Francisco Rodrigo Porto Cavalcanti ·
Sören Andersson
Editors

Optimizing Wireless Communication Systems

Springer

Editors
Francisco Rodrigo Porto Cavalcanti
Universidade Federal do Ceará
Fortaleza-CE
Campus do Pici, Bloco 910
Brazil
rodrigo@gtel.ufc.br

Sören Andersson
Ericsson AB
Isafjordsgatan 14E
SE-164 80 Stockholm
Sweden
soren.n.andersson@ericsson.com

ISBN 978-1-4419-0154-5 e-ISBN 978-1-4419-0155-2
DOI 10.1007/978-1-4419-0155-2
Springer Dordrecht Heidelberg London New York

Library of Congress Control Number: 2009931759

Printed on acid-free paper

Springer is part of Springer Science+Business Media (www.springer.com)

Optimizing Wireless Communication Systems

"To Eduarda and Renesa"

Foreword

In June 2000, GTEL (Wireless Telecommunications Research Group) at the Federal University of Ceará was founded by Professor Rodrigo Cavalcanti and his colleagues with the mission of developing wireless communications technology and impact the development of the Brazilian telecommunications sector. From the start, this research effort has been supported by Ericsson Research providing a dynamic environment where academia and industry together can address timely and relevant research challenges. This book summarized much of the research output that has resulted from GTEL's efforts. It provides a comprehensive treatment of the physical and multiple access layers in mobile communication systems describing different generations of systems but with a focus on 3G systems. The team of Professor Cavalcanti has contributed scientifically to the development of this field and built up an impressive expertise. In the chapters that follow, they share their views and knowledge on the underlying principles and technical trade-offs when designing the air interface of 3G systems.

The complexity of 3G systems and the interaction between the physical and multiple access layers present a tremendous challenge when modeling, designing, and analyzing the mobile communication system. Herein, the authors tackle this problem in an impressive manner. Their work is very much in line with the developments in 3GPP providing a deeper understanding of the evolution of 3G and also future enhancements. Two main themes are treated, resource management and transceiver designs. A common thread in both themes is the use of multi antenna systems or MIMO systems to enhance system performance.

Researchers or engineers active in wireless communications and interested in the design and optimization of current and emerging mobile systems are encouraged to share the results and insights of this comprehensive book.

KTH - Royal Institute of Technology
Stockholm, Sweden

Professor Björn Ottersten

Preface

Introduction

Mobile and wireless communication systems are a prominent communications technology with profound economical and social impacts in practically all parts of the world. The current state of wireless communication systems allows for a much wider scope of applications than what it used to be originally, that is, to be a mobile extension of the public switched telephone network. The convergence of mobile systems and the Internet has become a reality as new radio access technologies emerged with improved coverage, capacity, and latency. While the desire to develop and establish a truly mobile Internet dates back to the mid-1990s, it is only now that a significant increase in the volume of data is being witnessed by most cellular operators, not only in Europe and Japan, but also throughout North and Latin Americas. This book is about some of the underlying technological breakthroughs that allowed the evolution to the current state of development in wireless technology.

The focus of the book is on the two lower layers of the ISO/OSI layered model, that is, the physical and data link layers, including the link and media access control sublayers. These two layers are of specific importance in wireless systems, as opposed to many of its wired counterparts. This is fundamentally due to spectrum shortage, the broadcast nature of interference, and time variability of the wireless channel. As a consequence, much of the improvements in coverage, capacity, and latency of modern wireless systems are due to new approaches for tackling old problems in high-capacity radio communications in these two lower layers.

Intended Audience and Usage

This book is intended for researchers in the field of wireless communications, more specifically to the ones involved with the design and optimization of current and emerging wireless access technologies for mobile communications. Graduate students working in subjects such as radio resource management, OFDM, and MIMO, as well as in third-generation systems and beyond, will benefit from the state-of-the-art concepts, methods, examples, and case studies presented. Every chapter, in addition to having a clear ambition to address the state of the art of the corresponding subject, discusses basic concepts in the introductory sections and gives references for the interested reader to deepen his/her understanding. All chapters can be used independently as a complement to a graduate-level "advanced" wireless communications course, where each chapter can be subject to a directed study or a seminar. The book may also be of interest to the practitioner or to engineers involved in standardization efforts. The attention to technical details from standards is given in several chapters when performance results and case studies are presented. The idea is to demonstrate how advanced concepts can be adapted to be applicable in more realistic scenarios. Finally, almost every chapter of the book sheds light, directly or indirectly, on the subject of performance evaluation of wireless systems by means of system and link-level simulations. As the complexity of wireless systems grows, efficient and correct methods for modeling and performance simulations of these systems are becoming a fundamental discipline on their own.

Organization of the Book

Part I – Resource Allocation

Radio resource allocation (also known as radio resource management or RRM) has its roots in frequency reuse planning of first-generation cellular systems. Its fundamental goal is to increase spectrum efficiency. More efficient utilization of the radio spectrum plays such an important role because spectrum is simultaneously a very scarce and widely shared resource.

In the evolution of second- and third-generation systems, RRM became a discipline on its own, encompassing a variety of techniques such as power control, frequency hopping, dynamic channel allocation, and more advanced multi-antenna concepts, such as beamforming solutions as well as various transmit diversity schemes. Then, the emergence of packet-switched data services in third-generation systems and beyond has demanded a new set of RRM techniques able to handle mixed services scenarios. These included concepts borrowed from wired data networks, such as packet scheduling and congestion control, but that were reformulated and adapted to the wireless environment. More recently, highly configurable emerging radio access technologies, such as orthogonal frequency division multiplexing (OFDM)-based multiple access, have widened the scope of RRM. By means of advanced optimization approaches, radio resource allocation in time and frequency is now possible with fine granularity, increasing the efficiency potential of spectrum usage to unprecedented levels. This is mainly due to a clever exploitation of the multiuser diversity made available by these emerging systems.

Chapter 1 deals with power control. Transmission power is one fundamental resource whose optimization impacts directly on coverage and capacity. Power control has been a key technique since second-generation systems to achieve energy efficiency and interference management. This chapter focuses mostly on the latter. A basic introduction to some fundamentals in wireless communications is included. Basic propagation phenomena and modeling are first discussed. Then, a review about the fundamentals of power control is given along with classical algorithms, including analysis of convergence. A new approach to power control, based on game theory, is then presented, appropriated to emerging systems where multiple services with different quality-of-service (QoS) demands coexist. In particular a class of "opportunistic" distributed power control algorithms is derived for elastic data services, making it relevant to reconsider the supremacy of adaptive modulation and coding in current wireless systems. Finally a discussion about the use of channel prediction methods to improve the performance of existing algorithms is presented.

Chapter 2 presents an overview of RRM for the commercially most successful mobile communication system to date, that is, GSM, along with its packet-switched counterpart, EDGE. RRM has played a key role in the long-lasting life of GSM, which, dating back to more than 25 years now, is still able to cope with the majority of worldwide voice traffic. The chapter begins with a review on the fundamentals of the GSM/EDGE technology according to the respective 3GPP standards. Then, several RRM techniques are described as applied to GSM/EDGE along with performance results, using a detailed and realistic simulation model. These include power control, dynamic channel allocation, spatial division multiple access (SDMA), and management of multiple services by interference balancing. A discussion about large-scale modeling and simulation of wireless systems is also presented, including traffic modeling of data services.

Chapter 3 is a practitioner-oriented tutorial on HSPA deployment and optimization. HSPA is the key access technology currently behind the mobile broadband Internet expansion. The chapter serves a dual scope. First, a review about the HSPA standard is given. Both HSDPA and HSUPA are presented in aspects such as protocol stack, network architecture, channel structure, and physical layer procedures. A description of radio resource management fundamentals in HSPA is presented including aspects such as power allocation, mobility management, and related protocol aspects. Then the author describes several field results and real case studies leading to optimized broadband experience via HSPA. The chapter ends with suggestive guidelines for planning and dimensioning HSPA networks for the residential market.

Chapter 4 builds on the previous chapter to propose and analyze advanced congestion control mechanisms for HSPA, as well as for WCDMA (wideband code division multiple access) systems. While the baseline WCDMA/HSPA system can bring significant capacity improvements over GSM/EDGE, the growing demand for data services may rapidly press its spectrum efficiency to the limit. Quality-of-service management by means of congestion control is then proposed for dealing with multiple services competing for radio resources. Congestion control functions, in the form of admission control, load control, and packet scheduling, are responsible for keeping the network load at controlled levels and maintaining stability while ensuring QoS levels. Basic concepts and new methods are discussed and results showing the capacity benefits of employing congestion control demonstrate a significant impact. The proposed methods are shown to be fully automatic and scalable, able to cope with many services under different network loads. Case studies for WCDMA and HSPA are presented using realistic simulation scenarios composed of services such as World Wide Web access and voice-over-IP.

Chapter 5 addresses state-of-the-art OFDMA systems and corresponding resource allocation aspects. As previously mentioned, OFDMA opens up a new breed of RRM techniques due to the high flexibility and granularity with which frequency and time radio resources (i.e., subcarriers and time slots) can be allocated to multiple users. Advanced optimization techniques can then be employed to map radio resources to active connections in such a way as to fulfill network-level objectives such as maximization of the overall capacity or satisfaction of QoS levels. The chapter begins by establishing the system-level scenarios for RRM in OFDMA and their differences. Then a review of the key optimization and algorithmic approaches suitable for these problems is given. A new scheduling approach for OFDMA is proposed, based on the maximization of the user satisfaction ratio. A case study for 3GPP's long-term evolution (LTE) system is presented to illustrate the performance of the proposed methods and concepts. Finally, a new method for power allocation for OFDMA is presented along with results showing superior performance as compared to existing approaches.

Finally (for Part I), *Chapter 6* looks to the near future of wireless systems by dealing with the topic of multi-access networks. In this case, multiple radio access technologies cooperate to increase coverage and capacity. By means of a common core network infrastructure, complementary features of different radio access technologies can be combined to increase return of investment of existing networks while attending new demands for coverage and capacity. The chapter begins with a conceptual review about multi-access networks and the involved fundamental trade-offs. Then, concepts and methods for common radio resource management are exposed. These methods can be seen as an extension of conventional RRM methods for the multi-access case. Typical CRRM procedures include access selection and inter-system (or vertical) handovers. A case study involving a UMTS (Universal Mobile Telecommunication System) and a wireless local area (WLAN) joint network is explained and illustrated with simulation results.

Part II – Transceiver Architectures

The significant improvements at the physical layer have been instrumental for the increase of the wireless link capacity over the last decade. OFDM itself, already a popular modulation mechanism in fixed digital subscriber lines, has been combined with the use of multiple antennas at both ends of wireless links, in the so-called multiple input multiple output (MIMO) schemes. MIMO has changed the way wireless engineers face the fundamental capacity limits of the wireless channel by exploiting fading variability in favor of it. This fact also illustrates the major challenge – How can a wireless system be designed that allows for a practical implementation in the presence of such potentially fast fading propagation channels between and among the multitude of employed antennas? The main aspect to take into consideration is how to make such a system design both observable and controllable – the former important in order to generate the appropriate amount of radio network measurements and the associated signaling and the latter significant in the sense of keeping the interference levels under control on a system level.

The understanding and modeling of MIMO propagation channels have reached a rather mature level during the last decade; a remaining problem is, however, the computational complexity associated with using any of the available detailed MIMO models in system (or even link-) level simulations. There are also still many aspects to understand when it comes to including also antenna design – and modeling aspects for any realistic MIMO application – this is in particular the case on the user equipment side, mainly due to the fundamental restrictions originating from the size (in wavelengths) of handheld or portable devices.

Chapter 7 deals with a basic concept when analyzing wireless links by explaining the way wireless links can be modeled and have their performance efficiently evaluated. Modeling and simulating wireless systems is a complex task which starts with a good assessment of the physical layer behavior. The chapter discusses two main aspects. First, the authors discuss the approaches for dividing complex wireless system simulations into two independent, more tractable parts, namely link and system-level simulations. Then they focus on how to design reliable link-level simulators. Besides that, a software development framework is proposed for flexible and modular construction of link-level simulators. Several case studies are presented, involving the modeling and simulation of actual mobile systems, to illustrate the concepts.

Chapter 8 presents an overview of techniques related to the problem of equalization for wireless systems. The hereby desired recovery of coded symbols transmitted through a propagation channel is treated for the SISO scenario as well as for the SIMO (beamforming) case. Techniques for channel identification and tracking are discussed together with means to handle time-dispersive channels using either time- or frequency-domain techniques. Furthermore, case studies exemplify typical equalization solutions for wireless systems in use today. Finally, the chapter discusses the concept of – and principles for – turbo-equalization, that is, equalization structures that achieve near-optimal performance by jointly performing equalization and decoding.

Chapter 9 treats channel estimation for OFDM-based systems. Since the granularity in the time and frequency-domains are rather high, the concepts of frequency-domain interpolation and time-domain filtering are required for a practical implementation of a channel estimation algorithm. These concepts become very important for allowing as low a density as possible of the reference or pilot signals dedicated for aiding the channel estimation over the frequency band of interest and over time as the channel changes. As mandated by the OFDM systems currently emerging from ongoing standardization efforts (in, e.g., 3GPP LTE), good channel estimation performance and robustness as well as the associated system design aspects – like the desire to reduce the signaling overhead – becomes highly relevant for MIMO applications. Such aspects are discussed in that chapter and different channel estimators are evaluated and compared by means of simulations.

Chapter 10 brings the discussions in Chapter 9 further toward an application of channel estimation methods and the related channel state information and channel quality indicators to the problem of adapting modulation scheme and coding rate for a MIMO-OFDM system. This is commonly known as the problem of link-adaptation, and the problem to address is how to best select transmission parameters (like transmit antennas, bit rates, transmit power) for a certain estimated channel realization in order to optimally utilize the available system resources (like spectral efficiency or ultimately even energy consumption). The chapter outlines and discusses the different gains that can be achieved – diversity and multiplexing – and the relation between them. Furthermore, some hybrid MIMO transmission schemes are suggested and evaluated for various numbers of employed transmit antennas.

In *Chapter 11* the authors present an innovative space–time–frequency multiple-access (STFMA) MIMO wireless communication system combining space-domain and frequency-domain spreading by means of linear precoding, along with a time-domain block-spreading CDMA strategy. Precoding across space (transmit antennas) and frequency (subcarriers) provides robustness against deep channel fades while providing space and frequency diversities, while block-spreading enables multiple accessing. They utilize a tensorial algebra-based decomposition to model the received signal in the STFMA system. Thanks to the powerful identifiability properties of this tensor decomposition, blind signal detection based on multiuser detection is possible.

Chapter 12 finally addresses the problem of how to reduce the overhead signaling that is typically present for MIMO transmit schemes employing closed-loop channel state information feedback. The techniques that are discussed are mainly based on the concept of transmitter precoding, that is, the feedback from the user equipment consists of an index in a pre-designed transmit codebook, known at both transmitter and receiver sides; the codebook design is also described in the chapter in the form of illustrative examples. The transmitter then applies the so-indicated codebook vector of antenna weights, and this process is updated regularly. Clearly, the performance of closed-loop precoding schemes will suffer from high-speed terminals since the selected codebook index then quickly becomes outdated, and a possible remedy is then to switch to an open-loop transmit scheme, where mainly the modulation and coding rates are updated regularly and the potential additional transmit weights are designed to primarily generate diversity gains. Examples are provided, where the performance of different transmit schemes are compared by means of simulation studies.

Fortaleza, Brazil Francisco R. P. Cavalcanti
Stockholm, Sweden Sören N. Andersson

Acknowledgments

This book is the result of almost 10 years of research activities at the Wireless Telecommunications Research Group (GTEL) at the Federal University of Ceara (UFC), Brazil. GTEL was created in the year 2000 as a joint effort among UFC, Ericsson Brazil and Ericsson Research in Sweden. The present book – apart from all the M.Sc. and Ph.D. students that have been produced – can be seen as a direct and very successful result of this stimulation effort.

We are particularly thankful to Eduardo Oliva, Maria Valeria Marquezini, and Andrea Barros of Ericsson Brazil, who have managed the strategic, formal, and legal aspects of the research projects over the years with GTEL, as well as the management at Ericsson Brazil, Fernando Aragão and Trond Fidje, who have been supportive in the efforts involved in this research cooperation throughout the years.

We express our gratefulness also to the many students that have taken part in and contributed to the research efforts over the years, some of them chapter authors in this very book and turned into professors and industry experts now. Among them we would like to mention with special gratitude André Almeida, Charles Cavalcante, Emanuel Bezerra, Leonardo Sampaio, Tarcisio Maciel, Vicente Souza, Waltemar Sousa, Walter Cruz, and Yuri Silva. We would also like to acknowledge professors João Mota and João Romano for their various and invaluable contributions to the consolidation of GTEL since its foundation. Participating chapter authors from other institutions are thanked for their long-term and prolific partnership. One further and special thank you is to Mrs. Ana Carvalho for her endless dedication as executive secretary at GTEL.

Previously at Ericsson Research and currently with Ericsson's Business Unit Networks, a special thank you is passed on to Bo Göransson for his feedback over the years regarding the many detailed research issues of relevance to consider within the area of multiple antenna systems and to Henrik Asplund in the Propagation Group at Ericsson Research for his guidance on radio wave propagation aspects and channel modeling intricacies.

An expression of particular gratitude finally goes to the current and previous research managers Mikael Höök, Sverker Magnusson, and Sven-Olof Jonsson at

Ericsson Research who over the years have supported, directed, and reviewed the different projects' ambitions and results from the GTEL research cooperation.

Fortaleza, Brazil Francisco R. P. Cavalcanti
Stockholm, Sweden Sören N. Andersson

Contents

Contributors

Francisco R. P. Cavalcanti received a D.Sc. degree in electrical engineering from University of Campinas (UNICAMP), Brazil, in 1999. Upon graduation he joined the Federal University of Ceara (UFC) where he is an adjunct professor and holds the Wireless Communications Chair at the Teleinformatics Engineering Department. In 2000, he founded and since then has directed GTEL, a research institute based in Fortaleza, Brazil, focused on the advancement of wireless telecommunications technologies. He is also a program manager directing a program of research projects in wireless communications at GTEL sponsored by the Ericsson Research and Development Center in Brazil. He has published over 100 conference and journal papers in topics related to radio resource management, cross-layer algorithms, and transceiver architectures for wireless systems and networks. Prof. Cavalcanti also holds a "Leadership and Management" professional certificate from the Massachusetts Institute of Technology.
Wireless Telecom Research Group (GTEL), Campus do Pici - CP 6005, 60455-760, Fortaleza, CE, Brazil
e-mail: `rodrigo@gtel.ufc.br`

Sören Andersson received an M.Sc. EE and Ph.D. degrees, in Automatic Control from Linköping Institute of Technology, Sweden, in 1988 and 1992, respectively. During 1993 he was a postdoctoral research associate at Yale University. He then joined the Department for Access Technologies and Signal Processing – where research in advanced antenna systems for wireless networks was initiated – at Ericsson Research, Ericsson AB, Stockholm, Sweden in 1994. There he was active in research on adaptive antennas in cellular systems and was project manager for the research and subsequent field-trials carried out with respect to the application of adaptive antennas for GSM. Between 1998 and 2008 he managed Ericsson Research's activities in the area of antenna systems and propagation, and in 2008 he was appointed as an expert in multi-antenna systems. His research interests are in the general areas of statistical signal and array processing for wireless communications and radio access technologies, the hereby implicated product implementation requirements aspects, as well as radio network issues related to the application of advanced antenna solutions for wireless systems.
Ericsson AB, Ericsson Research, EAB/TU, Isafjordsgatan 14E, S-164 80, Stockholm, Sweden
e-mail: `soren.n.andersson@ericsson.com`

Alex P. da Silva received a B.Sc. in electrical engineering from Federal University of Ceará (UFC), Brazil, in 2004. During his graduate studies, he took part in the Double-Degree Program, receiving, also in 2004, a Generalist Engineer degree from École Centrale de Nantes (ECN) France. He received his M.Sc. degree in teleinformatics engineering from UFC, in 2007. Since 2003, he has been working in projects inside a technical cooperation between UFC and Ericsson of Brazil. These projects aim at proving solutions for radio resource management for 3G and multi-access networks and radio resource allocation for OFDMA-based systems. He is member of the Wireless Telecommunications Research Group (GTEL), Fortaleza, Brazil. His research interests include wireless communications and mobile networks, multi-access networks, and OFDMA-based systems.
Wireless Telecom Research Group (GTEL), Campus do Pici - CP 6005, 60455-760, Fortaleza, CE, Brazil
e-mail: alex@gtel.ufc.br

Aline O. Neves received a B.Sc. and an M.Sc. degree in electrical engineering from the State University of Campinas (UNICAMP), Brazil, in 1999 and 2001, respectively. She received her Ph.D. in 2005, also in electrical engineering, from the University René Descartes (Paris V), Paris, France. Recently, she has become an assistant professor at the Engineering, Modeling and Applied Social Science Center of the Federal University of ABC, Santo André, Brazil. Her research interests consist of equalization, channel estimation, source separation, and information theoretic learning.
Centro de Engenharia, Modelagem e Ciências Sociais Aplicadas, Universidade Federal do ABC, Rua Santa Adélia, 166, Santo André, SP, Brazil
e-mail: aline.neves@ufabc.edu.br

André L. F. de Almeida received a B.Sc. and an M.Sc. degree in electrical engineering from the Federal University of Ceará (UFC), Fortaleza, Brazil, in 2001 and 2003, respectively, and the double Ph.D. in sciences and teleinformatics engineering, respectively, from the University of Nice Sophia Antipolis (UNSA), France, and UFC, Brazil, in 2007. In 2002 he was a visiting researcher at Ericsson Research, Stockholm, Sweden, where he worked on MIMO channel measurements for indoor propagation modeling. He was a postdoctoral fellow with the I3S laboratory, CNRS, Sophia Antipolis, France, from January to December 2008. He is now a senior researcher with the Wireless Telecom Research Group (GTEL), Fortaleza, Brazil, where he has worked in transceiver architectures for wireless systems within the GTEL-Ericsson Research cooperation. Dr. Almeida is affiliated with the Department of Teleinformatics Engineering of the Federal University of Ceará. His main research interests lie in the area of signal processing for communications and include array processing, blind signal separation and equalization, multiple-antenna techniques, multicarrier and multiuser communications. Recent work of Dr. Almeida has focused on the development of tensor models for transceiver design in wireless communication systems.
Wireless Telecom Research Group (GTEL), Campus do Pici - CP 6005, 60455-760, Fortaleza, CE, Brazil
e-mail: andre@gtel.ufc.br

Carlos H. M. de Lima received a B.Sc. and an M.Sc. degree in electrical engineering from the Federal University of Ceará (UFC) in 2002 and 2004, respectively. Since then he has been working as a research scientist. From 2000 to 2005, he worked in the Wireless Telecommunications Research Group (GTEL), Fortaleza, Brazil. In 2005 he was a visiting researcher at Ericsson Research, Luela, Sweden, working on power control techniques for the enhanced uplink HSPA system. In 2006, he worked at Nokia Institute of Technology, Manaus, Brazil. Currently, he is pursing his D.Sc. in the Department of Electrical and Information Engineering, University of Oulu, Finland. He is also a member of the research staff of the Centre for Wireless Communications, Oulu, Finland.
Centre for Wireless Communications, University of Oulu, Erkki Koiso-Kanttilan katu 2S-door 90570, Finland
e-mail: carlos.lima@ee.oulu.fi

Charles Casimiro Cavalcante received a D.Sc. degree from the University of Campinas (UNICAMP) in São Paulo, Brazil in 2004. Dr. Cavalcante has been working on signal processing strategies for communications where he has several papers published and he has worked on funded research projects on the area. He has held a grant for Scientific and Technological Development from the Brazilian Research Council (CNPq) from 2004 to 2007. Since March 2007 he is a visiting professor at Teleinformatics Engineering Department of UFC and a researcher of the Wireless Telecommunications Research Group (GTEL) where he leads research on signal processing and wireless communications. His main research interests are in signal processing for communications, blind source separation, wireless communications, and statistical signal processing.
Wireless Telecom Research Group (GTEL), Campus do Pici - CP 6005, 60455-760, Fortaleza, CE, Brazil
e-mail: charles@gtel.ufc.br

Cibelly A. de Araújo received a B.Sc. degree in electrical engineering and a M.Sc. degree in teleinformatics engineering from the Federal University of Ceará (UFC), Fortaleza, Brazil, in 2006 and 2008, respectively. She is currently working toward her D.Sc. degree at the same institution. Since 2005, she has been a researcher at the Wireless Telecommunications Research Group, Fortaleza, Brazil. Currently, she is also a researcher within the technical cooperation between GTEL and Ericsson Research. Her research interests include cross-layer aspects for wireless communications, scheduling, link adaptation, and feedback reporting in MIMO-OFDM systems.
Wireless Telecom Research Group (GTEL), Campus do Pici - CP 6005, 60455-760, Fortaleza, CE, Brazil
e-mail: cibelly@gtel.ufc.br

Cristiano Magalhães Panazio received a B.Sc. and an M.Sc. degree in electrical engineering from the State University of Campinas (UNICAMP), Brazil, in 1999 and 2001, respectively. He received his Ph.D. in 2005, also in electrical engineering, from the Conservatoire National des Arts et Métiers (CNAM), Paris, France. In 2006, he became assistant professor at Escola Politécnica of the University of São Paulo. His research interests include equalization, multicarrier modulation, spread spectrum techniques, space–time receivers, and synchronization techniques.
Laboratory of Communications and Signals, Department of Telecommunications and Control, USP, São Paulo, Brazil
e-mail: `cpanazio@lcs.poli.usp.br`

Darlan C. Moreira received a Bachelor's degree in electrical engineering and the Master of Science degree in teleinformatics engineering from the Federal University of Ceará (UFC), Brazil, in 2005 and 2007, respectively. He is currently pursuing the Doctor's degree at the same institution. He is a member of the Wireless Telecommunications Research Group (GTEL), Fortaleza, Brazil, and since 2004, he has been working in projects within the technical cooperation between GTEL and Ericsson Research. In 2007 he was a visiting researcher at Ericsson Research, Stockholm, Sweden, working on channel quality measurement and reporting for 3GPP's long-term Evolution (LTE) wireless system. His research interests include cross-layer aspects of wireless communications, scheduling, and link adaptation in MIMO-OFDM systems.
Wireless Telecom Research Group (GTEL), Campus do Pici - CP 6005, 60455-760, Fortaleza, CE, Brazil
e-mail: `darlan@gtel.ufc.br`

Elvis M. G. Stancanelli received the B.Sc. degree in Electrical Engineering from the State University of Londrina (UEL), Brazil, in 2002. In 2001 and 2002, he took part in a project under technical cooperation between the University of São Paulo (USP), Brazil, and Ericsson Research. In July 2004, he received the M.Sc. degree in Electrical Engineering from the Polytechnic School of the University of São Paulo (EPUSP). At the same time, he joined the Wireless Telecommunications Research Group (GTEL), Fortaleza, Brazil, as researcher. Since 2004 he has been working in several projects within the technical cooperation between GTEL and Ericsson Research, where he developed link-level simulators for wireless standards such as GSM/EDGE, WCDMA, HSPA, and 3GPP's LTE. Currently, he is pursuing his D.Sc. degree in Teleinformatics Engineering at the Federal University of Ceará (UFC), Brazil. Some of his research interests are interference mitigation, diversity techniques, system modeling, and applied computational intelligence.
Wireless Telecom Research Group (GTEL), Campus do Pici - CP 6005, 60455-760, Fortaleza, CE, Brazil
e-mail: `emiguel@gtel.ufc.br`

Emanuel Bezerra Rodrigues received B.Sc. and M.Sc. degrees in electrical engineering from the Federal University of Ceará (UFC), Brazil, in 2001 and 2004, respectively. He worked in the Wireless Telecom Research Group (GTEL-UFC) from 2001 to 2007 participating in several research projects sponsored by the Ericsson Research Brazilian Branch. In 2004 he was a visiting researcher at Ericsson Research, Linköping, Sweden, working on congestion control techniques for the high-speed packet access system. He is currently doing his Ph.D. studies at the Signal Theory and Communications Department (TSC) of the Technical University of Catalonia (UPC), Spain. His main research interests are radio resource management, QoS control, and cross-layer optimization for mobile communication systems.
Technical University of Catalonia - UPC, Campus Nord, Jordi Girona 1-3, 08034 Barcelona, Spain e-mail: emanuel@tsc.upc.edu

Fabiano de S. Chaves received the B.S. degree in electrical engineering and the M.S. degree in teleinformatics engineering from Federal University of Ceará (UFC), Brazil, in 2003 and 2005, respectively. He was from 2002 to 2005 with the Wireless Telecommunications Research Group (GTEL), Fortaleza, Brazil. He is conducting his doctorate studies at University of Campinas (UNICAMP), Brazil, and is also member of the IEEE and of the Brazilian Telecommunications Society (SBrT). His research interests include distributed power control for communication systems, non-cooperative game theory, and interplays between signal processing and control methodologies.
Department of Communications, School of Electrical and Computer Engineering University of Campinas – UNICAMP, P.O. Box 6101, Campinas, 13083-852, São Paulo, Brazil
e-mail: fabiano@decom.fee.unicamp.br.

Gérard Favier received an engineering diploma from ENSCM (Ecole Nationale Supérieure de Chronométrie et de Micromécanique), Besançon, and ENSAE (Ecole Nationale Supérieure de l'Aéronautique et de l'Espace), Toulouse, the Engineering Doctorate and State Doctorate degrees from the University of Nice Sophia Antipolis, in 1973, 1974, 1977, and 1981, respectively. In 1976, he joined the CNRS (Centre National de la Recherche Scientifique) and now he works as a research director of CNRS at the I3S Laboratory, in Sophia Antipolis. From 1995 to 1999, he was the director of the I3S Laboratory. His present research interests include nonlinear process modeling and identification, blind equalization, tensor decompositions, and tensor approaches for wireless communication systems.
Laboratoire I3S/UNSA/CNRS, 2000 route des Lucioles, Les Algorithmes/Euclide B BP 121, Sophia Antipolis, France
e-mail: favier@i3s.unice.fr

Ícaro L. J. da Silva received a Bachelor degree in electrical engineering and a Master of science degree in teleinformatics engineering from the Federal University of Ceará (UFC), Brazil, in 2006 and 2009 respectively. Currently he is pursuing his M.Sc. degree in teleinformatics engineering from the same university. Since 2006 he has been with the Wireless Telecommunications Research Group (GTEL) where he is a researcher working on MIMO antenna systems and related issues such as space–time coding, spatial multiplexing and limited feedback. In 2009 he is a visiting researcher at Ericsson Research, Stockholm, Sweden, working in aspects such as MIMO precoding and limited feedback.
Wireless Telecom Research Group (GTEL), Campus do Pici - CP 6005, 60455-760, Fortaleza, CE, Brazil
e-mail: icaro@gtel.ufc.br

João Cesar M. Mota received his B.Sc. degree in physics from the Federal University of Ceará (UFC), Brazil, in 1978, the M.Sc. degree from Pontifícia Universidade Católica (PUC-RJ), Brazil, in 1984, and D.Sc. degree from the University of Campinas (UNICAMP), Brazil, in 1992, all in telecommunications engineering. Since August 1979, he has been in the UFC, and currently he is professor with the Teleinformatics Engineering Department. Dr. Mota worked in Institut National des Télécommunications and Institut de Recherche en Communications et Cybernetique de Nantes, both in France, as invited professor during 1996–1998 and spring 2006, respectively. He was general chairman of the 19th Brazilian Telecommunications Symposium – SBrT'2001 and the International Symposium on Telecommunications – ITS'2006. He is responsible for the international mobility program for engineering students of UFC. His research interests include digital communications, adaptive filter theory, and signal processing. He is member and counselor of the Sociedade Brasileira de Telecomunicações and member of the IEEE communications Society and IEEE Signal Processing Society. He is counselor of the IEEE Student Branch in UFC.
Wireless Telecom Research Group (GTEL), Campus do Pici - CP 6005, 60455-760, Fortaleza, CE, Brazil
e-mail: mota@gtel.ufc.br

João Marcos Travassos Romano received the degrees of engineer and MS in electrical engineering from the University of Campinas (UNICAMP), Brazil. He received his Ph.D. in automatic and signal processing from the University of Paris – XI in 1987. In 1988 he joined, as an associate professor, the School of Electrical and Computer Engineering (FEEC) at UNICAMP where he is currently professor. Since 1989, he is recipient of a Research Fellowship from CNPq, the national foundation for science and technology in Brazil. He has also been an invited professor at the René Descartes University in Paris and at the Communications and Electronic Laboratory in CNAM – Paris. Professor Romano served the Brazilian Communications Society (SBrT), a sister society of ComSoc-IEEE, as vice president (1996–2000) and president (2000–2004). Professor Romano has over 140 journal and conference publications. He has submitted four Brazilian patents and one international patent. He was the advisor of 23 M.Sc. and 12 D.Sc. These is at UNICAMP.
Department of Microwaves and Optics, School of Electrical and Computer Engineering, UNICAMP, PO Box 6101, 13083-852, Campinas, Brazil
e-mail: romano@dmo.fee.unicamp.br

Leonardo S. Cardoso received an electrical engineering degree from the Federal University of Ceará (UFC), Brazil, in 2003. He received his M.Sc. degree from the same institution in 2006. From 2001 to 2006 he worked in several projects under a technical cooperation between UFC and the Brazilian branch of Ericsson Research. Those projects aimed at studying solutions for radio resource management issues for 2/3G and multi-access networks. During the same period, he was member of the Wireless Telecommunications Research Group (GTEL), Fortaleza, Brazil. In 2006, he joined the Eurecom Institute, France, working in two projects that dealt with multi-access networks and real-time MIMO channel performance assessment. He also significantly contributed to the EMOS MIMO platform at the Eurecom institute. He is currently pursuing his Ph.D. at Supélec, France. His research interests include wireless communications, multi-access networks, cognitive radio, and random matrix theory.
Alcatel-Lucent Chair in Flexible Radio - Supélec, 3 rue Joliot-Curie, 91192 Gif-Sur Yvette Cedex, France
e-mail: `leonardo.cardoso@supelec.fr`

Mário I. J. Da Silva has been working in telecommunications since 1998. He has a B.Sc. engineering degree from the Federal University of Ceará (UFC) in Brazil. He also has a Masters degree from the Institut National des Telecommunications in France, and during his Masters he joined Motorola Labs in Paris, where he carried out research on UMTS physical layer performance. He subsequently began working as a radio design engineer in O2 UK and moved to O2 Ireland in 2001, where he works as a principal engineer. For the last 7 years, he has been involved in several projects on UMTS inclusive of the deployment of broadband over HSPA. He is currently working on radio and core optimization.
O2 Telefonica Ireland, 28/29 Sir John Rogerson's Quay, Docklands, Dublin 2, Ireland
e-mail: `mario.dasilva@o2.com`

F. Rafael M. Lima received a B.Sc. in electrical engineering and an M.Sc. in teleinformatics engineering from the Federal University of Ceará, UFC, Brazil, in 2005 and 2008, respectively. In 2008 he was a visiting researcher at Ericsson Research, Lulea, Sweden, working on packet scheduling techniques and QoS management for the 3GPPs long-term evolution (LTE) system. He is currently a researcher and a PhD candidate at the Wireless Telecom Research Group, GTEL, working in radio resource allocation for OFDMA-based systems. His research interests include radio resource management to WCDMA/HSDPA networks, packet scheduling, admission control, link adaptation, and load control.
Wireless Telecom Research Group (GTEL), Campus do Pici - CP 6005, 60455-760, Fortaleza, CE, Brazil
e-mail: `rafaelm@gtel.ufc.br`

Raimundo Abreu de Oliveira Neto received a Bachelor and Master of Science degrees in electrical engineering from the Federal University of Ceará (UFC), Brazil, in 2001 and 2004, respectively. From 2002 to 2008 he was with the Wireless Telecommunications Research Group (GTEL) where he has worked as a researcher for the technical cooperation between GTEL and Ericsson Research. Presently, he is senior engineer at Petrobras - Petroleo Brasileiro S/A. His research interests are power control, radio resource management, multi-access networks, and economic models for telecommunications.
Wireless Telecom Research Group (GTEL), Campus do Pici - CP 6005, 60455-760, Fortaleza, CE, Brazil
e-mail: neto@gtel.ufc.br

Renato da Rocha Lopes received a B.Sc. and an M.Sc. degree in electrical engineering from the University of Campinas (UNICAMP), Brazil, in 1995 and 1997, respectively. In 2003, he received the Ph.D. in electrical engineering from the Georgia Institute of Technology. Since then, he has been with the School of Electrical and Computer Engineering at UNICAMP, first as a post-doctoral fellow, then, since 2006, as an assistant professor. He is the recipient of several scholarships from the Brazilian government. His research interest spans the general area of communications theory, including MIMO systems, turbo receivers, channel estimation and equalization, and multiuser wireless communications.
Department of Communications, School of Electrical and Computer Engineering, UNICAMP, PO Box 6101, 13083-852, Campinas, Brazil
e-mail: rlopes@decom.fee.unicamp.br

Ricardo B. Santos received his B.Sc. in electrical engineering and M.Sc. in teleinformatics engineering from the Federal University of Ceará (UFC), Fortaleza, Brazil, in 2005 and 2008, respectively. Nowadays he is a researcher of Wireless Telecom Research Group (GTEL) working in radio resource allocation in OFDMA-based systems. His research interests include radio resource management to WCDMA/HSDPA networks, packet scheduling, admission control, power control, link adaptation, load control, and heuristic algorithms. In 2008 he was a visiting researcher at Ericsson Research, Lulea, Sweden, working on packet scheduling techniques and QoS management for the 3GPP's long-term evolution (LTE) system.
Wireless Telecom Research Group (GTEL), Campus do Pici - CP 6005, 60455-760, Fortaleza, CE, Brazil
e-mail: brauner@gtel.ufc.br

Rui Facundo Vigelis received his B.Sc. degree in electrical engineering in 2005 and Master of Science degree in teleinformatics engineering in 2006, both from the Federal University of Ceará (UFC) in Fortaleza, Brazil. Since September 2006 he is working toward a D.Sc. at UFC in advanced problems of communication systems. He has also worked on funded projects on the subject of OFDM-based wireless system in 2006 where he has published papers on channel estimation methods for wireless systems. His research interests include statistical signal processing, differential geometry, manifold learning, and wireless communications.
Wireless Telecom Research Group (GTEL), Campus do Pici - CP 6005, 60455-760, Fortaleza, CE, Brazil
e-mail: rfvigelis@gtel.ufc.br

Stefan Wänstedt joined Ericsson in 1999 and as a Senior Research Engineer at Advanced Wireless Ericsson Algorithm Research, Luleå, Sweden. The current focus of his work is on wireless IP optimization, including real-time services over cellular systems, in particular HSPA and LTE. Previous assignments focused on radio network performance measurements for cellular systems, including GPRS and voice quality models. He has also worked with projects related to streaming over WCDMA and EDGE. He holds a Ph.D. in geophysics from Luleå University of Technology.

Tarcisio F. Maciel received a B.Sc. and an M.Sc. degree in electrical engineering from the Federal University of Ceará, Fortaleza, Brazil, in 2002 and 2004, respectively. He received the Ph.D. in electrical engineering from the Technische Universität Darmstadt, Darmstadt, Germany, in 2008. In 1999, he attended the Technische Universität Hamburg-Harburg, Hamburg, Germany, as part of a 1-year sandwich graduation program. From 2001 to 2004 he was with the Wireless Telecom Research Group (GTEL), Fortaleza, Brazil, working in the research projects on radio resource management for wireless systems developed by GTEL in cooperation with Ericsson Research. From 2005 to 2008 he was with the Communications Engineering Lab, Darmstadt, Germany, where he developed his Ph.D. studies on resource allocation for systems with multiple antennas. Currently he is a professor of computer engineering at the Federal University of Ceará, Campus of Sobral, and a senior researcher at the Wireless Telecommunications Research Group. His main research interests are in the areas of wireless communication systems, resource allocation, adaptive antennas, and optimization.
Wireless Telecom Research Group (GTEL), Campus do Pici - CP 6005, 60455-760, Fortaleza, CE, Brazil
e-mail: maciel@gtel.ufc.br

Vicente A. de Sousa Jr. received a B.Sc. Electrical Engineer degree from the Federal University of Ceará (UFC), Fortaleza, Brazil, in 2001. During his graduate studies, he took part in a technical training of Motorola and Eldorado Institute. He received his M.Sc. degree from the UFC, in 2002. Between 2001 and 2006, he had been working in projects inside a technical cooperation between UFC and Ericsson of Brazil where developed solutions to smart antennas systems, radio resource management techniques for 3G networks and interworking of UMTS and WLAN systems. For the same period, he was member of the Wireless Telecommunications Research Group (GTEL), Fortaleza, Brazil. He is presently working toward his D.Sc. degree at UFC, Brazil. Sousa is also R&D coordinator of Nokia Technology Institute (INdT), Manaus, Brazil. His research interests include wireless communications and mobile networks, evolutionary computation, multi-access networks, and WiMAX systems.
Nokia Institute of Technology (INdT), Rua Torquato tapajós, 7200 - Colonia Terra Nova, 69093-415, Manaus, AM, Brazil
e-mail: vicente.souza@indt.org.br

Walter C. Freitas Jr. received a D.Sc. degree in teleinformatic engineering from Federal University of Ceará (UFC), Brazil, in 2006 and his B.Sc. and M.Sc. degrees in electrical engineering from the same university. During his studies, he was supported by the Brazilian agency FUNCAP and Ericsson. From 2005 to 2006, Dr. Walter was with the Nokia Institute of Technology as a senior researcher. Since 2006 he has been a project manager at GTEL working for the Ericsson–GTEL program of projects. From 2008 he is a professor at the Telinformatics Engineering Department at the Federal University of Ceará. His main area of interest concerns features development to improve the performance of the wireless communication systems, application of link adaptation techniques, OFDMA resource allocation, MIMO systems, and space–time coding.
Wireless Telecom Research Group (GTEL), Campus do Pici - CP 6005, 60455-760, Fortaleza, CE, Brazil
e-mail: walter@gtel.ufc.br

Yuri C. B. Silva received his B.Sc. and M.Sc. degrees from the Federal University of Ceará (UFC), Fortaleza, Brazil, in 2002 and 2004, respectively, and the Ph.D. from the Technische Universität Darmstadt, Germany, in 2008, all in electrical engineering. In 1999 he attended the Technische Universität Berlin, Germany, as part of a 1-year sandwich graduation program. From 2001 to 2004 he was with the Wireless Telecom Research Group (GTEL), Fortaleza, Brazil, working within the technical cooperation between GTEL and Ericsson Research. In 2003 he was a visiting researcher at Ericsson Research, Stockholm, Sweden, where he developed advanced radio resource management solutions for the GSM/EDGE standard. From 2005 to 2008 he was with the Communications Engineering Lab of the Technische Universität Darmstadt and currently he is a senior researcher at GTEL. His main research interests are in the areas of wireless communication systems, resource allocation, adaptive antennas, multicast services, and precoding techniques.
Wireless Telecom Research Group (GTEL), Campus do Pici - CP 6005, 60455-760, Fortaleza, CE, Brazil
e-mail: yuri@gtel.ufc.br

Acronyms

16-QAM 16-Quadrature amplitude modulation
3G Third generation
3GPP 3rd. Generation Partnership Project
4G Fourth generation
ABC Always best connected
AC Admission control
ACK Acknowledgement
AcVI Actual value interface
A-DPCH Associated dedicated physical channel
ADSL Asymmetric digital subscriber lines
ALS Alternating least squares
AM Acknowledged mode
AMC Adaptive modulation and coding
AMR Adaptive multirate
AP Access point
ARP Allocation/retention priority
ARQ Automatic repeat request
AS Access selection
AS Active set
ASBPC Autonomous SINR balancing power control
ATM Asynchronous transfer mode
AVI Average value interface
AWGN Additive white Gaussian noise
BB Branch-and-bound
BCCH Broadcast control channel
BEP Bit error probability
BER Bit error rate
BGR Benveniste–Goursat–Ruget theorem
BLAST Bell Labs layered space–time
BLEP Block error probability
BLER Block error rate

BPSK Binary phase-shift keying
BS Base station
BSC Base station controller
BSS Base station subsystem
BTCE Block-type channel estimation
BTS Base transceiver station
CAC Call admission control
CC Chase combining
CC Congestion control
CCCH Common control channel
CDMA Code-division multiple–access
CESM Capacity ESM
CFH Cyclic frequency hopping
CIR Carrier-to-interference ratio
CMA Constant modulus algorithm
CN Core network
CONFAC Contrained factor decomposition
CP Cyclic prefix
CPE Customer premise equipment
CPICH Common pilot channel
CQ Channel quantization
CQI Channel quality indicator
CRC Cyclic redundancy check
CRESM Cutoff rate ESM
CRRM Common radio resource management
CS Circuit-switched
CSE Circuit-switched equivalent
CSI Channel state information
CTA Coverage threshold algorithm
DBA Distributed balancing algorithm
DBLAST Diagonal Bell Labs layered space–time
DCA Dynamic channel allocation
DCCH Dedicated control channel
DCH Dedicated channel
DFE Decision-feedback equalizer
DL Downlink
DPC Distributed power control
DPCCH Dedicated physical control channel
DPCH Dedicated physical channel
DPDCH Dedicated physical data channel
DQPSK Differential quadrature phase-shift keying
DS Delay scheduler
DS-CDMA Direct-sequence code division multiple access
DTCH Dedicated traffic channel
DTX Discontinuous transmission

E-AGCH E-DCH access grant channel
E-DCH Enhanced dedicated channel
EDGE Enhanced data rate for GSM evolution
E-DPCCH Enhanced dedicated physical control channel
E-DPDCH E-DCH dedicated physical data channel
EESM Exponential ESM
EFLC Error feedback-based load control
EFR Enhanced full rate
EGC Equal gain combining
EGPRS Enhanced general packet radio service
EGT Equal gain transmission
E-HICH E-DCH hybrid ARQ indicator channel
eNB Enhanced Node B
EPC Evolved packet core
E-RGCH E-DCH relative grant channel
ERT Estimated RAN throughput algorithm
ESM Effective SINR mapping
E-TFC E-DCH transport format combination
ETSI European Telecommunications Standards Institute
ETU Extended typical urban
EUL Enhanced uplink
E-UTRAN Evolved UMTS terrestrial radio access network
FACCH Fast associated control channel
FDD Frequency division duplex
FDM Frequency division multiplexing
FDMA Frequency division multiple access
FEC Forward error correction
FER Frame erasure rate
FFT Fast Fourier transform
FH Frequency hopping
FIFO First-in-first-out
FIFS First-in-first-served
FIR Finite impulse response
FN Frame number
FP Frame Protocol
FSK Frequency shift keying
FSQP Feasible sequential quadratic programming
FSR Frame success rate
FTP File Transfer Protocol
G2 Alamouti space–time block code (STBC)
G3 3 transmitter antenna STBC
GA Genetic algorithm
GAP Generalized assignment problem
GASP Generalized access selection problem
GBR Guaranteed bit rate

GERAN GSM/EDGE radio access network
GGSN Gateway GPRS support node
GMSK Gaussian minimum shift keying
GPRS General packet radio service
GSM Global system for mobile communication
GW Gateway
H-ARQ Hybrid automatic repeat request
HLR Home location register
HMTS Hybrid MIMO transmit scheme
HSDPA High-speed downlink packet access
HS-DPCCH High-speed dedicated physical control channel
HS-DSCH High-speed downlink shared channel
HSN Hopping sequence number
HSPA High-speed packet access
HS-PDSCH High-speed physical downlink shared channel
HSUPA High-speed uplink packet access
HTTP Hypertext Transfer Protocol
ICI Inter-carrier interference
IEEE Institute of Electrical and Electronics Engineers
IETF Internet Engineering Task Force
IFFT Inverse fast Fourier transform
IIR Infinite impulse response
IMS IP multimedia subsystem
IMT International Mobile Telecommunications
IP Internet Protocol
IR Incremental redundancy
IRC Interference rejection combining
ISI Inter symbol interference
ITU International Telecommunication Union
JLC Jump-based load control
KPI Key performance indicator
KRST Khatri–Rao space–time
L2S Link-to-system-level
LA Link adaptation
LAC Link admission control
LBA Load balancing algorithm
LC Load control
LD Linear detection
LESM Logarithmic ESM
LF Limited feedback
LiESM Linear ESM
LL Link-level
LLC Link layer control
LMS Least mean square
LORAF Low-rank adaptive filter

LOS Line-of-sight
LS Least squares
LSDF Link-Level Software Development Framework
LTE Long-term evolution
LTI Linear time-invariant
LUBA Link utilization balancing algorithm
LuT Look-up table
MA Multi-access
MAC Medium access control
MAI Mobile allocation index
MAIO Mobile allocation index offset
MAL Mobile allocation list
MANET Mobile ad hoc network
MAP Maximum a posteriori
MAT Multi-antenna transmission
MCAS Modulation, coding, and antenna scheme
MCBS-CDMA Multi-carrier block-spread code division multiple access
MC-CDMA Multicarrier code division multiple access
MCDS-CDMA Multi-carrier direct-sequence code division multiple access
MCS Modulation and coding scheme
MCSE CSE maximization algorithm
MC-SSSMA Multi-carrier spread space spectrum multiple access
MIESM Mutual-information ESM
MIH Media-independent handover
MIMO Multiple-input multiple-output
MISO Multiple-input single-output
ML Maximum likelihood
MLSE Maximum-likelihood sequence estimation
MMSE Minimum-mean-square-error
MPF Multicarrier proportional fair
MR Maximum rate
MR59FR Multi-rate at 5.9 kbit/s with full rate
MRC Maximal ratio combining
MRT Maximal ratio transmission
MS Mobile station
MSC Mobile switching center
MSE Mean-squared error
MTDS-CDMA Multi-tone direct sequence
MTSI Multimedia telephony services over IMS
MUI Multi-user interference
MURPA Multiuser residual power allocation
NACK Negative acknowledgement
NAS Non-access stratum
NBAP Node B application part
NE Nash equilibrium

NLMS Normalized least-mean-square algorithm
NRT Non-real time
OF Orthogonality factor
OFDM Orthogonal frequency division multiplexing
OFDMA Orthogonal frequency division multiple access
OLPC Outer-loop power control
OOP Object-oriented programming
OPC Opportunistic power control
OPC-F Opportunistic power control with fairness
OQ-DPC-1 Opportunistic QoS distributed power control – 1
OQ-DPC-2 Opportunistic QoS distributed power control - 2
OSI Open systems interconnection
OSIC Ordered successive interference cancellation
OVSF Orthogonal variable spreading factor
PA Power allocation
PACE Pilot-assisted channel estimation
PARAFAC Parallel factor
PAST Projection approximation subspace tracking
PBCCH Packet broadcast control channel
PC Power control
PDCP Packet Data Convergence Protocol
PDTCH Packet data traffic channel
PDU Protocol data unit
PF Proportional fair
PhCH Physical channel
PHY Physical
PPC Partial phase combining
PS Packet-switched
PSC Packet scheduling
PSK Phase-shift keying
PSTN Public-switched telephone network
QAM Quadrature amplitude modulation
QBA Queue-based algorithm
QEGT Quantized equal gain transmission
QoS Quality-of-service
QP Quadratic programming
QPP Quadratic permutation polynomial
QPSK Quadrature phase shift keying
QSA Quantized signal adaptation
RA Rate adaptation
RAN Radio access network
RAT Radio access technology
RB Radio bearer
RBER Raw bit error rate
RF Radio frequency

RFH Random frequency hopping
RLC Radio link control
RLS Radio link set
RLS Recursive least squares
RM Rate maximization
RMA Rate maximization algorithm
RMSE Root mean square error
RNC Radio network controller
RR Round Robin
RRA Radio resource allocation
RRC Radio resource control
RRM Radio resource management
RT Real-time
RU Resource unit
RXLEV Received signal level
RXQUAL Received signal quality
SA Simulated annealing
SAC Session admission control
SACCH Slow associated control channel
SASP Strict version of the access selection problem
SAT Single-antenna transmission
SAW Stop-and-wait
SBA Satisfaction balancing algorithm
SBPS Service-based power setting
SDC Selection diversity combining
SDCCH Stand-alone dedicated control channel
SDPC Soft dropping power control
SDT Selection diversity transmission
SDU Service data unit
SEA Super exponential algorithm
SER Symbol error rate
SES Simple exponential smoothing
SF Spreading factor
SGSN Service GPRS support node
SHO Soft handover
SIC Successive interference cancellation
SIMO Single-input multiple-output
SINR Signal-to-interference-plus-noise ratio
SIP Session initiation protocol
SIP Signal-interference product
SIR Signal-to-interference ratio
SISO Single-input single-output
SISO Soft-input/soft-output
SL System-level
SM Spatial multiplexing

SMIRA Stepwise maximum-interference removal algorithm
SMS Short message service
SMT Single and multi-antenna transmission
SNDCP Sub-network-Dependent Convergence Protocol
SNR Signal-to-noise ratio
SORA Satisfaction-oriented resource allocation
SOVA Soft-output Viterbi algorithm
SQP Sequential quadratic programming
SRA Stepwise removal algorithm
ST Space–time
ST Subspace tracking
STBC Space–time block code
STF Space–time–frequency
STFMA Space–time–Frequency Multiple-Access
ST-LE Space–time Linear Equalizer
STM Space–time Multiplexing
STS Space–time Spreading
STTC Space–time Trellis Code
SVD Singular value decomposition
TB Transport block
TCH Traffic channel
TCP Transport control protocol
TDC Time delay compensation
TDD Time division duplex
TDL Tapped delay line
TDMA Time division multiple access
TF Transport format
TFCI Transport format combination indicator
TPC Transmit power control
TrCH Transport channel
TSTP Tensor-based space–time precoding
TTI Transmission time interval
UDP User Datagram Protocol
UDPC Up-down power control
UE User equipment
UHPD Users with highest packet delay
UL Uplink
ULA Uniform linear array
ULPT Users with lowest packet throughput
ULTR Users with lowest transmission rate
UM Unacknowledged mode
UMTS Universal Mobile Telecommunication System
URT Estimated user and RAN throughput algorithm
US Uncorrelated scattering
USF Uplink state flag

UTA Utility-based algorithm
UTRAN UMTS terrestrial radio access network
VBLAST Vertical Bell Labs layered space–time
VHO Vertical handover
VOFI Variable orthogonality factor interface
VoIP Voice over IP
VQ Vector quantization
VSER Vector symbol error rate
WAG WLAN access gateways
WCDMA Wideband code division multiple-access
WH Walsh–Hadamard
WIBRO Wireless broadband
WiMAX Worldwide interoperability for microwave access
WLAN Wireless local area network
WPF Weighted proportional fair
WSS Wide-sense stationary
WWW World Wide Web
ZF Zero-forcing
ZMCSCG Zero mean circularly symmetric complex gaussian

Part I
Resource Allocation

Chapter 1
Power Control for Wireless Networks: Conventional and QoS-Flexible Approaches

Fabiano de S. Chaves, Francisco R. P. Cavalcanti, Raimundo A. de Oliveira Neto, and Ricardo B. Santos

1.1 Introduction

Power control is an important functionality in radio resource management (RRM) of wireless communication systems, especially the cellular ones. This importance comes from the fact that the transmission power is an essential radio resource and must be employed in an efficient way. Power control techniques must attain two different objectives: minimize the interference in the wireless system and save energy.

In relation to the first objective, the transmitted power influences the communication quality in the receiver and, at the same time, generates interference for other links which use the same frequency band in the network. As a consequence, an adjustment in the transmission power of a link in order to increase its communication quality can potentially degrade the performance of the other links. Therefore, it is necessary to control the transmitted power so that the received power is the necessary minimum power in order to satisfy the quality requirements and, at the same time, to prevent the generation of unnecessary interference for the other links.

Concerning the second one, power control is essential for energy efficiency, since communications nodes using low power levels mean longer lifetime of batteries for user equipments (UEs) and more energy resources available for central nodes as base stations (BSs) in cellular systems. Therefore, power control serves both to manage the amount of interference in the system and to rationalize the use of energy resources, increasing the system capacity.

The importance of power control technique can be attested by the fact that it was standardized in third-generation wireless systems and therefore requires special attention. Power control has been the subject of attention of a large number of researches. However, few works compile in a systematic way the different approaches in this area. Therefore, this chapter provides a survey about the many facets of power control for wireless communication systems. The concepts are accompanied by results that illustrate potential gains and trade-offs involved in each approach.

F. Cavalcanti, S. Andersson (eds.), *Optimizing Wireless Communication Systems*,
DOI 10.1007/978-1-4419-0155-2_1, © Springer Science+Business Media, LLC 2009

This chapter is organized as follows. The first section presents some basic definitions and the model which describes the problem and its variables in a general framework. After that, a classification of the algorithms according to the type of communication infrastructure available is given. A detailed description of the main algorithms from the literature is provided. In addition, some recent improvements on such algorithms are presented. Moreover, advanced topics such as the application of game theory and prediction techniques aiming the improvement of power control algorithms are exposed. Finally, the conclusions of the chapter are presented.

1.2 Models and Basic Definitions

In the study of power control techniques, some basic definitions are necessary for the perfect understanding of the subject. These concepts are presented in this section.

1.2.1 Propagation Channel

The communication signal is irradiated through a physical medium which permits its transmission between distinct points. The transmission medium is formed by the interaction of electromagnetic radio waves with natural objects, such as atmospheric layers, clouds, fog, rain, trees, mountains, and man-made objects, e.g., buildings and cars. This physical medium of interest here, which can be time-variant, is called the radio propagation channel. When an electromagnetic wave travels through a propagation channel, the transmitted power is affected by a channel gain g. That is, if the transmitter emits a signal with power p_{t}, the received power p_{r} by the receiver will have a value given by

$$p_{\mathrm{r}} = p_{\mathrm{t}} \cdot g. \tag{1.1}$$

In wireless communication systems, most channel models assume that the channel gain g depends on three propagation effects: path loss, shadowing, and short-term fading. The channel gain is then composed of the multiplicative composition of each of these effects. Besides these propagation mechanisms, the broadcast nature of the radio channel leads to co-channel interference among multiple radio links sharing the same frequency bands. These effects are described in detail in the next subsections.

1.2.1.1 Path Loss

When a communication signal travels in space, its power is attenuated by a distance-dependent factor called path loss PL. There are several mathematical models for this phenomenon, depending on the propagation environment. The simplest one is the free-space path loss model. In this case, the received power p_r is proportional to $1/d^2$ (square inverse law) [4], where d is the distance between the transmitter and the receiver. For the non-free-space case, path loss is frequently assumed proportional to $1/d^{\alpha_{PL}}$ [40, 56], where α_{PL} is the path loss exponent which represents the rate in which the path loss increases with distance d. That is, the higher α_{PL}, more attenuating the propagation channel. Common values of α_{PL} vary from 2 to 6 [40, 56]. It is important to emphasize that α_{PL} depends on the specific environment (for instance, urban, rural, micro-cellular, the height of constructions) and the carrier frequency.

Other more accurate path loss models exist in the literature. They may be empirical (e.g., Okumura [36] and Hata [19]), semi-empirical (e.g., Cost-231 [5]), or deterministic, such as the ones based on ray-tracing [55]. In this chapter it is assumed that the path loss is only a function of distance, therefore calculated as

$$PL(d) = K_{PL} d^{-\alpha_{PL}}, \tag{1.2}$$

where K_{PL} represents the perceived path loss at the reference distance $d = 1$ in the same unit of d.

Sometimes it is preferable to express (1.2) in decibel scale (dB). This can be achieved by

$$\overline{PL}(d) = \overline{K}_{PL} - 10\alpha_{PL}\log_{10}(d), \tag{1.3}$$

where $\overline{K}_{PL}$ is the reference path loss K_{PL} in decibel scale.

1.2.1.2 Shadowing

The second phenomenon which affects the channel gain g is shadowing, caused by statistic fluctuations around a mean value. If a transmitter emits a communication signal, the path loss $PL(d)$ for all points positioned at a distance d of the transmitter is the same, according to (1.2); however, the channel gains g are different. Such phenomenon occurs due to shadowing. Shadowing provokes variations in the channel gain around its mean (given by the path loss). In cellular systems, shadowing can be modeled as a log-normal random variable.[1.1]

Among the many causes of shadowing, obstructions of communication signals due to large obstacles can be emphasized. Considering the carrier frequencies usually employed in cellular systems (1-2 GHz), the time scale of power variation due to shadowing is on the order of tens to hundreds of wavelengths. In academic literature, shadowing is also known as large-scale fading [40, 56].

[1.1] If the channel gain is expressed in dB scale, then the additional (over the path loss) shadowing component can be modeled by a zero-mean normal random variable with a given standard deviation. This standard deviation is a characteristic of the environment and typically ranges from 6 to 12 dB [40, 56].

1.2.1.3 Short-Term Fading

The third component of channel gain, denominated short-term fading, is caused by fast fluctuations in the amplitude of the communication signal in a short period of time or in a short displaced distance [40, 56]. The main cause of these fluctuations is the combination of different propagation paths of several replicas of the transmitted signal (multi-paths). These replicas arrive in the receiver in slightly distinct instants and with random amplitude and phase. This occurs due to reflection and scattering of the signal during its trajectory. Reflections happen when the signal reaches obstacles with larger size than the wavelength of the signal. On the other hand, scattering occurs when the objects' dimensions are in the same order of magnitude of the wavelength. When several replicas of the signal arrive in the receiver, their distinct phases add randomly both constructively and destructively, resulting in fast fluctuations on the amplitude of the received composite signal. Consequently, the signal power will vary rapidly too.

Beyond the addition of replicas of the signal, a second important factor which influences the short-term fading is the speed of the mobile station (MS) as well as of the objects around it in the propagation environment. The faster the MS moves, the faster it experiments the signal power variations in time.

These three main signal propagation mechanisms (path loss, shadowing, and fast fading) overlap in time and space. Figure 1.1 illustrates the superposition of propagation mechanisms as a mobile user covers a given distance.

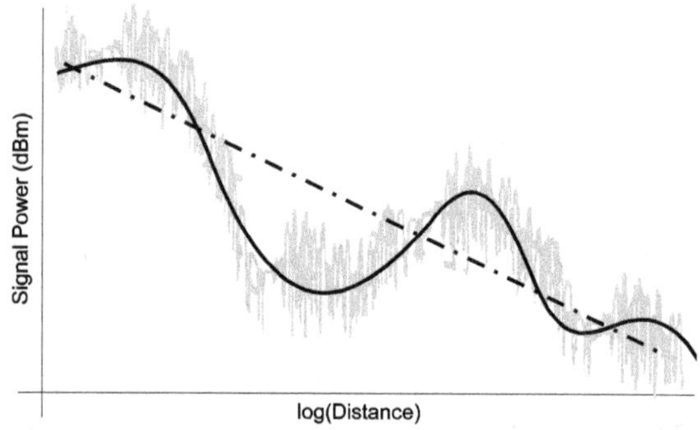

Fig. 1.1 Signal propagation mechanisms.

1.2.1.4 Co-channel Interference

One of the objectives of implementing power control algorithms is to control the excess of interference in the cellular system. This interference originates by the frequency reuse scheme in the system. This scheme permits the same frequency band to be used in different cells, according to a planned way, with the objective of increasing the capacity of the system. This is necessary because the available spectrum for each cellular operator is very limited for the user demand. Frequency reuse is a fundamental concept applied in all high-capacity cellular systems, independent of the specific multiple access scheme employed.

The most traditional approach to frequency reuse is based on a fixed channel allocation scheme, where adjacent cells are grouped into clusters of a suitable number of cells N, as seen in Fig. 1.2 for $N = 3$. The available channels in the frequency pool of the system are divided into N subsets of channels, and each subset is allocated to a cell in the cluster. After that, this pattern is replicated over all clusters. Cells with the same subset of channels are called co-channel cells and cause co-channel interference, due to the utilization of the same frequency bands (or, simply, channels), which enables the reception of non-desirable signals from other co-channel cells, as illustrated in Fig. 1.2.[1.2]

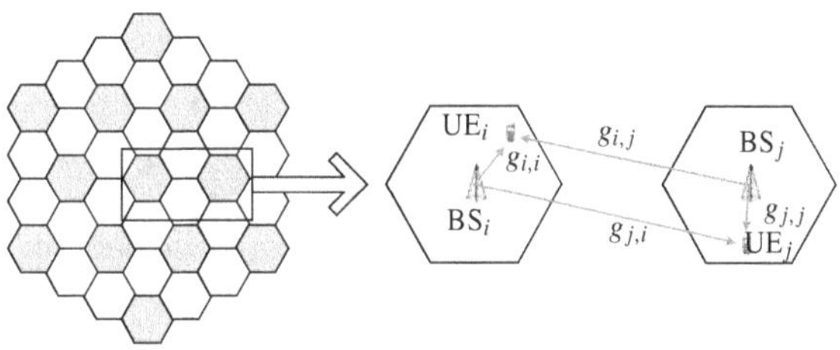

Fig. 1.2 Cell grid and co-channel interference generated by the frequency reuse.

The amount of co-channel interference affects the quality of the received signal. This interfering power will compete with the power of the desired signal. The result of this competition can be measured through the signal-to-interference-plus-noise ratio (SINR). The SINR in the ith link is calculated as

$$\gamma_i = \frac{g_{i,i} \cdot p_i}{I_i}, \tag{1.4}$$

where $g_{i,i}$ is the channel gain, p_i is the transmission power, and I_i is the co-channel interference plus noise in the ith link given by

$$I_i = \sum_{j \neq i}^{N} g_{i,j} \cdot p_j + \nu_i, \tag{1.5}$$

where N is the number of co-channel links and $g_{i,j}$ is the channel gain between transmitter j and receiver i, while ν_i is the noise power relative to ith link. Sometimes, for purposes of mathematical modeling, the noise power is neglected, and the SINR is therefore reduced to signal-to-interference ratio (SIR).

Performance measures like average and instantaneous data throughput, packet reception delay, and bit error rate (BER) can be considered as quality of service (QoS) measures. Since these performance metrics are related to the quality of communication links, which is commonly quantified by the SINR, the later is an indirect measure of QoS and an important quantity to be controlled.

[1.2] The hexagonal layout in Fig. 1.2 is illustrative as other cell arrangements are possible [57].

According to (1.4) and (1.5), the transmission powers directly affect the SINR. Therefore, transmission power control plays a key role in radio resource management (RRM), whose goal is to control the quality of the received signal. Requirements on the quality of the received signal can be different for distinct services. For instance, services requiring higher BERs will demand higher SINRs. In general, different levels of QoS requirements can be translated into different target SINR levels.

1.2.1.5 Classification of Power Control Algorithms

The power control algorithms can be classified according to the signalization and control architecture employed. This classification divides the algorithms into two groups: centralized and decentralized (or distributed) algorithms. In centralized schemes, a central controller has all information about the established connections and channel gains at every moment and controls all the transmitted powers of all mobile stations in the network [35]. Centralized power control requires extensive control signaling in the network and, therefore, is hard to be applied in practice. It can be used to determine superior bounds on the performance of decentralized algorithms. In its turn, a distributed implementation employs several decentralized controllers, where each one individually controls the power of the transmitters under its management, generally located in the same cell. In this case, the algorithm depends only on local information, such as SINR or channel gain of the specific user.

1.3 Centralized Power Control

The centralized scheme presented in this section was originally presented by Zander in [61]. It is based in the solution of a system of inequalities, where the variables are the transmission powers. Linear algebra elements are employed in its solution.

1.3.1 Problem Formulation

In this modeling, it is assumed that the noise power ν is null, therefore SIRs are considered instead of SINRs. Zander's algorithm has the objective of maximizing the minimum SIR of all co-channel links. In other words, the final goal is to find the maximum SIR that can be achieved in all co-channel links and the corresponding powers. This is equivalent to maximizing the balanced (equalized) SIRs. The solution consists in solving a system of inequalities using the Perron–Frobenius Theorem [11] (see Theorem 1.1 below). The following paragraphs describe the development of the Zander's algorithm .

Let $\mathbf{N} = \{1,\ldots,N\}$ be the set of N co-channel links in N different cells. The SIR of the ith link can thus be expressed by

$$\gamma_i = \frac{p_i}{\sum_{j=1}^{N} \frac{g_{i,j}}{g_{i,i}} \cdot p_j - p_i} = \frac{p_i}{\sum_{j=1}^{N} z_{i,j} \cdot p_j - p_i}, \quad \forall i \in \mathbf{N}, \tag{1.6}$$

where $z_{i,j} = \frac{g_{i,j}}{g_{i,i}}$ is the normalized channel gain in relation to link i. Let $\gamma_{\min}$ be the lowest SIR of all co-channel links. Therefore, for each link i,

$$\frac{p_i}{\sum_{j=1}^{N} z_{i,j} \cdot p_j - p_i} \geq \gamma_{\min}, \quad \forall i \in \mathbf{N}. \tag{1.7}$$

Converting (1.7) in matrix form,

$$\frac{1+\gamma_{\min}}{\gamma_{\min}} \mathbf{p} \geq \mathbf{Z}\mathbf{p}, \tag{1.8}$$

where $\mathbf{p} = [p_1,\ldots,p_N]^{\mathrm{T}}$ and $\mathbf{Z} = [z_{i,j}]$, $i,j \in \{1,...,N\}$ are the power vector and the normalized channel gain matrix, respectively. The intent is to find the positive power vector and the maximum balanced $\gamma_{\min}$ that satisfy (1.8). The Perron–Frobenius Theorem [11] is used to solve problems concerning non-negative matrix inequalities. The theorem is stated as follows:

Theorem 1.1 (Perron–Frobenius Theorem). *Given a non-negative irreducible matrix* $\mathbf{A}$,

- $\mathbf{A}$ *has exactly one real positive eigenvalue* λ^* *for which the corresponding eigenvector is positive.*
- *The minimum real* λ *such that the inequality* $\lambda \cdot \mathbf{b} \geq \mathbf{A}\mathbf{b}$ *has solutions for* $\mathbf{b} > 0$ *is* $\lambda = \lambda^*$.
- *The maximum real* λ *such that the inequality* $\lambda \cdot \mathbf{b} \leq \mathbf{A}\mathbf{b}$ *has solutions for* $\mathbf{b} > 0$ *is* $\lambda = \lambda^*$.

Notice that $\mathbf{Z}$ is a positive matrix. Then, the Perron–Frobenius Theorem is applicable to (1.8). Moreover, (1.8) is in the form $\lambda \cdot \mathbf{b} \geq \mathbf{A}\mathbf{b}$, with the following correspondences:

$$\lambda = \frac{1+\gamma_{\min}}{\gamma_{\min}}, \tag{1.9}$$

$$\mathbf{A} = \mathbf{Z} \quad \text{and} \quad \mathbf{b} = \mathbf{p}. \tag{1.10}$$

Applying the first and second propositions of the Perron–Frobenius Theorem to (1.8), it can be observed that there exists only one pair $(\lambda^*, \mathbf{p}^*)$, where λ^* is the minimum λ which satisfies $\lambda \cdot \mathbf{p} \geq \mathbf{Z}\mathbf{p}$, for $\mathbf{p} > 0$, with λ given by (1.9). According to (1.9), a minimum value of λ corresponds to a maximum value of γ_{min}. Therefore, the maximum value of γ_{min} which satisfies (1.8) is given by $\frac{1}{\lambda^*-1}$.

On the other hand, it is important to observe the other SIRs. For all links, the maximum SIR of all co-channel links γ_{max} can be written as

$$\frac{p_i}{\sum_{j=1}^{N} z_{i,j} \cdot p_j - p_i} \leq \gamma_{\text{max}}, \quad \forall i \in \mathbf{N}. \tag{1.11}$$

Expressing (1.11) in matrix form,

$$\frac{1+\gamma_{\text{max}}}{\gamma_{\text{max}}} \mathbf{p} \leq \mathbf{Zp}. \tag{1.12}$$

Now, applying the first and third propositions of the Perron–Frobenius Theorem to (1.12), it can be observed that there exists only one pair $(\lambda^*, \mathbf{p}^*)$ such that λ^* is the maximum λ which satisfies $\lambda \cdot \mathbf{p} \leq \mathbf{Zp}$ for $\mathbf{p} > 0$, with $\lambda = \frac{1+\gamma_{\text{max}}}{\gamma_{\text{max}}}$. In this case, a maximum value of λ corresponds to a minimum value of γ_{max}. Therefore, the minimum value of γ_{max} which satisfies (1.12) is given by $\frac{1}{\lambda^*-1}$. Since the maximum value of γ_{min} and the minimum value of γ_{max} are the same, determined by the eigenvector $\mathbf{p}^*$ relative to the eigenvalue λ^* of matrix $\mathbf{Z}$, the choice of power vector $\mathbf{p}^*$ maximizes the balanced SIR of all co-channel links, given as

$$\gamma^* = \frac{1}{\lambda^* - 1}. \tag{1.13}$$

The balanced SIR depends on the normalized channel gain matrix, that is, the instantaneous propagation conditions. Then it is possible that in some instants, the SIRs of all links can be found below a threshold for acceptable communication. This would be an extremely bad situation, which can be avoided by using some link-removal techniques. The balanced SIR will increase at the cost of penalizing some links with a temporary interruption of transmission. At the same time, it is desirable to reestablish acceptable communication by removing a minimum number of links. This demands a suitable method for the choice of links to be removed. In fact, it is convenient to classify each link as active or inactive (temporarily denied to transmit), since the link-removal procedure is periodic, and make decisions based on the quality of time-varying channels.

1.3.2 Stepwise Removal Methods

The stepwise removal algorithm (SRA) proposed in [61] removes links, one by one, until the SIR achieved in the remaining links is greater than or equal to a required threshold. SRA consists of two steps:

Step 1: Determine the maximum feasible SIR γ^* from the normalized channel gain matrix **Z** given by (1.13). If $\gamma^* \geq \gamma_{\min}$ (in this case, $\gamma_{\min}$ represents the minimum threshold for acceptable communication), utilize the eigenvector correspondent to the eigenvalue λ^* (Theorem 1.1) as power vector and stop. If $\gamma^* < \gamma_{\min}$, execute step 2.

Step 2: Remove the link n for which the sum of its row and column in the normalized channel gain matrix **Z**

$$\sum_{i=1}^{N} z_{n,i} + z_{i,n} \tag{1.14}$$

is maximized and thus forms a new square sub-matrix $\mathbf{Z}'$ with dimension $(N-1)$. Determine the new γ^* corresponding to $\mathbf{Z}'$. If $\gamma^* \geq \gamma_{\min}$ (power control is feasible), utilize the corresponding eigenvector as power vector, else repeat step 2 until γ^* becomes larger than $\gamma_{\min}$.

The row and column sums provide bounds on the dominant eigenvalue of matrix **Z**. This removal procedure seeks to maximize the lower bound for γ^* [11, 61].

Other stepwise link-removal methods use information of transmission powers in addition to the normalized channel gain matrix **Z**. The idea is that the larger the transmission power, the greater the interference it causes to other co-channel links. The stepwise maximum-interference removal algorithm (SMIRA), proposed in [26], removes the link which causes the highest total interference power or the one with the highest received interference power until the balanced SIR of the remaining links is larger than or equal to the minimum specified threshold. SMIRA is shown to outperform the SRA.

1.4 Distributed Power Control

In spite of the fact that centralized power control finds an optimum solution, its practical implementation in wireless systems is very difficult. This occurs because measuring all channel gains in real time (including the interfering ones) is very costly due to the signaling overheads involved. Therefore, distributed solutions are desirable for practical purposes. In this section are described algorithms which make the power control of each link the most independently possible of the channel gains of the other co-channel links. In a distributed implementation, each link controls its transmission power based only on measurements of its own signal quality.

1.4.1 Distributed Balancing Algorithm

In [60], Zander proposed a distributed version for his centralized algorithm, previously presented in Section 1.3. In this approach, the algorithm assumes an iterative rule executed individually for each co-channel transmitter, unlike the centralized version, whose solution is given instantaneously by the central controller. Considering $\mathbf{N} = \{1,\ldots,N\}$, the set of N co-channel links, Zander's distributed balancing algorithm (DBA) is given by the following iterative function:

$$p_i(k+1) = \xi_{\text{DBA}} p_i(k) \left(1 + \frac{1}{\gamma_i(k)}\right), \quad \forall i \in \mathbf{N}, \tag{1.15}$$

where ξ_{DBA} is a positive factor of proportionality which must be chosen adequately and $\gamma_i(k)$ is again the SIR of link i. DBA makes the SIRs converge to γ^* defined in (1.13) as a function of λ^*, the unique real positive eigenvalue of the normalized channel gain matrix for which the corresponding eigenvector is positive.

In relation to the transmission powers, they converge to a multiple of the eigenvector which corresponds to λ^*. Since the terms in (1.15) are all positive, the transmission power is an increasing sequence. The factor ξ_{DBA} can be used to avoid the uncontrolled power increase by adjusting it at each iteration k according to

$$\xi_{\text{DBA}}(k) = \frac{1}{||\mathbf{p}(k)||}, \tag{1.16}$$

where $||\mathbf{p}(k)||$ is the norm of the power vector $\mathbf{p}(k) = [p_1(k),\ldots,p_N(k)]^{\text{T}}$. However, this trick to limit the powers in the convergence process yields a loss of decentralization, since it requires the instantaneous powers of all co-channel links. Another option would be to hard-limit all transmission powers, but this clearly leads to a loss of optimization in the convergence process of DBA.

Example 1.1 (SIR Balancing Using DBA). Consider a set of seven co-channel links in a cellular system with frequency reuse 3. The path loss model is $\overline{PL}(d) = 128 + 38\log_{10}(d) + \chi$, where d is expressed in kilometers, and the shadowing component is also incorporated, a zero-mean normal random variable with standard deviation set to 6. Figure 1.3(a) shows the SIR convergence for the DBA. This figure illustrates that all SIRs converge for the same value 11 dB. This value can be found through the normalized channel gain matrix, using (1.13).

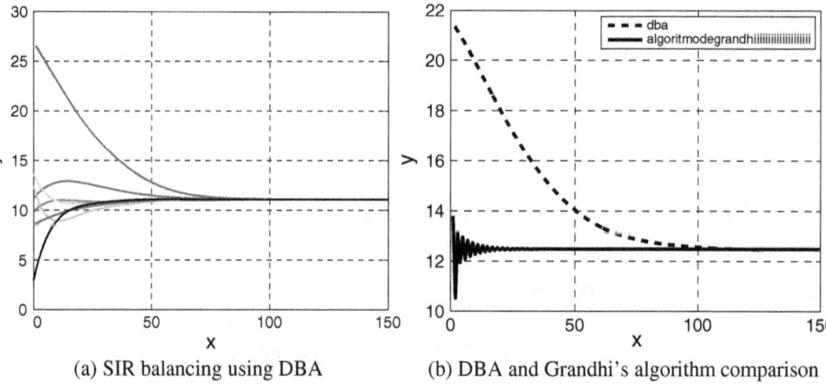

(a) SIR balancing using DBA (b) DBA and Grandhi's algorithm comparison

Fig. 1.3 SIR convergence in Examples 1.1 and 1.2.

1.4.2 Grandhi's Algorithm

Two terms influence the convergence of the iterative function of DBA, rewritten here in a different way in order to enlighten the discussion:

$$p_i(k+1) = \xi_{\mathrm{DBA}} p_i(k) + \xi_{\mathrm{DBA}} p_i(k) \frac{1}{\gamma_i(k)}. \tag{1.17}$$

The first term involves only the current transmitted power $p_i(k)$. Due to the presence of $\gamma_i(k)$ in the denominator, the second term is responsible for the convergence in the direction of the balanced SIR γ^*. In order to increase the convergence speed of the iterative process, the first term in (1.17) can be removed. The result is the Grandhi's algorithm, proposed in [16] and proved to be faster than DBA. Grandhi's algorithm can be expressed as

$$p_i(k+1) = \varepsilon_{\mathrm{G}} \frac{p_i(k)}{\gamma_i(k)}, \quad \forall i \in \mathbf{N}, \tag{1.18}$$

where ε_{G} is a positive constant of proportionality. It is important to emphasize that the SIRs also converge for the same value obtained by DBA, given by (1.13). The powers also converge to a multiple of the eigenvector corresponding to λ^*, in spite of the same problem concerning the increasing (or decreasing, in this algorithm) of the powers, which can be solved through the adjustment of the factor ε_{G} [16]:

$$\varepsilon_{\mathrm{G}}(k) = \frac{1}{\max\{\mathbf{p}(k)\}}, \tag{1.19}$$

where $\max\{\mathbf{p}(k)\}$ is the largest element of the power vector $\mathbf{p}(k)$. The adjustment of ε_{G} would also require some coordination among the co-channel links.

Example 1.2 (Comparison Between DBA and Grandhi's Algorithm). Consider a set of seven co-channel links in a cellular system with frequency reuse 3. The path loss model is the same as that of Example 1.1. Figure 1.3(b) shows the SIR convergence with the DBA and Grandhi's algorithm for the same link, with the same channel gains. As was expected, both attain the same balanced SIR; however, Grandhi's algorithm converges faster than DBA. While the first reaches the balanced SIR in 50 iterations, the last one converges in 130 iterations.

1.4.3 Distributed Power Control Algorithm

The algorithms presented so far are based only on SIR (null noise power) and therefore are idealized algorithms. Besides, there is no control over the balanced SIR because this depends on the normalized channel gain matrix, given by the propagation conditions.

This section presents an algorithm designed to work in the presence of noise, where it is possible to have certain control about the balanced SINR. This algorithm was proposed by Foschini and Miljanic in [9] and is referred to as distributed power control (DPC) algorithm. Its starting point is a differential equation, whose interpretation is the balancing of the SINR in each link i to a prescribed target value:

$$\gamma_i{}'(t) = -\beta_{\mathrm{F}}[\gamma_i(t) - \gamma^t], \quad \forall i \in \mathbf{N}, \tag{1.20}$$

where γ^t is the target SINR, $\gamma_i{}'(t)$ is the derivative of $\gamma_i(t)$ with respect to time, and β_{F} is a positive proportionality constant. According to the differential dynamics, the SINR evolves so that it converges to the target SINR by an amount proportional to the offset between both. Therefore, this dynamics will not stop unless $\gamma_i(t) = \gamma^t$. Different from the algorithms based on SIR whose resulting balanced SIR is dependent on the propagation conditions, the DPC algorithm will make the SINRs converge to a prescribed target SINR, provided that this target value is feasible. This latter aspect is discussed in detail in Section 1.5.

Substituting the SINR formula (1.4) in (1.20) results in

$$\left[\frac{g_{i,i}(t) \cdot p_i(t)}{I_i(t)}\right]' = -\beta_{\mathrm{F}}[\gamma_i(t) - \gamma^t]. \tag{1.21}$$

In a distributed implementation, the BS or MS can control only its own transmission power $p_i(t)$. The interference $I_i(t)$ and the channel gain $g_{i,i}(t)$ cannot be controlled. Therefore, considering only the temporal variation of transmission power $p_i(t)$, (1.21) becomes

$$p_i'(t) = -\frac{I_i(t)}{g_{i,i}(t)} \cdot \beta_{\mathrm{F}}[\gamma_i(t) - \gamma^t]. \tag{1.22}$$

On transforming (1.22) to discrete time, it becomes a difference equation:

$$p_i(k+1) - p_i(k) = -\beta_{\mathrm{F}} \cdot \frac{I_i(k) \cdot \gamma_i(k)}{g_{i,i}(k)} + \beta_{\mathrm{F}} \cdot \frac{\gamma^t \cdot I_i(k)}{g_{i,i}(k)}, \tag{1.23}$$

where k is the time index and $p_i(k)$, $g_i(k)$, $I_i(k)$, and $\gamma_i(k)$ are, respectively, the transmission power, the channel gain, the interference, and the SINR. Replacing $\frac{I_i(k)\cdot\gamma_i(k)}{g_{i,i}(k)}$ by $p_i(k)$ and $\frac{I_i(k)}{g_{i,i}(k)}$ by $\frac{p_i(k)}{\gamma_i(k)}$ and recombining the terms result in the DPC algorithm:

$$p_i(k+1) = p_i(k)\left[1 - \beta_{\mathrm{F}} + \beta_{\mathrm{F}} \cdot \frac{\gamma^t}{\gamma_i(k)}\right], \quad \forall i \in \mathbf{N}. \tag{1.24}$$

The positive proportionality constant β_{F} plays a key role in the stability and convergence of DPC. Assume that for a given channel gain matrix and a set of individual target SINR requirements $[\gamma_1^t, \ldots, \gamma_N^t]$, the power vector which makes the individual SINR requirements to be attained is $\mathbf{p}^* = [p_1^*, \ldots, p_N^*]^{\mathrm{T}}$. β_{F} can be set to $\beta_{\mathrm{F}} = 1$ in order to assure the convergence of $p_i(k)$ to the corresponding p_i^* in (1.24). This value of β_{F} is shown to be the largest possible value while any value of β_{F} in $(0,1]$ is called universal and also valid [9]. As can be observed in (1.24), DPC with low values of β_F is less responsive to changing conditions. In Section 1.5, the choice $\beta_F = 1$ is shown to give the fastest convergence.

1.4.4 Up–Down Algorithm

This is a simple power control algorithm also known as the fixed-step power control algorithm. In this scheme, at each power update period, the algorithm sets a power control command which increases or decreases the transmitted power by one step δ_{UD} or keeps it constant. The choice of this action is carried out based on the comparison between the actual SINR and a target SINR γ^t. Therefore, in this iterative process, the up–down algorithm will attempt to reach a target SINR γ^t.

Like DPC, the up–down algorithm will converge provided there exists a feasible solution. Due to its simple formulation, practical systems such as UMTS/WCDMA (universal mobile telecommunication system/wideband code division multiple access) use this algorithm with $\overline{\delta}_{UD} = 1$ dB [22].

1.5 Feasibility and Convergence Aspects of Distributed Power Control

In this section feasibility and convergence aspects of target tracking distributed power control are investigated. First, this section makes a particular analysis for the DPC algorithm, after which a more general framework is exposed.

By using the DPC algorithm, two questions arise. Will this algorithm get to balance the SINRs of all co-channel links for any target SINR γ^t chosen? Which value of β_F yields the highest convergence speed? For didactic purposes, the case of two co-channel links is considered in both analyses. However, the conclusions are valid to the general N co-channel link framework.

In order to answer the first question, it is important to establish a necessary and sufficient condition for the feasibility of a given target SINR. A target SINR γ^t is feasible when the following system of equations has positive solutions in p_1 and p_2:

$$\gamma^t = \frac{g_{1,1} \cdot p_1}{g_{2,1} \cdot p_2 + \nu_1} = \frac{g_{2,2} \cdot p_2}{g_{1,2} \cdot p_1 + \nu_2}. \tag{1.25}$$

Rewriting (1.25) in matrix form,

$$\mathbf{B} \cdot \mathbf{p} = \mathbf{n}, \tag{1.26}$$

where

$$\mathbf{B} = \begin{bmatrix} 1 & -\gamma^t \frac{g_{2,1}}{g_{1,1}} \\ -\gamma^t \frac{g_{1,2}}{g_{2,2}} & 1 \end{bmatrix}; \quad \mathbf{p} = \begin{bmatrix} p_1 \\ p_2 \end{bmatrix}; \quad \mathbf{n} = \begin{bmatrix} \gamma^t \frac{\nu_1}{g_{1,1}} \\ \gamma^t \frac{\nu_2}{g_{2,2}} \end{bmatrix}. \tag{1.27}$$

The resolution of (1.26) gives

$$p_1 = \frac{\gamma^t}{1-(\gamma^t \vartheta)^2} \left(\frac{v_1 \cdot g_{2,2} + v_2 \cdot \gamma^t \cdot g_{2,1}}{g_{1,1} \cdot g_{2,2}} \right), \tag{1.28}$$

$$p_2 = \frac{\gamma^t}{1-(\gamma^t \vartheta)^2} \left(\frac{v_2 \cdot g_{1,1} + v_1 \cdot \gamma^t \cdot g_{1,2}}{g_{1,1} \cdot g_{2,2}} \right), \tag{1.29}$$

where $\vartheta = \left(\frac{g_{1,2} \cdot g_{2,1}}{g_{1,1} \cdot g_{2,2}} \right)^{1/2}$. From (1.28) and (1.29), it can be observed that the system will have positive solutions when $\gamma^t \vartheta < 1$. Note that ϑ is the magnitude of the eigenvalue with highest absolute value of the matrix $\mathbf{C} = \frac{1}{\gamma^t}(\mathbf{I} - \mathbf{B})$, where $\mathbf{I}$ is the identity matrix. This criterion for the target SINR feasibility is also valid for N co-channel links [62].

Answering the second question, the convergence speed of the DPC algorithm is maximized when $\beta_\text{F} = 1$ [9]. In this case, the algorithm becomes

$$p_i(k+1) = p_i(k) \cdot \frac{\gamma^t}{\gamma_i(k)}, \quad \forall i \in \mathbf{N}. \tag{1.30}$$

Expressing (1.24) in matrix form,

$$\mathbf{p}(k+1) = \mathbf{D}\mathbf{p}(k) + \beta_\text{F}\mathbf{n}, \tag{1.31}$$

where $\mathbf{D} = (\mathbf{I} - \beta_\text{F}\mathbf{B})$. After $(k-1)$ iterations, $\mathbf{p}(k)$ can be expressed as

$$\mathbf{p}(k) = (\mathbf{I} + \mathbf{D} + \mathbf{D}^2 + \cdots + \mathbf{D}^{k-2})\beta\mathbf{n} + \mathbf{D}^{k-1}\mathbf{p}(1). \tag{1.32}$$

If the magnitude of all eigenvalues of $\mathbf{D}$ is lower than 1, then the series $(\mathbf{I} + \mathbf{D} + \mathbf{D}^2 + \cdots + \mathbf{D}^{k-2})$ will converge to $(\mathbf{I} - \mathbf{D})^{-1}$ [49]. Thus, assuming $\gamma^t \vartheta < 1$, i.e., the absolute value of each eigenvalue of $\mathbf{D}$ is strictly lower than 1,

$$\lim_{k \to \infty} \mathbf{p}(k) = (\mathbf{I} - \mathbf{D})^{-1}\beta_\text{F}\mathbf{n} = (\beta_\text{F}\mathbf{B})^{-1}\beta_\text{F}\mathbf{n} = \mathbf{B}^{-1}\mathbf{n} = \mathbf{p}, \tag{1.33}$$

which means that the powers will converge to the values given by the solution of (1.26), with the SINRs balanced in the value γ^t.

According to this, the lower the magnitude of the eigenvalue with highest magnitude, the higher the convergence speed; the optimum value of β_F is the one which minimizes the magnitude of the eigenvalue with highest magnitude of matrix $\mathbf{D}$. The eigenvalues of matrix $\mathbf{D}$ for the case of two co-channel links are $\lambda_1 = (1 - \beta_\text{F}) + \gamma^t \beta_\text{F} \gamma$ and $\lambda_2 = (1 - \beta_\text{F}) - \gamma^t \beta_\text{F} \vartheta$. Therefore, the fastest convergence is achieved with $\beta_\text{F} = 1$. Observe that this value of β_F makes $|\lambda_1| = |\lambda_2| = \gamma^t \vartheta < 1$. This is illustrated in Fig. 1.4. In [9], the optimality of $\beta_\text{F} = 1$ regarding the convergence speed of DPC is proved for the general case of N co-channel links.

The previous analysis of convergence and stability is restricted to the particular case of two co-channel links. A general analysis is considered in this section. The

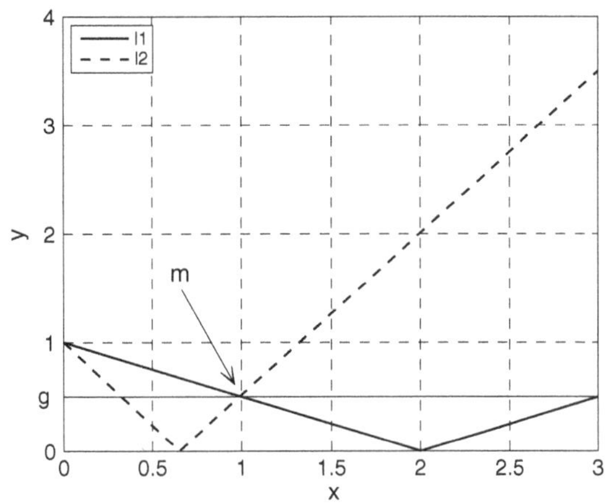

Fig. 1.4 Analysis of the values of β_F for two co-channel links.

standard power control framework developed in [58] is suitable for a broad class of distributed power control algorithms. It can be applied to the conventional distributed power control algorithms, characterized by prescribed SINR requirements of individual communication links, such as DBA, Grandhi's, and DPC algorithms.

The standard power control framework is established by identifying common properties of the interference constraints that permit a general proof of convergence to a unique fixed point. The power update process can be represented as an iterative function, as follows:

$$\mathbf{p}(k+1) = \boldsymbol{\zeta}(\mathbf{p}(k)), \tag{1.34}$$

where $\boldsymbol{\zeta}(\mathbf{p}(k)) = [\zeta_1(\mathbf{p}(k)),\ldots,\zeta_N(\mathbf{p}(k))]^{\mathrm{T}}$ is the iterative vector function associated with a general distributed power control problem with N co-channel links.

Definition 1.1. An iterative vector function $\boldsymbol{\zeta}(\mathbf{p})$ is said to be standard if it satisfies the following conditions[1.3,1.4]:

C-1 Monotonicity: If $\mathbf{p} \leq \mathbf{p}'$, then $\boldsymbol{\zeta}(\mathbf{p}) \leq \boldsymbol{\zeta}(\mathbf{p}')$.
C-2 Scalability: For all $\alpha > 1$, $\boldsymbol{\zeta}(\alpha\mathbf{p}) < \alpha\boldsymbol{\zeta}(\mathbf{p})$.

The interesting properties of standard iterative functions are stated in the following three theorems. Their proofs are found in [58].

Theorem 1.2. *If $\boldsymbol{\zeta}(\mathbf{p})$ is standard and a fixed point exists, then the fixed point is unique.*

Theorem 1.3. *If $\boldsymbol{\zeta}(\mathbf{p})$ is standard and a fixed point $\mathbf{p}^*$ exists, then any power vector $\mathbf{p}$ converges to $\mathbf{p}^*$.*

[1.3] In [58], $\boldsymbol{\zeta}(\mathbf{p})$ is called an interference function, since it represents the effective interference that transmitters must overcome. However, this section refers to it more generally as iterative function.

[1.4] The positivity property, present in the original definition of standard functions in [58], can be shown as a consequence of monotonicity and scalability [29].

Theorem 1.4. *If a fixed point* $\mathbf{p}^*$ *exists, then any power vector* $\mathbf{p}$ *converges to* $\mathbf{p}^*$ *under the totally asynchronous model.*

In order to have a clearer idea of the effective impacts of such theorems, some remarks follow. First, a fixed point $\mathbf{p}^*$ of a function $\boldsymbol{\zeta}(\mathbf{p})$ is such that $\boldsymbol{\zeta}(\mathbf{p}^*) = \mathbf{p}^*$. This means that if an iterative function meets a fixed point, it remains at this point. In the context of distributed power control, a fixed point of the power update iterative vector function corresponds to a system operating point. Therefore, according to Theorem 1.2, if a standard power update iterative vector function has a fixed point, only one system operating point exists.

Theorems 1.3 and 1.4 are related to the convergence of power updates to a fixed point. Theorem 1.3 assures the convergence of any power vector $\mathbf{p}$ to the fixed point $\mathbf{p}^*$. The proof of such theorem is based on a synchronous network operation, where all transmitters decide simultaneously for their own transmission powers at every power control actuation. The power update convergence in asynchronous network operation mode is addressed in Theorem 1.4.

Therefore, on the condition that there exists a fixed point, the class of standard power control algorithms has its convergence guaranteed in a network-wide synchronous or asynchronous mode. It is shown in [58] also that the continuity of the standard function and the introduction of an upper bound to the transmission power are sufficient conditions for the existence of a fixed point.

This important issue of convergence and stability of distributed power control algorithms is addressed in other parts of this chapter, where "non-conventional" approaches are treated. The theories of *type-II standard iterative functions* [51] and *two-sided scalable iterative functions* [51] are used in Sections 1.6.2 and 1.7.3, respectively, to prove convergence and stability of different algorithms.

1.6 Power Control for QoS-Flexible Services

Quality of service is a measure or a set of measures which indicates the degree of satisfaction of an individual communication link with the service provision or the efficiency of the network in managing the available resources. As QoS measures can be considered the average and instantaneous data throughput, packet reception delay, and BER. The relationship between these performance metrics and the SINR justifies the use of the later as an indirect measure of QoS at the physical layer.

In previous sections, the distributed power control problem is characterized by the single objective of meeting fixed prescribed SINR requirements of individual communication links. This power control framework is very suitable for services with strict QoS requirements, like voice communications with prescribed transmission and bit error rates. A fixed BER can be translated to a fixed target SINR depending on the modulation and coding scheme employed for transmission. All algorithms discussed in Section 1.4 seek to solve this QoS-fixed problem.

Emerging wireless networks are required to provide multiple services with distinct characteristics. Besides the traditional voice service with strict QoS requirement, some data services, such as world wide web (WWW) browsing and file download, tolerate larger variations of the link quality. For a fixed BER, variations in the link quality, i.e., in the SINR, are accommodated by changing the transmitter's data rate. This is usually accomplished by means of adaptive modulation and coding (AMC) [13]. Assuming the employment of AMC schemes, the relationship between individual link capacity C (in bits/s) and the SINR can be represented by a Shannon's channel capacity-like expression:

$$C = B\log_2(1+\kappa\gamma), \tag{1.35}$$

where B is the channel bandwidth and κ is a loss factor.

This mixed-service scenario requires a different treatment from the point of view of power control, since QoS-flexible applications allow, accordingly, more flexible approaches with the exploitation of the quality of communication links for an efficient resource allocation and QoS provision. In this section, some power control procedures suitable for exploitation of the flexible quality of communication links in data services are discussed.

1.6.1 Techniques of Target SINR Decreasing

Power control is said to be feasible if, given the individual SINR requirement of every communication link, there exists a feasible power vector which satisfies all SINR requirements. In Section 1.3, the Perron–Frobenius Theorem is used to obtain the maximum common SINR level for all communication links which results in feasible power control. Furthermore, algorithms are developed to remove those critical links which cause infeasibility. Moreover, the focus of those approaches are on fixed SINR requirements, i.e., on power control for QoS-fixed applications. In this section, communication links are assumed to be tolerant to flexible QoS (i.e., SINR) and the aim is to avoid infeasibility in a decentralized fashion. To this end, an adaptive target SINR is adopted for each link. The target SINR is adjusted according to the transmission power.

The essential idea of this approach is that communication links in worse propagation conditions have to use higher transmission powers to attain a given target SINR level. Consequently, they cause excessive interference to co-channel links. Therefore, it would be desirable to decrease the target SINR of such links in critical propagation conditions. This would also imply a reduced transmission rate for such link according to an adaptive modulation and coding mechanism. On the other hand, more favored links using low transmission powers could attain substantially higher levels of SINR with a slight increase in their powers without disturbing co-channel links with more interference.

The main target SINR decreasing approach, the soft dropping power control, is proposed in [59]. Transmission power update in soft dropping power control is given by

$$\overline{p}_i(k+1) = \overline{p}_i(k) + \overline{\gamma}_i^t(\overline{p}_i(k)) - \overline{\gamma}_i(k), \tag{1.36}$$

which has the same target tracking structure of DPC algorithm, but instead of a fixed target SINR, the SINR to be targeted in time instant $(k+1)$ is a function of transmission power $\overline{p}_i(k)$, that is $\overline{\gamma}_i^t(\overline{p}_i(k))$.

Figure 1.5 illustrates the mapping from transmission power to target SINR, where in addition to power constraints, lower and upper bounds for the target SINR are considered. Links where transmission power is below a specified minimum threshold $\overline{p}'_{\text{min}}$ are allowed to attempt a high-quality connection by targeting a prescribed maximum target SINR $\overline{\gamma}^t_{\text{max}}$. On the other extreme, links using power levels between a given maximum threshold $\overline{p}'_{\text{max}}$ and the maximum transmission power $\overline{p}_{\text{max}}$ will have as target SINR a minimum threshold $\overline{\gamma}^t_{\text{min}}$. Assuming $\overline{\gamma}_{\text{min}}$ as the minimum SINR level for acceptable communication, the relationship $\overline{\gamma}_{\text{min}} \leq \overline{\gamma}^t_{\text{min}} \leq \overline{\gamma}^t_{\text{max}}$ must hold, and every transmitter i aims for an acceptable target SINR. Finally, if the transmission power is between $\overline{p}'_{\text{min}}$ and $\overline{p}'_{\text{max}}$, the target SINR is defined as a linear function (in logarithmic scale) of transmission power. Note that the points $(\overline{p}'_{\text{min}}, \overline{\gamma}^t_{\text{max}})$ and $(\overline{p}'_{\text{max}}, \overline{\gamma}^t_{\text{min}})$ determine uniquely this linear function.

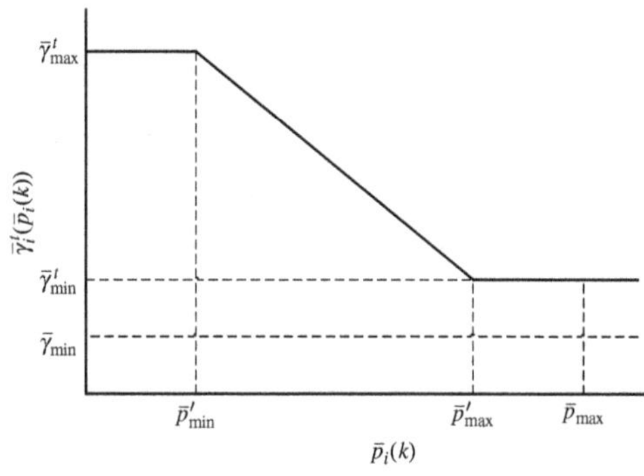

Fig. 1.5 Target SINR mapping of soft dropping power control.

In [59], soft dropping power control is shown to belong to the general framework of standard power control, discussed in Section 1.5. Therefore, its convergence to a unique fixed point is guaranteed, regardless of the initial values of transmission power and target SINR. Simulation results in [59] demonstrate performance improvements over fixed target SINR algorithms in a global system for mobile (GSM) communication environment. To be more precise, considering the maximum target SINR value of soft dropping $\overline{\gamma}^t_{\text{max}}$ as the SINR to be targeted by the fixed target algorithm, soft dropping provides lower power consumption with small deterioration in the average SINR level. The reduced level of co-channel interference allows more simultaneous transmissions with SINR levels above a minimum threshold for acceptable communication.

1.6.2 Opportunistic Power Control

Opportunistic power control offers an alternative vision for the distributed power control problem. In fact, conventional and opportunistic power controls have opposite philosophies. While the former is concerned with the satisfaction of strict SINR requirements by increasing the transmission power when the link condition is poor, in opportunistic power control, QoS requirements are not a concern and the transmission power is increased to transmit more information when channel gain is large and/or interference is low.

This alternative framework has its roots in concepts of opportunistic communications [17, 53], mainly the idea of scheduling the transmission according to channel quality. This fundamental concept is implemented through an opportunistic distributed power control [28, 51].

Opportunistic power control exploits the quality of communication links. Its strategy is simply to increase the transmission power when the effective interference decreases (or equivalently, when the effective channel gain becomes higher). The effective interference $I_i^e(\mathbf{p}_{-i})$ perceived at the receiver of a given link i is defined as the quotient between interference and channel gain of this link, that is,

$$I_i^e(\mathbf{p}_{-i}) = \frac{1}{g_{i,i}}\left(\sum_{j\neq i} g_{i,j}p_j + v_i\right), \quad \forall i \in \mathbf{N}, \tag{1.37}$$

where $\mathbf{N} = \{1,\dots,N\}$ is the set of N co-channel links. Moreover, the effective interference is such that

$$\gamma_i(\mathbf{p}) = \frac{p_i}{I_i^e(\mathbf{p}_{-i})}, \tag{1.38}$$

where $\gamma_i(\mathbf{p})$ is the SINR achieved at the receiver of link i.

Opportunistic algorithm is proposed in [51] with the introduction of the target signal-interference product (SIP). The idea is to keep the product of the signal power and the effective interference a constant Γ_i, that is $\Gamma_i = p_i I_i^e(\mathbf{p}_{-i})$. OPC updates transmission power in each link according to the following iterative function:

$$p_i(k+1) = \zeta_i(\mathbf{p}(k)) = \frac{\Gamma_i}{I_i^e(\mathbf{p}_{-i}(k))}, \quad \forall i \in \mathbf{N}, \tag{1.39}$$

where Γ_i is the target SIP of link i.

The analysis of convergence and stability of OPC cannot be carried out within the framework of standard power control. An opportunistic algorithm violates the monotonicity condition of standard functions, stated in Section 1.5, since its power iteration is decreasing with $I_i^e(\mathbf{p}_{-i})$. However, as for standard power control, the iterative functions of opportunistic power control algorithms present properties that permit a general proof of convergence to a unique fixed point. This new framework, proposed in [51], is called *type-II standard power control.*

Definition 1.2. An iterative vector function $\boldsymbol{\zeta}(\mathbf{p})$ is said to be type-II standard if it satisfies the following conditions:

C-1 Type-II Monotonicity: If $\mathbf{p} \leq \mathbf{p}'$, then $\boldsymbol{\zeta}(\mathbf{p}) \geq \boldsymbol{\zeta}(\mathbf{p}')$.
C-2 Type-II Scalability: For all $\alpha > 1$, $\boldsymbol{\zeta}(\alpha\mathbf{p}) > (1/\alpha)\boldsymbol{\zeta}(\mathbf{p})$.

The interesting properties of type-II standard iterative functions are stated in theorems below. Their proofs are found in [51].

Theorem 1.5. *If $\boldsymbol{\zeta}(\mathbf{p})$ is type-II standard and a fixed point exists, then the fixed point is unique.*

Theorem 1.6. *If $\boldsymbol{\zeta}(\mathbf{p})$ is type-II standard and a fixed point $\mathbf{p}^*$ exists, then any power vector $\mathbf{p}$ converges to $\mathbf{p}^*$.*

Theorem 1.7. *If a fixed point $\mathbf{p}^*$ exists, then any power vector $\mathbf{p}$ converges to $\mathbf{p}^*$ under the totally asynchronous model.*

Therefore, on the condition that there exists a fixed point, the class of type-II standard power control algorithms (OPC algorithm included) has its convergence guaranteed in a network-wide synchronous or asynchronous mode. It is also demonstrated in [51] that an upper bound to the transmission power is a sufficient condition for the existence of a fixed point.

The presented OPC algorithm is highly unfair, since only a few (in favored conditions) terminals are able to transmit. Unfairness is an intrinsic characteristic of opportunistic power control, since in opposition to a fixed SINR target tracking power control, it magnifies the near-far effect by assigning more power to links in favored conditions, and vice versa. In order to attenuate this effect, a fairness mechanism is introduced. The power levels of favored links are penalized, while terminals in poor propagation conditions have their powers increased. The opportunistic power control with fairness (OPC-F) algorithm [28] has the same update expression (1.39), but its target SIP Γ_i is given by

$$\Gamma_i = \frac{\rho_i}{[\widehat{g}_{i,i}(k)]^2}, \quad \forall i \in \mathbf{N}, \tag{1.40}$$

where ρ_i is a constant and $\widehat{g}_{i,i}(k)$ is the estimation of the average channel gain for terminal i calculated in a given sampling window. Convergence and stability analysis for OPC-F is analogous to that for OPC algorithm, leading to the same conclusions [28]. Both OPC and OPC-F algorithms have their performances compared with other opportunistic algorithms at the end of Section 1.7.3, in Example 1.7.

1.7 Power Control Games

In this section, the employment of game theory to the problem of distributed power control is discussed. Game theory is a mathematical branch dedicated to the analysis of interactions among interdependent rational decision makers. In distributed power control, every communication terminal defines its transmission power individually, and the decision of every transmitter is influenced by decisions of all other transmitters. This framework characterizes a non-cooperative game, where the decision makers (players) are the transmitters, whose decision variables are their own transmission powers. Non-cooperative game theory is suitable for the formulation of selfish decision strategies in problems where the decision makers present conflicting interests and are not allowed or able to negotiate their decisions.

In the following section, some fundamentals of non-cooperative game theory and some power control algorithms based on that framework are presented. This section also presents a novel class of opportunistic power control algorithms.

1.7.1 Non-cooperative Games

Games are mathematical representations of a particular class of optimization problems. The interested reader can find excellent texts about historical and technical aspects in game theory in [3, 10, 30–34]. In order to have a clear picture of this class of problems and the importance of a game theoretic approach, Example 1.3 is introduced.

Example 1.3 (Centralized Optimization). Consider the simultaneous minimization of two objective functions, $J_1(x_1,x_2) = x_1^2/2 + x_1x_2 + x_2^2 + x_2$ and $J_2(x_1,x_2) = x_1^2 + x_1x_2 + x_2^2 + 6x_2$ with respect to variables $x_1, x_2 \in \mathbb{R}$. In a centralized optimization process, a global objective function $J(x_1,x_2)$ can be composed of a weighted sum of the original ones. For simplicity, consider that there is no preference on the performance of one or other objective. Then, $J(x_1,x_2) = J_1(x_1,x_2) + J_2(x_1,x_2)$, and the problem can be stated as

$$\begin{aligned} &\min_{x_1,x_2} \left\{ J(x_1,x_2) = 3x_1^2/2 + 2x_1x_2 + 2x_2^2 + 7x_2 \right\} \\ &s.t. \quad x_1, x_2 \in \mathbb{R}. \end{aligned} \tag{1.41}$$

The solution of this problem is straightforward. First-order necessary conditions, $\nabla J(x_1,x_2) = \mathbf{0}$, yield two equations:

$$\begin{aligned} 3x_1 + 2x_2 &= 0, \\ 2x_1 + 4x_2 &= -7. \end{aligned} \tag{1.42}$$

These have the unique solution $(x_1^*, x_2^*) = (7/4, -21/8)$, which is a global minimum point of J, since the Hessian matrix

$$\nabla^2 J(x_1,x_2) = \begin{bmatrix} 3 & 2 \\ 2 & 4 \end{bmatrix} \tag{1.43}$$

is definite positive. The minimum value of the function is $J(x_1^*,x_2^*) = -9.19$. Individual performances are given by $J_1(x_1^*,x_2^*) = 1.20$ and $J_2(x_1^*,x_2^*) = -10.39$.

Example 1.3 illustrates a conventional optimization problem, where a central entity has the task of minimizing a global objective function with respect to its decision variables. Optimization theory provides strong analytical results and successful iterative algorithms for the treatment of problems of this type.

In many real world problems, however, a centralized optimization process is not feasible. Centralized power control for wireless networks, for instance, typically requires unacceptable signaling overheads and computational effort. In problems where centralized optimization is prohibitive, decentralized or distributed solutions become interesting alternatives. Example 1.4 helps to understand the differences between centralized and distributed optimization problems.

Example 1.4 (Distributed Optimization). Consider the same problem presented in Example 1.3, where the simultaneous minimization of two objective functions, $J_1(x_1,x_2) = x_1^2/2 + x_1x_2 + x_2^2 + x_2$ and $J_2(x_1,x_2) = x_1^2 + x_1x_2 + x_2^2 + 6x_2$ with respect to variables $x_1,x_2 \in \mathbb{R}$, must be carried out. In the framework of distributed optimization, there is no central entity with decision power over all variables. On the contrary, each decision variable is governed by a single entity, whose performance criterion is its own objective function. This problem can thus be stated as

$$\begin{aligned} &\min_{x_1} \left\{ J_1(x_1,x_2) = x_1^2/2 + x_1x_2 + x_2^2 + x_2 \right\} \\ &s.t. \quad x_1,x_2 \in \mathbb{R}, \end{aligned} \tag{1.44}$$

and, simultaneously,

$$\begin{aligned} &\min_{x_2} \left\{ J_2(x_1,x_2) = x_1^2 + x_1x_2 + x_2^2 + 6x_2 \right\} \\ &s.t. \quad x_1,x_2 \in \mathbb{R}. \end{aligned} \tag{1.45}$$

Observe the difficulty in finding the meaning of optimality in the sense of conventional optimization in this example. On the other hand, it is clear that the distributed optimization structure establishes a conflict between the two entities, since the individual performance is dependent on the decisions of both, and these decisions are made unilaterally, without information exchange or negotiation. This particular class of optimization problems can be represented within the framework of non-cooperative games.

A game is characterized by three basic elements: a set of players or decision makers, a set of possible actions or strategy space for each player, and a set of objective functions mapping action profiles into real numbers. In an attempt to establish a relationship with conventional optimization problems, these can be seen as one-player games, where the only decision maker optimizes a function by choosing proper values for its decision variables. Of course, this relationship is only for didactic purposes, since game theory is concerned with more general problems, where multiple players with conflicting objectives are involved in a decision process.

In Example 1.4, a two-player static non-zero-sum non-cooperative infinite game is enunciated. In opposition to zero-sum games, where the gains of a player represent exactly the losses of the other, in a non-zero-sum game the competition between the players is not so severe. A game is also classified as static or dynamic according to its decision-making structure. Games where players make their decisions without information about decisions of their opponents are static. This is equivalent to a simultaneous action process. On the other hand, a game with a sequential decision-making process and information transference is considered dynamic. A further classification as finite or infinite regards the set of possible decisions (strategy space) for each player. Since $x_1, x_2 \in \mathbb{R}$, the game of Example 1.4 is infinite. For a detailed material about classification of games and further aspects, the reader is encouraged to consult specialized texts as [3, 10]. This section restricts the discussion to the class of N-player static non-zero-sum non-cooperative infinite games, since it gives the basis for the development of distributed power control solutions.

In a non-cooperative game, in spite of the individual decision making, a desirable solution is one in which each player is satisfied with its performance. This means that the concept of equilibrium replaces the one of optimality. In general lines, an equilibrium solution is characterized by the absence of motivation for a unilateral deviation of any player. Since the interest of this section lies in problems without hierarchy or preferences among the players, the Nash equilibrium solution is appropriate.

1.7.1.1 Nash Equilibrium

Nash equilibrium (NE) concepts are widely employed in non-cooperative games, since they allow predictable and stable outcomes through self-optimization. Notions of uniqueness and stability of nash equilibrium (NE) solutions motivate the fundamental discussion about iterative algorithms.

For the sequence of presentation, it is important to establish the difference between strategy and solution. Roughly, a strategy is a rule of decision. Solutions or decisions result from adopted strategies. In Example 1.4, for instance, two possible strategies for Player 2 could be $\zeta_2 = x_1 + 3$ and $\zeta_2 = 5$. In the first case the decision of Player 2, x_2, depends on the choice of Player 1 according to ζ_2, while with the second strategy a fixed decision is made, that is $x_2 = 5$. Only in the case of fixed strategies, these have the same meaning of decisions. There is another aspect concerning the concept of strategy: the rule of decision can be deterministic, such as the examples above, or stochastic. Our focus is on pure (deterministic) strategies.

The study of pure-strategy NE solutions in static non-cooperative infinite games is based on concepts of reaction curves. The reaction curve of a player is the strategy which corresponds to the best response of the player with respect to any action of other players. The definition below makes this notion more precise.

Definition 1.3 (Reaction Curves). In an N-player non-zero-sum game, let the set of players be denoted by $\mathbf{N} = \{1,\dots,N\}$. Let $\{x_1,\dots,x_N\}$ be the set of decision variables associated with the N players, $\{\mathbf{X}_1,\dots,\mathbf{X}_N\}$ the set of strategy spaces, and $J_i(x_1,\dots,x_N)$, $i \in \mathbf{N}$, the set of cost functions. Assume that the minimum of the cost function of Player 1, $J_1(x_1,\dots,x_N)$, with respect to $x_1 \in \mathbf{X}_1$, can be attained for each $x_{-1} \in \mathbf{X}_{-1}$, where $x_{-1} \triangleq \{x_2,\dots,x_N\}$ and $\mathbf{X}_{-1} \triangleq \mathbf{X}_2 \times \cdots \times \mathbf{X}_N$. Then, the set $\zeta_1^{\mathrm{B}}(x_{-1}) \subset \mathbf{X}_1$ defined by

$$\zeta_1^{\mathrm{B}}(x_{-1}) = \{x_1^{\mathrm{B}} \in \mathbf{X}_1 : J_1(x_1^{\mathrm{B}},x_{-1}) \leq J_1(x_1,x_{-1}), \quad \forall x_1 \in \mathbf{X}_1\} \tag{1.46}$$

is called the *optimal response* or *rational reaction set* of Player 1. If the set $\zeta_1^{\mathrm{B}}(x_{-1})$ has a single element for every $x_{-1} \in \mathbf{X}_{-1}$, then it is called the *best response function* or *reaction curve* of Player 1. These definitions are also valid for Player $i = 2,\dots,N$ simply by replacing the index 1 by i.

Once in an intersection point of the reaction curves, each player is satisfied with its performance, since it is the best it can do. In fact, such a point is a pure-strategy NE solution. This relationship comes directly from the definitions of reaction curves (Definition 1.3) and Nash equilibrium solution (Definition 1.4).

Definition 1.4 (Nash Equilibrium Solution). A given N-tuple $\{x_1^{\mathrm{N}},\dots,x_N^{\mathrm{N}}\}$, with $x_i^{\mathrm{N}} \in \mathbf{X}_i$, $i \in \mathbf{N}$, is called a (pure) Nash equilibrium solution for a N-player non-zero-sum infinite game if

$$J_i(x_1^{\mathrm{N}},\dots,x_N^{\mathrm{N}}) \leq J_i(x_1^{\mathrm{N}},\dots,x_{i-1}^{\mathrm{N}},x_i,x_{i+1}^{\mathrm{N}},\dots,x_N^{\mathrm{N}}), \quad \forall i \in \mathbf{N}.$$

The concepts of reaction curves and NE solution are explored in Example 1.5, where the distributed optimization problem stated in Example 1.4 is addressed. A decentralized solution is obtained as the NE point.

Example 1.5. Since in the problem formulated in Example 1.4 the individual objective functions are twice-differentiable, the best response functions or reaction curves of players are given by ζ_1^{B} and ζ_2^{B} as follows:

$$\begin{aligned}
&\text{(Reaction curve – Player 1):} \quad \frac{\partial J_1(x_1,x_2)}{\partial x_1} = 0 \Rightarrow \zeta_1^{\mathrm{B}} = -x_2, \\
&\text{(Reaction curve – Player 2):} \quad \frac{\partial J_2(x_1,x_2)}{\partial x_2} = 0 \Rightarrow \zeta_2^{\mathrm{B}} = -\frac{(x_1+6)}{2}.
\end{aligned} \tag{1.47}$$

The intersection point of ζ_1^{B} and ζ_2^{B} gives the unique NE solution, that is $(x_1^{\mathrm{N}},x_2^{\mathrm{N}}) = (6,-6)$. The individual performances in the NE point are $J_1(x_1^{\mathrm{N}},x_2^{\mathrm{N}}) = 12$ and $J_2(x_1^{\mathrm{N}},x_2^{\mathrm{N}}) = 0$.

In general, the effectiveness of NE solutions is dependent on three fundamental aspects: existence, uniqueness, and stability. Obviously, the existence of NE points means that there are solutions where all players are satisfied with their performances. However, the multiplicity of these points can lead to ambiguous outcomes.

To be more precise, consider a two-player non-cooperative infinite game with two NE points given by (x_1^{N1}, x_2^{N1}) and (x_1^{N2}, x_2^{N2}). Since there is no reason for players to prefer one particular equilibrium solution over the other(s) and their decisions are made independently, an interchanged solution (x_1^{N1}, x_2^{N2}) or (x_1^{N2}, x_2^{N1}), which is not an equilibrium point can happen. This justifies the search for unique NE solutions.

The definition of NE solutions in static non-cooperative infinite games as common intersection points of the best response functions of the players characterizes a unique NE solution as the unique solution of the fixed point equation. The study of stability of a fixed point solution gives rise to the notion of iterative algorithms. A given NE solution is said to be stable if after any deviation of one or more players, this solution is restored in an iterative (rational) sequence of moves of the players, governed, of course, by their reaction curves. In order to illustrate this dynamics, Fig. 1.6 shows the reaction curves and the iterative process of convergence of a stable NE solution (Fig. 1.6(a)) and an unstable NE solution (Fig. 1.6(b)) in a two-player non-cooperative infinite game.

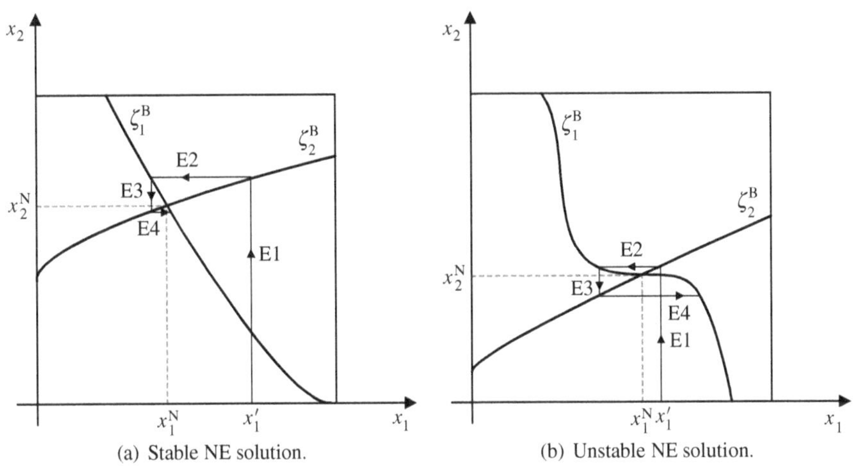

Fig. 1.6 Convergence of Nash equilibrium solutions under a unilateral deviation.

In Fig. 1.6(a) and (b), ζ_1^B and ζ_2^B represent the best response functions of Player 1 and Player 2, respectively, and (x_1^N, x_2^N) is a NE point. Assume a unilateral deviation of Player 1, which decides for $x_1' \neq x_1^N$. Events E1, E2, E3, and E4 illustrate the first steps of the sequential process of decisions after the deviation from the NE point. Given the choice x_1' of Player 1, E1 represents the decision process of Player 2 according to its best response function ζ_2^B, that is, the decision of Player 2 is given by $\zeta_2^B(x_1')$. In the sequence, Player 1 acts in response to the previous decision of Player 2, reaching $\zeta_1^B(\zeta_2^B(x_1'))$, and so on. In Fig. 1.6(a), this sequence of decisions will converge to (x_1^N, x_2^N), the stable NE point. On the contrary, in Fig. 1.6(b), a deviation from the NE point results in a sequence of best responses that do not restore the equilibrium.

The guaranteed convergence to the unique fixed point is a very important property of the stable (and unique) NE solution in a static non-cooperative infinite game. It means that regardless of the initial conditions (first decisions of players), if every player adopts its best response function as decision strategy, the iterative process of decisions will converge to the NE solution. Strong analytical results define conditions for the desirable characteristics of existence, uniqueness, and stability of NE solutions [3, 10]. These ideas form the basis for the development of several distributed power control algorithms; some of them are discussed in next section.

To finish this section, we return to the discussion about centralized and distributed optimization. It is important to be careful in making comparisons between solutions coming from the two optimization processes. It is expected that centralized solutions are better than distributed ones. The problem considered in Examples 1.3, 1.4, and 1.5, where the task is the minimization of two objective functions, follows this general rule. Centralized optimization gives $J_1(x_1^*, x_2^*) = 1.20$ and $J_2(x_1^*, x_2^*) = -10.39$ as individual performances, while the NE solution results in $J_1(x_1^{\mathrm{N}}, x_2^{\mathrm{N}}) = 12$ and $J_2(x_1^{\mathrm{N}}, x_2^{\mathrm{N}}) = 0$. In this comparison, however, the important structural differences between the optimization processes must also be taken into account. In centralized optimization, a central entity has unrestricted knowledge and decision power over all variables. On the other hand, distributed optimization makes use of restricted information and individual decision making. Distributed solutions, such as those provided by game theory, are valuable for problems where centralized optimization is prohibitive due to aspects such as lack of global information or high computational complexity.

1.7.2 Game-Based Distributed Power Control Algorithms

In distributed power control, transmitter terminals define their transmission power individually, and the decision of each transmitter influences the performance of all others. This general operational structure characterizes a non-cooperative game, where the decision makers (players) are the transmitters, whose decision variables are their own transmission powers. The third element of a game, i.e., the set of objective functions, defines the individual goals of players and can also induce desirable network behaviors.

The distributed power control approaches derived within the framework of non-cooperative games are essentially based on concepts of Nash equilibrium (NE). A variety of objective functions is considered and represents the diversity of interests which can be involved. The objectives of the power control procedure depend on the applications. Services with strict QoS requirements, like voice communications with prescribed fixed transmission and bit error rates, require the meeting of fixed SINR levels for individual communication links. On the other hand, data applications which tolerate larger delays, such as WWW browsing, allow the exploitation of different approaches, where instead of a single pre-defined set of QoS requirements, multiple simultaneous objectives may be considered such as maximization of data throughput and energy efficiency.

Next, some of the solutions derived from non-cooperative games are discussed. Because of the basic differences of power control for QoS-fixed and QoS-flexible applications, they are addressed separately.

1.7.2.1 QoS-Fixed Applications

The development of a game theoretic solution to distributed power control in the context of QoS-fixed applications retakes the power control problem with fixed SINR requirements discussed in Section 1.4. In this new framework, each player decides its own transmission power level aiming to meet the prescribed target SINR. Since this is a traditional target tracking problem, the game can be formulated as

$$\min_{p_i} \left\{ J_i(p_i, \mathbf{p}_{-i}) = \left(\gamma_i^t - \gamma_i(p_i, \mathbf{p}_{-i})\right)^2 \right\}, \quad \forall i \in \mathbf{N}, \tag{1.48}$$

where $\mathbf{N} = \{1, \ldots, N\}$ is the set of N players or interfering communication links. Player i has p_i as transmission power, γ_i^t as target SINR, and $\gamma_i(p_i, \mathbf{p}_{-i})$ as the SINR, expressed in (1.4). The vector of powers excluding the ith player is denoted as $\mathbf{p}_{-i}$.

The development of a NE solution for game (1.48) requires the investigation of its existence, uniqueness, and stability. The existence of an equilibrium in this game is guaranteed by Theorem 1.8.

Theorem 1.8. *For each $i \in \mathbf{N}$ let $\mathbf{P}_i$ be a closed, bounded, and convex subset of a finite-dimensional euclidian space, and the cost functional $J_i : \mathbf{P}_1 \times \cdots \times \mathbf{P}_N \longrightarrow \mathbb{R}$ be jointly continuous in all its arguments and strictly convex in p_i for every $p_j \in \mathbf{P}_j$, $j \neq i$. Then, the associated* N*-player non-zero-sum game admits a Nash equilibrium in pure strategies.*

Proof. See [3], Chapter 4, pp. 173–174. □

Without loss of generality, strategy spaces in game (1.48) are assumed to be $\mathbf{P}_i = [p_{\text{min}}, p_{\text{max}}]$ and therefore are closed, bounded, and convex subsets of a finite-dimensional euclidian space, for all $i \in \mathbf{N}$. The satisfaction of remaining conditions stated in Theorem 1.8 can be easily verified by explicitly expressing the objective function $J_i(p_i, \mathbf{p}_{-i})$ in terms of all transmission powers. Next, properties of uniqueness and stability are discussed. For this purpose, the best response function or NE strategy of each player in (1.48) must be derived.

Proposition 1.1. *The best response function or NE strategy of each player in (1.48) is given by*

$$\zeta_i^B = \max(p_{\text{min}}, \min(\widetilde{p}_i, p_{\text{max}})), \quad \forall i \in \mathbf{N}, \tag{1.49}$$

where $\widetilde{p}_i$ is the unconstrained minimizer of the objective function $J_i(p_i, \mathbf{p}_{-i})$, that is,

$$\begin{aligned}\widetilde{p}_i &= \arg\min_{p_i \in \mathbb{R}} J_i(p_i, \mathbf{p}_{-i}) \\ &= \frac{\gamma_i^t}{g_{i,i}} \left(\sum_{j \neq i} g_{i,j} p_j + \nu_i \right),\end{aligned} \tag{1.50}$$

where $g_{i,j}$ is the channel gain between transmitter j and receiver i, and ν_i represents the noise power at ith receiver.

Proof. Objective function $J_i(p_i, \mathbf{p}_{-i})$ is a continuous quadratic function of p_i. Then, it is easy to verify that $\partial J_i(p_i, \mathbf{p}_{-i})/\partial p_i = 0$ gives $\widetilde{p}_i$, which is the minimizer of $J_i(p_i, \mathbf{p}_{-i})$, since $\partial^2 J_i(p_i, \mathbf{p}_{-i})/\partial p_i^2 > 0$. However, if $\widetilde{p}_i \notin \mathbf{P}_i$, it cannot be the best response of Player i, since it is not a feasible solution. In this case, also because $J_i(p_i, \mathbf{p}_{-i})$ is quadratic in p_i, if $\widetilde{p}_i < p_{\min}$, the best response of Player i is $p_{\min}$. Analogously, if $\widetilde{p}_i > p_{\max}$, the best response of Player i is $p_{\max}$. □

Theorem 1.9. *The game formulated in (1.48) has a unique and stable NE solution.*

Proof. From Theorem 1.8, it is known that there exists an equilibrium point in game (1.48). Let $\mathbf{p}^{\mathrm{N}} = [p_1^{\mathrm{N}}, \ldots, p_N^{\mathrm{N}}]^{\mathrm{T}}$ be the Nash equilibrium in this game. By definition, $\mathbf{p}^{\mathrm{N}}$ results from the common intersection of best responses given by (1.49), that is $\mathbf{p}^{\mathrm{N}} = \boldsymbol{\zeta}^{\mathrm{B}}(\mathbf{p}^{\mathrm{N}})$, where $\boldsymbol{\zeta}^{\mathrm{B}}(\mathbf{p}^{\mathrm{N}}) = \left[\zeta_1^{\mathrm{B}}(\mathbf{p}^{\mathrm{N}}), \ldots, \zeta_N^{\mathrm{B}}(\mathbf{p}^{\mathrm{N}})\right]^{\mathrm{T}}$. Uniqueness and stability proofs come from the fact that the best response $\boldsymbol{\zeta}^{\mathrm{B}}(\mathbf{p})$ is a standard function, as shown in [58]. From the discussion in Section 1.5, a standard function makes any feasible initial point converge to the unique fixed point in synchronous or asynchronous decision process. □

Therefore, the NE strategy given in (1.49) leads to a unique and stable NE solution. This decision strategy is exactly the well-known distributed power control (DPC) algorithm presented in Section 1.4.3.

Another approach, based on submodular games, also recovers the DPC algorithm. In this case, the game formulation is given by

$$\min_{p_i \in \mathbf{P}_i(\mathbf{p}_{-i})} \; p_i, \quad \forall i \in \mathbf{N}, \tag{1.51}$$

where $\mathbf{P}_i(\mathbf{p}_{-i})$ is the set of feasible power levels for Player i that depends on the power allocation of all other players, i.e.,

$$\mathbf{P}_i(\mathbf{p}_{-i}) = \{p_i \geq 0 : \gamma_i(p_i, \mathbf{p}_{-i}) \geq \gamma_i^t\}. \tag{1.52}$$

Sets $\mathbf{P}_i(\cdot)$ present the following property: $\mathbf{p}_{-i} < \mathbf{p}'_{-i} \Rightarrow \mathbf{P}_i(\mathbf{p}_{-i}) \supset \mathbf{P}_i(\mathbf{p}'_{-i})$. This characterizes submodular sets [2]. Therefore, the game formulated in (1.51) is called submodular game. The best response of each player is the DPC algorithm and makes any feasible power vector converge to the unique NE point.

1.7.2.2 QoS-Flexible Applications

Most approaches of distributed power control for QoS-flexible applications consider multiple objectives. Because of this, the objective function of each player in a power control game can be seen as a satisfaction measure, where revenues or utilities and losses or costs can be combined. The satisfaction in a communication link is often related to the trade-off between a measure of QoS and the energy consumption. In the following, some such approaches are discussed.

Consider a wireless system where each terminal transmits L_{inf} information bits in frames (packets) of $L_{tot} > L_{inf}$ bits. A fixed rate r (bits/s) is assumed for each terminal. Then, a utility function which accounts for the amount of information successfully received per unity of energy expended (bits/Joule) is given below:

$$J_i(p_i, \mathbf{p}_{-i}) = \frac{L_{inf} FSR_i r}{L_{tot} p_i}, \quad \forall i \in \mathbf{N}, \tag{1.53}$$

where FSR_i is the frame success rate (FSR) of communication link i, the probability of correct reception of a frame at the receiver. Assuming perfect error detection and no error correction, FSR can be expressed as $FSR_i = (1 - BER_i)^{L_{tot}}$, where BER_i is the bit error rate (BER). The utility function (1.53) presents a mathematical anomaly. In case of no transmission ($p_i = 0$), the best strategy for the receiver is to make a guess for each bit, i.e., $FSR_i = 2^{-L_{tot}}$, resulting in infinite utility. Since BER is a monotonically decreasing function of SINR, FSR is monotonically increasing with SINR. Then, FSR_i can be substituted in (1.53) with a specific function of SINR, according to properties of the system such as modulation and receiver structure. Therefore, in order to avoid the degenerate solution, FSR in (1.53) is replaced by an *efficiency function*, defined as

$$f_e(\gamma_i) = (1 - 2BER_i)^{L_{tot}}, \tag{1.54}$$

and closely follows the behavior of FSR. The resulting utility function is

$$J_i(p_i, \mathbf{p}_{-i}) = \frac{L_{inf} f_e(\gamma_i) r}{L_{tot} p_i}, \quad \forall i \in \mathbf{N}. \tag{1.55}$$

Making a guess for each bit yields $f_e(\gamma_i) = 0$. Then, conveniently, in case of no transmission one obtains zero utility.

The power control game can, therefore, be stated as

$$\max_{p_i \in \mathbf{P}_i} J_i(p_i, \mathbf{p}_{-i}), \quad \forall i \in \mathbf{N}, \tag{1.56}$$

where $J_i(p_i, \mathbf{p}_{-i})$ is given in (1.55) and $\mathbf{P}_i = [0, p_{\max}]$. This utility function is quasi-concave in transmission power. The investigation of existence, uniqueness, and stability of the NE solution in the game stated in (1.56) is carried out in [15, 44]. The existence of an equilibrium in this game is guaranteed by Theorem 1.10.

Theorem 1.10. *A Nash equilibrium exists in game (1.56) if, for all* $i = 1,\dots,N$,

1. $\mathbf{P}_i$ *is a non-empty, convex, and compact subset of some euclidian space* $\mathbb{R}^N$ *and*
2. $J_i(\mathbf{p})$ *is continuous in* $\mathbf{p}$ *and quasi-concave in* p_i.

Proof. See [44], Appendix I. Results obtained from [6, 7, 12] are used in the proof of existence of a fixed point. □

The proof of uniqueness and stability of the NE solution in game (1.56) is analogous to that of game (1.48). The key aspect is to realize that the best response function in game (1.56) is standard. The best response function is given as

$$\zeta_i^{\mathrm{B}} = \min(\widetilde{p}_i, p_{\max}), \quad \forall i \in \mathbf{N}, \tag{1.57}$$

where $\widetilde{p}_i$ is the unconstrained maximizer of the utility function in (1.56), that is,

$$\widetilde{p}_i = \arg\max_{p_i \in \mathbb{R}} J_i(p_i, \mathbf{p}_{-i}) = \frac{\widetilde{\gamma}_i^t}{g_{i,i}} \left(\sum_{j \neq i} g_{i,j} p_j + \nu_i \right), \tag{1.58}$$

where $\widetilde{\gamma}_i^t$ is the target SINR which solves

$$f_e'(\widetilde{\gamma}_i^t)\widetilde{\gamma}_i^t - f_e(\widetilde{\gamma}_i^t) = 0, \tag{1.59}$$

with $f_e'(\cdot)$ denoting the first-order derivative of $f_e(\cdot)$.

At the NE point, a terminal either attains its utility maximizing SINR $\widetilde{\gamma}_i^t$ or it fails to do so and transmits at maximum power $p_{\max}$. When the configuration is such that all terminals use the same modulation technique and the same packet length, they have the same efficiency function. Therefore, in this case, if every terminal is able to achieve its utility maximizing SINR, this value is the same for all terminals. Although this power control solution is similar to the conventional fixed target algorithms for QoS-fixed applications, $\widetilde{\gamma}^t$ is derived from the particular efficiency function, while the target SINR in conventional algorithms is determined by subjective measures of quality.

Since the power control solution for game (1.56) is a NE point, no terminal can increase its utility through an individual effort. However, it is known that an incremental decrease in the transmission power of every terminal in (1.56) leads to a solution that *"Pareto dominates"* the original NE. The following definition establishes the meaning of Pareto dominance and optimality [44].

Definition 1.5. A power vector $\widetilde{\mathbf{p}}$ *Pareto dominates* another vector $\mathbf{p}$ if, for all $i \in \mathbf{N}$, $J_i(\widetilde{\mathbf{p}}) \geq J_i(\mathbf{p})$ and for some $i \in \mathbf{N}$, $J_i(\widetilde{\mathbf{p}}) > J_i(\mathbf{p})$. Furthermore, a power vector $\mathbf{p}^*$ is *Pareto optimal* (*efficient*) if there exists no other power vector $\mathbf{p}$ such that $J_i(\mathbf{p}) \geq J_i(\mathbf{p}^*)$ for all $i \in \mathbf{N}$ and $J_i(\mathbf{p}) > J_i(\mathbf{p}^*)$ for some $i \in \mathbf{N}$.

Therefore, the NE solution to game (1.56) is not efficient. As mentioned, it is expected that a solution obtained from a distributed decision process be less efficient than solutions obtained through cooperation among the decision makers or as a result of centralized optimization. However, decentralized decisions can be compatible with overall system efficiency if pricing mechanisms are employed properly.

Pricing is a cost imposed on the individual expenditure of resources with the aim of preventing damaging effects for the overall system. It induces efficient resource utilization rather than the aggressive competition of purely non-cooperative games, while maintaining the non-cooperative (distributed) nature of the resulting solution.

The game (1.56) can thus be reformulated as a non-cooperative power control game with (linear) pricing as follows:

$$\max_{p_i \in \mathbf{P}_i} J_i(p_i, \mathbf{p}_{-i}) - a_p \varepsilon_i p_i, \quad \forall i \in \mathbf{N}, \tag{1.60}$$

where $J_i(p_i, \mathbf{p}_{-i})$ is the utility of game (1.56), given by (1.55); a_p and ε_i are positive scalars; and $\mathbf{P}_i = [p_{\min}, p_{\max}]$, with $p_{\min}$ derived from $\gamma \geq 2\ln(L_{tot})$ [44]. The pricing factor a_p must be tuned such that the individual interest of terminals, represented by the net utility to be maximized, leads to the best possible improvement in overall network performance.

Although there is similarity between games formulated in (1.56) and (1.60), they cannot be solved in the same way. In game (1.56), the quasi-concavity of utility functions allows the establishment of the existence of a NE solution. Game (1.60), however, does not have quasi-concave utility functions. Moreover, analytical techniques based on convexity and differentiability are no longer applicable to prove NE existence. In this case, the existence of equilibria is assured by *supermodularity theory* .

Definition 1.6. A game with strategy spaces $\mathbf{P}_i \subset \mathbb{R}$ for all $i \in \mathbf{N}$ is *supermodular* if, for each i, $J_i(p_i, \mathbf{p}_{-i})$ has non-decreasing differences in $(p_i, \mathbf{p}_{-i})$, that is, for all $\mathbf{p}_{-i} > \mathbf{p}'_{-i}$ the quantity $J_i(p_i, \mathbf{p}_{-i}) - J_i(p_i, \mathbf{p}'_{-i})$ is non-decreasing in p_i.

The set of Nash equilibria of a supermodular game is non-empty and has a largest element and a smallest element [44, 52]. The game formulated in (1.60) with a pricing factor a_p is shown in [44] to be a supermodular game. Furthermore, a totally asynchronous algorithm is developed to generate a sequence of powers that converge to the smallest NE, i.e., the NE with minimum total transmission powers, which is the one that yields the highest net utility. This algorithm performs the maximization of the net utility given in (1.60), which requires the satisfaction of a condition similar to (1.59), but that contains a term which depends explicitly on the power of each terminal:

$$f_e'(\tilde{\gamma}_i^t)\tilde{\gamma}_i^t - f_e(\tilde{\gamma}_i^t) - a_p \varepsilon_i p_i^2 = 0. \tag{1.61}$$

The implementation of a power control procedure that provides Pareto-dominant solutions requires a central coordination. The central node announces the pricing value and each terminal uses the mentioned algorithm to obtain the smallest NE. The pricing factor is increased and announced to all terminals until the utility of at least one terminal begins to decrease. At this point, $a_p = a_{p,\mathrm{BEST}}$, and the solution is Pareto dominant. Example 1.6 compares the solutions of games (1.56) and (1.60).

Example 1.6. Consider a generic spread spectrum wireless system with a central communication node and stationary terminals that transmit $L_{inf} = 64$ information bits in frames of $L_{tot} = 80$ bits. A fixed rate $r = 10$ kbits/s is assumed for each terminal and the channel bandwidth is $B = 1$ MHz. Maximum transmission power is set to 2 W and the noise power at the receiver is $\nu = 5 \times 10^{-15}$ W. Path gain for each communication link is obtained using the simple path loss model $\overline{PL}(d) = 120 + 40\log_{10}(d)$, where distance d from the transmitter to the central receiver node is given in kilometers. The efficiency function $f_e(\gamma_i) = (1 - \mathrm{e}^{-0.5\gamma_i})^{L_{tot}}$ approximates the probability of successful frame reception for non-coherent frequency shift keying (FSK) modulation. Parameter ε in the game with pricing (1.60) is set to 1.

Figure 1.7 shows the distribution of utilities and transmission powers with distance between transmitter and receiver for power control games with and without pricing. In Fig. 1.7(a) and (b), utilities and powers are dependent on the terminal location or, more precisely, on the channel quality. Without pricing ($a_p = 0$), the utility is maximized at the same SINR, $\tilde{\gamma} = 12.42$, for all terminals. As the pricing factor is increased, the equilibrium is shifted to points where terminals attain lower SINR levels, expend less power, and achieve higher utilities. The benefits in terms of utility are entirely due to the reduction of power. At the equilibrium with pricing, SINRs are no longer equal for all communication links; transmitters closer to receiver attain higher SINRs than transmitters far away.

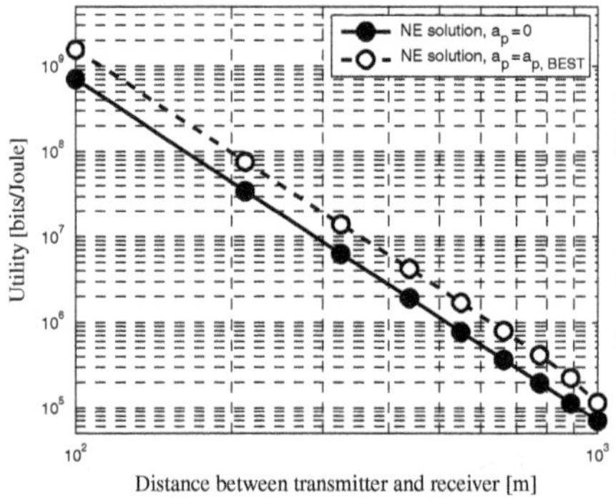

(a) Distribution of utilities with distance.

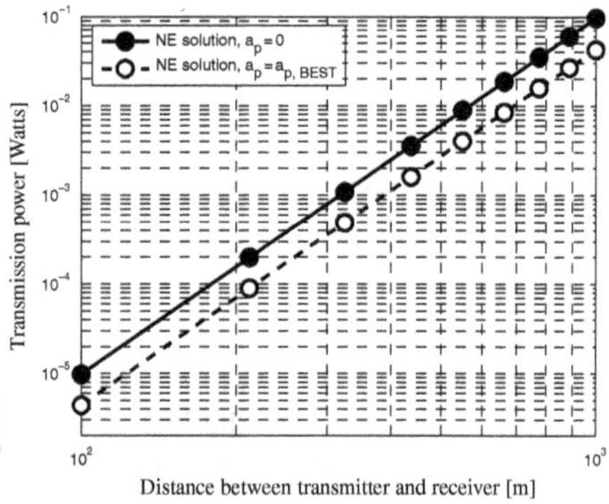

(b) Distribution of powers with distance.

Fig. 1.7 Distribution of utilities and powers with distance between transmitter and receiver for power control games with and without ($a_p = 0$) pricing.

In distributed power control approaches discussed above for QoS-flexible applications, objective functions are dependent on specific system configurations, such as the modulation scheme and the packet length. Aspects such as channel gain variations are not considered, and the resulting algorithms are not convenient for practical implementation, since they require the periodic numerical solution of (1.59) and (1.61) for each terminal. There exist approaches more appropriate for practical implementation and that are not constrained to specific system configurations.

Shannon's channel capacity-like expressions are suitable for representing revenues of terminals as a function of the SINR. On the other hand, the pricing mechanism is usually employed as a linear function of the transmission power. As an example of such configuration, consider the approach of [1] for the uplink transmission in a single-cell code division multiple access (CDMA) cellular system, where the distributed power control game is formulated as

$$\min_{p_i \geq 0} \left\{ J_i(p_i, \mathbf{p}_{-i}) = a_{p,i} p_i - a_{s,i} \ln(1+\gamma_i) \right\}, \quad \forall i \in \mathbf{N}, \tag{1.62}$$

where $a_{p,i}$ and $a_{s,i}$ are "user-specific" positive parameters, which define the importance level given, respectively, to the power-saving objective and to the achievement of high levels of SINR. Since the objective function $J_i(p_i, \mathbf{p}_{-i})$ in (1.62) is twice-differentiable in p_i, and its second-order derivative is positive, the best response function of each terminal is given by

$$\zeta_i^{\mathrm{B}} = \begin{cases} \dfrac{1}{g_i} \left[q_i - \dfrac{r}{B} \left(\sum_{j \neq i} g_j p_j \right) \right] & \text{if } \sum_{j \neq i} g_j p_j \leq \dfrac{B}{r} q_i \\ 0 & \text{else.} \end{cases}, \tag{1.63}$$

where g_i is the channel gain between transmitter i and the base station, B denotes the channel bandwidth, and r is the transmission rate. Parameter q_i is expressed as

$$q_i = \frac{a_{s,i} g_i}{a_{p,i}} - \frac{r}{B} \nu, \tag{1.64}$$

with ν representing the noise power at the receiver.

It can be observed from (1.63) and (1.64) that the resulting transmission power of terminal i is proportional to the difference between $(a_{s,i}/a_{p,i})g_i$ and (r/B) $(\sum_{j \neq i} g_j p_j + \nu)$, if the value of the difference is positive (for negative values, transmission is interrupted). Therefore, for a given spreading factor (B/r), the transmission power level is defined according to the channel gain and interference plus noise power, but also depends on the choice of parameters $a_{p,i}$ and $a_{s,i}$. In agreement with the definition of such parameters, the lower the quotient $(a_{p,i}/a_{s,i})$, the more power-saving the power allocation. On the other hand, higher SINRs are obtained with higher values of this quotient. In [1], it is shown that the set of fixed point equations has a unique NE solution. Furthermore, the stability of the NE is investigated with the proposition of two asynchronous power control algorithms.

The general framework of power control games for QoS-flexible applications in wireless communication systems is characterized by a combination between a utility and a cost for the composition of an individual satisfaction measure. This framework allows flexible operation, where distinct individual priorities can be contemplated with the adjustment of some parameters. However, approaches derived within this framework are not appropriate for dealing with prescribed levels of QoS. Next, a class of algorithms which combines efficiency on the resource utilization with the fulfillment of essential QoS constraints is addressed.

1.7.3 A Class of Opportunistic Algorithms

The key aspect in distributed power control for QoS-flexible applications is the exploitation of the quality of communication links for implementing an efficient utilization of power resources. This is accomplished by opportunistic power control, addressed in Section 1.6.2, and non-cooperative power control games, discussed in Section 1.7.2.2. In both frameworks, each terminal achieves a level of QoS according to its channel quality and its preference for high QoS or low power consumption. However, the mentioned frameworks do not take into account operational constraints which are common to any practical communication system.

Besides the natural limitation on the transmission power, there exist constraints on the quality of the signal at the receiver, such as a minimum level of SINR for an acceptable communication. Furthermore, the finite number of adaptive modulation and coding schemes leads to a capacity upper bound. This means that above a given SINR level $\gamma_{\max}$, no improvement on the QoS is obtained. Therefore, it is desirable to maintain the SINR levels of terminals inside the region of operation $[\gamma_{\min}, \gamma_{\max}]$.

In the aforementioned power control frameworks, prescribed SINR requirements are not a concern. Then, in order to manage the QoS of individual terminals in an opportunistic fashion while satisfying SINR thresholds, one can consider a target tracking power control, where the target SINR is defined as an increasing function of effective channel gain and assumes values in the interval $[\gamma_{\min}, \gamma_{\max}]$. The ratio between the channel gain of a given communication link and the interference perceived at the receiver defines the effective channel gain $g_i^e(\mathbf{p}_{-i})$ of this link:

$$g_i^e(\mathbf{p}_{-i}) = \frac{g_{i,i}}{\sum\limits_{j \neq i} g_{i,j} p_j + \nu_i}, \quad \forall i \in \mathbf{N}. \tag{1.65}$$

Then, the effective channel gain is such that

$$\gamma_i(\mathbf{p}) = g_i^e(\mathbf{p}_{-i}) p_i, \tag{1.66}$$

where $\gamma_i(\mathbf{p})$ is the SINR achieved at the receiver of link i.

The dynamics of a typical wireless communication system is illustrated in Fig. 1.8 and helps to characterize the target SINR function. In this figure, straight lines L1 and L2 represent distinct propagation conditions of a given communication link in successive time instants k and $(k+1)$, respectively. Due to (1.66), effective channel gains $g_i^e(\mathbf{p}_{-i}(k))$ and $g_i^e(\mathbf{p}_{-i}(k+1))$ are the slopes of straight lines L1 and L2. The aspect to be pointed out is that a decreasing function of transmission power is an increasing function of effective channel gain, since for $\mathbf{p}_{-i}(k) > \mathbf{p}_{-i}(k+1)$ one has $g_i^e(\mathbf{p}_{-i}(k)) < g_i^e(\mathbf{p}_{-i}(k+1))$. Therefore, a decreasing continuous function of transmission power $f_i(p_i)$ can be adopted as target SINR for all $i \in \mathbf{N}$. This approach generalizes the soft dropping power control addressed in Section 1.6.1.

This adaptive target SINR leads to an energy-efficient scheme of QoS provision, since its response to an improvement on the effective channel gain is the increase of the SINR level to be targeted with the expenditure of less power. Assuming feasible

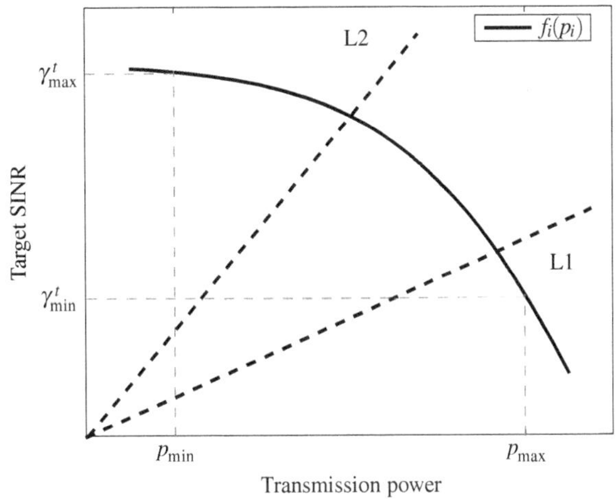

Fig. 1.8 Target SINR as a decreasing continuous function of transmission power with power and QoS constraints.

power control, operation in the specific SINR interval $[\gamma_{\min}, \gamma_{\max}]$ can be accomplished by designing $f_i(p_i)$ such that a terminal transmitting at maximum power targets the minimum SINR threshold, while the maximum SINR threshold is defined as target for a transmitter using the minimum power level.

Therefore, a target tracking power control game, where the target SINR is a decreasing function of power, can be stated as follows:

$$\min_{p_i} \left\{ J_i\left(p_i, \mathbf{p}_{-i}\right) = \left(f_i(p_i) - \gamma_i(\mathbf{p})\right)^2 \right\}, \ \forall i \in \mathbf{N}, \tag{1.67}$$

where $f_i(p_i)$ is the adaptive target SINR and $\gamma_i(\mathbf{p})$ is the actual SINR. For this quadratic cost function, the best response of Player i, ζ_i^{B}, satisfies

$$f_i(\zeta_i^{\mathrm{B}}) = g_i^e(\mathbf{p}_{-i})\zeta_i^{\mathrm{B}}, \quad \forall i \in \mathbf{N}. \tag{1.68}$$

Re-arranging the terms from (1.68),

$$\zeta_i^{\mathrm{B}} = \frac{f_i(\zeta_i^{\mathrm{B}})}{g_i^e(\mathbf{p}_{-i})} = f_i^{-1}(g_i^e(\mathbf{p}_{-i})\zeta_i^{\mathrm{B}}), \tag{1.69}$$

where $f_i^{-1}(\cdot)$ is the inverse function of $f_i(\cdot)$. Then, the following relationship holds for the best response:

$$f_i(\zeta_i^{\mathrm{B}}) = g_i^e(\mathbf{p}_{-i}) f_i^{-1}(g_i^e(\mathbf{p}_{-i})\zeta_i^{\mathrm{B}}). \tag{1.70}$$

Calculating ζ_i^{B} requires the resolution of (1.68) for a specific function $f_i(\cdot)$. However, in order to analyze a class of power control algorithms characterized by the family of decreasing continuous functions, ζ_i^{B} must be expressed in terms of a generic $f_i(p_i)$.

Proposition 1.2. *The best response function ζ_i^B in game (1.67) can be expressed, without loss of generality, as*

$$\zeta_i^B = \frac{f_i(p_i)}{g_i^e(\mathbf{p}_{-i})}, \quad \forall i \in \mathbf{N}, \tag{1.71}$$

where, necessarily,

$$f_i(p_i) = g_i^e(\mathbf{p}_{-i}) f_i^{-1}(g_i^e(\mathbf{p}_{-i}) p_i). \tag{1.72}$$

Proof. The representation of ζ_i^{B} comes directly from (1.69) and (1.70). □

Existence, uniqueness, and stability of the NE solution of game (1.67) are proved by the theory of *two-sided scalable* iterative functions. Two-sided scalability is a general framework for distributed power control established in [51]. It encompasses standard (conventional) and type-II standard (opportunistic) frameworks as particular cases and identifies common properties of iterative functions which permit a general proof of convergence to a unique fixed point.

Definition 1.7. An iterative vector function $\boldsymbol{\zeta}^{\mathrm{B}}(\mathbf{p}) = \left[\zeta_1^{\mathrm{B}}(\mathbf{p}), \ldots \zeta_N^{\mathrm{B}}(\mathbf{p})\right]^{\mathrm{T}}$ is said to be two-sided scalable if it presents the following property: for all $\alpha > 1$, $(1/\alpha)\mathbf{p} \leq \mathbf{p}' \leq \alpha\mathbf{p}$ implies

$$\frac{1}{\alpha}\boldsymbol{\zeta}^{\mathrm{B}}(\mathbf{p}) < \boldsymbol{\zeta}^{\mathrm{B}}(\mathbf{p}') < \alpha\boldsymbol{\zeta}^{\mathrm{B}}(\mathbf{p}).$$

Theorem 1.11. *The iterative vector function given by the best response defined in Proposition 1.2 is two-sided scalable.*

Proof. Theorem 1.11 is proved in [41]. □

The interesting properties of two-sided scalable functions are stated in the following three theorems. Their proofs are found in [51].

Theorem 1.12. *If $\boldsymbol{\zeta}^B(\mathbf{p})$ is two-sided scalable and a fixed point exists, then the fixed point is unique.*

Theorem 1.13. *If $\boldsymbol{\zeta}^B(\mathbf{p})$ is two-sided scalable and a fixed point $\mathbf{p}^N$ exists, then any power vector $\mathbf{p}$ converges to $\mathbf{p}^N$.*

Theorem 1.14. *If a fixed point $\mathbf{p}^N$ exists, then any power vector $\mathbf{p}$ converges to $\mathbf{p}^N$ under the totally asynchronous model.*

According to Theorems 1.11–1.14, on the condition that there exists a fixed point, the class of distributed power control algorithms which correspond to the NE strategy of game (1.67) has its convergence guaranteed in a network-wide synchronous or asynchronous mode. However, a fixed point may not exist. Sufficient conditions for the existence of a fixed point are established in the following:

Corollary 1.1. *[51] Given a two-sided scalable iterative function* $\boldsymbol{\zeta}^B(\mathbf{p})$*, if* $\boldsymbol{\zeta}^B(\mathbf{p})$ *is continuous and* $\boldsymbol{\zeta}^B(\mathbf{p}) \leq \mathbf{p}_{\max}$ *for all* $\mathbf{p}$*, then a fixed point exists.*

Corollary 1.1 establishes that the continuity of the two-sided scalable function and the introduction of an upper bound to the transmitter power are sufficient conditions for the existence of a fixed point. To complete the analysis of convergence for this class of power control algorithms, consider Proposition 1.3.

Proposition 1.3. *[51] If the iterative functions* $\boldsymbol{\zeta}^B(\mathbf{p})$ *and* $\widetilde{\boldsymbol{\zeta}}^B(\mathbf{p})$ *are two-sided scalable, then* $\boldsymbol{\zeta}^B_{\min} = \min\left(\boldsymbol{\zeta}^B(\mathbf{p}), \widetilde{\boldsymbol{\zeta}}^B(\mathbf{p})\right)$ *and* $\boldsymbol{\zeta}^B_{\max} = \max\left(\boldsymbol{\zeta}^B(\mathbf{p}), \widetilde{\boldsymbol{\zeta}}^B(\mathbf{p})\right)$ *are also two-sided scalable.*

The constrained iterative function $\boldsymbol{\zeta}^{\mathrm{B}}_c(\mathbf{p})$, given by

$$\boldsymbol{\zeta}^{\mathrm{B}}_c(\mathbf{p}) = \max\left(\mathbf{p}_{\min}, \min\left(\mathbf{p}_{\max}, \boldsymbol{\zeta}^{\mathrm{B}}(\mathbf{p})\right)\right), \tag{1.73}$$

where the power vector is constrained to lie within $[\mathbf{p}_{\min}, \mathbf{p}_{\max}]$, is thus demonstrated to be also two-sided scalable by repeatedly applying Proposition 1.3. Therefore, the class of distributed power control algorithms given by the iterative function in Proposition 1.2 converges to a unique NE point in a network-wide synchronous or asynchronous mode.

Algorithms which fall into the discussed class come from the derivation of the NE strategy (best response function) in game (1.67) for a specific function $f_i(p_i)$, i.e., they come from the resolution of (1.68) for ζ_i^{B}. Algorithms developed in [43] and [42] belong to this class.[1.5]

In [43], a decreasing exponential function of transmission power, expressed as

$$f_i^1(p_i) = 10^{(\alpha_1/10)} p_i^{-\beta_1}, \tag{1.74}$$

is adopted as target SINR. Parameters α_1 and β_1 are defined such that $f_i^1(p_i)$ meets the points $(p_{\min}, \gamma^t_{\max})$ and $(p_{\max}, \gamma^t_{\min})$, as discussed in Fig. 1.8. The resulting algorithm is called opportunistic QoS distributed power control – 1 (OQ-DPC-1). It is given by the following iterative function:

$$\overline{\zeta}_i^{\mathrm{B}}(\mathbf{p}) = \frac{1}{1+\beta_1}\left(\alpha_1 - \overline{g}_i^e(\mathbf{p}_{-i})\right), \tag{1.75}$$

where $\overline{(\cdot)}$ denotes the decibel value of $(\cdot)$.

The adaptive target SINR adopted in [42] is a decreasing sigmoid function of transmission power:

[1.5] Power control algorithms in [43] and [42] were originally developed within a restricted noncooperative game framework, where in order to assure global convergence and stability, the particular target SINR functions were required to be analytic, decreasing, and differentiable functions of transmission power.

$$f_i^2(p_i) = \frac{\alpha_2}{1+\beta_2 p_i^2}, \tag{1.76}$$

with parameters α_2 and β_2 defined such that $f_i^2(p_i)$ meets the points $(p_{\min}, \gamma_{\max}^t)$ and $(p_{\max}, \gamma_{\min}^t)$. It results in the opportunistic QoS distributed power control - 2 (OQ-DPC-2) algorithm, expressed by the following iterative function:

$$\begin{aligned} \zeta_i^{\mathrm{B}}(\mathbf{p}) = & \sqrt[3]{\frac{\alpha_2}{2\beta_2}\frac{1}{g_i^e(\mathbf{p}_{-i})} + \sqrt{\left(\frac{\alpha_2}{2\beta_2}\frac{1}{g_i^e(\mathbf{p}_{-i})}\right)^2 + \frac{1}{(3\beta_2)^3}}} \\ & + \sqrt[3]{\frac{\alpha_2}{2\beta_2}\frac{1}{g_i^e(\mathbf{p}_{-i})} - \sqrt{\left(\frac{\alpha_2}{2\beta_2}\frac{1}{g_i^e(\mathbf{p}_{-i})}\right)^2 + \frac{1}{(3\beta_2)^3}}}. \end{aligned} \tag{1.77}$$

Example 1.7 compares the discussed OQ-DPC algorithms with the opportunistic OPC and OPC-F algorithms, addressed in Section 1.6.2. Under realistic conditions, computer simulations show that OQ-DPC algorithms are more effective than the conventional opportunistic approaches in terms of throughput and fairness.

Example 1.7. Consider a generic spread spectrum wireless system with a central communication node and multiple transmitter terminals. Path loss for each terminal is modeled as $\overline{PL}(d) = 129 + 35\log_{10}(d) + \chi$ [dB], where d is the terminal-cell site distance in kilometers and χ is a zero-mean normal random variable with standard deviation set to 7 dB. Time-variant Rayleigh fading is implemented following Jake's model [23] with 50 Hz maximum Doppler spread. Power control rate is 1.5 kHz and the maximum transmission power is limited to 21 dBm, with a dynamic power range of 70 dB. The relationship between individual link capacity C_i and SINR is given by a Shannon's channel capacity-like expression,

$$C_i(\gamma_i(\mathbf{p})) = B\log_2(1 + \kappa\gamma_i(\mathbf{p})), \quad \forall i \in \mathbf{N}, \tag{1.78}$$

where $B = 1.25$ MHz is the channel bandwidth and $\kappa = 0.5$ is a loss factor. This is in accordance with the adoption of an efficient AMC scheme with adaptive processing gain. The SINR region of operation imposed by practical limitations is $[-19.7, -7.78]$ dB, which corresponds to the throughput range [9.6, 144] kbps. Figure 1.9 shows the performances of the presented OQ-DPC algorithms and particular settings of OPC and OPC-F algorithms in 5 s of network operation.

Overall system throughput (sum of all users' throughput) and outage, i.e., the probability of achieved SINR falling below the minimum SINR threshold $\Pr\{\gamma < \gamma_{\min}\}$, are plotted in Fig. 1.9(a) and (b), respectively. Overall system throughput is a measure of efficiency in the use of power resources, while outage is related to fairness. OPC and OPC-F algorithms are set to provide the maximum overall system throughput. The blind operation of OPC and OPC-F with respect to QoS of individual links explains their poor performance, since they can waste power resources in links where throughput is already saturated, while in other links the power level is not enough to establish an acceptable transmission. OQ-DPC-2 is characterized not only by high-throughput levels for all system loads, but also by rapidly increasing

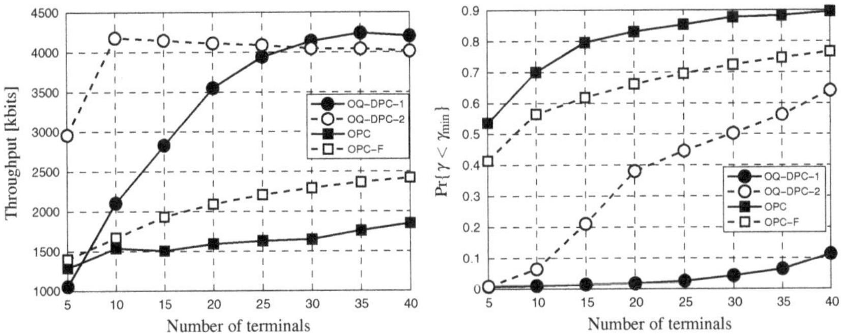

(a) Overall system throughput.

(b) Outage.

Fig. 1.9 Overall system throughput and outage in 5 s of network operation for OQ-DPC-1, OQ-DPC-2, and particular settings of OPC and OPC-F algorithms.

outage with system load. On the other hand, OQ-DPC-1 provides low outage for all considered loads and increasing throughput with system load, achieving throughput similar to OQ-DPC-2 for high-loaded systems.

This example illustrates the importance of associating the fulfillment of essential QoS constraints with the efficiency on the resource utilization. Operational limitations of practical systems, such as power constraints and the finite number of adaptive modulation and coding schemes, impose a region of operation in terms of the quality of the signal at the receiver. While opportunistic OPC and OPC-F algorithms provide an efficient power allocation but are blind with respect to QoS, OQ-DPC algorithms combine both aspects. Because of this, OQ-DPC algorithms outperform the opportunistic ones.

1.8 Prediction of Channel State Information

In distributed power control, transmission power is updated by using some channel state information, usually the measured signal-to-interference-plus-noise ratio (SINR), which contains information of channel gain and interference. Two main problems affect the quality of SINR measurements and consequently lead to degradation of power control performance. The first one is the uncertainty caused by fading, nonlinear effects, and mismatched models for interference power dynamics. The second problem is the round-trip delay in the control loop. Despite the inaccuracy of SINR measurements, most power control algorithms assume perfect knowledge of this quantity. In general, it is supposed that in the interval between two power control iterations, the variation of channel gain and interference power is not significant.

The employment of signal processing techniques in the power control problem can be helpful in dealing with uncertain and delayed measurements. In this section, some approaches toward the filtering, prediction, or estimation of fundamental quantities in the power control procedure are discussed.

1.8.1 Taylor's Series

The distributed power control (DPC) algorithm [9] presented in Section 1.4.3 is a classical solution, originally derived from a differential dynamics that makes the SINR of each communication link evolve toward a given target SINR. The setting for fast convergence has the parameter $\beta = 1$, and the algorithm reduces to

$$p_i(k+1) = \frac{\gamma^t}{\gamma_i(k)} p_i(k) = \frac{I_i(k)}{g_i(k)} \gamma^t, \quad \forall i \in \mathbf{N}, \tag{1.79}$$

where $\mathbf{N} = \{1, \ldots, N\}$ is the set of N co-channel links and k is the discrete time index. For each link i, p_i is the transmission power, γ^t denotes the target SINR, and the actual SINR is represented by γ_i, as follows:

$$\gamma_i(k) = \frac{g_i(k)}{I_i(k)} p_i(k), \tag{1.80}$$

where channel gain and interference are, respectively, denoted by g_i and I_i.

In the derivation of DPC algorithm, it is assumed that path gain and interference do not vary between two consecutive iterations. In order to observe the effects of channel gain and interference variation on the dynamics of DPC algorithm, consider the expression for $\gamma_i(k+1)$ according to (1.80), with $p_i(k+1)$ given by (1.79):

$$\gamma_i(k+1) = \left(\frac{g_i(k+1)}{g_i(k)} \cdot \frac{I_i(k)}{I_i(k+1)} \right) \gamma^t. \tag{1.81}$$

Since the aim is to achieve $\gamma_i(k+1) = \gamma^t$, the time variation of channel gain and interference causes the disturbance given by the term inside parenthesis in (1.81).

The effect of this disturbance can be attenuated by using prediction. It is clear that $g_i(k+1)$ and $I_i(k+1)$ are not available at current time instant k. However, if predicted values of channel gain and interference, $\widehat{g}_i(k+1)$ and $\widehat{I}_i(k+1)$, replace $g_i(k)$ and $I_i(k)$ in DPC update expression (1.79), they will also replace them in (1.81). Therefore, accurate prediction of channel gain and interference, i.e., $\widehat{g}_i(k+1) \approx g_i(k+1)$ and $\widehat{I}_i(k+1) \approx I_i(k+1)$, makes the SINR $\gamma_i(k+1)$ tend to the target γ^t.

This analysis is found in [37, 38], where a simple prediction method based on Taylor's series is proposed to improve DPC algorithm. Let $f(x)$ be a continuous and differentiable function. Neglecting high-order terms of the series, $f(x)$ can be expanded as

$$f(x) \approx f(x_0) + f'(x_0) \cdot (x - x_0), \tag{1.82}$$

where $f'(\cdot)$ is the first-order derivative of $f(\cdot)$ and the equivalent discrete time form is

$$f(k+1) \approx 2f(k) - f(k-1). \tag{1.83}$$

This expression can be used to predict path gain and interference. The improved DPC algorithm is, therefore, given by

$$p_i(k+1) = \frac{\widehat{I}_i(k+1)}{\widehat{g}_i(k+1)}\gamma^t = \frac{2I_i(k) - I_i(k-1)}{2g_i(k) - g_i(k-1)}\gamma^t, \quad \forall i \in \mathbf{N}. \tag{1.84}$$

The classical DPC algorithm (1.79) and its prediction-based version (1.84) are considered in Example 1.8. This numerical example illustrates the improved performance due to prediction.

Example 1.8. Consider a set of co-channel base stations in downlink (base station to mobile terminal) transmission, which comprises a central cell and one layer of interferers. Mobile terminals are uniformly distributed over the cell area. Path loss for each terminal is modeled as $\overline{PL}(d) = 120 + 40\log_{10}(d) + \chi$ [dB], where d is the cell site-terminal distance in kilometers and χ is a zero-mean normal random variable with standard deviation set to 6 dB. Rayleigh fading is implemented following Jake's model [23] with 20 Hz maximum Doppler spread. The power control rate is 1 kHz and the target SINR is 5 dB.

Figure 1.10 illustrates the evolution in time of the SINR of a mobile terminal in the central cell. Both algorithms, the conventional and the improved DPC, are observed under the same channel gain conditions in Fig. 1.10(a) and (b), respectively. It is clearly observable that DPC with prediction of channel gain and interference is more efficient than the conventional one in stabilizing the SINR around the target value. The advantage of employing the prediction method is confirmed by the values of mean squared error between the actual and the target SINRs.

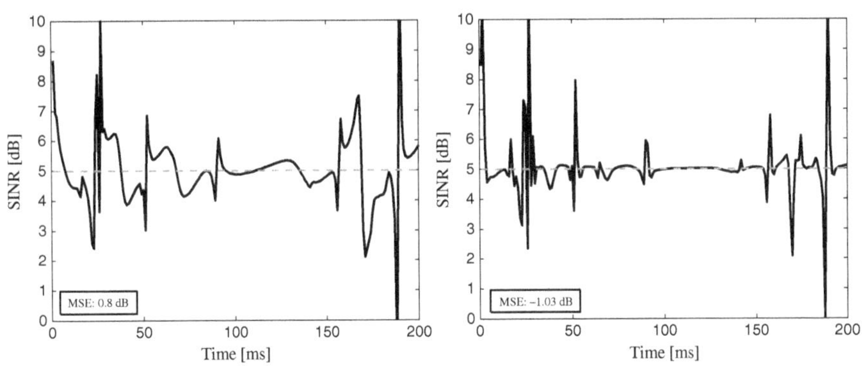

(a) Conventional DPC algorithm. (b) DPC algorithm with prediction.

Fig. 1.10 Evolution in time of the SINR using channel and interference prediction based on Taylor's series.

1.8.2 Other Prediction Approaches

The prediction task in power control can be carried out by traditional signal processing techniques. The simple structure of linear transversal filters is suitable for accomplishing predictive power control in a distributed fashion. Let $x(k+1)$ denote the quantity to be predicted and $\widehat{x}(k+1)$ the predicted value of $x(k+1)$. Then,

$$\widehat{x}_i(k+1) = \sum_{j=0}^{M-1} w_{i,j}(k)x_i(k-j), \quad \forall i \in \mathbf{N}, \tag{1.85}$$

where $w_{i,j}(k)$ is the jth coefficient of the filter for communication link i.

Basically, adaptive prediction uses past and current data to adjust the filter coefficients, so that past predictions closely match observed data. Subsequently, these parameters are used to generate future predictions. The well-known least mean square (LMS) and recursive least squares (RLS) adaptive algorithms are commonly used to determine the filter coefficients [20].

Applications of linear adaptive prediction to power control can be found in [14, 24, 25, 54]. As a general rule, predictive power control performs better than the conventional one. Furthermore, in a comparison between LMS and RLS, the former presents faster convergence, while the latter exhibits superior tracking performance.

Another possible approach is based on the Kalman filter, which is a fundamental tool for analyzing and solving a broad class of prediction and estimation problems. Consider the representation of a linear discrete time dynamic system, with $\mathbf{x}$ as the state vector and $\mathbf{y}$ as the measurement vector:

$$\mathbf{x}(k+1) = \mathbf{A}(k)\mathbf{x}(k) + \mathbf{v}_x(k), \tag{1.86}$$

$$\mathbf{y}(k) = \mathbf{C}(k)\mathbf{x}(k) + \mathbf{v}_y(k), \tag{1.87}$$

where k denotes the discrete time index; matrices $\mathbf{A}(k)$ and $\mathbf{C}(k)$ are known and have appropriate dimensions; $\mathbf{v}_x(k)$ and $\mathbf{v}_y(k)$ are denoted as process noise and measurement noise, respectively; and the initial state $\mathbf{x}(0)$ is unknown. Assume that the unknown disturbances $\{\mathbf{x}(0), \{\mathbf{v}_x\}, \{\mathbf{v}_y\}\}$ are zero-mean random variables with known second-order statistics. In this general context, let the objective be the estimation of a linear combination of the states, that is $\mathbf{s}(k) = \mathbf{L}(k)\mathbf{x}(k)$, by using current and past measurements $\{\mathbf{y}\}$.

The well-known Kalman filter is the solution to the problem of minimizing the expected filtered error energy, i.e.,

$$\min_{\widehat{\mathbf{s}}(k)} E\left\{\widetilde{\mathbf{s}}^H(k)\widetilde{\mathbf{s}}(k)\right\}, \tag{1.88}$$

where $\widetilde{\mathbf{s}}(k) = \mathbf{s}(k) - \widehat{\mathbf{s}}(k)$ is the filtered error and the linear discrete time dynamic system is described by (1.86) and (1.87). The interested reader can find a complete treatment on Kalman filtering in [18, 47].

In the power control problem, variables which carry information about the channel state are measured or estimated at the receiver, such as channel gain and interference, and can be predicted by using the Kalman filter. The use of predicted values of these quantities can attenuate effects of round-trip delay. On the other hand, performance degradation due to uncertainties on channel state information can be minimized with estimates provided by the Kalman filter. Details on the application of Kalman filtering to power control in wireless communication systems can be found in [27, 46].

Finally, H_∞ filtering appears as an alternative to the conventional filtering tools. In this approach, disturbances $\{\mathbf{x}(0), \{\mathbf{v}_x\}, \{\mathbf{v}_y\}\}$ in the linear discrete time dynamic system described by (1.86) and (1.87) are unknown, but deterministic. Furthermore, the optimization criterion is no longer the minimization of a quadratic function of the filtered error. The aim in this framework is to minimize the "worst-case" energy gain from disturbances to the filtered error. Therefore, H_∞ filtering is closely related to robustness.

Since only in particular problems the explicit minimization of the "worst-case" energy gain from disturbances to the filtered error is possible, it is common to consider the following sub-optimum problem:

$$\max_{\mathbf{x}(0),\mathbf{v}_x(k),\mathbf{v}_y(k)} \frac{\sum_{k=0}^{L} ||\mathbf{s}(k) - \widehat{\mathbf{s}}(k)||^2}{||\widetilde{\mathbf{x}}(0)||^2_{\boldsymbol{\Pi}^{-1}} + \sum_{k=0}^{L} \left(||\mathbf{v}_x(k)||^2_{\mathbf{Q}_x^{-1}(k)} + ||\mathbf{v}_y(k)||^2_{\mathbf{Q}_y^{-1}(k)} \right)} < \theta^2, \quad (1.89)$$

where $||\cdot||$ denotes the euclidian norm, such that $||\mathbf{v}_y(k)||^2_{\mathbf{Q}_y^{-1}(k)} = \mathbf{v}_y^H(k)\mathbf{Q}_y^{-1}(k)\mathbf{v}_y(k)$; $\widetilde{\mathbf{x}}(0) = \mathbf{x}(0) - \widehat{\mathbf{x}}(0)$; and the prescribed θ^2 gives the energy gain upper bound. Thus, the lower the value of θ^2, the more restrictive the filter with respect to the energy gain from disturbances to the filtered error. In other words, parameter θ is responsible for the robustness level of the H_∞ filter. The derivation and a complete treatment of the H_∞ filter can be found in [18, 45, 47].

The employment of the H_∞ filter in the place of conventional filtering methods becomes attractive when uncertainties on the adopted models are significant and/or disturbances are not statistically well characterized. In other words, the H_∞ filter is robust and can be preferable to the Kalman filter, the optimum linear solution in the sense of the minimum error variance, if disturbances are not gaussian random variables or if their statistics are unknown. Examples of applications of H_∞ filtering to the distributed power control problem can be found in [39, 48], where the effective channel gain and the interference power must be predicted or estimated. In [39], H_∞ estimation is shown to outperform the Kalman filtering.

1.9 Conclusions and Topics for Future Research

Power control, as an essential technique for wireless communication systems, is a broad area of research and has received much attention from academy and industry. In this chapter, the main topics on power control were covered:

- Classical theory of centralized and distributed power control for QoS-fixed applications, including the presentation of feasibility and convergence analysis, as some of the main algorithms available in the literature
- Alternative frameworks and distributed power control algorithms based on concepts of opportunistic communications for QoS-flexible services
- Employment of non-cooperative game theory to the distributed power control problem for QoS-fixed and QoS-flexible applications, including the study of several algorithms
- Overview of some approaches for prediction of channel state information toward an improved power control

Power control for QoS-fixed services, such as voice communications, can be considered a consolidated research field. Since the early 1990s extensive studies have been yielding distinct approaches and algorithms accompanied by strong analytical results concerning feasibility and convergence. On the other hand, power control for QoS-flexible applications is an evolving research topic. The interest on this subject is more recent, motivated by the demand for data services in wireless communication networks. Although fundamental analytical results and some suitable approaches and algorithms exist, there is room for further investigation.

In a more general context, the scientific community has been working in concepts of *cognitive wireless networks* [8, 21]. Such systems are expected to present properties like decentralized optimization, pro-active or predictive operation, self-adaptation to different operation conditions, capacity of learning and robustness. Distributed power control must play a key role in a cognitive communication system. Its approach from the viewpoint of automatic control theory, as done in [50] for QoS-flexible services, seems to be a step on the right direction. Automatic control is suitable for dealing with multiple objectives. Furthermore, it is able to provide adaptive, decentralized, and robust solutions. Therefore, the employment of feedback control methodologies to the distributed power control problem can be a promising research field.

References

1. Alpcan, T., Basar, T., Srikant, R.: CDMA uplink power control as a noncooperative game. Wireless Networks **8**, 659–670 (2002)
2. Altman, E., Altman, Z.: S-modular games and power control in wireless networks. IEEE Transactions on Automatic Control **48**(5), 839–842 (2003)
3. Basar, T., Olsder, G.J.: Dynamic Noncooperative Game Theory, 2nd edn. SIAM (1998)
4. Chiller, J.: Mobile Communications. Addison-Wesley (2000)

5. Damosso, E.: Digital Mobile Radio Towards Future Generation Systems. European Commission (1999)
6. Debreu, G.: A social equilibrium existence theorem. Proceedings of National Academy of Science **38**, 886–893 (1952)
7. Fan, F.: Fixed point and minima theorems in locally convex topological linear spaces. Proceedings of National Academy of Science **38**, 121–126 (1952)
8. Fitzek, F.H.P., Katz, M.D.: Cognitive Wireless Networks: Concepts, Methodologies and Visions Inspiring the Age of Enlightenment of Wireless Communications. Springer (2007)
9. Foschini, G.J., Miljanic, Z.: A simple distributed autonomous power control algorithm and its convergence. IEEE Transactions on Vehicular Technology **42**(4), 641–646 (1993)
10. Fudenberg, D., Tirole, J.: Game Theory. MIT Press (1991)
11. Gantmacher, F.R.: The Theory of Matrices. Chelsea Publishing (1960)
12. Glicksberg, I.L.: A further generalization of the Kakutani fixed point theorem with application to Nash equilibrium points. Proceedings of American Mathematics Society **3**, 170–174 (1952)
13. Goldsmith, A.J., Chua, S.G.: Adaptive coded modulation for fading channels. IEEE Transactions on Communications **46**(5), 595–602 (1998)
14. Gombachika, H.S.H., Tafazolli, R., Evans, B.G.: A comparative study of predictive transmit power schemes for S-UMTS. Wireless Networks **11**, 215–222 (2005)
15. Goodman, D., Mandayam, N.: Power control for wireless data. IEEE Personal Communications **7**(2), 48–54 (2000)
16. Grandhi, S.A., Vijayan, R., Goodman, D.J.: Distributed power control in cellular radio systems. IEEE Transactions on Communications **42**(2–4), 226–228 (1994)
17. Hanly, S., Tse, D.N.C.: Multi-access fading channels: Part II: Delay-limited capacities. IEEE Transactions on Information Theory **44**(7), 2816–2831 (1998)
18. Hassibi, B., Sayed, A.H., Kailath, T.: Indefinite-Quadratic Estimation and Control: A Unified Approach to H_2 and H_∞ Theories. SIAM (1999)
19. Hata, M.: Empirical formula for propagation loss in land mobile radio services. IEEE Transactions on Vehicular Technology **29**(3), 317–325 (1980)
20. Haykin, S.: Adaptive Filter Theory, 4th edn. Prentice Hall (2001)
21. Haykin, S.: Cognitive radio: Brain-empowered wireless communications. IEEE Journal on Selected Areas in Communications **23**(2), 201–220 (2005)
22. Holma, H., Toskala, A.: WCDMA for UMTS: Radio Access for Third Generation Mobile Communications, 3rd edn. Wiley (2004)
23. Jakes, W.C.: Microwave Mobile Communications, 2nd edn. Wiley (1974)
24. Lau, F.C.M., Tam, W.M.: Novel predictive power control in a CDMA mobile radio system. Proceedings of IEEE Vehicular Technology Conference **3**, 1950–1954 (2000)
25. Lau, F.C.M., Tam, W.M.: Achievable-SIR-based predictive closed-loop power control in a CDMA mobile system. IEEE Transactions on Vehicular Technology **51**(4), 720–728 (2002)
26. Lee, T.H., Lin, J.C., Su, Y.T.: Downlink power control algorithms for cellular radio systems. IEEE Transactions on Vehicular Technology **44**(1), 89–94 (1995)
27. Leung, K.K.: Power control by interference prediction for broadband wireless packet networks. IEEE Transactions on Wireless Communications **1**(2), 256–265 (2002)
28. Leung, K.K., Sung, C.W.: An opportunistic power control algorithm for cellular network. IEEE Transactions on Networking **14**(3), 470–478 (2006)
29. Leung, K.K., Sung, C.W., Wong, W.S., Lok, T.M.: Convergence theorem for a general class of power control algorithms. IEEE Transactions on Communications **52**(9), 1566–1574 (2004)
30. Luce, R.D., Raiffa, H.: Games and Decisions. John Wiley & Sons (1957)
31. Nash, J.: Equilibrium points in N-person games. Proceedings of National Academy of Science **36**, 48–49 (1950)
32. Nash, J.: Non-cooperative games. Annals of Mathematics **54**, 286–295 (1951)
33. Neumann, J.V.: Zur theorie der gesellschaftsspiele. Mathematische Annalen **100**, 295–320 (1928)
34. Neumann, J.V., Morgenstern, O.: Theory of Games and Economic Behavior, 1st edn. Princeton University Press (1944)

35. Novakovic, D.M., Dukic, M.L.: Evolution of the power control techniques for DS-CDMA toward 3G wireless communication systems. IEEE Communications Surveys, pp. 2 –15 (2000)
36. Okumura, T., Ohmori, E., Fukuda, K.: Field strength and its variability in VHF and UHF land mobile service. Review of the Electrical Communication Laboratory **16**(9–10), 825–873 (1968)
37. de Oliveira Neto, R.A., de S. Chaves, F., Cavalcanti, F.R.P., Maciel, T.F.: A new distributed power control algorithm based on a simple prediction method. Lecture Notes in Computer Science **3124**, 431–436 (2004)
38. de Oliveira Neto, R.A., de S. Chaves, F., Cavalcanti, F.R.P., Maciel, T.F.: New distributed power control algorithms for mobile communications. Journal of the Brazilian Telecommunications Society **20**(2), 65–71 (2005)
39. Qian, L., Gajic, Z.: Variance minimization stochastic power control in CDMA systems. IEEE Transactions on Wireless Communications **5**(1), 193–202 (2006)
40. Rappaport, T.S.: Wireless Communications. Prentice-Hall (1996)
41. de S. Chaves, F., Cavalcanti, F.R.P., de Oliveira Neto, R.A., Santos, R.B.: Opportunistic distributed power control with adaptive QoS and fairness for wireless networks. Wireless Communications and Mobile Computing, published online, DOI: 10.1002/wcm.753, 1–14 (2009)
42. de S. Chaves, F., Cavalcanti, F.R.P., Santos, R.B., de Oliveira Neto, R.A.: Opportunistic distributed power control with QoS guarantee in wireless communication systems. Proceedings of IEEE Workshop on Signal Processing Advances in Wireless Communications, pp. 1–5 (2007)
43. de S. Chaves, F., de Sousa Jr., V.A., de Oliveira Neto, R.A., de Lima, C.H.M., Cavalcanti, F.R.P.: Performance of energy efficient game theoretical-based power control algorithm in WCDMA. Proceedings of IEEE International Symposium on Personal, Indoor and Mobile Radio Communications, pp. 1–5 (2006)
44. Saraydar, C.U., Mandayam, N.B., Goodman, D.J.: Efficient power control via pricing in wireless data networks. IEEE Transactions on Communications **50**(2), 291–303 (2002)
45. Shaked, U., Theodor, Y.: H_∞-optimal estimation: A tutorial. Proceedings of IEEE Conference on Decision and Control **2**, 2278–2286 (1992)
46. Shoarinejad, K., Speyer, J.L., Pottie, G.J.: Integrated predictive power control and dynamic channel assignment in mobile radio systems. IEEE Transactions on Wireless Communications **2**(5), 976–988 (2003)
47. Simon, D.: Optimal State Estimation: Kalman, H_∞, and Nonlinear Approaches. John Wiley & Sons (2006)
48. Sorooshyari, S., Gajic, Z.: Autonomous dynamic power control for wireless networks: User-centric and network-centric consideration. IEEE Transactions on Wireless Communications **7**(3), 1004–1015 (2008)
49. Strang, G.: Linear Algebra and Its Applications. Harcourt (1988)
50. Subramanian, A., Sayed, A.H.: Joint rate and power control algorithm for wireless networks. IEEE Transactions on Signal Processing **53**(11), 4204–4214 (2005)
51. Sung, C.W., Leung, K.K.: A generalized framework for distributed power control in wireless networks. IEEE Transactions on Information Theory **51**(7), 2625–2635 (2005)
52. Topkis, D.M.: Supermodularity and Complementarity. Princeton University Press (1998)
53. Tse, D.N.C., Hanly, S.: Multi-access fading channels: Part I: Polymatroid structure, optimal resource allocation, and throughput capacities. IEEE Transactions on Information Theory **44**(7), 2796–2815 (1998)
54. Virtej, I., Kansanen, O., Koivo, H.: Enhanced predictive fast power control for 3G systems. Proceedings of IEEE Vehicular Technology Conference **4**, 2864–2868 (2001)
55. Walfisch, J., Benoni, H.L.: A theoretical model of UHF propagation in urban environments. IEEE Transactions on Antennas and Propagation **36**(12), 1788–1796 (1988)
56. Yacoub, M.D.: Fundamentals of Mobile Radio Engineering. CRC Press (1993)
57. Yacoub, M.D.: Wireless Technology: Protocols, Standards, and Techniques. CRC Press (2001)
58. Yates, R.D.: A framework for uplink power control in cellular radio systems. IEEE Journal on Selected Areas in Communications **13**(7), 1341–1347 (1995)

59. Yates, R.D., Gupta, S., Rose, C., Sohn, S.: Soft dropping power control. Proceedings of IEEE Vehicular Technology Conference **3**, 1694–1698 (1997)
60. Zander, J.: Distributed cochannel interference control in cellular radio systems. IEEE Transactions on Vehicular Technology **41**(3), 305–311 (1992)
61. Zander, J.: Performance of optimum transmitter power control in cellular radio systems. IEEE Transactions on Vehicular Technology **41**(1), 57–62 (1992)
62. Zander, J., Kim, S.L.: Radio Resource Management for Wireless Networks. Artech House Publishers (2001)

Chapter 2
Radio Resource Management Performance for the GSM/EDGE Radio Access Network

Yuri C. B. Silva, Tarcisio F. Maciel, and Francisco R. P. Cavalcanti

2.1 Introduction

This chapter presents a broad study on the potential of applying RRM techniques to the global system for mobile communication (GSM)/enhanced data rates for GSM evolution (EDGE) system. Even though the presented results are focused on the GSM/EDGE system, the principles employed by most of the considered RRM techniques can be applied to other radio access networks (RANs). We thus hope that the reader will also learn about RRM strategies and adapt these concepts to the RAN of his/her own interest.

The provision of multiple services is one of the key features of GSM/EDGE, which has been the focus of several studies, such as [22, 24, 25, 37]. Additionally, some of the themes that have been covered by recent research include power control [36, 41, 45], dynamic channel allocation [20, 46, 47, 49, 60], multi-antenna techniques [18, 19, 34, 48, 50, 51], among others.

In this chapter, the RRM techniques are placed within the context of GSM/EDGE and results are presented, which indicate the achievable gains in terms of capacity and/or quality of service (QoS). The results demonstrate that, by using appropriate RRM techniques, the GSM/EDGE radio access network capacity remains competitive with other emerging access technologies, thus allowing for substantial operational cost reductions for the already deployed infrastructure.

The remaining of this chapter is organized as follows: Section 2.2 briefly describes the architecture of the GSM/EDGE radio access network, along with its protocol stack and standard RRM functionalities. Section 2.3 presents the RRM techniques considered in the scope of this chapter, which are power control, dynamic channel allocation, management of multiple services, and multi-antenna techniques. Next, in Section 2.4, some aspects concerning the simulation and modeling of GSM/EDGE networks are discussed. The achieved simulation results are presented in Section 2.5 for the studied RRM techniques. Finally, trends and directions for the further evolution of RRM in GSM/EDGE networks are discussed in Section 2.6.

F. Cavalcanti, S. Andersson (eds.), *Optimizing Wireless Communication Systems*,
DOI 10.1007/978-1-4419-0155-2_2, © Springer Science+Business Media, LLC 2009

2.2 Fundamentals of RRM in GSM/EDGE

The second generation of cellular systems was marked by a transition from analog-to-digital radio communications. GSM emerged in this context, with its phase 1 specification and initial deployment dating back to the early 1990s. GSM had a significant role in unifying the previously diverging European standards. The ubiquity of GSM, which facilitated international roaming among operators, the creation of the low-cost short message service, the support for circuit-switched data connections, as well as further improvements of the technology, such as the introduction of more efficient speech codecs, led to widespread GSM availability throughout the world, reaching the expressive mark of over 3 billion subscribers by the end of 2007 [27].

The provision of data services was improved with the introduction of the general packet radio service (GPRS) in 1997, which added support for packet-switched connections and provided four different coding schemes, with rates ranging from 8 to 20 kbit/s. In 1999, the enhanced GPRS (EGPRS) was introduced and then adopted as the packet system of the GSM/EDGE radio access network, which is the focus of this section. In the following section, an overview of GSM/EDGE is presented along with some of its standard functionalities.

2.2.1 GSM/EDGE Radio Access Network Overview

The GSM/EDGE radio access network (GERAN) represents the evolution of the GSM system for providing improved packet data transmission. The GPRS and EGPRS are radio technologies that provide packet-switched connections between MS and BS, while the GERAN is composed of several network elements that are interconnected through standard interfaces. The two main elements of the radio access network are

- **Base station subsystem (BSS)**: It comprehends the base transceiver station (BTS), or simply BS, which is the onsite base station, and the base station controller (BSC), which is a controlling unit responsible for a group of BTSs.
- **Core network**: It has functionalities such as mobility management, authentication, charging, among others, and also provides access to networks outside of the cellular system. In the case of circuit-switched connections, the mobile switching center (MSC) is the main element of the core network, providing accessibility to the conventional public-switched telephone network (PSTN). In the case of packet-switched connections, the main elements are the service GPRS support node (SGSN) and the gateway GPRS support node (GGSN). The former performs routing and delivery of packets within the cellular system and the latter provides connectivity to external data packet networks, such as the Internet.

With the purpose of maintaining compatibility with the existing GSM infrastructure, the EDGE technology has many parameters in common with GSM, including the sharing of the same frequency spectrum. EGPRS introduces some improvements with regard to GPRS, such as the 8-PSK (phase-shift keying) modulation and additional modulation and coding schemes (MCSs). Due to this improved physical layer, EGPRS provides high data rates, reaching 384 kbit/s or more when multiple timeslots are reserved to a single MS.

Through the adequate configuration of parameters from the protocol layers of GSM/EDGE, it is possible to provide multiple services. The data transmission of EGPRS may be adjusted, for example, to support applications with different quality-of-service (QoS) requirements, such as World Wide Web (WWW), file transfer protocol (FTP), and streaming of audio/video files.

The standardization of the GSM/EDGE radio access network is coordinated by the 3rd. Generation Partnership Project (3GPP). The standards have undergone some major revisions, with Rel-8 being the latest release as of 2008.

The GSM/EDGE network is already well established, in terms of technical maturity as well as market deployment. This, however, has not stopped the development of new techniques for improving its performance and providing capacity gains.

2.2.1.1 Channel Structure

The GSM/EDGE system implements multiple access through frequency division as well as through time division. Each frequency carrier has a cyclic time structure associated, which is composed of hyperframes, superframes, multiframes, frames, and timeslots [7], as it can be seen in Fig. 2.1.

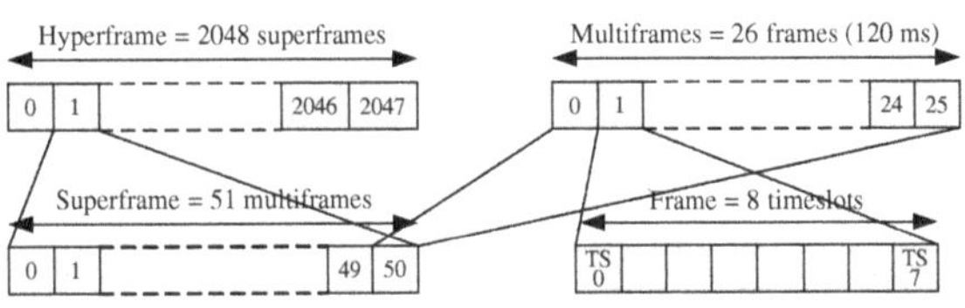

Fig. 2.1 GSM/EDGE frame structure.

Besides the displayed frame structure, which is employed for traffic channels, there is an alternative signaling frame structure that defines a multiframe with 51 frames and a superframe with 26 multiframes.[2.1]

The basic time unit is the timeslot, which is equivalent to roughly 0.577 ms. A sequence of eight timeslots defines a time division multiple access (TDMA) frame, and a group of four frames composes a TDMA radio block.

Among the 26 frames of the multiframe structure in Fig. 2.1, the 13th and the last frame are reserved for control and other functionalities. Therefore, the other 24 frames may be employed for traffic, i.e., six radio blocks.

A physical channel is defined by the pair timeslot/frequency. A logical channel, on the other hand, corresponds to the information flow between a BS and an MS.

[2.1] A superframe still contains 26 × 51 frames in total.

The logical channels may be divided into traffic and control channels. Next, some of the most relevant logical channels are presented:

- **Traffic channel (TCH)**: This is a circuit-switched traffic channel used for voice as well as circuit-switched data transmission.
- **Packet data traffic channel (PDTCH)**: This is a packet-switched traffic channel used for data transmission.
- **Broadcast control channel (BCCH)**: This is a downlink control channel that distributes general information to the MSs concerning the system configuration. The information may include number of common control channels, possible combinations of control channels, whether support for packet-switched traffic is enabled, among others. There is also the packet broadcast control channel (PBCCH), which is the corresponding channel for data MSs.
- **Common control channel (CCCH)**: It corresponds to a set of common control channels that are used for implementing access management functions.
- **Dedicated control channel (DCCH)**: It corresponds to a set of dedicated control channels that are used for measurements, signaling, among other functionalities. The main circuit-switched dedicated control type channels are the slow associated control channel (SACCH) and fast associated control channel (FACCH), which provide connection-specific signaling information concerning the channels they are associated to, and the stand-alone dedicated control channel (SDCCH), which can be used for signaling during call setup. These channels can be used both in the uplink and downlink.

The SACCH, for example, is important for transmitting information related to signal level and signal quality measurements. The reporting periods have a duration of 480 ms (104 TDMA frames) and they are employed, e.g., by the power control (PC), handover, and link adaptation (LA) algorithms discussed later in this chapter.

The actual radio transmission requires that the different logical channels be mapped onto the physical channels. A physical channel may be composed of only control channels, two half-rate traffic channels plus control channels, or a full-rate traffic channel plus control channels. The combination of traffic and control channels is possible by either employing the previously described reserved frames, in the case of SACCH, or by stealing slots from traffic channels, in the case of FACCH. A more detailed description of the possible channel combinations can be found in [6].

2.2.1.2 Protocols

This section presents an overview of the protocol stack of the GSM/EDGE network for packet data transmission. Figure 2.2 shows how the user plane[2.2] protocol layers are organized among the different network elements in the 3GPP standards.

The focus of this chapter lies on the radio link between the MS and the BS, which is supported by the following three protocol layers:

[2.2] There are protocol stacks for the user plane and control plane. The former refers to the actual data transmission and the latter is used for control and signaling.

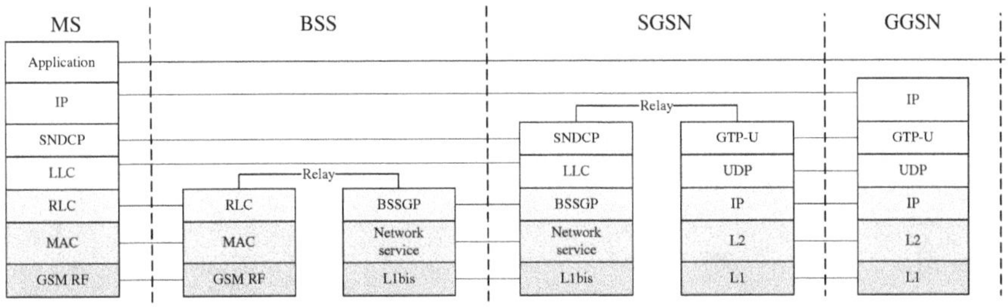

Fig. 2.2 Packet-switched user plane protocols of the GSM/EDGE network.

- **Link layer control (LLC)**: It offers a reliable and secure logical link between the MS and the SGSN for superior layers. One of its main functionalities consists of performing the segmentation of packets arriving from higher layers.
- **Radio link control (RLC) and medium access control (MAC)**: These protocols provide services for the transfer of information over the physical layer. Among their functionalities are the error-correcting procedures enabled through the selective retransmission of erroneous blocks. The RLC function offers a reliable radio link to the higher layers, while MAC treats issues such as channel allocation and the multiplexing/scheduling of MSs.
- **GSM RF or physical layer**: It provides data transfer services over the physical channel between the BS and the MS. Among its functionalities are the coding of data and the detection/correction of transmission errors in the physical medium.

The data transmission process can be briefly described as follows. The packets that arrive from the internet protocol (IP) and sub-network-dependent convergence protocol (SNDCP) layers are segmented into LLC layer frames. The LLC frames are segmented into RLC/MAC blocks as they are being requested by the system. The RLC/MAC blocks are transmitted over the four bursts of a TDMA radio block, where the term burst corresponds to the transmission of data during the time of a timeslot.

The RLC/MAC blocks are composed of header and data fields, which have variable lengths depending on the current MCS. EGPRS has nine MCSs, with the first four employing Gaussian minimum shift keying (GMSK) modulation and the remaining ones 8-PSK. The lowest MCSs transport a smaller amount of information data per block, but are more robust to variations in the link quality.

The MCSs are organized into different families, each with a certain base payload [4]. In the case of retransmissions, only an MCS of the same family may be chosen. MCSs 7–9 transmit two RLC/MAC blocks per TDMA radio block, while the others transmit only one block. A summary of the main MCS parameters is presented in Table 2.1. The interested reader can refer to [3, 5] in order to obtain a detailed description of each protocol in the GSM/EDGE protocol stack.

Table 2.1 MCS parameters (with data rate in kbit/s and payload in bytes per TDMA radio block).

MCS	**1**	**2**	**3**	**4**	**5**	**6**	**7**	**8**	**9**
Modulation	⊢	GMSK		⊣	⊢		8-PSK		⊣
Code rate	0.53	0.66	0.85	1.0	0.37	0.49	0.76	0.92	1.0
Data rate	8.8	11.2	14.8	17.6	22.4	29.6	44.8	54.4	59.2
Payload	22	28	37	44	56	74	112	136	148
Family	C	B	A	C	B	A	B	A	A

2.2.2 Link Adaptation

The link adaptation (LA) mechanism of EGPRS, which is described in [8], tries to provide the best possible quality to the MS through the modification of the current MCS. This adaptation occurs according to the availability of link quality measurements and it intends to exploit the channel diversity and maximize data rates by suitably selecting an MCS according to the channel state.

Ideally, LA could be employed on a per-block basis, i.e., a new MCS would be selected for each radio block (20 ms) [38, 39]. In practice, however, the standard LA mechanism uses the same link quality estimations used by power control, which are periodically reported to the BS each 480 ms [9].

Figure 2.3, based on results presented in [22], shows how LA behaves according to the SIR. For a given SIR, the MCS providing the highest data rate should be selected.

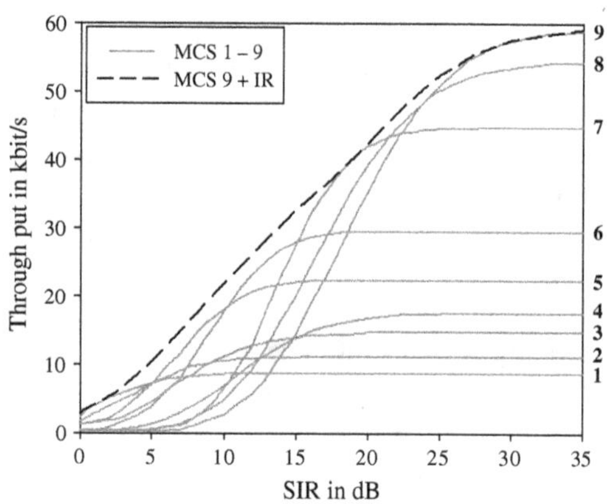

Fig. 2.3 Link adaptation with pedestrian mobility (3 km/h).

Another mechanism, which may replace or act together with LA, is the incremental redundancy (IR), which is also shown in Fig. 2.3. With IR the amount of redundancy is increased for each additionally required retransmission. IR improves the reception of retransmissions by combining the retransmitted blocks with the data already available at the receiver. More details on IR can be found in [23, 40].

2.2.3 Frequency Hopping

The frequency hopping (FH) technique consists of periodically changing the transmission frequency with the purpose of introducing diversity. The diversity effect may include both frequency and interference diversity, which are illustrated in Fig. 2.4 for two co-channel MSs.

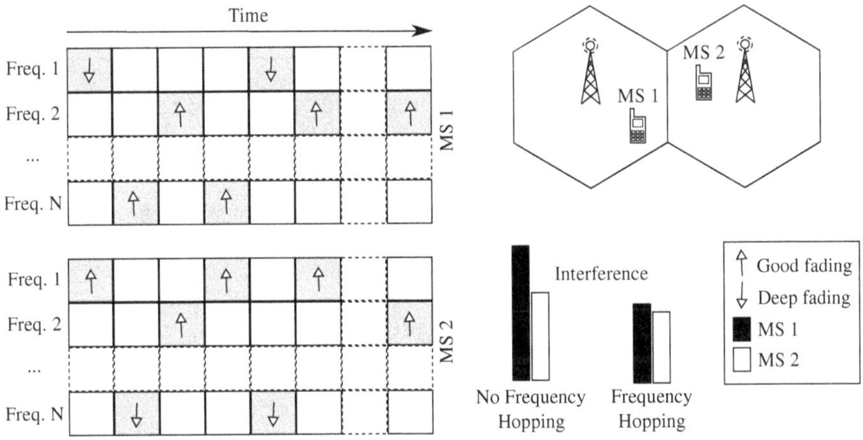

Fig. 2.4 Frequency hopping.

Because a different frequency is used after each hop, the MSs perceive different fading gains at each time, as indicated by the arrows in Fig. 2.4. Consequently, the users' links do not remain for a long time in a deep fading and a more reliable communication might be achieved.

Regarding interference, suppose some MSs perceive strong interference while other MSs perceive weak interference (or no interference at all), as illustrated in Fig. 2.4. Without frequency hopping, the links of some users would be subject to high interference for a long time. By hopping across different frequencies, the set of co-channel interferers seen by a user changes after each hop and the MSs would experience periodically changing interference profiles, so that almost the same average interference would be perceived by all MSs.

Thus, the main benefit of frequency hopping is to average out the fading and interference effects, thus allowing the use of more aggressive reuse patterns, such as 1/3 and 1/1.[2.3] The types of frequency hopping are described as follows:

[2.3] This notation indicates the cluster size of the frequency reuse pattern, i.e., the group of cells/sectors within which there can be no reuse of the available frequencies.

- **Random frequency hopping (RFH)**: It performs the hopping in an unordered fashion, according to a pseudo-random sequence determined based on system parameters and the algorithm presented in [6].
- **Cyclic frequency hopping (CFH)**: It performs the hopping in an ordered fashion, according to a previously established cyclic sequence.

RFH provides both frequency and interference diversity, while CFH presents frequency diversity, but not interference diversity, since the co-channel MSs hop over the same frequencies. The main system parameters involved with the frequency hopping algorithms are presented in Table 2.2.

Table 2.2 Description of the frequency hopping parameters.

Parameter	Description
MAL	Mobile allocation list containing the frequencies available for allocation
MAIO	Mobile allocation index offset indicating the offset within the MAL
N_{freq}	Number of frequencies per mobile allocation list
FN	TDMA frame number currently in use
HSN	Hopping sequence number allocated to each sector
MAI	Mobile allocation index referencing the frequency of MAL to be used

The frequency hopping algorithm proposed in [6] can be described as

$$\begin{cases} \text{MAI} = (\text{FN} + \text{MAIO}) \bmod (N_{\text{freq}}), & \text{if HSN} = 0, \\ \text{MAI} = (S + \text{MAIO}) \bmod (N_{\text{freq}}), & \text{if HSN} \neq 0, \end{cases} \tag{2.1}$$

where the first equation relates to CFH and the second to RFH. The S variable corresponds to the generation of the pseudo-random sequence, which is a function of the FN, HSN, N and the hopping table defined in [6]. The standard defines 64 possible orthogonal hopping sequences. The orthogonality is assured for MSs that have the same HSN but different MAIOs, i.e., they never become co-channel interferers.

The MAIO allocation to MSs entering the system can be done either at random, in the case of RFH, or it can follow a certain allocation algorithm, such as the one defined in Section 2.3.2, in the case of CFH. More details on the implementation of frequency hopping for GSM/EDGE can be found in [6, 42].

2.3 Advanced Radio Resource Management for GSM/EDGE

Since the deployment of the first GSM-based networks, the demand for mobile communication has increased enormously. Conventional GSM/EDGE networks have reached their capacity limits and, in order to serve the growing demand for mobile communication, solutions to increase system capacity are required.

Spectrum is a scarce resource whose use is granted and regulated by estate institutions, such that a capacity expansion through the acquisition of new frequency bands may become very expensive. In this context, techniques to optimize the usage of radio resources, i.e., RRM techniques, gain importance as alternative to enhance the system capacity without needing additional spectrum.

This section describes several advanced RRM techniques contextualized in a GSM/EDGE network. The RRM techniques discussed here are not necessarily an integral part of the GSM/EDGE standard, but might be incorporated as part of proprietary solutions.

2.3.1 Power Control

Power control (PC) is a well-established RRM technique which aims at, mainly, reducing interference levels in a wireless network and conserving battery power of terminals. A detailed discussion about PC has already been provided in Chapter 1 of this book. In this section, some particular aspects of PC in the context of GSM/EDGE networks are detailed.

PC is well described in the standard for the uplink of GSM/EDGE networks [9, 10]. In the downlink, PC is not a mandatory feature. However, a downlink PC algorithm can also be implemented at the BSs as long as the restrictions imposed by the standard are respected. The power characteristics of base and mobile stations are described in [9, 10] and among the standard restrictions are, for example, the usage of discrete power levels in steps of 2 dB and dynamic power ranges limited to 30 dB and to 10 dB for voice and data services, respectively.

The standard PC algorithm for the uplink in the GSM/EDGE network is a variable-step up-down power control (UDPC) algorithm with some allowed step sizes (all in integer multiples of 2 dB) [9, 10]. However, arbitrarily large power adjustments, such as 30 dB at once, are not foreseen. The up-down algorithm described in Chapter 1 is the PC algorithm that can more easily be related to the standard algorithm considered in the uplink of the GSM/EDGE network. In this chapter, the up-down algorithm of Chapter 1 is referred to just as UDPC.

In order to apply PC, link measurements are required. Indeed, the actuation frequency and performance of PC depend directly on the availability of such measurements at the BS.

For the voice service, two standard measurements performed by the MS are the received signal level (RXLEV) and the received signal quality (RXQUAL). RXLEV values are measured in the range of -110 to -48 dBm for each TDMA frame within one SACCH multiframe and are mapped afterward to one of 64 possible levels. Average RXLEVs are then reported by the MS to its serving BS. For the RXQUAL, the GSM/EDGE standard states that its value must be related to the BER before decoding, also termed raw bit error rate (RBER). RBER values can be estimated, e.g., as part of the channel equalization or decoding processes. For example, a method to estimate RBER values consists of comparing a reencoded version of a correctly decoded frame with the originally received frame, which is required due to the data interleaving done across the half-bursts of the frame. Based on the estimated RBER values, an average RBER over one SACCH multiframe should be computed, mapped to one of eight possible discrete values, and reported by the MS to its serving BS, where it can be used to perform, e.g., downlink PC, LA, and handovers. Average RXLEV and RXQUAL values become available at the BS at the end of the subsequent SACCH multiframe, thus imposing a delay of 480 ms between measurement and availability of the measured values at the BS. Thus, it should be noted that PC actuates at a very low frequency in the GSM/EDGE network with two power adjustments each second. Some particular cases in which this frequency might be higher are also defined by the standard.

For data services, the GSM/EDGE standard states that the link quality measurements should be related to the bit error probability (BEP) within each burst of the radio blocks received within one SACCH multiframe. The BEP of the four bursts of a radio block are used to calculate the mean BEP (MEAN_BEP) and the coefficient of variation of the BEP (CV_BEP) of the block, as described in [9]. MEAN_BEP and CV_BEP values are computed for all the correctly decoded radio blocks within the duration of one SACCH multiframe. The average of the MEAN_BEP and CV_BEP values are calculated from these values, mapped to 32 and 8 values, respectively, and reported back by the MS to its serving BS, where they can be used to perform, e.g., downlink PC, LA, and handovers. Since data is interleaved across the bursts of a block, the BEP of each burst should be estimated using, e.g., the same method described for the voice service, i.e., the comparison of a reencoded version of a correctly decoded block with the originally received one.

For data services in the GSM/EDGE network, PC is a particularly challenging task because of the bursty nature of data traffic. The size of data packets may vary in a broad range of values and such packets require quite different amounts of time to be transmitted. For small packets, iterative PC may not have enough time to converge to the target SINR before the packet is completely transmitted. Moreover, depending on the adopted scheduling discipline the MS transmitting on the shared channel may change each 20 ms, thus also affecting the convergence of PC algorithms.

In the GSM/EDGE network, LA has priority over PC and the dynamic power range available for PC is limited to 10 dB for data services instead of the 30 dB used for the voice service. The reduced dynamic power range leads to higher average interference levels [45].

One reason for a higher minimum transmit power is to ensure a higher reliability in the reception of the uplink state flag (USF) transmitted within the downlink RLC blocks. This 3-bit flag is stored in the header of RLC blocks sent on the downlink and is used to coordinate channel accesses in the uplink. The USF is decoded by all MSs sharing a downlink channel via TDMA and indicates which MS gets access to the uplink channel during the next radio block. Thus, in order to enable scheduling in the uplink, all MSs must be able to reliably decode the USF independently of their positions within the cell. Throughout this chapter, every time the transmit power range or the transmit antenna pattern are modified in the downlink, there is a risk that the USF might not be decoded correctly. This is a practical GSM/EDGE issue that the RRM algorithms need to overcome.

PC and LA have somewhat conflicting objectives. The former usually aims at providing just the minimum required quality to the links, for example, a target SINR, thus reducing power consumption at MSs and the overall interference in the system. The latter aims at providing the highest possible data rate according to the current link quality while employing maximum transmit power. In order to avoid concurrence between PC and LA, previous works on PC for data services in the GSM/EDGE network considered high target SINR values lying outside the interval in which LA works [45]. However, with such high target SINR values, only a very small number of links benefit from PC, while most of them transmit at full power.

EGPRS can be considered an energy-efficient service, since it maximizes throughput for a fixed transmit power by LA and, consequently, MSs' data sessions will probably take shorter times in average. However, maximizing instantaneous rates might not always be the best policy, especially when considering mixed-service scenarios where one service may be experiencing an excess quality while the other services are below their QoS limit. This subject (co-existence of multiple services) will be further explored in this chapter.

Different downlink PC algorithms are considered later in Section 2.5.2 of this chapter, which involve non-standard features including power adjustments of up to 30 dB at once and higher actuation frequency, such as one power adjustment at each 120 or 20 ms. Regarding data services, the impact of increasing the dynamic power range of PC from 10 to 30 dB is also investigated.

2.3.2 Dynamic Channel Allocation

The channel allocation procedure consists, essentially, of distributing a finite number of channels among the several base and mobile stations within a cellular network. An efficient channel allocation algorithm may lead to benefits for the system, be it in terms of reduced blocking rates or QoS improvements. Classically, algorithms may be classified as fixed or dynamic [33], even though there are also those which combine characteristics of both, which are called hybrid.

Dynamic channel allocation (DCA) assumes that there is a central channel pool, from which the channels may be allocated on-demand, i.e., there is no fixed distribution of the channels among the cells. This higher flexibility allows that the fluctuations in the offered traffic and co-channel interference be treated with a higher efficiency. DCA thus requires that the network be capable of offering all frequencies within each cell, as well as providing reliable measurements of the parameters used by the DCA algorithms (e.g., number of MSs and radio link quality).

The DCA algorithms may be classified according to the metric they optimize [59] or to their degree of centralization [15]. The first criterion considers characteristics such as adaptability to traffic and interference, as well as channel reusability. The second form of classification concerns the degree of centralization of the algorithm, e.g., centralized, distributed, or locally distributed. Locally distributed algorithms allow the exchange of information among nearby BSs, aiming at the improvement of the quality estimation procedure and consequently of the allocation decisions.

The centralized approach has a rather high implementation cost, mainly due to the excessive signaling. The fully distributed strategy may also not be adequate, since it does not provide the necessary means for reliable interference estimation. A locally distributed algorithm represents the most feasible approach, allowing the exchange of information among the neighboring BSs in order to achieve more precise interference estimates. For these reasons, and since high interference tight reuse patterns provide the highest capacity potential, here we focus on a locally distributed interference adaptive algorithm.

The application of DCA to an actual system must take into account the characteristics and practical limitations of the cellular network. Some works have proposed and evaluated DCA algorithms specifically for GSM/EDGE, such as the dynamic frequency and channel allocation [46, 47].

It is worth mentioning that the use of DCA in this context is not compatible with random hopping. Due to the fact that the MSs are constantly hopping over the frequencies in an unordered fashion, the channel selection procedure becomes irrelevant, since with each hop the set of co-channel interferers may change completely. It is therefore required that frequency hopping be disabled or that a coordinated cyclic hopping be implemented, in which the groups of co-channel interferers hop over the same frequencies.

2.3.2.1 Measurements and SIR Estimation

The GSM/EDGE cellular network does not count with direct SIR measurements. The SIR must be inferred based on the measurement report mechanisms available in the network [9].

Each MS monitors the power levels arriving from nearby BSs. This information is accumulated and periodically reported to the BS to which the MS is connected. The measurements are done for the BCCH channel and the report period is of 104 TDMA frames (480 ms). For each frame a different BS is measured. The number of BCCH carriers and the measurement order are parameters defined by the system. As an example, suppose that the BCCH list contains 32 elements, then there will be (104/32) received power measurements for each carrier, i.e., three or four samples per BS.

After the measurements within the report period are concluded, the MS has to organize the data and prepare the report to be sent to the BS. The samples of each carrier are averaged, and among all measured carriers only the six with the highest received power levels are included within the report. The transmission occurs through the SACCH control channel during the next 480 ms. The total delay until the report is available at the BS, including the measurement and transmission times, is therefore of roughly 1 s.

The original purpose of this measurement mechanism would be to aid in the handover decisions, indicating which cells offer the best signal quality to the MS. Nevertheless, it may also be employed to produce channel SIR estimates.

The reports are based on measurements of the BCCH channel, which is always active. In order to obtain a more realistic SIR estimation, the actual channel activity of the BSs should be taken into account. However, this channel activity information is not globally available, it has to be shared among the BSs [46]. In practical terms this information exchange could be feasible, representing a signaling increase within the backbone of the cellular network. Note that if power control is employed in the system, it would also be necessary to share the power adjustments of each BS, since the BCCH measurements are done for full power.

2.3.2.2 Channel Selection and Admission Control

The considered DCA algorithm prioritizes the channel presenting the best SIR. In the case when several channels perceive no interference, or when the estimated SIR is the same, the choice is done at random among them.

The admission control corresponds to an optional stage of the DCA algorithm. Differently from the case with RFH, with DCA the MS will be subject to the same interference profile for a certain period of time, therefore it is important to guarantee that the channel is offering a minimum acceptable quality.

Even though the channel selection procedure prioritizes the channel perceiving the best SIR, high-load situations may occur, for which even the best channel would not be able to offer a satisfactory quality to the MS. In such cases, blocking the MS might be a better option than letting it enter the system, since it is expected that its QoS will probably not be satisfied. Besides, the additional interference that would be introduced in the system is avoided.

The admission criterion can be based on a minimum SIR threshold, which may be defined based on link-level simulation results. In the case of the enhanced full rate (EFR) speech codec, for example, a minimum SIR of 8 dB is required in order to assure that the frame erasure ratio will be kept at acceptable levels [24].

Another aspect that may be taken into account by the admission control is related to the impact that the introduction of a new MS might have over the quality of the MSs already allocated in the system. This "impact test" corresponds to performing an estimate of how the SIR of the co-channel MSs would be degraded after admitting a new MS. In case the admission would result in the SIR of any of the co-channel MSs being reduced to a value below the threshold, blocking would be activated. Note that the complexity for implementing such estimate might be prohibitive in practical terms.

2.3.3 Management of Multiple Services

The support of multiple services, such as web-browsing, e-mail, audio/video streaming, among others, is one of the main features of the third generation of cellular systems and beyond. Since these services have different characteristics and requirements, the use of efficient RRM techniques is therefore essential to ensure their QoS levels and to optimize system capacity.

In systems with multiple services the available radio resources may be either segregated or shared, i.e., different frequency groups may be reserved for the services or it may be allowed that they have access to the total set. The isolated approach is not very efficient, since for situations of asymmetric load the overloaded service does not have access to the channels reserved to the other services, even though they may be unoccupied. The sharing of channels avoids problems of this nature, but has the side-effect of creating a situation in which the different services cause interference among each other, which may have certain implications on the capacity of interference-limited systems.

Since the services have different QoS requirements and different traffic patterns, their combination within interference-limited scenarios implies that the more demanding service will limit system capacity, even though the other services still may perceive sufficient QoS. In [54], a method has been proposed for balancing the QoS in code division multiple access (CDMA) systems, based on a power allocation methodology that allows the joint service capacity to be increased.

A more general definition for QoS balancing was presented in [22, 25], which was called per-service capacity balancing. The referred work has demonstrated that the system capacity is maximized when the per-service capacities are reached for the same load. In the case of interference-limited systems, it has also been shown that the capacity balancing may be achieved through an interference balancing process, such as the service-based power setting (SBPS) technique for mixed-service GSM/EDGE networks [24], which consists of applying offsets in the transmission powers of the different services.

2.3.4 Multi-antenna Techniques

The application of multi-antenna techniques to mobile communication systems is capable of providing capacity gains as well as the improvement of the quality of service perceived by the subscribers. The spatial filtering realized through the use of narrow beams, which can be either selected from a fixed set (switched fixed beams) or adaptively steered toward the desired MSs (adaptive beamforming), is able to significantly reduce the co-channel interference levels. This approach can be considered a more flexible and evolved form of the sectorization that is employed as a baseline in most mobile communication systems. The interference reduction allows for the implementation of more aggressive frequency reuse patterns, such as 1/3 or 1/1, thus resulting in higher spectral efficiencies.

The switched fixed beams, as well as adaptive beamforming based on direction-of-arrival estimation techniques, are more adequate to macrocellular environments with low angular spread and strong line-of-sight [55]. In the case of indoor or micro-cellular environments presenting strong multi-path components, the adaptive arrays must rely on the estimation of the complex channel coefficients in order to adapt to the current channel conditions [11].

The gains provided by multi-antenna techniques, however, are not limited to those of the spatial filtering functionality. Techniques that take advantage of the spatial diversity associated with the presence of multiple antennas may also be applied, such as the maximum ratio combining (MRC) or the interference rejection combining (IRC).

In the context of the application of adaptive antennas to multiple services in GSM/EDGE networks there are some practical aspects that should be taken into account, mainly with regard to the data service over EGPRS. For the voice service there would be no problem with replacing the sector antennas with antenna arrays, other than the restriction that they may not be applied to the BCCH carriers, for the same reason that other techniques such as FH and PC may not be used, which is to guarantee that all MSs have uninterrupted access to the broadcast channel.

The results section carries out the evaluation of some multi-antenna strategies for scenarios where both voice and data share the same frequency spectrum (see Section 2.5.4). It assesses, among other things, the performance of a strategy for which the adaptive antennas are applied only to the voice service, in order to avoid the problems associated with their use within EGPRS. Since all MSs hop over the same set of frequencies, the interference reduction for the voice MSs is also reflected upon the data MSs, thus bringing benefits for both. The performance of the strategy combining different antennas for the different services is also compared to the fully multi-antenna case, i.e., with both services employing multiple antennas.

2.4 Simulation and Modeling of GSM/EDGE Networks

Studying the performance of modern wireless networks, such as GSM/EDGE, is a complex task. Due to the large number of variables and mechanisms involved, a pure analytical study is not feasible and computer simulations are applied to investigate the system's characteristics of interest [31, 35].

In this chapter, the strategy of dividing the simulations into link and system levels is employed. These two types of simulators are then connected through an appropriate interface. This is a complexity reduction strategy, which is discussed in more details in Chapter 7 of this book. We also employ dynamic system simulations, which include channel and traffic variabilities with time. In the following, we describe several simulation models, which are used later to assess the performance of RRM techniques in a GSM/EDGE network. In spite of considering here a GSM/EDGE network, many of the models introduced in this section are quite general and can be used to evaluate the performance of other modern wireless networks.

2.4.1 Cellular Grid, Frequency Reuse, and Mobility Models

In this chapter, the cellular network is modeled as a macrocellular system composed of tri-sectored cells organized in 1/3 or 1/1 uniform frequency reuse patterns [59]. Figure 2.5 illustrates a cellular grid with a 1/3 frequency reuse.

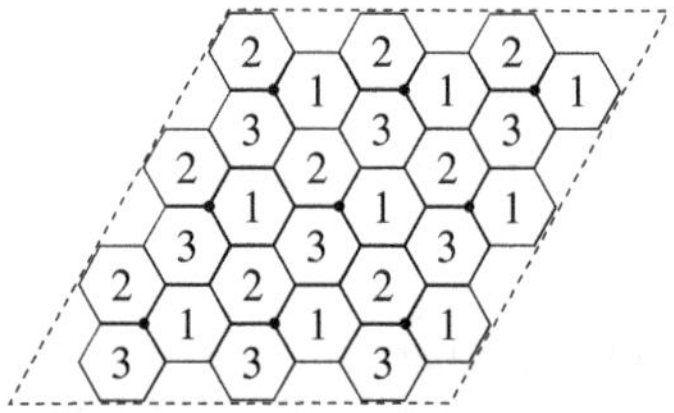

Fig. 2.5 Uniform cellular grid with 1/3 frequency reuse.

In Fig. 2.5, the sectors with same number employ the same set of channel frequencies and are co-channel interferers. A number of MSs are distributed over the area covered by the cellular grid and can freely move within it. Each cell sector in Fig. 2.5 is assumed to use a typical sector antenna, such as that presented in [53].

MSs' mobility can be modeled according to a random-walk (Markovian) mobility pattern [16, 17]. Only pedestrian mobility is considered in this chapter, which assumes an average speed of 3 km/h. The current MS' speed and direction of movement are uniformly distributed within [1 km/h, 5 km/h] and $[0, 2\pi]$, respectively. These are held until the MS walks a distance of 5 m, after which new speed and direction are sorted. Other mobility models, including that of vehicular mobility, can be found in [16, 17].

Because MSs move over the grid shown in Fig. 2.5 and because only a limited area is covered, MSs could eventually leave the coverage area. Additionally, due to the geographic distribution of the co-channel sectors in Fig. 2.5, sites on the border of the grid perceive less interference than those in middle of the grid, which is termed a border effect. In order to allow infinite mobility over a limited region and to avoid border effects, which are undesired, a wrap-around technique is usually considered. Wrap-around techniques are usually based on cell replication or on a geometric model which yields homogeneous average interference levels in the whole grid. Herein, the wrap-around technique described in [59] is employed, which consists of bending the grid in order to form a torus surface, as illustrated in Fig. 2.6. Note that the described grid and mobility models can be directly applied to other wireless networks.

2.4.2 Propagation Models

As it has been discussed in Chapter 1, radio communication is affected by large- and small-scale fading, which ultimately result from reflection, refraction, and diffraction of the transmitted radio waves [44, 56]. These effects are assumed in the

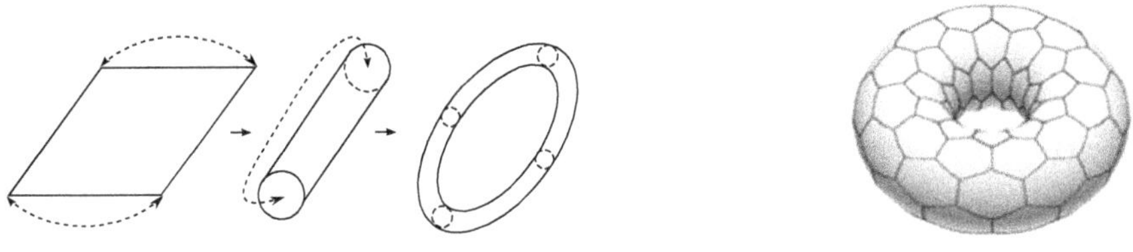

(a) Torus model (b) Torus mapping

Fig. 2.6 Infinite mobility model.

GSM/EDGE simulations considered in this chapter and their models are described in the sequel.

There are different average path loss models, which are adequate for different propagation scenarios [44, 56]. Herein, the Okumura–Hata model is employed, which applies to urban and suburban environments where the average building height is approximately uniform [53]. Considering this model and denoting by d the distance in km between a BS and an MS, by f_c the system central carrier frequency in MHz, and by h_{BS} the BS height in meters and measured with respect to the average rooftop of buildings, the average path loss L_{pl} is given by

$$L_{pl} = 40(1 - h_{BS} \times 4 \times 10^{-3}) \log(d) - 18 \log(h_{BS}) + 21 \log(f_c) + 80 \text{ in dB}. \quad (2.2)$$

As the MS moves in the coverage area, large obstacles such as buildings may obstruct the propagation path between the BS and MS causing fluctuations on the received signal power, i.e., shadowing the received signal. Shadowing is usually modeled by a lognormal random variable with standard deviation σ_{sf} [44, 56]. Because shadowing relates with the position of large obstacles in the coverage area, it is position-dependent and spatially correlated. The modeling of the spatially correlated shadowing can be accomplished by sorting independent lognormal shadow samples for a rectangular grid composed of points uniformly separated by a shadowing decorrelation distance d_{sf} [59]. This model is termed a shadowing map and for each BS in the network such a shadowing map is created. Then, the shadowing $L_{sf}(x,y)$ associated with an arbitrary position (x,y) in the grid can be obtained through linear interpolation.

Besides the spatial correlation of shadowing captured by the above model, it is also worth modeling the spatial correlation of the shadowing perceived in the links between different BSs and the same MS. This kind of correlation occurs, for example, when the MS moves within a tunnel. This inter-BS shadowing correlation can be modeled with help of an additional shadowing map, which is associated with all MSs. In order to obtain the shadowing sample $L_{sf}^{(b,m)}(x,y)$ for the link between a BS b and an MS m, the shadowing samples $L_{sf}^{(b)}(x,y)$ and $L_{sf}^{(m)}(x,y)$, obtained respectively from the BS' and MSs' shadowing maps, are combined as

$$L_{sf}^{(b,m)}(x,y) = \sqrt{1-\rho_{sf}} L_{sf}^{(b)}(x,y) + \sqrt{\rho_{sf}} L_{sf}^{(m)}(x,y), \quad (2.3)$$

where $0 \leq \rho_{\mathrm{sf}} \leq 1$ is a coefficient which controls the amount of shadowing correlation among BSs [59].

Multi-path fading leads to deep fluctuations in the received signal power. It has been modeled in the link-level simulations, where time-correlated fast fading is generated using the well-known Jakes' model [30].

2.4.3 Link Quality Measurements

Considering the presented radio propagation models, the measures considered in the system simulations to evaluate the communication links between BSs and MSs are discussed in the following.

The power P_{r} received by the MS depends on the transmit power P_{t} used by the BS and on the particular state of the link between BS and MS. The received power P_{r} of an MS can be expressed as

$$P_{\mathrm{r}} = P_{\mathrm{t}} + G_{\mathrm{ant}} - L_{\mathrm{pl}} - L_{\mathrm{sf}} \text{ in dBm}, \tag{2.4}$$

where the antenna gain G_{ant}, the average path loss L_{pl}, and the shadow fading L_{sf} can be obtained considering the relative position of the BS and MS. Additionally, extra gains/losses, such as cabling losses at the BS or additional antenna gains at the MS, can be easily added/subtracted in (2.4). For simplicity, such additional gains/losses are not considered here.

Co-channel sectors share the same frequencies and generate interference. A common measure of the link quality corresponds to its SINR. Assume that MS i is served by the sector i and let N_{ci} denote the number of interfering co-channel sectors. Denoting by $p_{\mathrm{r}i,j}$ the power received by MS i from sector j, and by ν the average noise power, the SINR γ_i of MS i is given by

$$\gamma_i = \frac{p_{\mathrm{r}i,i}}{\sum\limits_{j=1,\, j\neq i}^{N_{\mathrm{ci}}} p_{\mathrm{r}i,j} + \nu}. \tag{2.5}$$

For interference-limited scenarios, the average noise power ν can be neglected and the SINR in (2.5) reduces to the SIR.

In the link-level simulations, curves such as bit error rate (BER), block error rate (BLER), or frame erasure rate (FER) as functions of the SINR (or SIR) are obtained by averaging the link performance over a long period of time. In this way, the mean characteristics of mechanisms like data interleaving, fast fading, and fast interference variations are captured into the link-level results. In the system-level simulations, measures of the SINR (or SIR) can be easily obtained and can be mapped afterward into BER, BLER, or FER values using an adequate link-level curve.

Considering voice services, the average SINR of the eight half-bursts composing a voice frame is mapped to an FER value. Considering data services, the average SINR of the four bursts composing a radio block is mapped to a BLER value. Based on the FER, in the case of the voice service, or on the BLER, in the case of data services, a random test is used to determine whether a transmission has been successful.

For the simulations considered in this chapter, the RXQUAL, MEAN_BEP, and CV_BEP values shortly described in Section 2.3.1 cannot be measured as described in the standard because they depend directly on RBER. Instead of the standard measurements, the average SINR $\bar{\gamma}_s$ over one reporting period is considered as link quality measurement. It is computed by averaging the mean SINR of each block within the reporting period and it is sent back to the BS by the MS. This model is applied here to both voice and data services and a reporting period equal to one SACCH multiframe is considered by default.

2.4.4 Traffic Models

Voice and data services have very different traffic patterns, thus requiring elaborate traffic models adequate to their peculiarities. They define the activity of the MSs and how the traffic is generated.

Dynamic arrival and departure of MSs' voice calls or data sessions are considered in the system. The arrival process is modeled in the dynamic simulations through Poisson processes with specific arrival rates for each traffic type. In both voice and data traffic cases, the interval between consecutive arrivals of new MSs in the system is modeled by a negative exponentially distributed random variable [43].

In the modeling of the voice service, two aspects are taken into account: the duration of the call and the speech activity during a call. The first aspect is modeled through an exponential distribution with a 120 s mean. The activity model, however, depends on the use or not of discontinuous transmission (DTX). In the case in which DTX is disabled, the BS continuously transmits speech frames, even during the periods in which the speaker is silent, thus generating interference during the whole call. In the case in which DTX is used as an interference reduction mechanism, voice activity must be modeled.

A slow voice activity detector is considered, which fits well in to the global system for mobile communication (GSM) voice service whose minimum talking/silent periods are of one SACCH multiframe [9]. The adopted voice activity model considers a two-state Markov chain for simulating the transition between active and silent states [26].

The state transition probabilities from active-to-silent $P_{a \to s}$ and from silent-to-active $P_{s \to a}$ can be determined from the equations:

$$P_{a \to s} = 1 - \exp(-T_{rep}/T_a) \quad \text{and} \quad P_{s \to a} = 1 - \exp(-T_{rep}/T_s), \tag{2.6}$$

respectively, where T_{rep} represents the duration of a reporting period and T_a and T_s correspond to the mean duration of the active and silent stages. This leads to a mean voice activity of $T_a/(T_a + T_s)$. Typically, a mean voice activity of 60% is considered [22]. For the GSM/EDGE network considered in this chapter, T_{rep} corresponds by default to the duration of an SACCH multiframe and during silent periods DTX disables the transmission of voice frames, thus reducing interference in the system. Note that by adapting the values of T_{rep}, T_a, and T_s the presented voice traffic model can be easily employed in other wireless networks.

The data traffic model of the WWW interactive service is fairly different from that of speech, having a strong bursty traffic characteristic. The adopted WWW model considers the arrivals of packets within a session, the times between packets, and the packet lengths [32]. It is a simplified version of the WWW model presented in [53], which additionally considers packet calls. In [53], there are random variables modeling the number of packets per packet call and the inter-arrival time between such packets. The traffic model of [32] concentrates packets belonging to a packet call into a single large packet. Table 2.3 presents a summary of the parameters, distributions, and values of the WWW traffic model.

Table 2.3 world wide web (WWW) traffic model.

Parameter	**Value**
Sessions	
Distribution of the number of packet calls per session	Geometric
Mean number of packet calls per session	10
Packet calls	
Distribution of the reading time between packet calls	Truncated Pareto
Mean reading time between packet calls μ_{TP}	10 s
Pareto distribution parameters α_{TP}, k_{TP}, and m_{TP}	1.4, 3.45, and 120 s
Number of packets per packet call	1
Packets	
Distribution of packet sizes	Lognormal
Mean packet size	4,100 bytes
Standard deviation of packet size	30,000 bytes
Maximum packet size	100,000 bytes

2.4.5 Evaluation Metrics

Key performance indicators in wireless networks are mainly related to quality and capacity measures. Other measures, such as blocking rate and channel reallocation rate, may also be relevant in certain situations. The analyses presented in this chapter are focused on the relative performance of the different RRM techniques, rather than on absolute figures. In this section, some evaluation metrics and requirements are defined.

The voice quality is expressed in terms of the FER, which takes into account the percentage of lost speech frames with regard to the total number of speech frames.

The data quality, on the other hand, is defined as the mean packet bit rate r during the session of an MS.

The individual QoS criteria, which determine the satisfaction of each MS, are defined taking as reference the values presented in [24]. The following services are considered: enhanced full rate (EFR), multi-rate at 5.9 kbit/s with full rate (MR59FR), and World Wide Web (WWW). The QoS requirements are given by

$$\begin{cases} \text{Voice (EFR):} & FER^{\text{req}} \leq 1\%, \\ \text{Voice (MR59FR):} & FER^{\text{req}} \leq 0.6\%, \\ \text{Data (WWW):} & r^{\text{req}} \geq 10\ \text{kbit/s}. \end{cases} \tag{2.7}$$

The global QoS criteria of the system are defined as the percentual of satisfied users of each service. For voice a 95% satisfaction is required and for data either 95% or 90%, depending on the scenario.

Besides the QoS requirements, which apply to interference-limited systems, there is also the blocking criterion for voice. The blocking limit tolerated by the system is of 2%, where the blocking rate is defined as the relation between the number of blocked MSs and the number of births that occurred during the period for collecting the statistics.

The system capacity C is expressed in terms of spectral efficiency, i.e., the offered load per cell normalized by the amount of utilized spectrum. The spectral efficiency is expressed as $C = r_{\text{tot}}/(B_{\text{tot}}N_{\text{cell}})$, where r_{tot} indicates the total load offered to the system in bit/s, B_{tot} represents the frequency bandwidth available for traffic, and N_{cell} corresponds to the number of cells.

In the case of the voice service, the offered load is measured in Erlangs, while for the data service it is measured in terms of the transmission rate (bit/s). The spectral efficiency units are therefore measured in Erl/MHz/cell and bit/s/Hz/cell, for the voice and data services, respectively.

The network capacity limit $C^{\max}$ corresponds to the maximum load that can be offered to the system before the QoS or blocking requirements are violated. In some situations the capacity is also expressed in its normalized form, which consists of dividing the capacity value by the maximum service capacity, i.e., $C_{\text{norm}} = C/C^{\max}$.

Although the required QoS values introduced in this section are meant for a GSM/EDGE network, the same framework can be applied to other wireless networks with different QoS requirements.

2.5 RRM Performance in GSM/EDGE

In this section, the performance of different RRM techniques is evaluated in a GSM/EDGE network by means of dynamic system level simulations employing the models described in the previous section. It should be noted that the performance assessment through simulations involves simplifications, which are required in order to make this task mathematically and computationally tractable.

Due to such simplifications, the absolute performance results obtained through the modeling and simulation of wireless networks might differ from those observed in real systems. However, the relative analyses conducted by means of simulative studies are valid since all the model parameters are kept consistent across the different scenarios. Moreover, a significant part of the performance gains observed in simulation studies can be usually obtained in real systems. In this way, important insight on the performance of RRM strategies applied to wireless networks is obtained, which is of crucial strategic importance in decision-making processes involved in the design and operation of these systems.

Therefore, in this section it is important to focus on the relative performance gains obtained by the RRM techniques compared to reference scenarios. It is also worth mentioning that there exists extensive literature on RRM applied to GSM/EDGE and other wireless networks. It is not our intent to be exhaustive in covering the existing literature, so that we restrict ourselves to providing only key references concerning each topic.

2.5.1 Overall scenario

In this chapter, a GSM/EDGE network covering an urban macrocellular scenario is considered. Because frequency spectrum is a limited and expensive resource, tighter frequency reuse patterns like 1/3 and 1/1 are usually pursued in the GSM/EDGE networks. Tight frequency reuses allow for obtaining higher spectral efficiencies. They lead, however, to additional co-channel interference that must be efficiently handled by means of RRM techniques. To investigate the performance of voice and data services, simulations considering different RRM techniques have been done for cellular grids implementing 1/3 and 1/1 frequency reuses and the obtained spectral efficiency values have been compared with those obtained in reference scenarios. The most relevant parameter values are shortly summarized in Table 2.4.

A central carrier frequency $f_c = 2{,}000$ MHz and a BS height of 15 m above the rooftop of buildings are considered. A system bandwidth of 2.4 MHz is taken into account, which corresponds to 12 GSM carriers. RFH will be often employed and in each simulation a total number of 10,000 calls or sessions are simulated. In most of the considered cases, an interference-limited system is considered, i.e., noise power is negligible and SINR values are equivalent to SIR values.

In the system, a shadowing standard deviation σ_{sf} value of 6 dB, a shadowing decorrelation distance d_{sf} of 110 m, and an inter-BS shadowing correlation factor ρ_{sf} of 0.4 are considered [58].

2.5.2 Power Control

In this section, the performance of some of the PC algorithms discussed in Chapter 1 is evaluated in the downlink of a GSM/EDGE network. The voice service

Table 2.4 Simulation parameters.

Parameter	Value
	System
Cell type	Tri-sectored
Frequency reuse pattern	1/3 or 1/1
Frequency of operation	2,000 MHz
System bandwidth	2.4 MHz (12 GSM carriers of 200 kHz)
# of transceivers per sector	4 in frequency reuse 1/3, 12 in the frequency reuse 1/1
# of timeslots per carrier	8
Frequency hopping	Random or cyclic
Transmission direction	Downlink
Maximum transmit power	35 dBm
MS mobility	Random-walk pedestrian model (3 km/h avg. speed)
# of simulated calls or sessions	10,000
Satisfaction degree	95% of voice calls, 90 or 95% of WWW sessions
	Propagation effects
Average path loss	$128.15 + 37.60\log(d)$ in dB, cf. (2.2)
Shadowing standard deviation	6 dB
Shadowing decorrelation distance	110 m
Inter-BS shadowing correlation factor	0.4
Fast fading	Considered at the link-level
	Services
Average voice call duration	120 s
Voice codec	EFR, MR59FR
Voice call average FER	1% with EFR, 0.6% with MR59FR
Voice blocking limit	2%
Discontinuous transmission	Enabled, disabled
Mean active voice period	1.1 s
Mean silent voice period	0.7 s
Mean voice activity	≈60%
WWW traffic	Modeled according to [32], cf. Section 2.4
Link adaptation	Enabled in ideal or non-ideal mode
Average session throughput	10 kbit/s

performance analysis considers two voice codecs, namely the EFR and the MR59FR. WWW traffic is considered as the data service. The system performance with and without PC is investigated for these services. In this section, a fraction of 95% of the MSs should be satisfied either for the voice or the WWW service.

The following PC algorithms are considered in this section: the one proposed in [21], which is termed here autonomous SINR balancing power control (ASBPC); the one proposed in [13, 57], which is termed here soft dropping power control (SDPC); and the up-down power control (UDPC) algorithm. All these algorithms have been previously discussed in Chapter 1.

In this section, the reported $\bar{\gamma}_S$ is represented by γ for simplicity of notation. The three considered PC algorithms are closed-loop algorithms based on the reported

SINR values. The algorithms share a similar mathematical structure, given by

$$P^{(k+1)} = P^{(k)} - c_{\mathrm{PC}} \varepsilon_{\gamma}^{(k)} \text{ in dBm}, \tag{2.8}$$

where $P^{(k)}$ represents the transmit power at iteration k, c_{PC} is a PC feedback constant, and $\varepsilon_{\gamma}^{(k)}$ is the feedback error signal of PC [28].

For all algorithms, the time delay compensation (TDC) technique discussed in Chapter 1 is employed to avoid instabilities due to measurement and report delays [28]. Considering the delay of one period associated with reporting the average SINR γ back to the BS, (2.8) becomes

$$P^{(k+1)} = P^{(k)} - c_{\mathrm{PC}} \varepsilon_{\gamma}^{(k-1)} \text{ in dBm}. \tag{2.9}$$

As a consequence, the power adjustment at iteration $k+1$ is performed based on outdated information, since the effect of the power adjustment associated with iteration k is captured only in the SINR made available at the BS at iteration $k+2$. In order to avoid this, the effect of the power adjustment done at iteration k is predicted and introduced in the feedback error signal under the assumption that the referred PC command was successful. Thus, considering TDC, the PC iteration becomes

$$P^{(k+1)} = P^{(k)} - c_{\mathrm{PC}} \varepsilon_{\gamma}^{k} + P^{(k)} - P^{(k-1)} \text{ in dBm}. \tag{2.10}$$

Considering TDC, the iterative formulation of the ASBPC, SDPC, and UDPC algorithms of Chapter 1 can be written as

$$p^{(k+1)} = p^{(k)} \left(1 - \beta_{\mathrm{F}} + \beta_{\mathrm{F}} \frac{\gamma_{\mathrm{t}} p^{(k-1)}}{\gamma^{(k)} p^{(k)}} \right) \text{ in W}, \tag{2.11}$$

$$P^{(k+1)} = P^{(k)} - \beta_{\mathrm{SD}} \left(\Gamma^{(k)} + P^{(k)} - P^{(k-1)} - \Gamma_{\mathrm{t}}(P^{(k)}) \right) \text{ in dBm, and} \tag{2.12}$$

$$P^{(k+1)} = P^{(k)} - \delta_{\mathrm{UD}} \operatorname{sign}(\Gamma^{(k)} + P^{(k)} - P^{(k-1)} - \Gamma_{\mathrm{t}}) \text{ in dBm}, \tag{2.13}$$

respectively.

For the ASBPC and SDPC algorithms, $\varepsilon_{\gamma}^{(k)}$ of (2.8) and (2.10) relates to the difference between the measured SINR γ and the target SINR γ_{t}, while the constant c_{PC} relates to the amount of the difference between measured and target SINRs that is compensated at each PC iteration. For the UDPC, $\varepsilon_{\gamma}^{(k)}$ relates to the sign of the difference between the measured and target SINRs, which is limited to either -1, 0, or $+1$, and the constant c_{PC} relates to the power step δ_{UD} introducing a fixed compensation of the feedback error signal. For more details on the formulation of power control iterations as control systems with feedback error signals refer to [28].

In particular for the voice service, several target SINR Γ_{t} values for the PC algorithms have been tested and the value resulting in the best spectral efficiency figures has been selected. For UDPC and ASBPC, Γ_{t} values 2 dB lower or higher than the selected ones resulted in worse system performance. The target SINR values considered in this section are higher than the 8 and 4 dB required with the EFR and

MR59FR codecs to achieve average FERs of 1 and 0.6%, respectively. Such differences are due to the particular models described in Section 2.4. In fact, an outer loop PC would be required in order to determine the most adequate target SINR for each scenario [29]. However, such an outer loop PC is not considered here.

In the SDPC case, a fixed target SINR is not specified, but a value for the parameter β_{SD} controlling the relationship between demanded power and target SINR. For the SDPC, β_{SD} value has been found experimentally, as suggested in [13], with $\beta_{SD} = 0.6$ providing the best results.

For the data service, a target SINR of 35 dB has been considered in order to avoid the concurrence between PC and LA. In order to investigate the impact of the actuation frequency of PC, different reporting periods have also been considered. Such parameter settings will be discussed in more detail together with the results obtained in the corresponding cases. The most relevant power control parameters are listed in Table 2.5.

Table 2.5 Power control parameters.

Parameter	Value
PC algorithms	ASBPC, SDPC, and UDPC
Maximum transmit power	35 dBm
Minimum transmit power	5 dBm
Power levels	Discrete in steps δ_{UD} of 2 dB
Power control time-step	1 iteration at each 480, 120, or 20 ms
Time delay compensation (TDC)	Enabled
β_F for ASBPC	1 (fastest convergence) [21, 28]
β_{SD} for SDPC	0.6 (defined experimentally, as in [13])
Minimum target SINR $\Gamma_{t,min}$ for SDPC	8 dB for EFR, 4 dB for MR59FR, 6 dB for WWW

In this section, scenarios implementing tight frequency reuses of 1/3 and 1/1 are considered and PC is employed to manage the co-channel interference and improve the system performance. RFH and pedestrian mobility, cf. Table 2.4, are considered.

2.5.2.1 Power Control Performance for the Voice Service

Initially, the performance of PC for the voice service considering the EFR codec is investigated. In order to determine the system capacity, simulations with increasing offered loads have been conducted until reaching the QoS limits given in Table 2.4. Figure 2.7 presents the percentual of satisfied MSs as a function of the voice spectral efficiency in Erl/MHz/cell considering the EFR codec and the 1/3 frequency reuse. In the cases in which PC is applied, DTX is also enabled.

In Fig. 2.7, the capacity limits by interference and by blocking are also shown. The system capacity is limited by interference whenever the blocking rate is below 2% but the fraction of satisfied MSs is lower than 95%, which corresponds to having more than 5% of the MSs perceiving average FER higher than the values specified

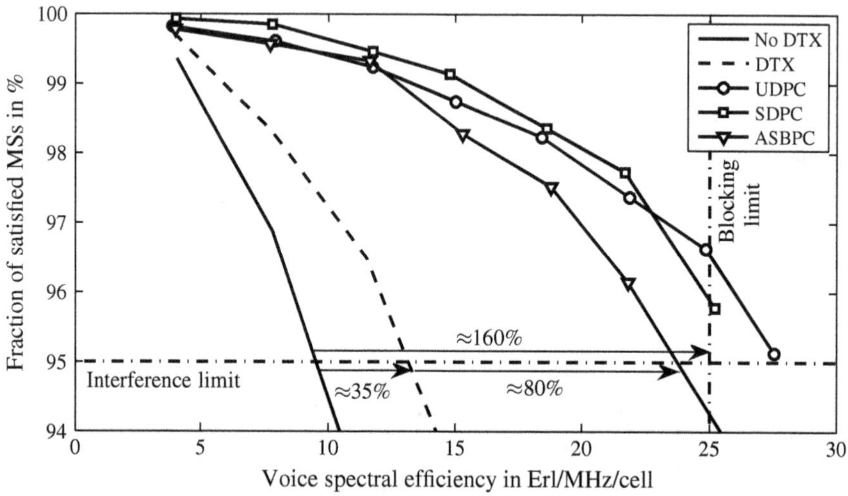

Fig. 2.7 PC performance with EFR voice codec and reuse 1/3 ($\Gamma_t = 16$ dB for UDPC and ASBPC).

in Table 2.4. Oppositely, the system capacity is limited by blocking if more than 2% of the arriving calls are blocked due to unavailability of free channels while the ongoing calls perceive acceptable average FER values. Given the number of channels per sector and the target blocking rate, the blocking limit of the system can be easily calculated using standard traffic engineering methods [44, 56]. The capacity limits by blocking and interference are drawn for all figures in this section, but are labeled only in Fig. 2.7.

In Fig. 2.7, it can be seen that the system capacity without PC and DTX is considerably lower than when these two features are enabled. Indeed, by enabling DTX a considerable amount of unnecessary interference is eliminated from the system and a capacity gain of more than 35% is obtained. By using ASBPC, a voice spectral efficiency of more than 23.5 Erl/MHz/cell is obtained, which represents an additional gain superior to 80% compared to the case with DTX only. Compared to the case in which both PC and DTX are disabled, the obtained spectral efficiency is about 1.5 times higher, which shows that ASBPC can substantially improve the system capacity. Such a high-capacity improvement comes from the reduction of the co-channel interference performed by the PC algorithms, which employ only as much power as required to attain the target QoS levels.

In spite of being able to perform larger power adjustments, the ASBPC provided lower spectral efficiency figures than the UDPC. This result might be explained by the selection of $\beta_F = 1$, which according to [21, 28] results in the fastest convergence. However, this value of β_F is not necessarily optimal and might lead to instability of the ASBPC algorithm, as discussed in [28]. Thus, it is possible that lower values of β_F could lead to better spectral efficiencies. Anyway, the optimization of the parameter β_F has not been performed for the ASBPC algorithm. In this case, the UDPC works in a more stable fashion leading to higher spectral efficiency values.

Indeed, it performs as good as the SDPC algorithm, which is allowed to make larger power adjustments like the ASBPC but used a more conservative value for β_{SD} and involves a self-regulation of the target SINR [13, 28, 57]. The UDPC and SDPC algorithms outperform here the ASBPC algorithm providing spectral efficiency gains higher than 160% compared to the case without PC and DTX. As it can be seen in Fig. 2.7, the UDPC and SDPC algorithms reach the blocking limit of 2% while there are more than 95% of satisfied MSs in the system, i.e., in the considered scenario PC turns the system into a blocking limited system.

Because the system capacity became limited by blocking, voice calls perceive QoS levels higher than the target ones. This suggests tightening the frequency reuse in order to make more channels available per sector and to potentially support higher number of voice calls. The frequency reuse tightening increases the number of available channels per sector, but it incurs in a considerable increase of co-channel interference due to the smaller reuse distance. In Fig. 2.8, the fraction of satisfied MSs against the voice spectral efficiency is shown for a GSM/EDGE network implementing a 1/1 frequency reuse and considering the EFR codec.

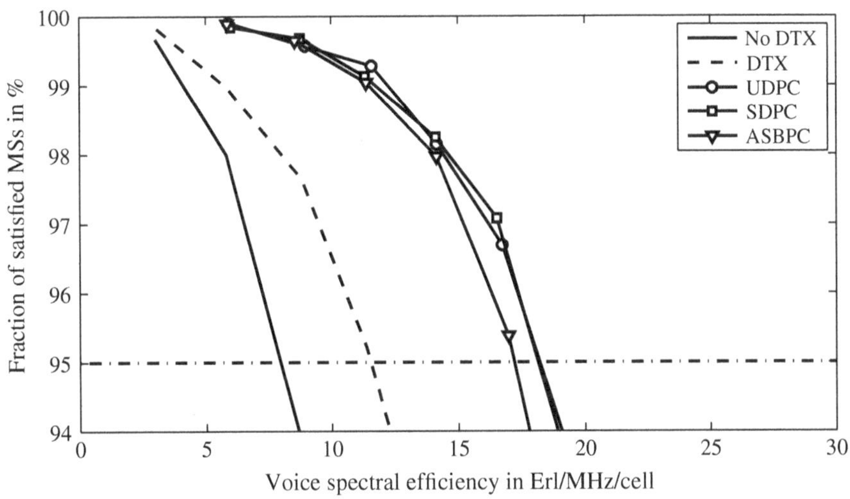

Fig. 2.8 PC performance with EFR voice codec and reuse 1/1 ($\Gamma_t = 18$ dB for UDPC and ASBPC).

It can be seen in Fig. 2.8 that the frequency reuse tightening results in a too large increase of the average co-channel interference and causes an overall reduction of the system spectral efficiency varying between 15 and 25%. The system might support higher voice loads if more robust voice codecs, such as the MR59FR, are used by the MSs. Such a codec allows the system to operate with acceptable voice quality at considerably lower SINR levels.

In the following, the spectral efficiency of the system considering the MR59FR codec is evaluated.

Figure 2.9(a) shows the fraction of satisfied MSs against the achieved voice spectral efficiency considering the MR59FR codec and a 1/3 frequency reuse. In this figure, it can be seen that considerably higher voice loads are supported in the system and, consequently, higher spectral efficiency values are achieved. Moreover, it can be seen that the use of DTX without PC is already sufficient to bring the system to a capacity limitation by blocking. Compared to the case without DTX and PC, the obtained spectral efficiency gains are of about 25%. Using PC in this scenario would only improve the QoS levels perceived by the MSs resulting in average FER per call considerably lower than the target values given in Table 2.4. Such a case is shown in Fig. 2.9(a) for the UDPC algorithm only, which provides a satisfaction level much higher than that achieved with DTX and no PC with more than 99% of satisfied MSs. In this case, a tightening of the frequency reuse might also be advisable. The results, analog to Fig. 2.8 but considering the MR59FR codec, are shown in Fig. 2.9(b).

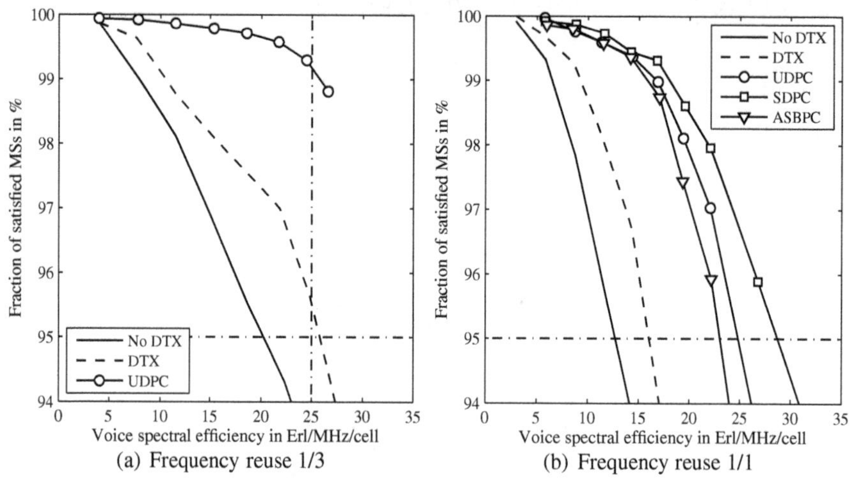

Fig. 2.9 PC performance with MR59FR codec.

Differently from Fig. 2.8, where a tightening of the frequency reuse resulted in too much interference to be managed, in Fig. 2.9(b) the use of a more robust codec allows a larger amount of interference to be supported. In Fig. 2.9(b) the spectral efficiency value achieved with the ASBPC at the capacity limit is almost the same as that achieved in the 1/3 frequency reuse with the EFR codec, in spite of the additional interference of the 1/1 frequency reuse. The same is valid for the UDPC, which achieves about 25 Erl/MHz/cell as in Fig. 2.7. In Fig. 2.9(b), it can be seen that SDPC outperforms the UDPC algorithm and achieves a spectral efficiency value of about 28.5 Erl/MHz/cell. Thus, a spectral efficiency gain of about 15% is obtained with acceptable QoS for voice calls. Moreover, no blocking has been verified since a larger number of channels are available in the 1/1 scenarios.

Additional capacity gains can be obtained by increasing the actuation frequency of the PC. In order to do this, alternative schemes have been proposed in [2, 52], which suggest to modify the PC signaling mechanisms in order to improve PC performance. The first approach consists of modifying the SACCH channel structure so that PC commands and link quality measurements could be transmitted in each SACCH burst, i.e., at each 120 ms. This corresponds to multiplying the rate of the power adjustments by 4. A second approach is to transmit PC commands and link quality measurements in-band, resulting in a 24-fold improvement in the PC actuation frequency. Nevertheless, it is worth mentioning that both strategies might reduce the protection of the control information sent over the SACCH and TCH channels, since some more bits would be allocated for PC purposes [52]. Even considering such improvements, the PC actuation frequency in the GSM/EDGE network is still too low to compensate for fast fading, especially for fast moving MSs. As a comparison, the highest PC actuation frequency, i.e., 50 Hz considering one iteration at each 20 ms, is still 30 times lower than that considered in WCDMA systems, which corresponds to 1.5 kHz. According to [29], even when operating at such frequency, wideband code division multiple-access (WCDMA) fast PC is not able to perfectly compensate for fast fading of MSs with speeds superior to 50 km/h.

Table 2.6 shows the performance of the SDPC algorithm considering different reporting periods, i.e., different actuation frequencies for the PC. SDPC is considered because it performed as good as or outperformed the ASBPC and UDPC algorithms. A frequency reuse of 1/3 is used with the EFR service because better spectral efficiency values have been achieved in this case. For the same reason, the 1/1 frequency reuse is selected when considering the MR59FR codec.

Table 2.6 Performance of the SDPC algorithm with reduced reporting periods (higher actuation frequency).

	EFR, 1/3			MR59FR, 1/1		
Reporting period in ms	480	120	20	480	120	20
Capacity in Erl/MHz	25	25	25	28.5	35	31.3
% of satisfied MSs	96	98	98	95	95	95

In Table 2.6, it can be seen that either QoS (with EFR) or capacity gains (with MR59FR) can be obtained by increasing the actuation frequency of the PC. For the MR59FR, a spectral efficiency gain of about 20% is achieved by performing PC adjustments at each 120 ms instead of at each 480 ms. However, adjusting transmit powers at each 20 ms does not enhance spectral efficiency. The reason for the latter result is that an increase of the actuation frequency leads to a higher variance of the interference in the system, which is harder to be tracked and compensated by the PC. Comparing the 35 Erl/MHz/cell supported in the 1/1 frequency reuse with MR59FR and power adjustments at each 120 ms, a spectral efficiency gain superior to 170% is obtained compared to the scenario without PC and DTX.

2.5.2.2 Power Control Performance for the WWW Service

In the GSM/EDGE network, the LA works on the same time basis as PC, i.e., with one adjustment at each 480 ms. Such adjustments can be made based on the reported average MEAN_BEP and average CV_BEP measured during one SACCH frame. Similarly to the voice service, a delay of one reporting period is involved before measurements become available at the BS.

If the target SINR of the PC algorithm is set outside of the range of SINR values covered by the LA, the performance of PC is strongly limited and no considerable capacity gain is obtained through PC, as it has been shown in [37]. This occurs mainly because PC will only apply for a few connections, whose link quality is really high, for instance, with SINR values above 35 dB. In [37], it has also been shown that an increase in the actuation frequency of LA leads to capacity gains of about 20%, which are similar to those previously shown for the voice service when increasing the actuation frequency of PC. Anyway, PC control does not provide any considerable capacity gain in this case too.

PC can provide capacity gains to the WWW service if the dynamic power range is extended from 10 to 30 dB and if the target SINR of the PC algorithm is allowed to lie inside the range covered by the LA, i.e., if the prioritization of LA over PC and the minimum transmit power constraint for the reliable reception of the USF are disabled.

In order to show this, the SDPC algorithm is considered in the sequel for the WWW service. A minimum target SINR of 6 dB has been considered, which corresponds to the QoS requirement of 10 kbit/s according to the link-level curves in Fig. 2.3. As before, the value 0.6 has been used for the parameter β_{SD}. In order to reduce the impact of the scheduling algorithm over the PC, the well-known first-in-first-served (FIFS) scheduling discipline is used, which maintains a channel allocated to the same MS for the total time needed to transmit its current packet. Considering these assumptions, Fig. 2.10 shows the capacity and QoS gains achieved with PC for the WWW service.

In Fig. 2.10(a), the fraction of satisfied MSs against the data spectral efficiency in bit/s/Hz/cell is shown. As it can be noted, PC can provide spectral efficiency gains of about 30% if a 10 dB dynamic power range is assumed and the prioritization of LA over PC is ignored. Additional gains of 15% are obtained by increasing the dynamic power range from 10 to 30 dB.

Figure 2.10(b) shows the 10th percentile of the MSs' average packet throughput, denoted by $r_{10\%}$, against the data spectral efficiency. In this figure, it can be seen that for low offered loads the average throughput achieved by 90% of the MSs is higher when PC is not used. Oppositely, for higher loads the average throughput perceived by 90% of the MSs is higher than when PC is disabled. This effect is more accentuated for a 30 dB dynamic power range. Consequently, the curves for the average throughput of 90% of the MSs with and without PC cross each other. The reasons for this crossing are as follows.

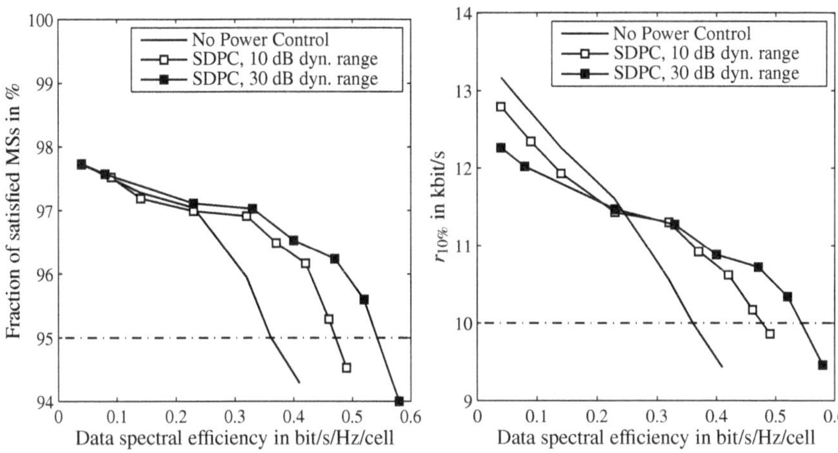

(a) Fraction of satisfied MSs (b) Average packet throughput

Fig. 2.10 PC performance for the WWW service in frequency reuse 1/3.

For lower loads, a smaller number of data sessions are present in the system and LA maximizes the throughput of each connection providing high rates, while PC reduces the rates of the MSs to attain only the target values. Because for low offered loads there are enough resources (timeslots) to serve the connections without incurring excessive co-channel interference, as it can be seen in Fig. 2.10(a), LA can operate at peak rates, and the rate reductions promoted by PC in order to attain the target SINRs could be eventually seen as unnecessary. In the truth, an outer loop PC should ideally be employed in this case to adjust the target SINR of the MSs and give them the higher rates that they could achieve without compromising system performance, i.e., without leading to excessive co-channel interference. When the load increases, co-channel interference increases and, in order to keep it under control, PC reduces transmit powers and consequently lowers the rates of several connections, which does not occur when considering LA without PC. The increased co-channel interference levels are efficiently combated by PC so that higher offered loads can be supported while maintaining the QoS levels at acceptable values and, consequently, better throughput values are attained by the MSs. This result is even better when considering a wider dynamic power range, i.e., 30 dB.

In total, the capacity gains provided by PC for the WWW service are modest (about 50% only) compared to those achieved for the voice service. Moreover, these gains are only achieved by violating some restrictions imposed by the standard, which suggests that PC is not very efficient for interactive data services. This is due to several aspects such as the bursty nature of data services, the MSs multiplexing performed by scheduling algorithms, and the competition between LA and PC, among other reasons.

2.5.3 Dynamic Channel Allocation

In this section, the different stages of the dynamic channel allocation algorithm are evaluated taking into account both ideal situations as well as those for which practical restrictions limit the efficiency of the algorithm.

The evaluation of DCA has been performed considering a GSM/EDGE cellular system in a macrocellular environment employing a 1/1 frequency reuse pattern. Such reuse pattern implies that all hopping frequencies are available within each sector, which corresponds to a scenario adequate to the application of DCA, since there exists the flexibility of distributing the channels freely among the cells. The voice service is considered in the evaluations using the EFR codec and considering DTX. In order to assess the impact of DCA on the system performance, scenarios employing either RFH or DCA combined with CFH are considered. RFH and CFH have been described in Section 2.2.3.

The evaluation begins with the simulation of the DCA algorithm under ideal circumstances (DCA ideal), i.e., with perfect SIR estimation. Some admission control alternatives are also considered. Finally, practical restrictions are introduced and some solutions for improving DCA performance are evaluated.

Figure 2.11 shows the performance of the DCA ideal algorithm with regard to RFH, in terms of quality of service and blocking rate. The value in dB represents the minimum SIR criterion of the admission control, and the term "impact" (DCA impact) refers to the impact test described in Section 2.3.2 with a 10 dB threshold.

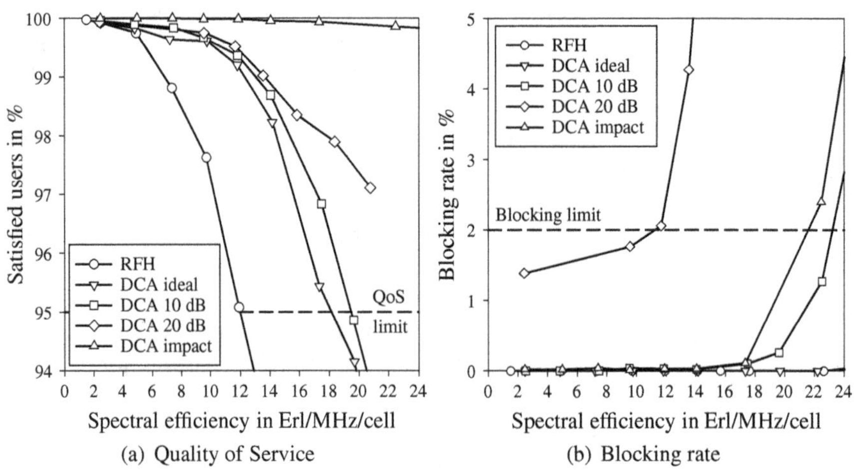

Fig. 2.11 Performance of ideal DCA and admission control.

Observing Fig. 2.11(a), it can be seen that the use of DCA provides significant capacity gains when compared to RFH. By applying the admission control with 10 and 20 dB thresholds, which indicate that users with SIRs below these levels are blocked, the quality of service is additionally improved. An even higher gain is provided by the impact test.

The admission control algorithm, even though presenting better quality of service to the MSs within the system, has its negative side, which is to increase the blocking error rate. This behavior can be seen in Fig. 2.11(b), where the presented metric includes the blocking of incoming calls as well as the dropping of ongoing calls. For a 10 dB threshold the blocking rate is significantly increased when compared to the case without admission control, but the interference still limits the offered load. For a 20 dB threshold, blocking becomes excessive, limiting capacity too early. Since, according to link-level results, an 8 dB SIR would be enough to guarantee minimum quality, the use of a threshold with such a high margin is not recommended. The best situation for the admission control would be the one in which the interference and blocking limits would meet at the same load, which might occur for a threshold slightly above 10 dB.

The impact test provides excellent QoS levels, almost reaching 100% satisfaction even for high offered loads. The blocking rate limits capacity at about 22 Erl/MHz/cell, which is still better than the other strategies. A threshold lower than 10 dB could provide better results, in the sense of equilibrating interference and blocking. It is important to emphasize, however, that this impact test is highly idealized, being its practical implementation probably not feasible after all restrictions are taken into account.

The measurement report mechanism, described in Section 2.3.2, is also used for performing SIR estimation in a practical system. As previously mentioned, two aspects might contribute to the imprecision of the estimates: the report delay and the number of reported cells.

Figure 2.12(a) shows how the DCA performance is compromised when assuming non-ideal SIR estimation (see the DCA report curve). The system capacity with dynamic channel allocation becomes inferior to the case with RFH, dropping from approximately 18 to 9 Erl/MHz/cell. The distribution (in logarithmic scale) of the mean SIR estimation error per user can be seen in Fig. 2.12(b), where the error is defined as the difference between the actual value and the estimated value. Since all values are negative, it may be concluded that the SIR is being overestimated.

The impact of faster report updates on the DCA performance is depicted by the DCA fast-rep curve, which assumes a delay corresponding to half the actual delay perceived within the system. It can be seen that the reduction of the delay has only a slight impact on the system capacity and that the distribution of the SIR estimation error was practically unaltered. Since the DCA algorithm employs cyclic hopping, which does not provide interference diversity after each hop, the interference profile perceived by the users barely changes within a measurement reporting period.

The use of extended reports, i.e., with a larger number of reported cells, presented good capacity results, as it can be seen in Fig. 2.12(a). By using measurement reports with 10 reported BSs (DCA ext-rep) instead of 6, the DCA capacity once again surpassed that of RFH, approaching the ideal performance and reaching a spectral efficiency of 15 Erl/MHz/cell. The improved DCA performance is due to the more

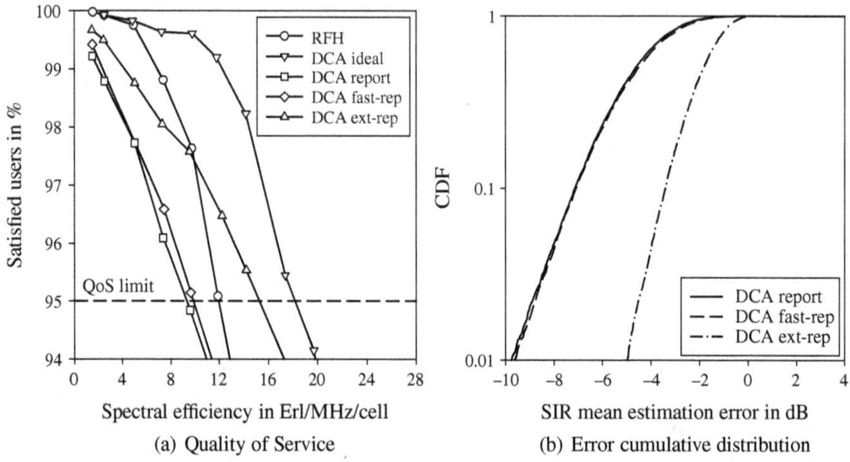

Fig. 2.12 Impact of the measurement reports on the DCA performance

precise SIR estimates (see Fig. 2.12(b)), which are achieved by the larger number of reported BSs, since more interferers are taken into account.

Table 2.7 presents a summary of the spectral efficiency achieved by the different algorithms. Note that the capacity is limited by interference for almost all cases, except for those marked with *, which are limited by blocking.

Table 2.7 Capacity results of the different algorithms.

Strategy	Spectral efficiency	Strategy	Spectral efficiency
RFH	12.0	DCA ideal	18.2
DCA report	9.3	DCA 10 dB	19.5
DCA fast-rep	9.8	DCA 20 dB	11.3*
DCA ext-rep	15.3	DCA impact	21.6*

The DCA algorithm is also analyzed for data services. Since the channels may be shared by several users, the selection procedure first prioritizes the empty channels with the best SIR, in order to avoid excessive queuing delays. If there are no empty channels, the selection is based exclusively on the SIR criterion. The ideal DCA algorithm has been assumed and a data user is deallocated if there is currently no other packet at the BS to be transmitted, which means that several channel reassignments may take place during the session of a data user.

Figure 2.13 shows the performance of RFH and DCA for a WWW service considering a FIFS scheduling discipline. The QoS is expressed in terms of the previously described $r_{10\%}$ rate metric. It can be seen that DCA provides significant capacity gains, reaching almost double the capacity of RFH, which demonstrates the potential of DCA for data services. The gains are mainly due to the bursty nature of WWW, which leads to high reassignment rates and allows the DCA algorithm to act quite often. Notice that lower gains are expected if practical constraints are taken into account, such as measurement reports and channel release timers, as well as in the case of non-bursty data traffic, such as download.

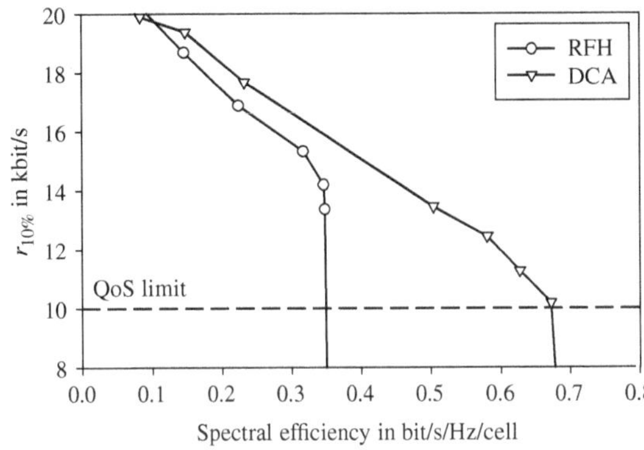

Fig. 2.13 DCA performance for the WWW data service.

2.5.4 Multi-antenna Techniques for Multiple Services

In order to provide capacity enhancements for GSM/EDGE networks with a mix of a voice and a data service, this section evaluates three strategies concerning the application of multi-antenna techniques, which are single-antenna transmission (SAT) with both services using a single sector antenna; single- and multi-antenna transmission (SMT) with a single sector antenna applied to the data service and switched fixed beams applied to voice; and multi-antenna transmission (MAT) with both services using switched fixed beams.[2.4]

For the cases in which switched fixed beams are applied, four-element uniform linear antenna arrays are considered for each sector of the BSs. In a sector, the antenna array is configured to steer four narrow beams covering the sector area. The voice service using the EFR codec and the WWW service are considered. For the voice service, DTX and PC are applied. For the WWW service, IR is employed considering only the MCS-9 of EGPRS.

The performance of the individual services may be seen in Fig. 2.14. The parameters are defined as voice service spectral efficiency (C_V), data service spectral efficiency (C_D), and the maximum offered spectral efficiency supported by the voice ($C^{max}_{V,SA}$), and data ($C^{max}_{D,SA}$) services with a single antenna.

It can be seen that the antenna arrays provide significant capacity gains, due to their spatial filtering and interference reduction characteristics. The capacity gains

[2.4] Note that MAT disregards the USF issue. A possible solution for a multi-antenna EGPRS system is the beam-based scheduling algorithm proposed in [12].

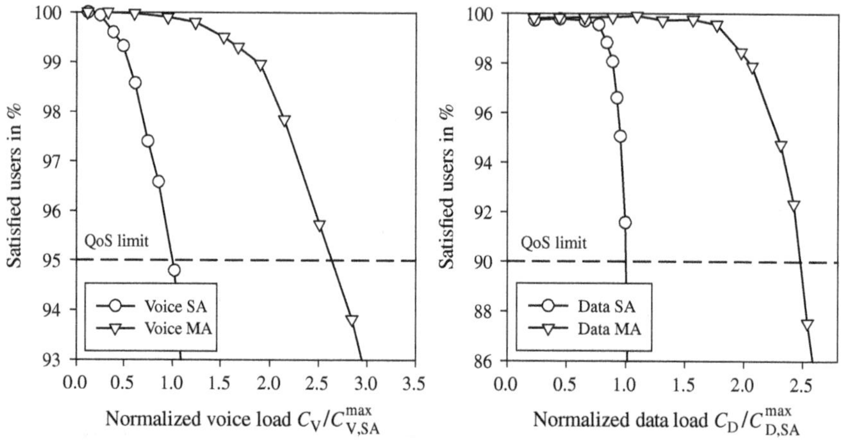

Fig. 2.14 Performance of the individual services: voice (left) and data (right).

may also be seen as quality of service gains, in case it is preferred to maintain the same load and offer better quality to the MSs, which would be adequate to the data service, since it would lead to higher transmission rates.

Figure 2.15 shows the values of the mean interference per MS for each service, with a single antenna (SA) and multiple antennas (MA). The mean interference is measured from the simulation of the individual services. Since we are simulating an interference-limited network that employs RFH, it may be assumed that the interference distribution will not significantly vary with the service mix. Therefore, out of simplicity, the individual services are taken as reference for calculating the power offset, corresponding to a mix of 100% of one service and 0% of the others.

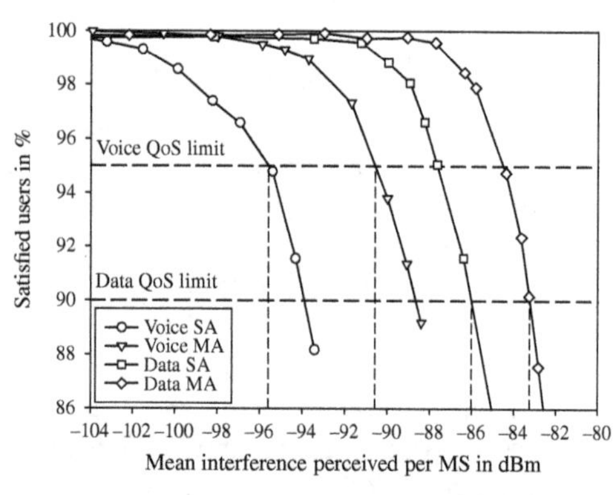

Fig. 2.15 Maximum interference supported by each service.

The application of multiple antennas results in an increase of the maximum interference supported by the services. The reason for this behavior might be the higher interference diversity introduced by the switched fixed beams.

As it has been discussed in Section 2.3.3, the system capacity is maximized when the maximum interference levels are balanced, which may be accomplished by adjusting the transmission power of the service which is more robust to interference (SBPS technique). The determination of the power offset may be done based on the interference levels supported by each service. From Fig. 2.15, a rough estimate of the power offset of each service/antenna configuration may be calculated as the absolute difference between the maximum interference levels, e.g., the difference between voice SA and data SA for the SAT configuration. This leads to offsets of approximately 9 dB (SAT), 4 dB (single and multi-antenna transmission (SMT)), and 6 dB (MAT). Note that the actual offset obtained from Fig. 2.15 for the MAT case is of roughly 7 dB, but it has been verified through simulations that the ideal offset for maximizing capacity is about 1 dB lower.

Next, the joint simulation of voice and data assumes that the frequency spectrum is shared among the services. The antenna strategies are initially evaluated for fixed voice load scenarios, in order to assess the achievable data capacities.

Figure 2.16 shows the data performance of the system for a scenario in which the normalized voice load is fixed at 0.75, i.e., 75% of the total supported voice load. The graphic is presented in terms of quality of service as a function of the data load. For each antenna strategy, quality curves are presented for voice (VQoS) and data (DQoS). The points at which the QoS requirements are met are indicated as well.

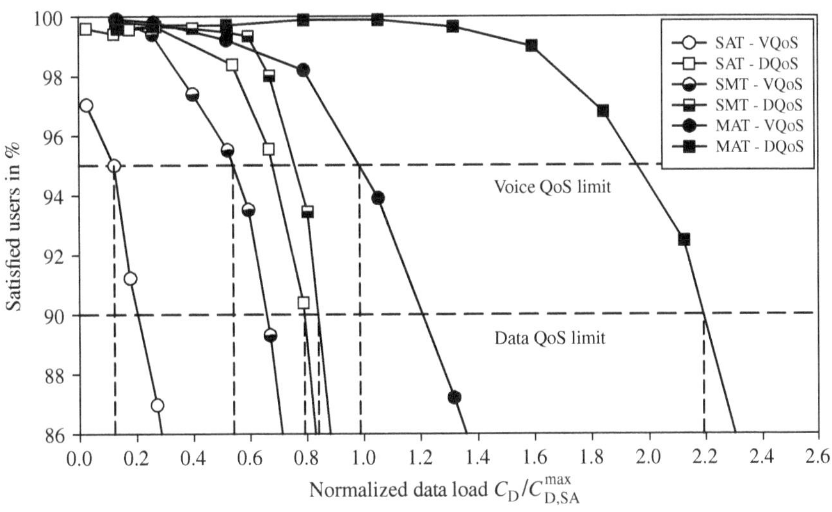

Fig. 2.16 Data service capacity for the different antenna strategies.

For the case of the strategy employing only single-antenna transmission (SAT), it can be seen that the voice service limits the data capacity right after the introduction of a few MSs, resulting in a capacity of less than 20% of the load supported by the data service individually.

The application of multi-antenna transmission only for the voice service (SMT) provides a reasonable increase in terms of data capacity, since the random frequency hopping algorithm allows the interference reduction of switched fixed beams to be perceived by all services. The multi-antenna transmission for both services (MAT) also provides improvements on the data capacity, as it can be seen in Fig. 2.16.

For all considered strategies, the voice service always limits system capacity, due to its more restrictive QoS requirements. The data service, which has less restrictive QoS requirements than voice, would be able to support higher offered loads if the voice restriction could be disregarded. This indicates that the system is unbalanced and that the SBPS concept may be applied.

The application of the power offset to each of the antenna strategies presents quite satisfactory results, as it can be seen in Fig. 2.17. The total system capacity considerably increases, and the QoS requirements are reached for similar offered loads, confirming the initial expectations regarding SBPS.

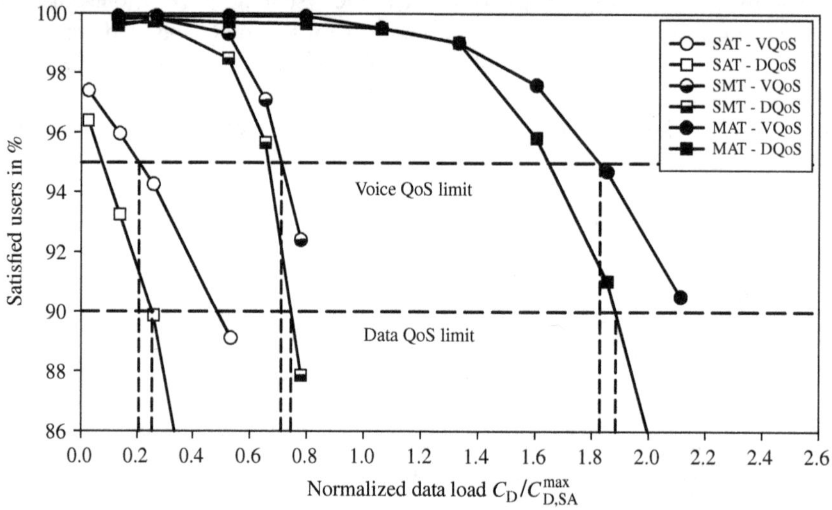

Fig. 2.17 Application of the power offset for the different antenna strategies.

Next, both voice and data loads are jointly varied for each service mix in order to obtain the capacity regions. The capacity region corresponds to the area within a data load vs. voice load graphic, for which the QoS requirements of both services are satisfied. Combinations of voice and data outside this region are not supported by the system. The left hand side of Fig. 2.18 shows the capacity regions – both simulated and theoretical – for the evaluated antenna strategies.

The extreme points correspond to the individual capacities of the isolated services. In [22] it has been shown that, for interference-limited scenarios, the capacity region of a balanced system is linear, i.e., it consists of a straight line connecting the individual capacity extremes. These regions are indicated as (Theory - SBPS).

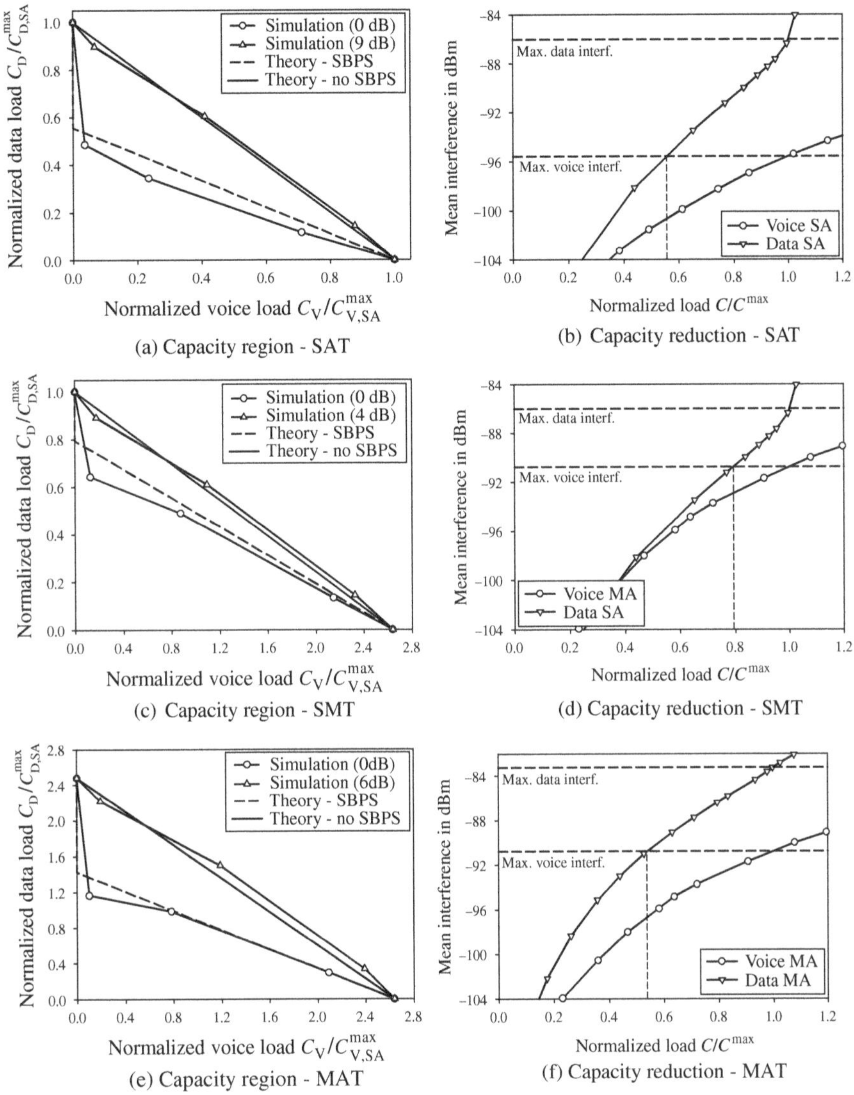

Fig. 2.18 Capacity regions (*left-hand side*) and capacity reduction (*right-hand side*) for the mix of voice and data services.

It can be seen that the application of the previously estimated power offsets has been fairly efficient, approximating the theoretical capacity regions for all cases. This result confirms that the application of antenna arrays, for one or more services, requires different power offsets in order to perform the interference balancing.

The system capacity prior to the SBPS application, mainly for the MAT case, has a much worse performance than the balanced system. As soon as a few voice users are introduced, the data capacity suffers a sudden capacity drop, and from then on it starts to decrease linearly with the increase in voice capacity.

This capacity drop is caused by the unbalanced interference distribution. Since the data service supports interference levels much higher than the voice service, the introduction of the latter in the system requires that the interference be decreased until the point at which it becomes possible to offer the minimum QoS requirements for voice. The larger the difference between the maximum interference supported by the services, the larger the sudden capacity drop effect.

The theoretical curves of the unbalanced system may be obtained from the results that relate load and interference, such as the ones shown on the right-hand side of Fig. 2.18. In these results, the load is normalized per service, i.e., the unit load corresponds to the maximum load supported by voice as well as by data. The maximum interference limits are those determined based on the quality requirements of each service (see Fig. 2.15).

Since the voice service is the most sensitive to interference, the crossing of the data curve with the maximum voice interference indicates how much the data load has to be reduced in order to satisfy the interference requirements of the voice service. The theoretical curve (Theory - no SBPS) may then be drawn as the maximum data capacity dropping until the estimated reduction point, followed by a straight line connecting it to the maximum voice load point.

Figure 2.19 shows the simulated capacity regions for each antenna strategy in a same graphic, with and without the application of SBPS. These figures provide a clearer view of the performance of the antenna strategies for multiple services. The application of the SMT technique, besides providing significant gains for the voice service itself, also provides gains to the data service. The capacity region area is expanded and certain voice load situations, for which it was previously prohibitive to add data users, now have a larger tolerance for data users. The application of MAT, on the other hand, results in a significant increase of the whole capacity region of the system making it possible to support high offered loads for both services.

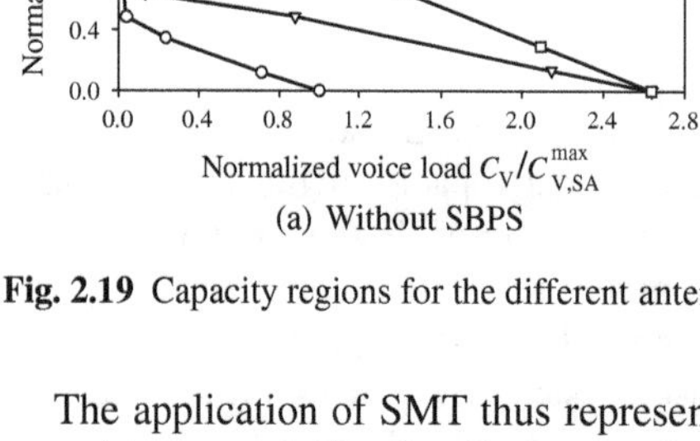

(a) Without SBPS

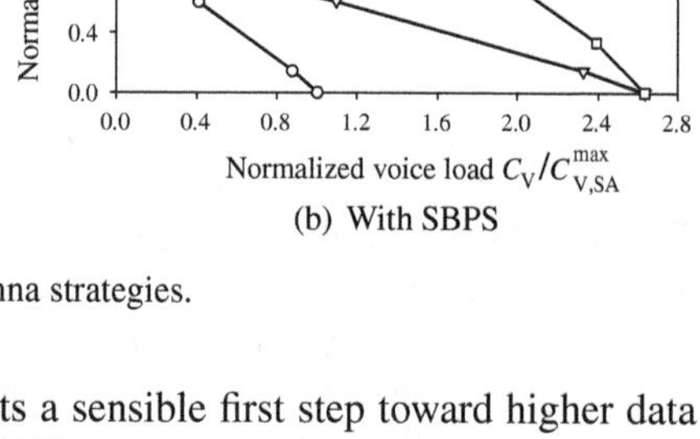

(b) With SBPS

Fig. 2.19 Capacity regions for the different antenna strategies.

The application of SMT thus represents a sensible first step toward higher data capacities, especially when the system is still dominated by voice users.

2.6 Conclusions and Research Directions

This chapter has shown how the efficient application of adequate RRM techniques can provide significant gains to the GSM/EDGE system. The techniques were employed to different services and considered different scenarios.

The PC technique is an efficient method for improving the capacity of both voice and data services. Due to the distributed nature of the considered algorithms, low implementation complexity is required and they may be implemented using the standard measures available on GSM/EDGE DCA, on the other hand, requires a certain cooperation among BSs in order to achieve reliable SIR estimates of the available channels. The increased complexity of DCA results in significant gains with regard to standard mechanisms such as random hopping, especially when applied in combination with admission control. Finally, the multi-antenna techniques, which require the deployment of additional hardware to the BSs, present the most promising capacity improvements, for both single and multi-service scenarios.

In Fig. 2.20, a summary of the spectral efficiencies achieved by the different RRM techniques is presented for both voice (EFR) and data (WWW) services considering a 1/1 frequency reuse. The best algorithm of each RRM technique is taken into account and the baseline corresponds to a simple random hopping scenario. It can be seen that the proposed techniques can provide considerably higher spectral efficiencies than the baseline case. The multi-antenna (MA) strategy provides the highest gains for both voice and data services, followed by the DCA and PC algorithms.

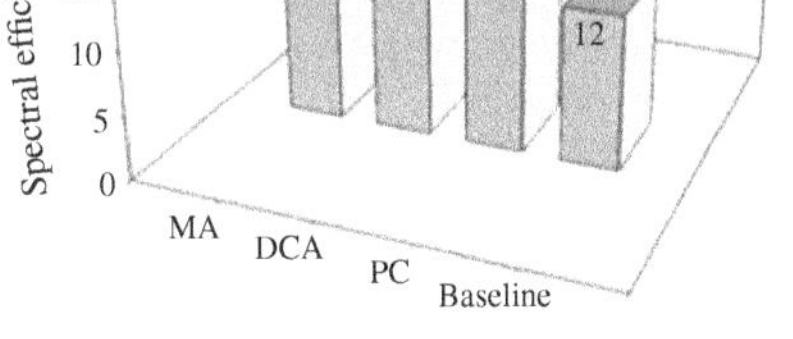

(a) Voice (EFR) service

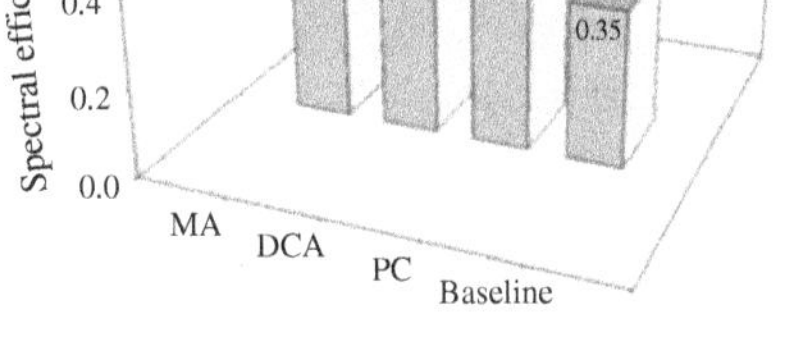

(b) Data (WWW) service

Fig. 2.20 Spectral efficiency achieved by the RRM algorithms assuming a 1/1 reuse.

A rough comparison of GSM/EDGE with a baseline high-speed downlink packet access (HSDPA) system with 5 MHz bandwidth, which is shown in Chapter 3 to be of about 0.7 bit/s/Hz/cell, indicates that RRM can potentially boost the GSM/EDGE capacity to approach and even surpass that of HSDPA. Notice, however, that this comparison is just illustrative of the potential capacity of GSM/EDGE with advanced features, as the HSDPA capacity figure does not consider similar improvements.

The further evolution of GSM/EDGE, as pointed out by [14], could include dual-antenna terminals, multi-carrier data services, mobility enhancements, high-order modulations, among other features. The evolved EDGE technology is expected to reach peak theoretical user data rates of 1.2 Mbit/s in the downlink and 473 kbit/s in the uplink [1]. Besides these system modifications, further studies on radio resource management techniques might still be able to boost system performance, especially with regard to the co-existence of multiple services.

References

1. 3G Americas: The case for evolved EDGE. Tech. rep. (2008). URL http://www.3gamericas.org
2. 3GPP: Performance of alternative fast power control schemes. Tech. rep., 3GPP, TSG GERAN Adhoc 2 Tdoc 0075/00 (2000). URL http://www.3gpp.org
3. 3GPP: GSM/EDGE radio access network (GERAN) overall description; stage 2. Tech. rep., 3rd. Generation Partnership Project, TS 43.051 V7.0.0 (2007). URL http://www.3gpp.org
4. 3GPP: General packet radio service (GPRS); overall description of the GPRS radio interface; stage 2. Tech. rep., 3GPP, TS 43.064 V7.9.0 (2008). URL http://www.3gpp.org
5. 3GPP: General packet radio service (GPRS); service description; stage 2. Tech. rep., 3GPP, TS 23.060 V8.2.0 (2008). URL http://www.3gpp.org
6. 3GPP: Multiplexing and multiple access on the radio path. Tech. rep., 3GPP, TS 45.002 V7.7.0 (2008). URL http://www.3gpp.org
7. 3GPP: Physical layer on the radio path; general description. Tech. rep., 3GPP, TS 45.001 V7.7.0 (2008). URL http://www.3gpp.org
8. 3GPP: Radio link control/medium access control (RLC/MAC) protocol. Tech. rep., 3GPP, TS 44.060 V8.1.0 (2008). URL http://www.3gpp.org
9. 3GPP: Radio subsystem link control. Tech. rep., 3GPP, TS 45.008 V7.11.0 (2008). URL http://www.3gpp.org
10. 3GPP: Radio transmission and reception. Tech. rep., 3GPP, TS 45.005 V8.1.0 (2008). URL http://www.3gpp.org
11. Alexiou, A., Haardt, M.: Smart antenna technologies for future wireless systems: trends and challenges. IEEE Communications Magazine **42**(9), 90–97 (2004)
12. Alm, M., Craig, S.: Adaptive antenna systems and EGPRS data protocol aspects. In: Proc. of the IEEE Vehicular Technology Conference, vol. 1, pp. 54–58 (2004)
13. Almgren, M., Eriksson, H., Wallstedt, K.: Power control in a cellular system. In: Proc. of the IEEE Vehicular Technology Conference, vol. 2, pp. 833–837 (1994)
14. Axelsson, H., Björkén, P., de Bruin, P., Eriksson, S., Persson, H.: GSM/EDGE continued evolution. Ericsson Review (1), 20–29 (2006)
15. Blogh, J., Hanzo, L.: Third generation systems and intelligent wireless networking, 1st edn. Wiley (2002)
16. Bratanov, P.I., Bonek, E.: Mobility model of vehicle-borne terminals in urban cellular systems. IEEE Transactions on Vehicular Technology **52**(4), 947–952 (2003)
17. Camp, T., Boleng, J., Davies, V.: A survey of mobility models for ad hoc network research. Wireless Communications and Mobile Computing **2**(5), 483–502 (2002)

18. Cavalcanti, F.R.P., de Sousa Jr., W.M., Silva, Y.C.B., Maciel, T.F.: Combined performance of packet scheduling and smart antennas for data transmission in EGPRS. In: Proc. of the IEEE Vehicular Technology Conference, vol. 2, pp. 797–801 (2002)
19. Dam, H., Berg, M., Andersson, S., Bormann, R., Frerich, M., Ahrens, F., Henß, T.: Performance evaluation of adaptive antenna base stations in a commercial GSM network. In: Proc. of the IEEE Vehicular Technology Conference, vol. 1, pp. 47–51 (1999)
20. Eriksson, H.: Capacity improvement by adaptive channel allocation. In: Proc. of the IEEE Global Telecommunications Conference, pp. 1355–1359 (1998)
21. Foschini, G.J., Miljanic, Z.: A simple distributed autonomous power control algorithm and its convergence. IEEE Transactions on Wireless Communications **42**(4), 641–646 (1993)
22. Furuskär, A.: Radio resource sharing and bearer service allocation for multi-bearer service, multi-access wireless networks – methods to improve capacity. Ph.D. thesis, Royal Institute of Technology, Stockholm, Sweden (2003)
23. Furuskär, A., Bladsjö, D., Eriksson, S., Frodigh, M., Jäverbring, S., Olofsson, H.: System performance of the EDGE concept for enhanced data rates in GSM and TDMA/136. In: Proc. of the IEEE Wireless Communications and Networking Conference, vol. 2, pp. 752–756 (1999)
24. Furuskär, A., de Bruin, P., Johansson, C., Simonsson, A.: Mixed service management with QoS control for GERAN – the GSM/EDGE radio access network. In: Proc. of the IEEE Vehicular Technology Conference, vol. 4, pp. 2635–2639 (2001)
25. Furuskär, A., Zander, J.: Multiservice allocation for multiaccess wireless systems. IEEE Transactions on Wireless Communications **4**(1), 174–184 (2005)
26. Goodman, D.J.: Efficiency of packet reservation multiple access. IEEE Transactions on Wireless Communications **40**(1), 170–176 (1991)
27. GSM Alliance and Wireless Intelligence: Global mobile market Q4 2007 (2008). URL `http://www.gsmworld.com/documents/20_year_factsheet.pdf`
28. Gunarsson, F.: Power control in cellular systems: analysis, design and estimation. Ph.D. thesis, Linköping University, Linköping, Sweden (2000)
29. Holma, H., Toskala, A.: WCDMA for UMTS, 2nd edn. Wiley (2002)
30. Jakes, W.C.: Microwave mobile communications, 1st edn. Wiley (1974)
31. Jeruchim, M.C., Balaban, P., Shanmugan, K.S.: Simulation of communication systems: modeling, methodology and techniques, 2nd edn. Springer (2000)
32. Johansson, C., Verdier, L.D., Khan, F.: Performance of different scheduling strategies in a packet radio system. In: Proc. of the IEEE International Conference on Universal Personal Communications, vol. 1, pp. 267–271 (1998)
33. Katzela, I., Naghshineh, M.: Channel assignment schemes for cellular mobile telecommunication systems: a comprehensive survey. IEEE Personal Communications Magazine **3**(3), 10–31 (1996)
34. Lau, B.K., Berg, M., Andersson, S., Hagerman, B., Olsson, M.: Performance of an adaptive antenna system in EGPRS networks. In: Proc. of the IEEE Vehicular Technology Conference, vol. 4, pp. 2354–2358 (2001)
35. Law, A.M., Kelton, W.D.: Simulation modeling and analysis, 3rd edn. McGraw-Hill (2000)
36. Maciel, T.F., de Oliveira Neto, R.A., de Sousa Jr., W.M., Cavalcanti, F.R.P., Silva, Y.C.B.: Estimating the QoS enhancement using power control in EGPRS. In: Proc. of the IEEE International Telecommunications Symposium (2002)
37. Maciel, T.F., Silva, Y.C.B., Cavalcanti, F.R.P., Cardoso, L.S.: Interference management for mixed-services through power control and service-based power setting. In: Proc. of the IEEE Vehicular Technology Conference, vol. 3, pp. 1640–1644 (2004)
38. Molkdar, D., Featherstone, W., Larnbotharan, S.: An overview of EGPRS: the packet data component of EDGE. Electronics and Communication Engineering Journal **14**(1), 21–38 (2002)
39. Molkdar, D., Lambotharan, S.: Link performance evaluation of EGPRS in LA and IR modes. In: Proc. of the IEEE Vehicular Technology Conference (2000)

40. van Nobelen, R.: Towards higher data rates for IS-136. In: Proc. of the IEEE Vehicular Technology Conference, vol. 3, pp. 2403–2407 (1998)
41. de Oliveira Neto, R.A., Maciel, T.F., Cavalcanti, F.R.P., Chaves, F.: Distributed power control with tracking of fast fading and interference. In: Proc. of the IEEE Vehicular Technology Conference, vol. 2, pp. 989–993 (2004)
42. Olofsson, H., Näslund, J., Sköld, J.: Interference diversity gain in frequency hopping GSM. In: Proc. of the IEEE Vehicular Technology Conference, vol. 1, pp. 102–106 (1995)
43. Papoulis, A., Pillai, S.U.: Probability, random variables and stochastic processes, 4th edn. McGraw-Hill (2002)
44. Rappaport, T.S.: Wireless communications: principles and practice, 1st edn. Prentice Hall Communications Engineering and Emerging Technologies. Prentice Hall (1999)
45. Rodríguez, R., Martínez, J., Romero, J.: Downlink power control performance in (E)GPRS networks. In: Proc. of the IEEE Vehicular Technology Conference, vol. 2, pp. 1125–1128 (2002)
46. Salmenkaita, M., Gimenez, J., Tapia, P.: A practical DCA implementation for GSM networks: dynamic frequency and channel assignment. In: Proc. of the IEEE Vehicular Technology Conference, vol. 4, pp. 2529–2533 (2001)
47. Salmenkaita, M., Gimenez, J., Tapia, P., Fernandez-Navarro, M.: Optimizing the GSM/EDGE air interface for multiple services with dynamic frequency and channel assignment. In: Proc. of the IEEE Vehicular Technology Conference, vol. 4, pp. 2215–2219 (2002)
48. Silva, Y.C.B., Maciel, T.F., Cavalcanti, F.R.P., Silva, E.B.: Performance comparison of same cell reuse strategies: SDMA and channel allocation tiering (CHAT). In: Proc. of the IEEE Vehicular Technology Conference, vol. 4, pp. 1978–1982 (2004)
49. Silva, Y.C.B., Silva, E.B., Maciel, T.F., Cavalcanti, F.R.P., Cardoso, L.S.: Combined performance analysis of signal level-based dynamic channel allocation and adaptive antennas. Lecture Notes in Computer Science – Service assurance with partial and intermittent resources **3126**, 92–103 (2004)
50. de Sousa Jr., W.M., Cavalcanti, F.R.P., Maciel, T.F., Silva, Y.C.B.: On the performance of EGPRS using same cell reuse. In: Proc. of Virginia Tech/MPRG Symposium on Wireless Personal Communications, pp. 227–238 (2002)
51. de Sousa Jr., W.M., Cavalcanti, F.R.P., Silva, Y.C.B., Maciel, T.F.: System-level performance of space–time scheduling for non real-time traffic in a 3G network. In: Proc. of the IEEE Vehicular Technology Conference, vol. 2, pp. 1207–1211 (2002)
52. Tidestav, C., Eriksson, M.: Fast power control for voice in GERAN. In: Proc. of the IEEE Personal, Indoor and Mobile Radio Communications, vol. 1, pp. C–134–138 (2001)
53. UMTS: Selection procedures for the choice of radio transmission technologies of the UMTS. Tech. rep., UMTS, UMTS 101.112 V3.2.0 (1998)
54. Wang, L., Aghvami, A.: Optimal power allocation based on QoS balance for a multi-rate packet CDMA system with multimedia traffic. In: Proc. of the IEEE Global Telecommunications Conference, vol. 5, pp. 2778–2782 (1999)
55. Winters, J.H.: Smart antennas for wireless systems. IEEE Personal Communications Magazine **5**(1), 23–27 (1998)
56. Yacoub, M.D.: Fundamentals of mobile radio engineering, 1st edn. CRC Press (1993)
57. Yates, R.D., Gupta, S., Rose, C., Sohn, S.: Soft dropping power control. In: Proc. of the IEEE Vehicular Technology Conference, vol. 3, pp. 1694–1698 (1997)
58. Zander, J.: Performance of optimum transmitter power control in cellular radio systems. IEEE Transactions on Vehicular Technology **1**, 57–62 (1992)
59. Zander, J., Kim, S.L., Almgren, M., Queseth, O.: Radio resource management for wireless networks, 1st edn. Artech House Publishers (2001)
60. Zhang, Y., Soong, B.H.: Performance evaluation of GSM/GPRS networks with channel reallocation scheme. IEEE Communications Letters **8**(5), 280–282 (2004)

Chapter 3
Performance Optimization in Practical HSPA Networks for Wireless Broadband Access

Mário I. J. Da Silva

3.1 Introduction to Broadband Wireless Access

Over the last few years, there has been a substantial growth in broadband Internet access worldwide driven by intensifying competition and by regulatory measures to improve competitor access to local loops. In the fixed broadband business, asymmetric digital subscriber lines (ADSL) remain the key technology, followed by cable networks.

On the fixed wireless side, worldwide interoperability for microwave access (WiMAX) technology (based on IEEE 802.16-2004) has been expected to capture a significant share of wireless broadband market and attract a large number of DSL or other wireline subscribers. However, up to 2007, WiMAX together with other fixed wireless access technologies correspond to less than 2% of the total number of fixed broadband subscriptions.

On the other hand, major mobile network operators have started a move to offset their eroding voice revenues by investing in data services. The introduction of high-speed packet access (HSPA) allowed these operators to enter the mobile broadband market. HSPA is a data-oriented 3G standard complementary to Universal Mobile Telecommunication System (UMTS) and released by the standardization body 3rd. Generation Partnership Project (3GPP). Since its first deployment in 2005, the number of HSPA networks has been increasing every year. The price to the end-user data usage has also dropped due to flat rate charging, which in turn is driving a significant increase in data traffic.

The growth in mobile broadband traffic has been significant and according to many forecasts will continue for an undetermined period of time. By 2012, mobile broadband is expected to account for 25% of the total broadband subscriptions in comparison with 3% in 2005. The fixed and mobile subscription forecasts are highlighted in Fig. 3.1. HSPA is expected to be the major access technology driving the growth of mobile data subscriptions by then. Some fundamental technical characteristics and performance of HSPA are to be discussed over the rest of this chapter aiming at an optimized deployment of wireless broadband access networks.

F. Cavalcanti, S. Andersson (eds.), *Optimizing Wireless Communication Systems*,
DOI 10.1007/978-1-4419-0155-2_3, © Springer Science+Business Media, LLC 2009

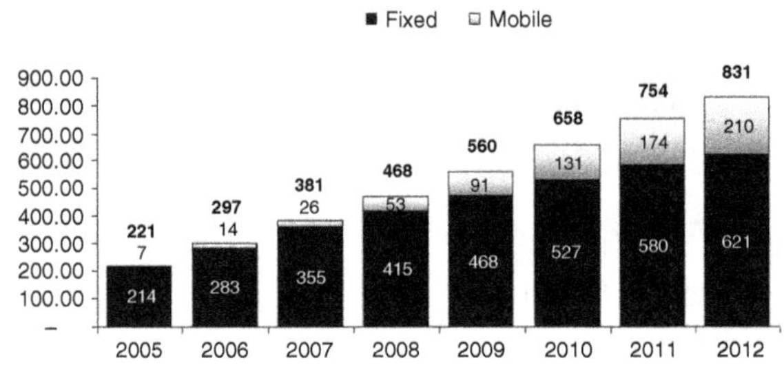

Fig. 3.1 Fixed and mobile broadband Internet subscriptions forecast (millions of subscribers, [12]).

3.2 System Overview

UMTS is a third-generation mobile system which uses the WCDMA technology. UMTS networks can be divided into core network (CN) and UMTS terrestrial radio access network (UTRAN). In the UTRAN side, two different operation modes are available in the specifications: frequency division duplex (FDD) and time division duplex (TDD). This chapter deals only with the FDD mode.

UMTS Release 99 was recognized as the first completed specification and was subsequently deployed in several markets. In this release, the system was designed to operate using a dedicated channel (DCH) for most of the services such as speech, video telephony, and packet radio bearers. Release 4 followed around a year later in 3GPP, but it did not have major additions from the FDD perspective. However, anticipating the requirement for improved packet performance, a feasibility study was carried out for high-speed downlink packet access (HSDPA) [1]. Finally, HSDPA was introduced in standards by 3GPP Release 5 in early 2002. high-speed uplink packet access (HSUPA) was subsequently added to Release 6 in late 2006. It is worth noting that the 3GPP standardization work carries on, and further releases such as Release 7 are already available.

3.2.1 HSPA Standard: Principles and Key Features

HSPA in this chapter refers to both HSDPA and HSUPA. As previously described, these technologies were specified as part of different UMTS standard releases, but they complement each other when considering a complete wireless broadband solution. The key benefits for these technology deployments are as follows:

- **Higher downlink bit rates, up to 14 Mbps**: This reduces download times and enables more advanced audio and video content and larger file transfers such as high-resolution streaming applications.
- **Latency reduction**: Shorter round-trip delay on the air interface. The benefits are better Internet browsing end-user experience and new services enabled with very high requirements on response times such as gaming applications to be carried over the HSPA bearer.

- **Higher uplink bit rates, up to 5.7 Mbps**: Reducing upload times and potentially enabling more peer-to-peer traffic and improved user experience in applications that require high upload speeds such as web photo storage and video uploads.
- **Capacity enhancements**: Higher system capacity in comparison to UMTS Release 99, which reduces the network cost per megabyte.

3.2.2 The Downlink Component of HSPA

HSDPA functionality is briefly described in the following sections. A detailed description of HSPA functionality is presented in [14, 15, 20, 27].

3.2.2.1 HSDPA Concepts

HSDPA delivers improved downlink performance by introducing new functionality for high-speed packet data transmission such as shared channel, fast hybrid automatic repeat request (H-ARQ) with soft-combining, shorter TTI (transmission time interval), and high-order modulation and fast packet scheduling.

Shared Channel

The HS-DSCH (high-speed downlink shared channel) is the fundamental transport channel added by HSDPA. The high-speed physical downlink shared channel (HS-PDSCH) is used to carry the HS-DSCH. HS-PDSCH uses a fixed spreading factor of 16, instead of the dedicated channels variable spreading factor used in Release 99. It achieves high data rates by allocating multiple codes to a single user. HS-PDSCH can be flexibly shared among UEs (user equipment) in the time domain (via time multiplexing) as well as in the code domain (via code multiplexing). For time multiplexing, the basic time unit of HSDPA is the TTI of 2 ms. In Fig. 3.2 is shown an example of time–code allocation plane for four users in HSPA.

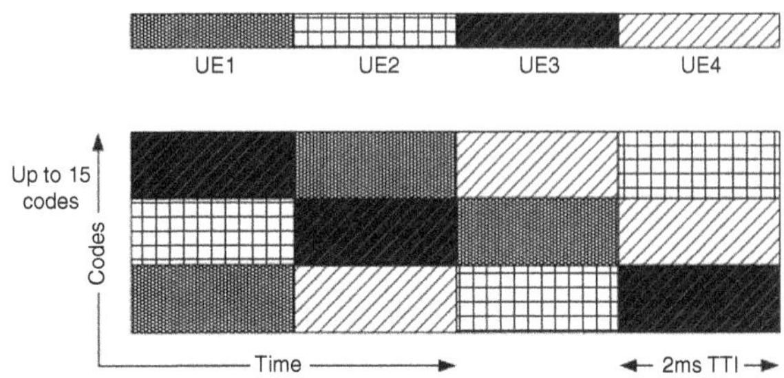

Fig. 3.2 Time and code multiplexing in HSDPA.

Fast Link Adaptation

The HS-PDSCH uses fast link adaptation rather than fast power control to deliver the improved downlink performance. It adjusts transmission parameters (coding rate, modulation type, and number of codes) according to instantaneous channel conditions based on downlink channel quality indicator (CQI) sent by the UE. This is done on a per TTI basis.

Fast H-ARQ with Soft-Combining

The H-ARQ mechanism employed in HSDPA uses layer 1 retransmission and is terminated in the Node B, thus allowing a rapid response to retransmission requests. This then decreases error rates for the RLC (radio link control) layer and reduces retransmission delays. Multiple parallel H-ARQ processes allow continuous transmission to a single user.

H-ARQ uses a soft-combining mechanism to recover the error-free packets as efficiently as possible, i.e., even if the retransmitted packets are corrupted, their combination can yield an error-free packet. The two soft-combining mechanism options are *chase combining (CC)* and *incremental redundancy (IR)* . In the first, the basic principle is to send a number of repeated transmissions of each coded data packet which allows the decoder to combine multiple received copies of the coded packet weighted by the signal-to-noise ratio (SNR) prior to decoding. The second is another implementation of the H-ARQ technique wherein instead of sending simple repeats of the entire coded packet, additional redundant information is incrementally transmitted if the decoding fails in the first attempt.

Shorter TTI (2 ms)

The HSDPA radio frame is a sub-frame of the 10-ms DCH frame. It has a length of 2 ms, i.e., equivalent to three UMTS time-slots. Hence, HSDPA TTI could be reduced to 2 ms, whereas typical WCDMA's packet-switched DCH TTI varies from 10 to 40 ms. This allows the network to allocate resources more efficiently as the user data transmission occupies HS-PDSCH for a shorter time. It also improves the round-trip delay for the packet. For more details about the HS-PSDSCH sub-frame structure the reader may refer to [2].

High Order Modulation

A 16-quadrature amplitude modulation (16-QAM) scheme is introduced in the downlink, in addition to the legacy QPSK from DCH Release 99. The 16-QAM allows for twice the peak data rate compared to QPSK. It is useful mainly in good radio channel conditions because both magnitude and phase information must be accurate to correctly discern the position of the symbol in the resulting constellation. The support for both QPSK and 16-QAM is mandatory in most of the HSDPA-related UE categories.

Fast Packet Scheduling

The scheduler is a key element to the overall system performance, especially in a highly loaded network. Packet scheduling controls to which users the shared-channel transmission is allocated. This decision can also be based on the received downlink channel quality from the UE (CQI), which each user device transmits as often as 500 times per second. The Node B packet scheduling occurs every TTI which allows utilization of information on the instantaneous channel conditions for each user. This has the benefits of multi-user selection diversity, i.e., to give the shared channel to the "best" user at a given time. Figure 3.3 illustrates an example of the channel quality variation for two users, where the best user is scheduled resulting in higher overall system capacity.

Fig. 3.3 Fast packet scheduling multi-user gain.

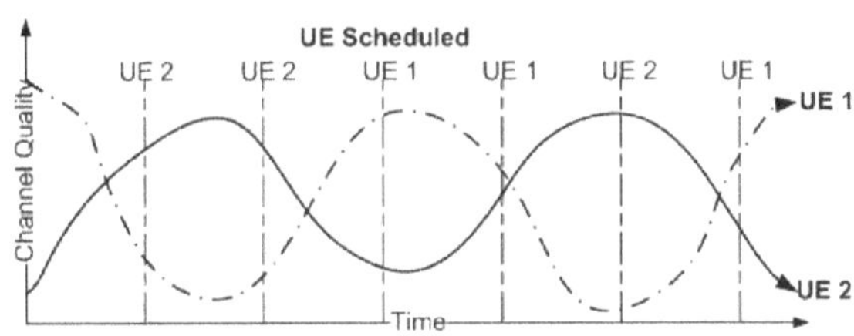

There are different packet scheduling algorithms that trade between fairness (i.e., similar transmission opportunities to all users in a cell) and total cell throughput. An overview of scheduling algorithms for wireless communications can be found in [19] and a comparison of packet scheduling algorithm performance in HSPA networks is provided, for instance, in [15, 16]. Three popular options are as follows:

- **round robin (RR)**: Cyclically assigns the channel to users, i.e., all users will be allocated equal amount of time but the system will not benefit fully from the channel quality variance.
- **Max C/I ratio**: Assigns the channel to the user with the best channel quality (carrier to interference ratio). It maximizes system capacity at the cost of the lowest degree of fairness.
- **proportional fair (PF)**: Assigns the channel to the user with the best relative channel quality, which is a combination of channel quality and the level of fairness desired. This definition of relative quality leads to many variations on the PF concept apart from its original proposal in [26] (see, e.g., [17, 23–25] for some of these variations). Depending on the specific version, the scheduler may take into account past resource allocations, the current level of performance of

a given user, and its instantaneous and average channel quality. The final goal of a PF scheduler is to achieve a balance between the maximization of total cell throughput and fairness, allowing all users to achieve a minimum quality of service.

HSDPA Architecture

HSDPA is implemented under the existing WCDMA architecture. The UTRAN contains the Node B, which is 3GPP terminology to describe the WCDMA base station; the radio network controller (RNC), which is responsible for control of several Node Bs and interface with the core network; and the corresponding interfaces.

The legacy WCDMA layer 2 protocols, i.e., RLC and MAC-d, are still presented in HSDPA. The key difference is the move of the packet scheduling from RNC to Node B and the layer 1 retransmissions in HSDPA. In order to support this, a new MAC-hs (MAC high speed) is introduced in both Node B and UE. The MAC-hs frame is composed of several MAC-d frames. It is worth noting that the RLC layer still has a role in the error control at the RNC level, i.e., L2 retransmission. Finally, the HS-DSCH frame protocol (FP) handles the data transport between RNC and the Node B [9]. The HSDPA protocol stack is shown in Fig. 3.4.

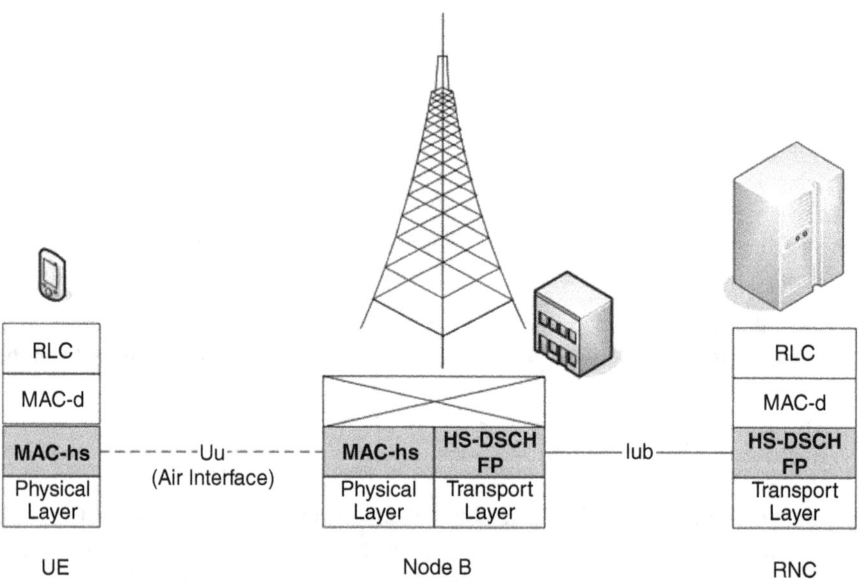

Fig. 3.4 HSDPA protocol stack.

HSDPA Channel Structure

In UMTS there are three types of channel: logical, transport, and physical. The logical channel refers to the type of information to be transferred, whereas the transport channel consists of characteristics required for the data transfer of the air interface. For instance, the dedicated traffic channel (DTCH), which carries user data, and dedicated control channel (DCCH), which carries user control data, can be both mapped into the HS-DSCH. The physical channel is the transmission resource in the air interface. For more information refer to [3, 10].

In the physical layer, the HS-PDSCH is used to carry the HS-DSCH, i.e., user data. One channelization code of fixed spreading factor SF=16 is adopted for HS-PDSCH. Multi-code transmission is allowed, which translates to an UE being assigned multiple channelization codes in the same HS-PDSCH sub-frame, the maximum number of codes depending on the UE capability.

Along with the HS-PDSCH channel, two physical channels are also introduced: HS-SCCH (high-speed shared control channel) and HS-DPCCH (high-speed dedicated physical control channel). HS-SCCH is a fixed-rate downlink physical channel (60 kbps, SF=128) that informs the user that data will be sent on the HS-PDSCH and in which format the data are coming. If code multiplexing is used in order to schedule more than one user per TTI, then more than one HS-SCCH will be necessary. The uplink HS-DPCCH carries acknowledgment information and the CQI of the user.

3.2.3 The Uplink Component of HSPA

HSUPA was introduced in 3GPP Release 6 to improve the uplink performance. It is also known in the industry as enhanced uplink (EUL) and by the UMTS standards as enhanced dedicated channel (E-DCH). An overall description of HSUPA is presented in [15, 21, 22, 28].

3.2.3.1 HSUPA Concepts

HSUPA introduces new functionalities in the uplink that are similar to the ones used by HSDPA, for instance fast scheduling, shorter TTI, and fast H-ARQ with soft-combining.

Fast Scheduling

The fast scheduler is a key functionality in HSUPA. It controls the bit rates and the time that the UE is allowed to transmit. The total received power (or UL interference) at the Node B is the limiting factor in the uplink.

Fast scheduling allows for a more relaxed interference threshold. Without fast scheduling, a conservative admission control policy would have to be implemented in order to reserve a greater interference margin in the system.

The uplink scheduling is capable of scheduling multiple users in parallel. This is more essential than in HSDPA as in most cases a single UE will not be able to utilize the full cell capacity due to limited UE transmit power. A more detailed explanation is provided in the following sections.

Shorter TTI

There are two options for HUSPA TTI duration: 10 ms (mandatory) and 2 ms (optional). A shorter TTI of 2 ms is supported by HUSPA to allow a significant reduction in overall data transmission delays and provide the means for the other features to adapt rapidly to the radio conditions and load. However, this is optional and will depend on the UE category.

Fast H-ARQ with Soft-Combining

A fast hybrid ARQ with soft-combining is adopted in the Node B, for similar reasons as in HSDPA, allowing the Node B to rapidly request retransmission of erroneously received data, which implies increased robustness and reduced retransmission delays.

Prior to decoding, the Node B combines information from the original transmission with that of later retransmissions, which is generally known as soft-combining mechanism. A negative acknowledgement (NACK) and acknowledgement (ACK) are sent from the Node B to the UE for every packet received. A key difference from the HSDPA scheme is that the ACK/NACK can be received from multiple cells in the active set. If at least one ACK is received, the UE assumes that the packet was successfully received, otherwise the packet is retransmitted. The mobility procedure is explained in the next section.

The uplink modulation based on QPSK is unchanged in Release 6. The increased bit rate is achieved by using multi-code transmission and a smaller spreading factor (SF=2).

HSUPA Architecture

HSUPA is also implemented under the existing WCDMA architecture. Similar to HSDPA, the Node B is responsible for packet scheduling for HSUPA and layer 1 retransmissions.

A new MAC entity (MAC-e) is added to handle H-ARQ retransmissions, scheduling, MAC-e multiplexing/demultiplexing, and E-DCH TFC selection. In the RNC, a new MAC entity (MAC-es) is added to provide in-sequence delivery (reordering) and to handle combining of data from different Node Bs in case of soft-handover.

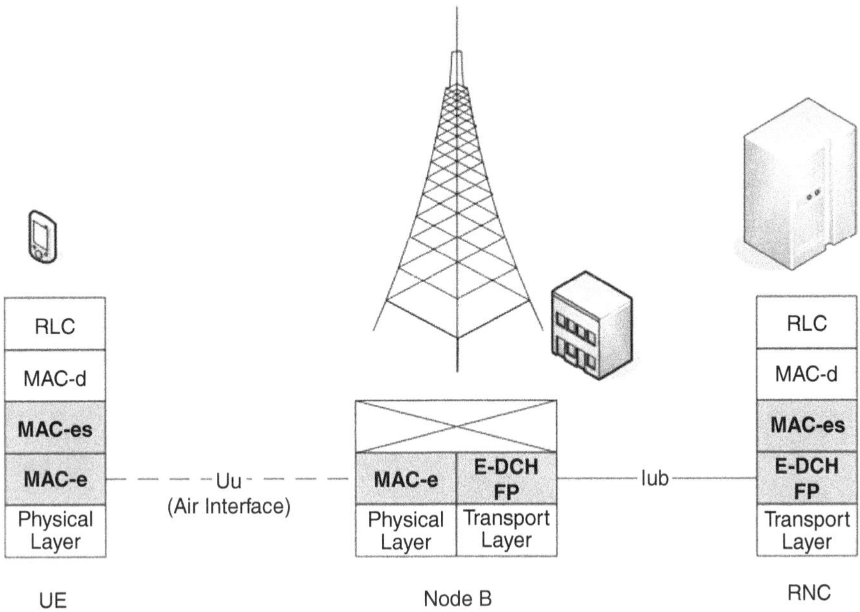

Fig. 3.5 HSUPA protocol stack.

The E-DCH FP handles the data transport between the RNC and the Node B [6]. The HSUPA protocol stack is shown in Fig. 3.5.

HSUPA Channel Structure

The HSUPA channel mapping among logical, transport, and physical channels is described in [3, 10].

The E-DCH is the transport channel in HSUPA which carries the user data. This is mapped to the E-DCH dedicated physical data channel (E-DPDCH) and the enhanced dedicated physical control channel (E-DPCCH) in the uplink. A brief description of the new physical channels introduced with HSUPA is presented below (see also [7]):

- **E-DPCCH**: used to carry the control information associated with the E-DCH
- **E-DPDCH**: used to carry the E-DCH transport channel, i.e., user data
- **E-DCH access grant channel (E-AGCH)**: informs the UE about its absolute limitation of the maximum amount of uplink resources the UE may use
- **E-DCH relative grant channel (E-RGCH)**: informs the UE to increase or decrease the resource limitation compared to the previously used value
- **E-DCH hybrid ARQ indicator channel (E-HICH)**: carries H-ARQ NACK/ACK information on the downlink direction, i.e., informs the UE whether a particular Node B has received the uplink packet correctly or not

3.2.4 Customer Premises Equipment

Customer premise equipment (CPE) terminology comes from the fixed broadband sector and refers to any device and associated equipment at a subscriber's premises that allow the connection to a broadband service provider. In the case of HSPA the device will connect to one of the nearest radio base stations (Node B).

In early UMTS deployments, handsets were the only devices available; connection to the personal computer can be done by data cable or via Bluetooth wirelessly. Nowadays, besides handsets, users may choose among PCMCIA data cards or plug and play USB modems. Another interesting device is the 3G wireless router, which can distribute an HSPA connection to several computers through cable or Wi-Fi. The location of the CPE should be preferably close to windows to reduce indoor penetration losses. Figure 3.6 illustrates some HSPA CPEs.

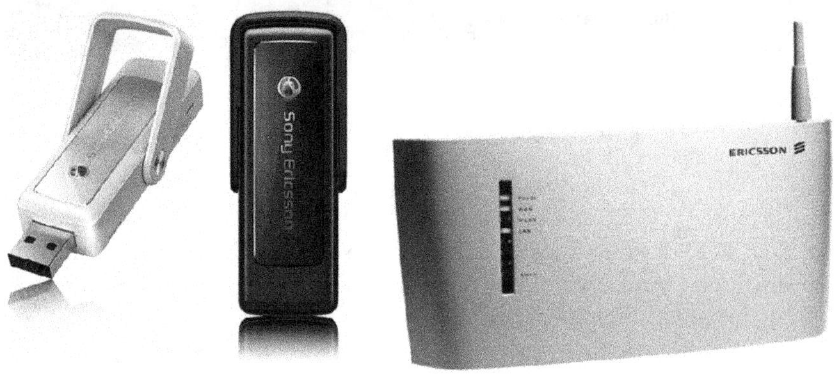

Fig. 3.6 Examples of HSPA CPEs.

3.2.4.1 HSDPA Devices

HSDPA devices are divided into 12 different categories according to the maximum bit rate supported [8]. The peak data rate of 14 Mbps is achieved in a category 10 UE with 15 parallel codes, which also supports 16-QAM modulation. Most of HSDPA devices commercially available are also backward compatible with UMTS and GSM/GPRS/EDGE networks.

3.2.4.2 HSUPA Devices

Similar to HSDPA, HSUPA devices are divided into six different categories based on the number of parallel codes, minimum spreading factor, TTI length, and peak bit rate supported [8]. The number of parallel codes ranges from 1 to 4. For instance, the peak bit rate of 5.7 Mbps is achieved in UE category 6 with minimum SF=2, 4 parallel codes and a TTI of 2 ms. It is worth noting that when four codes are transmitted in parallel, two codes shall be transmitted with SF2 and two with SF4.

3.2.5 HSPA Radio Resource Management Fundamentals

The purpose of this section is to present the fundamental aspects and key radio resource management (RRM) functions in HSPA.

3.2.5.1 HSPA RRC States

In HSPA, there are five different types of connection states: idle mode, where there is no connection established, and four different types of connected states or RRC (radio resource control) states depending on the UE activity [11]. These different states indicate the level of UE connection and the transport channels that can be used by the UE. RRC states are illustrated in Fig. 3.7.

The Cell-DCH state is when a dedicated physical channel is allocated to the UE in uplink and downlink. In case of HSPA the UE will also monitor the HS-SCCH channel in case data are scheduled in the HS-PDSCH. When all dedicated channels have been released, the UE goes into the Cell-FACH state. In this state the UE

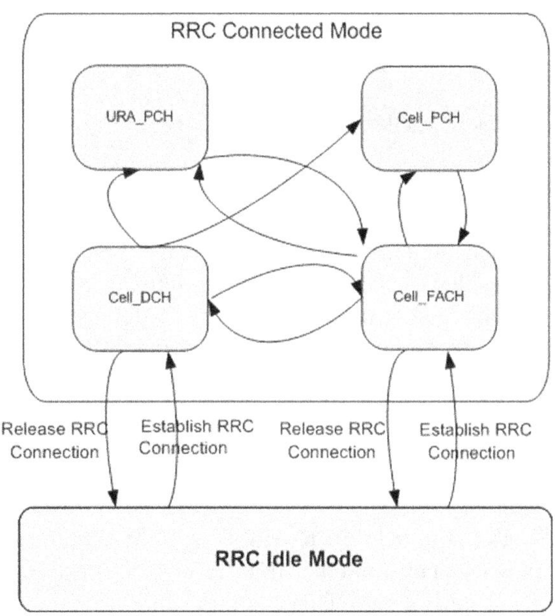

Fig. 3.7 HSPA RRC states.

continuously listens to the FACH, and a small volume of data can be transmitted using common transport channels.

In Cell-PCH and URA-PCH states, the UE is monitoring the PCH (paging channel). In that case, no dedicated channel is allowed to the UE, and no data transfer is possible. The difference between cell-PCH and URA-PCH is that the location of the UE is known on a cell and UTRAN registration area (URA), respectively. An URA contains many cells and has been specified to lessen signaling when a location update is needed. One advantage of these states over Cell-FACH state is conservation of the UE battery.

It is worth noting that not all the states and transitions are supported by the UE or network equipment. For instance, a number of commercial UEs might not support either Cell-PCH or URA-PCH.

3.2.5.2 HSDPA RRM Fundamentals

In order to achieve high performance, HSDPA requires an efficient allocation of radio resources. Typical RRM features in HSDPA include scheduling, admission control, and mobility management. The function of giving access to the shared channel to various users is performed by the scheduler in the Node B, whereas functions such as admission control and mobility management are carried out by the RNC.

The maximum achievable bit rate on radio bearers transmitted on the HS-DSCH transport channel depends on the following factors: number of users in queue, buffer occupancy, number of available SCCH codes, radio conditions (based on reported CQIs), UE capability (i.e., UE category), available power for HS-PDSCH, and available number of HS-PDSCH codes.

Packet Scheduler

The Node B scheduler indicates which UE (or UEs) to transmit to in the next TTI. The number of UEs to be scheduled in the same TTI will depend on the number of available HS-SCCH codes, as each user requires at least one HS-SCCH code. Depending on the scheduling algorithm, radio conditions might also be used to prioritize users. In order to help the scheduling decisions in the Node B, the UE will send periodically the CQI on the uplink. The CQI report estimates transport block size (TBS) that can be transmitted to the UE using a certain assumed HS-PDSCH power and a block error probability less than 10% [4]. It is important to know that the algorithms used to predict the TBS (transport block size) are specific to UE manufacturers. An example of how the UE can derive the reported CQI is presented in [20].

The reported CQI provides the Node B with a measure of the UEs' perceived channel quality and the UE receiver performance. For the purpose of estimation of the channel quality indicator, the UE assumes a total received HS-PDSCH power as the sum of the received power of the P-CPICH (primary common pilot channel), a measurement power offset signaled by higher layers during call set-up, and a reference power adjustment depending on the UE category [4]. The total received power is evenly distributed among the HS-PDSCH codes of the reported CQI value.

The UE capability will determine the upper limits of parameters such as TBS or codes. The UE category is provided during the call set-up stage to the Node B. As previously mentioned, the reported CQI is calculated using a reference power which might differ from the available power in the Node B. So, for each HS-PDSCH transmission, it is the Node B that determines the final TBS, modulation type, number of allocated HS-PDSCH codes, and HS-PDSCH transmission power.

Downlink Transmit Power

Downlink transmit power is a shared resource between HS-DSCH and DCH. The percentage of power available for HSDPA has a considerable bearing on the performance and capacity. For instance, if the DCH power utilization is high due to speech users (e.g., 50% of the total power), the operator might consider the deployment of a second carrier for HSDPA to minimize degradation in performance.

There are different options for power allocation for HSDPA. For instance, assuming that a fixed power is allocated for HSDPA, DCH (e.g., speech) bearers would not be able to use the HSPDA power even if there are no HSDPA users in the cell. Another alternative is dynamic power allocation, i.e., HSDPA uses the remaining power after common channels and DCH are considered, as illustrated in Fig. 3.8. The available HS-PDSCH power can be calculated by the Node B on a regular basis (e.g., 2 or 10 ms). CQI reports received by the Node B can be adjusted to the available power.

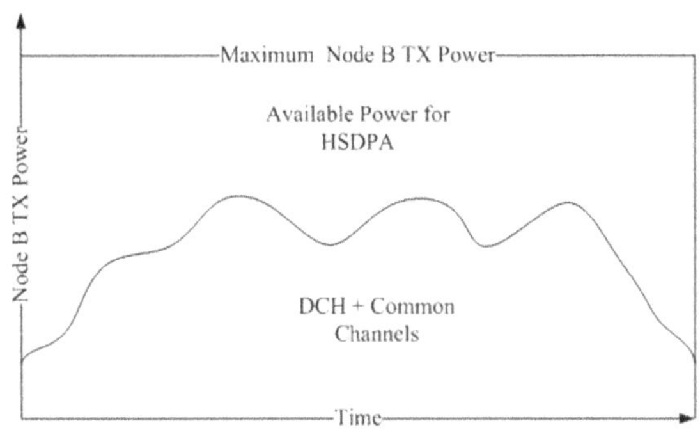

Fig. 3.8 Example of dynamic power sharing between HSDPA and non-HSDPA channels.

Code Allocation

HSDPA code allocation can also be static or dynamic. In the static case, a number of codes are reserved for HSDPA, while only codes unused by DCH can be allocated in the dynamic case. An algorithm will periodically check whether there are available codes free in the code tree and if so, they will be added to the HS-DSCH allocation.

CQI Adjustment

In order to protect the system against inaccurate CQI reports from the UE, CQI adjustment was implemented by some of the UMTS manufacturers. This is not mandatory in the 3GPP standards for UMTS. The CQI adjustment algorithm in the Node B processes the ACKs and NACKs received from the UE to determine if the UE is overestimating or underestimating the channel quality. Then the algorithm will adjust the CQI accordingly. Fast link adaptation in HSDPA always aims to achieve a block error rate of 10% for the initial transmissions, i.e., excluding retransmissions.

HSDPA Mobility

HSDPA mobility management is performed by the RNC. Unlike DCH, the HS-DSCH channel does not support soft-handover, i.e., it is established and maintained in only one cell. The cell associated with the serving HS-DSCH radio link is defined as the HS-DSCH serving cell. The procedure to change the serving cell is called HS-DSCH serving cell change, and it aims to keep the best cell in the active set as the HS-DSCH serving cell. HSDPA physical channels in the serving cell and non-serving cell are shown in Fig. 3.9.

The HS-DSCH serving cell change procedure uses the reporting event 1d, i.e., change of best cell. This measurement basically reports the best serving HS-DSCH cells to the RNC based on measurements of the reference channel, i.e., the P-CPICH. The selection of candidate cells for serving HS-DSCH cell is based on the two measurements below [5]:

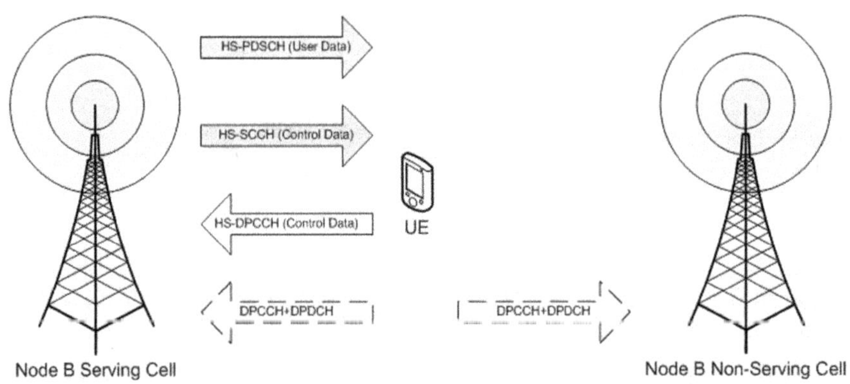

Fig. 3.9 HSDPA physical channels in the serving and non-serving cells.

- **P-CPICH E_c/I_0**: the received energy per chip for the P-CPICH divided by the power density in the band
- **P-CPICH RSCP**: the received signal code power measured on the P-CPICH

Let us assume E_c/I_0 is our measurement quantity. Then the event Change of best cell (event 1d) is triggered when [11]

$$(E_c/I_0)_{NOT\_BEST} \geq (E_c/I_0)_{BEST} + H_{1d}/2. \tag{3.1}$$

The variables in the formula are defined as follows: $(E_c/I_0)_{BEST}$ is the CPICH E_c/I_0 measurement of the serving cell; $(E_c/I_0)_{NOT\_BEST}$ is CPICH E_c/I_0 measurement of the non-serving cells in the active set; H_{1d} is the hysteresis parameter for the event 1d. It is an extra margin to limit the amount of event-triggered reports.

The report is triggered only after the conditions for the event 1d have existed for the specified time. It means that the event is not reported until it has been within the range for the time given by the time-to-trigger parameter. This aims at limiting the measurement signaling load. Figure 3.10 illustrates the hard-handover (i.e., cell change) procedure when there are two cells in the UEs active set.

The HSDPA serving cell change procedure can be either synchronous or asynchronous. Asynchronous reconfiguration implies that the involved Nodes B will obey the reconfiguration message as soon as it is received, and any untransmitted data in the MAC-hs buffer of the source Node B will be discarded, i.e., the RLC retransmission will be required in this case. In the synchronous case, an activation time is defined in the reconfiguration message, which specifies when the source Node B shall stop the MAC-hs scheduling on the HS-DSCH on the serving cell and when the target Node B shall start the MAC-hs scheduling on the HS-DSCH in the target cell. This allows the Node B an attempt to transmit any remaining data before

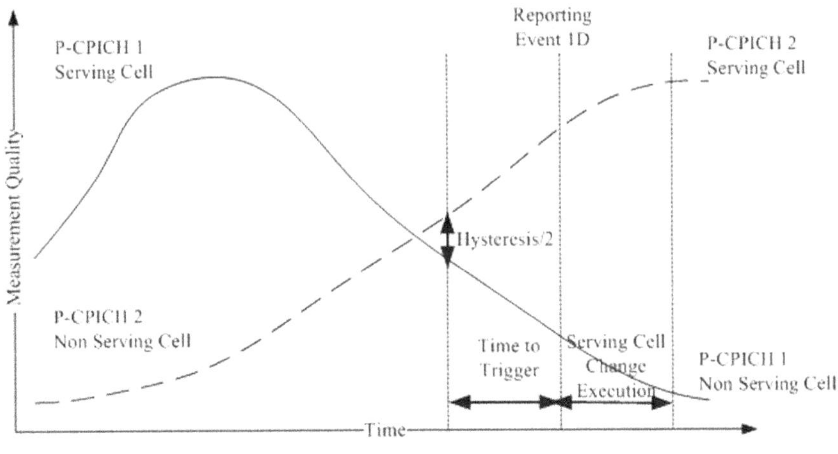

Fig. 3.10 A P-CPICH becomes better than the previously best P-CPICH.

the cell change occurs, which should minimize packet losses and RLC retransmissions. Therefore, synchronous reconfigurations are preferred for HS-DSCH serving cell change.

3.2.5.3 HSUPA RRM Fundamentals

Like HSDPA, the scheduling in HSUPA is done in the Node B, whereas admission control and mobility management are carried out in the RNC. The uplink radio resource controlled by the scheduler is the interference or uplink load.

A maximum noise rise threshold is set in the cell by the RNC to protect the existing users from service degradation caused by increasing interference. Each user contributes a certain amount to the total interference, therefore the maximum noise rise threshold can be seen as a common resource shared among the UEs. The Node B periodically measures the total interference and provides the scheduler with an estimation of the interference headroom available for the E-DCH channel. This headroom is the potential interference power that may be consumed on top of the thermal noise, the interference created by the DCH channels, and the inter-cell interference created by other sources (users connected to other cells, external interferers, and so on).

There are two different ways to configure the data flow: scheduled or non-scheduled modes. In the non-scheduled mode, the UE is configured once at call set-up and data flows independently; it could, for example, be used to carry voice over IP (VoIP) bearers minimizing delay. In the scheduled mode, the UE is required to adjust the bit rates periodically.

The Node B controls the uplink load dynamically through absolute grant and relative grant messages. Based on these messages, the UE updates the serving grant which represents the maximum E-DPDCH to DPCCH power ratio the UE may use in the next transmission. Expressing the serving grant as a maximum power ratio is motivated by the fact that the fundamental quantity the scheduler is trying to control is uplink interference, which is directly proportional to transmission power. The E-DPDCH transmission power is defined relative to the DPCCH to ensure the E-DPDCH is also affected by the power control commands.

The serving grant is used as basis for selecting the appropriate E-DCH transport format combination (E-TFC), thereby determining the data rate to be used for uplink transmission. The UE, E-TFC selection is performed given capability and power constraints according to [10]. In summary, the higher the power offset, the higher the data rate.

The absolute grant allows the Node B scheduler to directly adjust the granted rate of UEs under its control. The absolute grant is used to initialize the serving grant. The relative grants are used to incrementally adjust the UEs serving grants. For the serving E-DCH radio link set, it consists of transmissions of an "UP" or "DOWN" command. For the non-serving radio links, the relative grants are a means to avoid overload situations by transmitting a "DOWN" on E-RGCH.

If the UE receives a "DOWN" command, the serving grant goes one step down in the serving grant table. For the "UP" command there are different steps based on the threshold "3-index-step threshold" and "2-index-step threshold" which are configured by higher layers. Figure 3.11 shows an example of how the UE updates its serving grant based on the reception of relative grant "UP." The serving grant table definition is presented in [10]. It is worth noting that the corresponding index in the SG table is defined as SGLUPR.

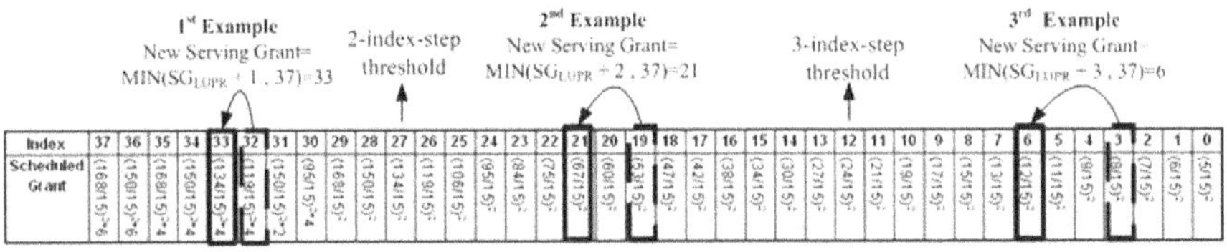

Fig. 3.11 UE received a serving relative grant "UP."

The UE has also a mechanism to inform the Node B whether the UE could use more resources or not. This is done through the "happy bit," which is a single bit field that is passed from MAC to the physical layer for inclusion on the E-DPCCH.

HSUPA Mobility

HSUPA mobility procedure works in the same way as HSDPA, i.e., the serving cell change procedure is performed in the same way as for HSDPA. The key difference is the support for soft-handover. In soft-handover situations, data are received independently in several Nodes Bs. The RNC, through the MAC-es layer, is responsible for reordering and macro-diversity selection [10]. HSUPA physical channels in the serving and non-serving cells are illustrated in Fig. 3.12.

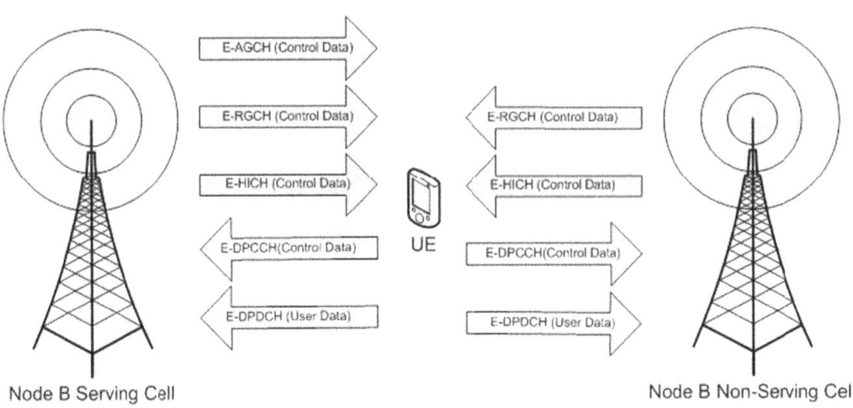

Fig. 3.12 HSUPA physical channels in the serving and non-serving cells.

Table 3.1 HSPA and DCH comparison.

Feature	DCH	HSDPA	HSUPA
Standard	Release 99 (R99)	Release 5	Release 6
Direction	UL/DL	DL	DL
Channel type	Dedicated (DCH)	Shared (HS-DSCH)	Dedicated (E-DCH)
Multi-code transmission	No*	Yes	Yes
TTI	10 or 20 ms	2 ms	2 or 10 ms
Adaptive modulation	No	Yes	No
Scheduler	At the radio network controller node	At the Node B	At the Node B
Error correction	RLC (layer 2)	ARQ (layer 1)	ARQ (layer 1)
Soft-handover	Yes	No	Yes
Peak bit rate	384 kbps	14 Mbps	5.7 Mbps

*Higher bit rates were supported by the specifications based on multiple parallel code transmission, but these were never implemented by any of the key manufacturers.

Based on the specifications [11], the E-DCH active set can be a subset of the DCH. A new event 1j was created for HSUPA, and it is triggered when a P-CPICH that is not included in the E-DCH active set but included in DCH active set becomes better than a P-CPICH that is in the E-DCH active set.

Table 3.1 summarizes the key differences among HSDPA, HSUPA, and the original WCDMA standard based on conventional DCH.

3.3 HSDPA Performance

The purpose of this section is to present different performance aspects related to HSDPA in the field and from simulation results.

3.3.1 HSDPA Simulations

HSDPA simulation is a good method for initial capacity estimation and, especially, for performance comparison of different RRM techniques. In the following example, a UE with five codes and support to both 16-QAM and QPSK modulation schemes was assumed. The traffic was considered to be 100% WWW (World Wide Web) type. The criteria for calculating the system capacity was to achieve a 90% satisfaction factor among the users. The user was considered satisfied if it was not blocked by the admission control and reached an average bit rate not smaller than 64 kbps. Table 3.2 shows a summary of the key assumptions used in this simulation.

Table 3.2 Key simulation assumptions.

Parameter	Value
Environment	Indoor
Number of cells	27
Cell radius	500 m
Node B transmit power	20 W
User speed	3 km
Number of codes reserved for HSDPA	5
Number of H-ARQ parallel process	6
Maximum number of H-ARQ retransmissions	5
Modulation	16-QAM and QPSK
WWW satisfaction threshold	90%
WWW session throughput satisfaction threshold	64 kbps

The HSDPA system capacity obtained from the simulation results based on the assumptions above was 1,116 Mbps/cell. Figure 3.13 shows the bit rate distribution by the Node B scheduler. This is an indication of the HSDPA performance across the cells.

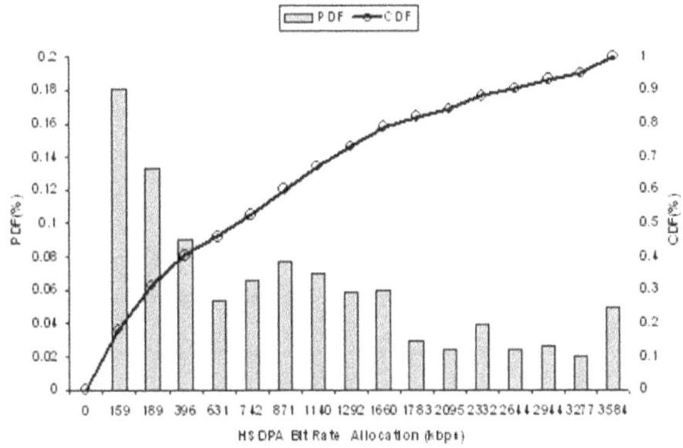

Fig. 3.13 HSDPA bit rate distribution from simulation.

3.3.2 HSDPA Field Trials: Bit Rates and Coverage

In this section, the performance results in terms of achievable bit rates obtained in the field are presented. The results were gathered from field trials. The selected area corresponds to a cluster of 21 sites with HSDPA activated in an urban area, as illustrated in Fig. 3.14.

Fig. 3.14 P-CPICH RSCP outdoor coverage plot for the trial cluster.

The drive testing was carried out in off-peak hours in a commercial UMTS network. The file transfer protocol (FTP) was the application used for this test; files were continuously downloaded from an internal FTP server. On the network side, 16-QAM modulation and the CQI adjustment feature were enabled for the trials. Three different scenarios were chosen according to Table 3.3.

Table 3.3 Test scenarios.

	Scenario A	Scenario B	Scenario C
HS-PDSCH Codes	5 codes	10 codes	10 codes
Uplink Bearer	DCH (up to 384 kbps)	HSUPA (to 1.4 Mbps)	DCH (up to 384 kbps)
HSDPA UE category	Category 6	Category 8	Category 8
UE type	Type 2 (equalizer)	Type 3 (equalizer and Rx diversity)	Type 3 (equalizer and Rx diversity)
Downlink Maximum bit rate (MBR)	3.6 Mbps	4 Mbps	4 Mbps

The results were obtained through drive testing, so the mobility aspect is included in the results. Figure 3.15 shows the measured UE P-CPICH RSCP distribution across the drive test route. In this case, the 95th percentile of the UE P-CPICH RSCP

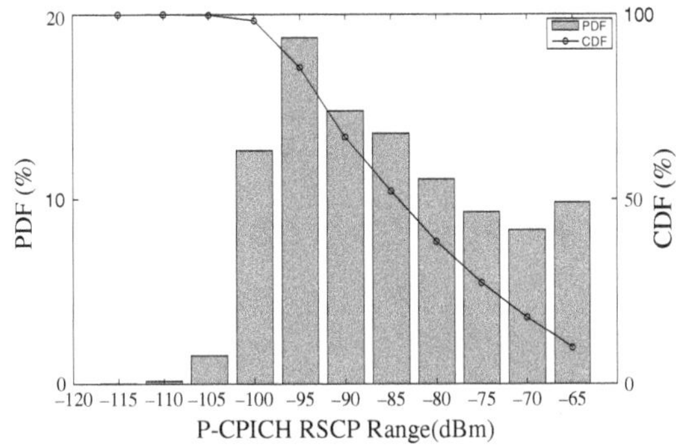

Fig. 3.15 Measured P-CPICH RSCP distribution for trial cluster.

distribution was around −98 dBm. As can be seen, this percentile corresponds to an area of relatively good coverage.

In Fig. 3.16 it is possible to see the single user bit rate distribution for the two scenarios. In Scenario A, 77% of the samples have a throughput greater than 1 Mbps. In Scenario B, approximately 96% of the measurement samples have an application throughput greater than 1 Mbps. This significant improvement can be attributed to several factors: double number of HS-PDSCH codes (10 codes), HSUPA, receiver diversity, and higher bit rate support UE category.

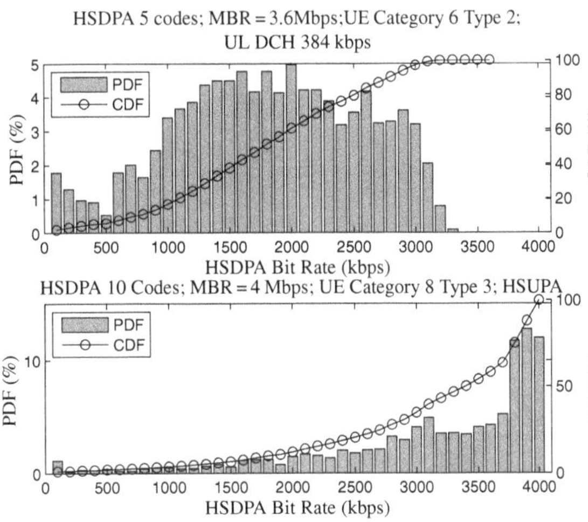

Fig. 3.16 HSDPA bit rate distribution for trial cluster.

Figure 3.17 shows the average bit rate versus RSCP on the move. As expected, the bit rate declines with a weaker radio signal. Finally, the average bit rates achieved in both mobile and stationary cases are presented in Fig. 3.18 for the different

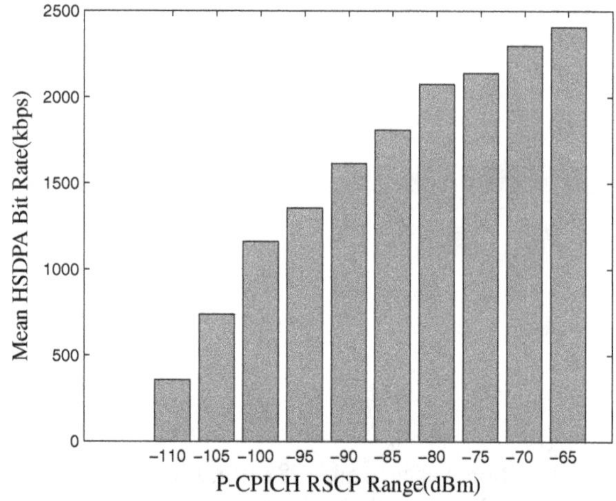

Fig. 3.17 HSDPA bit rate versus P-CPICH RSCP on moving for UE category 6.

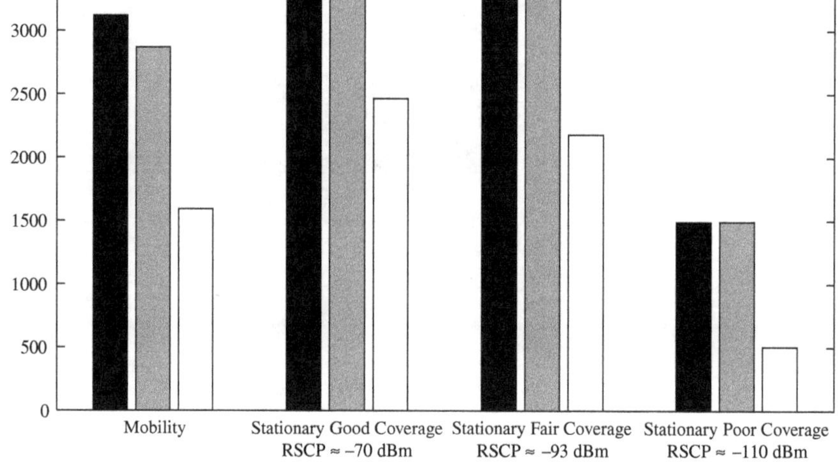

Fig. 3.18 HSDPA bit rates for mobile and stationary UEs.

scenarios. As can be seen, combined HSDPA 10 codes and HSUPA features provided the highest downlink bit rates for all the different scenarios, i.e., stationary and on the move.

3.3.2.1 CQI Adjustment Evaluation

An evaluation of this functionality was carried out using drive testing resulting in an average NACK rate of 10% when CQI adjustment was turned on and 24% when it was turned off. CQI adjustment gives a significant improvement in the average NACK rate. From the results it can be concluded that this specific UE overestimates the channel quality and hence reported CQI values are too high. The Node B will then assign higher bit rates to the UE which leads to a NACK ratio higher than the expected 10.

With CQI adjustment ON, the radio resources are being used more efficiently as a block error rate of 10% provides a better spectrum-efficient transfer of data over the air interface. It is also likely that different UE types will estimate the channel quality differently and hence report different CQIs for the same actual channel quality. Different UEs might also have different receiver capabilities and which also might lead to a difference in reported CQI value. CQI adjustment functionality is required to minimize the negative system impact due to inaccurate CQI reports. It also presumably evens out the potential difference in UE performance if they estimate the CQI differently.

Figure 3.19 shows an example of the impact of the CQI adjustment feature. It consists of the reported and used CQI distributions measured in the field when CQI adjustment is turned on. In this example, it can be seen that the UE overestimates

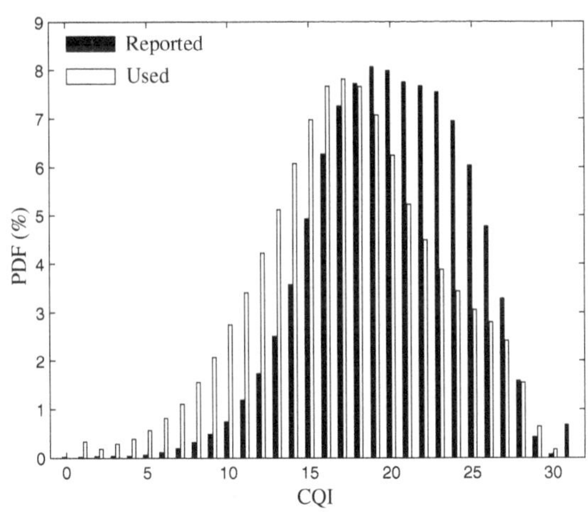

Fig. 3.19 Comparison between reported and used CQI for CQI adjustment ON.

the CQI when in good radio conditions (high CQI values) and underestimates under poor radio (i.e., low CQI values).

3.4 HSDPA Field Trials: Mobility Issues

HS-DSCH serving cell change causes a break in the data flow, i.e., RNC stops sending downlink data before cell change. From the application performance perspective this switch does not drop the connection; however, bit rates reduce during the cell change execution. Figure 3.20 shows an example of the performance during mobility. During the serving cell change procedure the user bit rates drop to lower values.

The potential reasons behind bit rate dips during HSDPA serving cell change procedure are as follows:

- Activation time: If the activation time is set far enough ahead to ensure that all data from the MAC-hs buffer have been transmitted, a significant delay can be introduced. If it is too short an RLC retransmission might be required as packets are dropped in the RNC.
- Degradation on P-CPICH E_c/I_0 due to lack of dominant server, which leads to low CQI reported by one UE (leading to lower bit rates allocated).
- Lack/low volume of data in the MAC-hs buffer in the target Node B just after serving cell change.
- UE delays on starting the decode of the HS-SCCH on the target Node B.

Optimization of serving cell change parameters such as event 1d threshold and activation time could potentially improve the bit rates in soft-handover areas, but would not eliminate this issue.

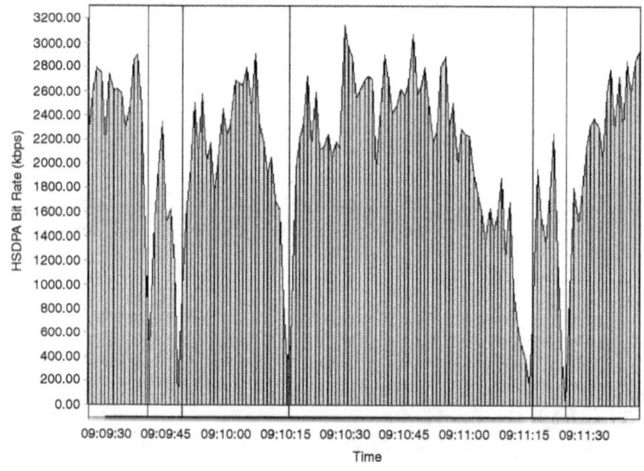

Fig. 3.20 HSDPA bit rate during serving cell change.

3.5 HSUPA Results: Field Trials

The increased data rates for HSUPA come from the new features previously described, plus the implementation of multi-codes transmission in the uplink. Obviously these bit rates are also dependent on several other factors, such as the maximum allowed uplink noise rise, the UE category, DCH R99 uplink loading, number of users in the queue, radio conditions, and network support.

The results presented below correspond to a HSUPA category 5 (2 Mbps). However, at the time these trials were carried out, the maximum bit rate (MBR) was limited to 1.4 Mbps. This trial was carried on a commercially deployed network, but the test was performed during off-peak hours (lowest traffic level). In this case, maximum allowed uplink noise rise was not the limiting factor.

Figure 3.21 shows the average user throughput distribution under mobility. The distribution graph shows a fairly stable user throughput with around 86% of the samples greater than 500 kbps.

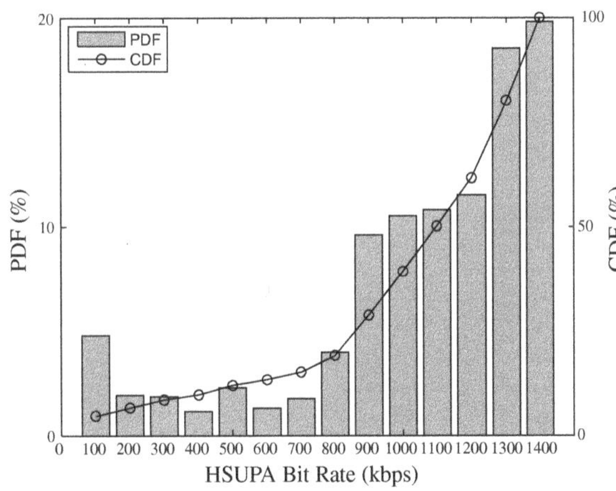

Fig. 3.21 HSUPA bit rates distribution for the trial cluster.

Figure 3.22 shows a comparison between the user HSUPA and DCH (at 384 kbps) uplink bit rate. The comparison is done in the same range of P-CPICH RSCP distribution. As can be seen, HSUPA provides higher bit rates in similar radio conditions.

Finally, the HSUPA mobility and stationary results are summarized in Figure 3.23. As can be seen, HSUPA could deliver average uplink bit rates greater than 500 kbps in all scenarios.

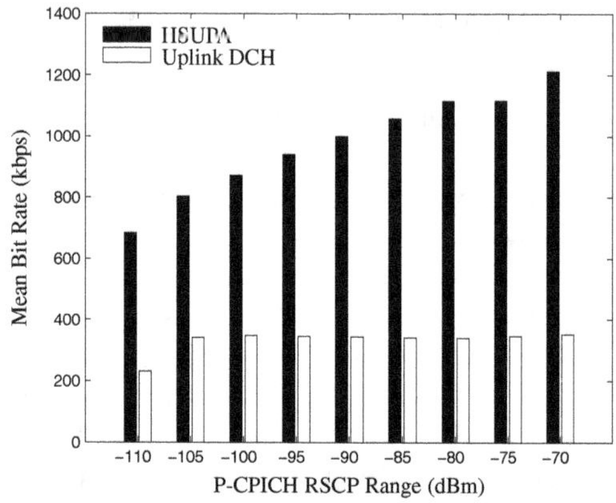

Fig. 3.22 Uplink bit rates versus P-CPICH RSCP.

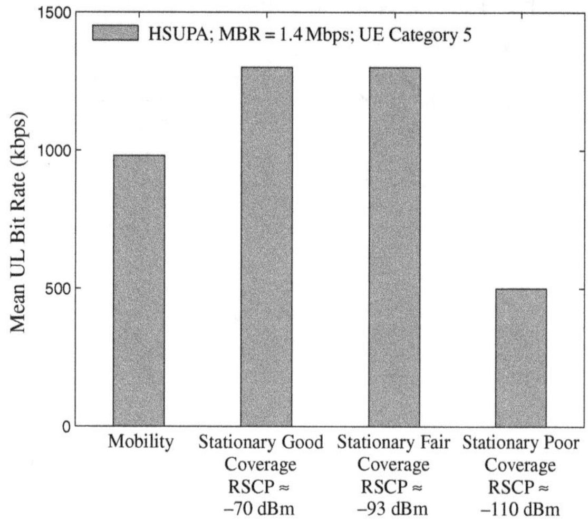

Fig. 3.23 HSUPA bit rates in mobility and stationary cases.

3.6 Applications Performance over HSPA

TCP/IP is a set of protocols developed to connect a number of different networks designed by different vendors together to the Internet. It uses the IP address, which uniquely identifies the connection of a host to the Internet. These complete suites of dominant protocols in the Internet include internet protocol (IP), transport control protocol (TCP), and user datagram protocol (UDP).

Other protocols were developed for doing specific tasks such as transferring files between computers, sending mail, and web browsing. Those are known as application layer protocols and are located over IP (layer 3) and UDP/TCP (layer 4) in the protocol stack. In this section, an introduction to the performance of these protocols over HSDPA will be covered.

3.6.1 TCP/IP Introduction

Traditionally there are two different transports: UDP and TCP. UDP is non-connection oriented and TCP is connection oriented. UDP is suitable for real-time multimedia applications (e.g., VoIP) due to the delay constraints of these types of services. This section focuses on key TCP characteristics, but more details about the TCP/IP can be found in [18].

TCP establishes a virtual path between the source and the destination and uses the services of IP to deliver individual segments to the receiver, but it controls the connection itself. TCP is responsible for verifying the correct delivery of data from client to server. Data can be lost in the intermediate network. TCP adds support to detect errors or lost data and to trigger retransmission until the data are correctly and completely received.

The connection establishment in TCP is called the three-way handshake. A client program sends a connection request (SYN) to a particular TCP server, including the initial TCP sequence number. The server acknowledges the connection (SYN, ACK) and displays the next sequence number that it expects to receive from the client together with the TCP window size to be used by the client. Finally the client sends a third segment acknowledging the receipt of the second segment and sends the client's receiver window size.

After the TCP connection is established data transfer can start. Client and server send and acknowledge data in both directions.

3.6.2 TCP Performance

TCP was not originally designed for wireless networks. Lost packets are attributed to congestions, and this results in severe reduction in bit rates as a precautionary mechanism. There are three key features in TCP that can have a significant impact on performance: flow control, error control, and congestion control.

3.6.2.1 Congestion Control and Retransmission

The flow control regulates the amount of data a source can send before receiving an acknowledgment from the destination. The maximum size of this window in one end is calculated as the minimum between the receiver's advertised window (RWND) and the sender's congestion window (CWND).

Error control provides reliability since it has a mechanism to detect and correct erroneous, out of order, or lost segments. In the TCP Reno implementation, a retransmission takes place if a retransmission timer expires or the sender receives three duplicate ACKs, as illustrated in Fig. 3.24.

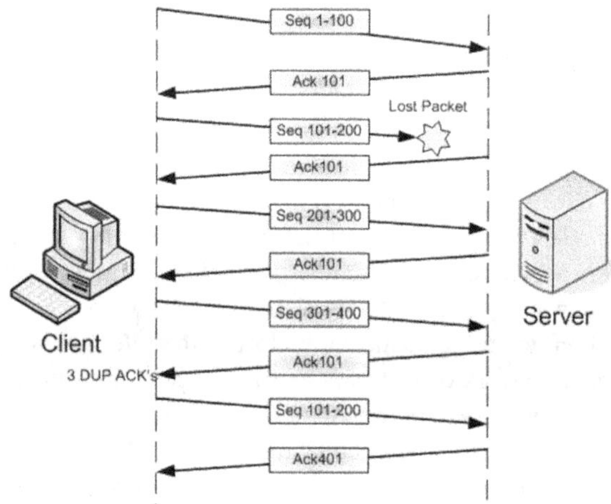

Fig. 3.24 TCP fast retransmission (three duplicate ACKs).

Congestion control is the mechanism to ensure that the sender is not overloading the network with packets. The congestion control actions taken by TCP are as follows:

- At the beginning of a new TCP connection the congestion window (CWND) is typically initialized to one segment.
- The sender begins the "slow start" phase. TCP slow start operates by increasing the CWND exponentially each time an ACK is received until it reaches a certain threshold (slow start threshold).
- Once the slow start threshold is reached, the sender moves to the congestion avoidance phase, which increases CWND additively instead of exponentially.
- The sender will transmit up to the minimum of the CWND and the advertised window (RWND).
- A new TCP slow start phase starts after a retransmission time out (RTO). RTO is dynamically corrected according to the measured round-trip time. Its actual calculation is based on the smoothed RTT and its deviation.
- A new congestion avoidance phase starts after receiving the third duplicate ACK. The CWND is reduced by half when a retransmission takes place.

Figure 3.25 shows an example of how congestion control works in TCP and its influence on the CWND size.

A simple formula to estimate the maximum achievable bit rate by an individual TCP flow is presented below. This should be seen in practical terms as an upper bound on the performance of TCP as the formula is not expected to apply under

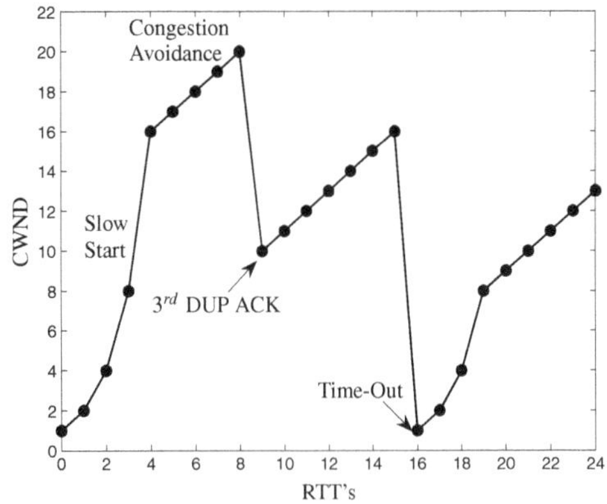

Fig. 3.25 TCP congestion control Example.

a number of situations where congestion avoidance does not fully control the TCP performance. The maximum TCP throughput achievable by an individual flow is given by [27]

$$BW = \frac{K \times MSS}{RTT \times \sqrt{p}} \quad [\text{Bps}], \tag{3.2}$$

where K is a constant typically equal to 0.93; MMS is the TCP segment size being used by the connection in bytes; RTT is the end-to-end round-trip time of the TCP connection (in seconds), and p is the packet loss rate for the path.

3.6.2.2 TCP in Wireless Systems

TCP was originally designed for wired networks. For instance, packet losses are assumed by TCP to be due to network congestion, and TCP stack uses the flow control mechanism to slow down the data transmission allowing the network to recover from the congestion situation. This assumption is reasonable for wired system where error rates are very low; however, this is not the case in the wireless system. The key factors that affect the TCP performance in wireless system are highlighted below:

- **Higher packet losses**: Packet error rate due to transmission over the air interface is typically higher than wired system. However, the typical TCP implementations, such as TCP Reno, do not have means to differentiate congestion from transmission errors.
- **Mobility**: User's mobility can introduce break in transmissions, i.e., data can be lost when the user moves from a serving cell to another. In this case, the performance will be further degraded due to the triggering time outs and shrinking TCP sender window.

There are a number of proprietary solutions to enhance TCP performance over wireless focused on minimizing the effect of packet losses in the TCP by fast error detection and recovery and also to adjust TCP algorithms for handling better the mobility aspects [13, 19]. A brief description of some key applications and how their performances are affected by TCP over HSPA are provided in the following sections.

3.6.3 Internet Applications

This section reviews some of the main applications of mobile data and explains how they perform in HSPA.

3.6.3.1 Domain Name System (DNS)

Domain name system (DNS) is used to map a name to an IP address. It consists of two types of messages: query and response. A DNS message is probably the first message sent from the user end device when access to the World Wide Web is requested.

The response times will vary depending on the RRC states of the UE. When the UE is in the idle mode, typically the first message can take from 1.5 to 2 s. If the first query failed, then DNS typically waits double the time before repeating the query and so on until reaching the maximum number of re-attempts. The DNS time out settings is configured in the personal computer or laptop. Figure 3.26 shows the average time for the DNS response messages for different RRC states.

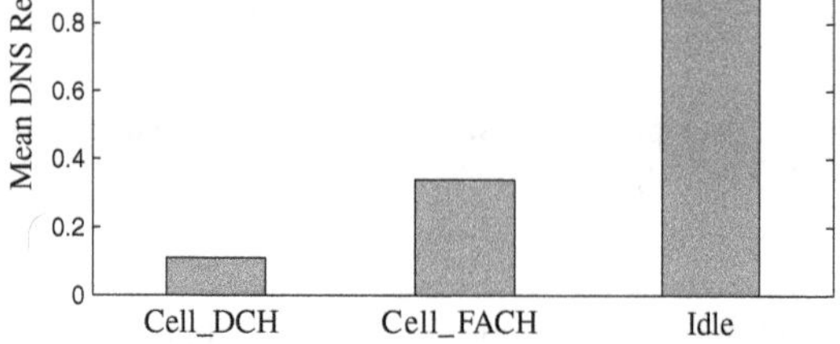

Fig. 3.26 Mean DNS response times.

As can be seen, the DNS response time is nearly 2 s when the UE comes from idle mode. In order to minimize the possibility of a time out and consequently an excessive delay, two approaches might be considered: first option is to extend the time that the UE stays in Cell_FACH and the second would be to activate either Cell_PCH or URA_PCH mode. The first option has the drawback that the UE will consume more battery in Cell_FACH than in the other two modes. In contrast Cell_FACH, the Cell_PCH and URA_PCH are not supported by all the UE manufacturers.

3.6.3.2 Hypertext Transfer Protocol

HTTP (Hypertext Transfer Protocol) is a protocol used mainly for access to information on the World Wide Web. Typically, an HTTP client initiates a request by establishing a TCP connection to port 80 on a host. An HTTP server listening on that port waits for the client to send a request message. Upon receiving the request, the server sends back a response.

HTTP messages are usually short, but TCP connection overhead (three-way handshake) might be considerable. HSDPA has implicitly improved latency and bit rate compared with DCH, resulting in better web browsing user experience. Figure 3.27 shows the download times distribution for both DCH and HSDPA on the move for a medium-size HTTP page (275 kB).

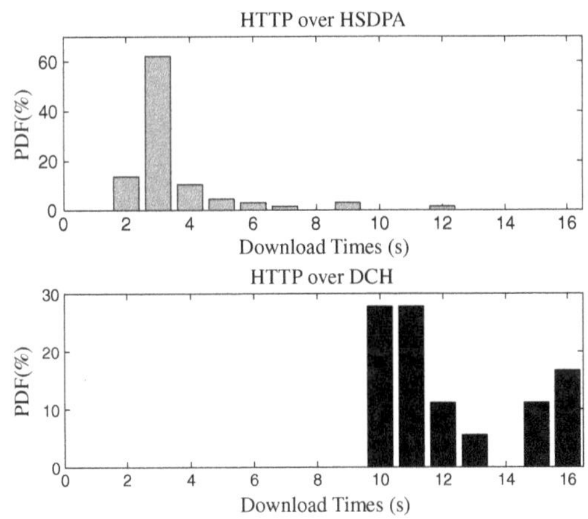

Fig. 3.27 HSDPA and DCH (R99) download times comparison for HTTP page.

3.6.3.3 File Transfer Protocol

FTP is a standard protocol for exchanging files from a host to another one that operates over TCP/IP. It needs two TCP connections, and port 21 is used for control connection, while port 20 is used for data connections. FTP client and server programs, which run in different machines, communicate to each other regardless of which operating systems are involved.

FTP is mainly throughput driven, and although HSDPA can provide high bit rates in the field as presented in the previous section, it is important to ensure that this is not limited by TCP settings.

Figure 3.28 shows the average and peak throughput achieved in a lab environment for different TCP receiver window sizes. As can be seen, the best average application throughput for a UE category 6 was achieved using a 64 Kbyte TCP receiving window. Lower TCP receiver window sizes (e.g., 16 Kbytes) can limit the FTP download bit rates that the user will be able to achieve.

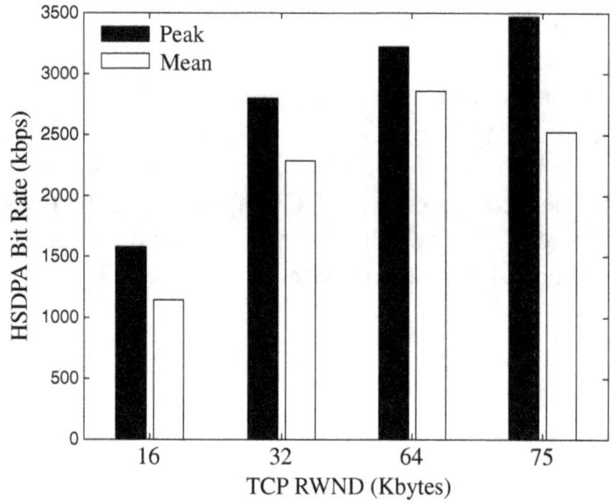

Fig. 3.28 TCP impact on average and peak FTP bit rates (lab measurement).

3.6.3.4 Latency-Sensitive Applications

Ping (Packet Internet Groper) is an application designed to diagnose network problems, which can also be used for determining the RTT (round-trip time) for an IP datagram to travel between two machines (host to router, router to router, etc.). It uses the ICMP (Internet Control Message Protocol), which sends "echo request" packets to the target machine and listens for "echo response" replies. Ping estimates the round-trip time and records packet loss statistics.

HSDPA has initially enhanced the UMTS round-trip time. Further improvements were followed with the introduction of HSUPA. Figure 3.29 shows the measured RTT (ping size=32 B) for the three different combinations on the downlink and uplink: DCH/DCH, HSDPA/DCH, and HSDPA/HSUPA.

As can be seen the HSDPA/HSUPA provides the lowest RTT. This should improve the user experience in applications that have high requirements on response times such as gaming and VoIP.

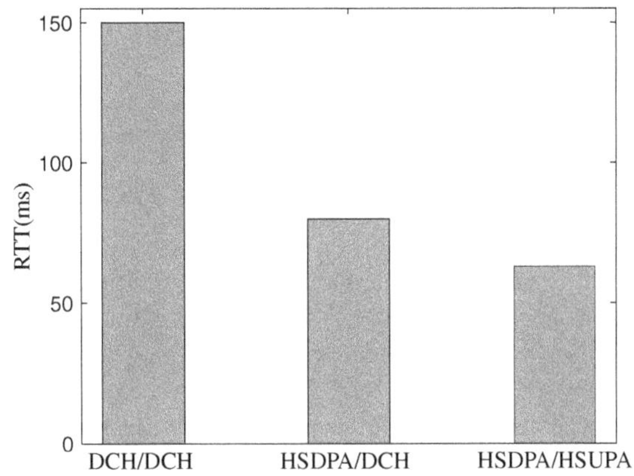

Fig. 3.29 RTT time comparison for ping size=32 B.

3.6.3.5 Other Applications

Peer-to-peer (or "P2P") file-sharing applications are growing in popularity. It uses the connectivity among network users and the cumulative bandwidth to share contents rather than conventional centralized resources. This type of application could potentially generate high volumes of data in both uplink and downlink directions, increasing the demand for capacity. Although the majority of the traffic is concentrated on the downlink, in certain scenarios the uplink can be more impacted as it uses dedicated radio resources, and it has less radio capacity than the downlink. The P2P file-sharing application could cause more network traffic than any other applications. Therefore, the volumes, priority, and traffic handled should be considered by the operators, especially in the case of flat rate price plans. For instance, in order to avoid further investment in network infrastructure, the broadband provider could reduce the fraction of resources available for this type of traffic during the peak hours, so it would not compromise the experience of other HTTP users.

Video streaming is another popular application to watch live or pre-stored video content. The streaming applications traditionally run over UDP/IP protocol stack and the UE buffers a number of seconds of the streamed traffic before starting to play the video. The quality of the video depends on the encoding rate, with high-quality videos requiring higher encoding rate.

HSDPA can handle video streaming applications but the performance will vary depending on the video encoding rate, buffering period, and cell load.

In earlier implementations of HSDPA, bit rates were not guaranteed, i.e., it was a best effort bearer with no service differentiation. Nevertheless, with the traffic growth trend, applications like video streaming will require a certain quality-of-service (QoS) level (e.g., a guaranteed bit rate) in order to avoid degraded user perception. Traffic differentiation and QoS management is possibly based on the QoS management mechanisms available in the 3GPP standards. This feature is expected to be actually implemented by manufacturers as the traffic growth pushes operators into service differentiation as a means to keep customers satisfied with their perceived QoS.

3.6.4 End-to-End Performance Troubleshooting – Case Study

HSPA performance depends on a large number of factors. Many are related to the radio network (e.g., radio link quality, number of users, available capacity), while others depend on the core network, IP backbone, CPE type, and PC settings, among others.

The term "end-to-end performance" in the context of this section means that radio aspects and core network nodes are considered as part of the investigation.

The end-to-end performance issues can be classified into different categories: degraded throughputs, session drops, and poor connectivity, among others.

In general, data performance issues are more complex to investigate than speech ones. The traditional radio performance management counters are usually not enough and complementary measurements are required in specific nodes. In this section, an end-to-end performance troubleshooting case study regarding degraded throughput is described.

3.6.4.1 Case Study

Initially, it was observed in tests at the operator's laboratory that HSDPA performance was not satisfactory: the measured average user FTP download throughputs was mostly less than 900 kbps for both HSDPA category 6 (up to 3.6 Mbps) and category 12 (up to 1.8 Mbps) UEs. The reason for the poor performance was unknown and therefore this was the starting point of the end-to-end (e2e) performance troubleshooting.

The complete wireless network configuration is complex and requires a structured methodology to be able to identify cross-domain issues. A simplified diagram of the e2e chain composed of UTRAN and CN elements is shown in Fig. 3.30.

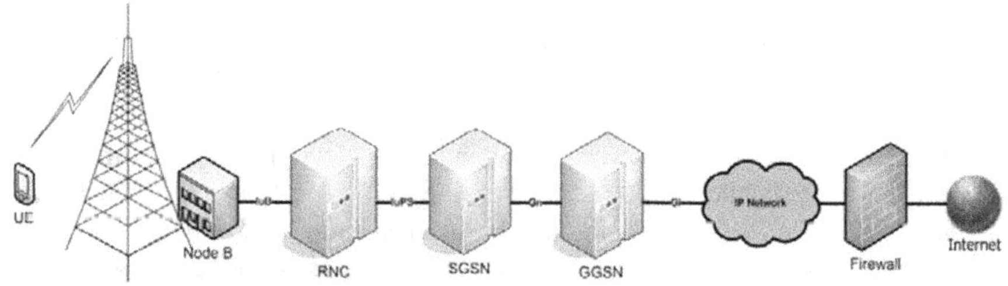

Fig. 3.30 Simplified network diagram.

The methodology used in this case was to exclude the well-performing nodes as candidates for the bottleneck, based on data collected from the key interfaces and specific test case results. To assist with evaluating the problem, several sources of data were used such as performance management counters, UE logs, IP sniffers, protocol analyzers, and internal system traces.

This investigation can be divided into three separate steps. The first step was to assess the HSDPA performance at the Node B as radio access is of concern. Subsequently, in the second step the investigation was extended across different domains. Finally the last step focused on a particular segment of the network once the problem was narrowed down. The three steps are outlined in more detail below.

HSDPA Node B and Radio Performance

A number of preliminary checks were carried out such as configuration and parameter inconsistencies, TCP settings, home location register (HLR) provision (e.g., maximum bit rate), UE benchmarking. However, none of the above mentioned was responsible for the degraded performance. The next step was to evaluate the Node B and radio performance. In order to do so, three test scenarios were chosen: in the first two, UE category is 12 while in the third one it is 6. Also, in the first scenario a single FTP download session was conducted while in the last two multiple sessions were considered.

A number of HSDPA key performance indicators (KPIs) derived from the Node B counters were pre-selected for this exercise, as highlighted in Table 3.4.

Table 3.4 HSDPA KPIs.

KPI	Description
Reported CQI	Measures the average CQI reported by the UE
Used CQI	Measures the average CQI allocated by the Node B
NACK rate (%)	Measures the number of NACKs divided by the sum of ACKs and NACKs received by the Node B. It can be considered as a BLER indicator for HSDPA
HSDPA MAC-hs efficiency (%)	Measures the percentage of time that the HSDPA sub-frames were transmitted containing user data during the period where the user buffer in the Node B was not empty
HSDPA transmission ratio (%)	Measures the percentage of time that HSDPA is effectively transmitting during the entire measurement period
Cell MAC-hs throughput (kbps)	Measures the cell-average DSCH throughput during the measurement period
User MAC-hs throughput (kbps)	Measures the user-average DSCH throughput during the measurement period
HSDPA Iub limiting ratio (%)	Measures the percentage of time that the bit rates were limited by the Iub bandwidth
HSDPA Iub frame losses (%)	Measures the percentage of frames lost in the Iub interface

The test consisted of continuous data transferring during the period of measurement. The HSDPA KPI results are presented below for the different scenarios.

Radio evaluation

The troubleshooting started with the CQI reports. These values range from 0 to 30, where 30 is the best channel quality indicator. For the three scenarios, the reported and used CQI were on average greater than 26 and 23 for UE category 12 and 6, respectively, as shown in Fig. 3.31. This basically proved that the user was in good radio conditions and was scheduled with a high transport block size. It is worth noting that NACK rate was around 10%, which was also a normal behavior, as shown in Fig. 3.32.

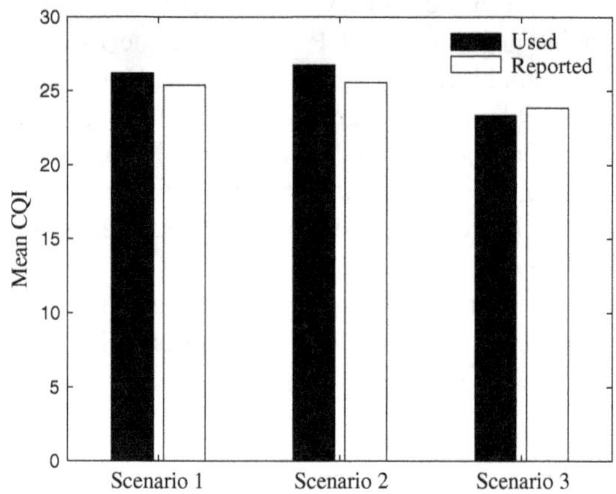

Fig. 3.31 Reported and used CQI.

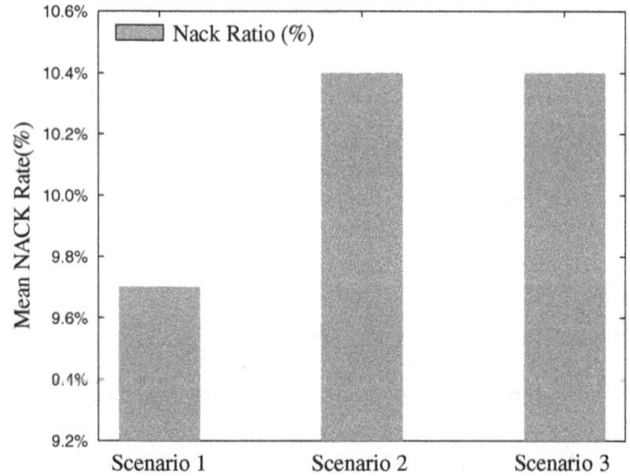

Fig. 3.32 NACK ratio.

Scheduler Efficiency

Figure 3.33 shows the MAC-hs efficiency and transmission ratio. MAC-hs efficiency was around 100% for all scenarios, which indicated that the Node B is sending data when data are available in the UE buffer as per normal operation. The transmission ratio for Scenario 1, however, was smaller than 60%. This was increased by approximately 37% once more FTP downloads were initiated (Scenario 2), suggesting that the UE was not fully utilizing the radio capability. This is due to lack of data in the user buffer, i.e., the Node B was not receiving enough data.

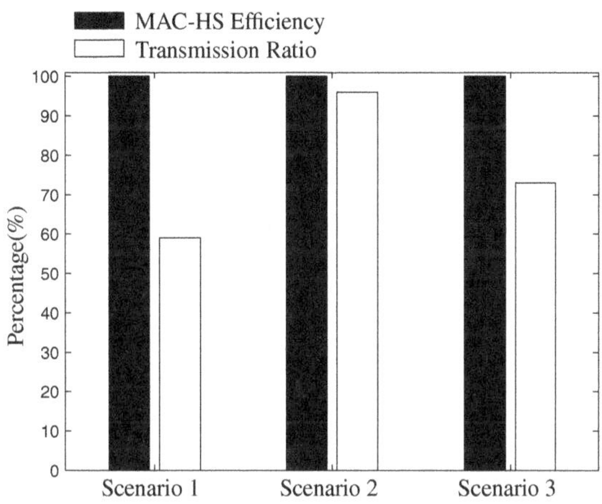

Fig. 3.33 Scheduler efficiency.

Cell and User Throughputs

Figure 3.34 presents the cell and user throughputs. There is basically no difference between them as this test was carried out in a controlled environment. This is just to show that when the Node B has data in the user buffer, it can deliver higher throughputs. This is more evident in Scenario 3 where the throughput in the air interface reaches 2,500 kbps for a HSDPA category 6 UE.

Iub Evaluation

Finally, the last two HSDPA KPIs refers to the Iub bandwidth and performance. It was found that the Iub limiting ratio is practically null, leading to the conclusion that the Iub bandwidth was not the bottleneck (no HSDPA frame was lost in the Iub for all the scenarios). In summary, it can be concluded that the Iub was not the reason for the poor performance.

Fig. 3.34 User and cell Throughputs.

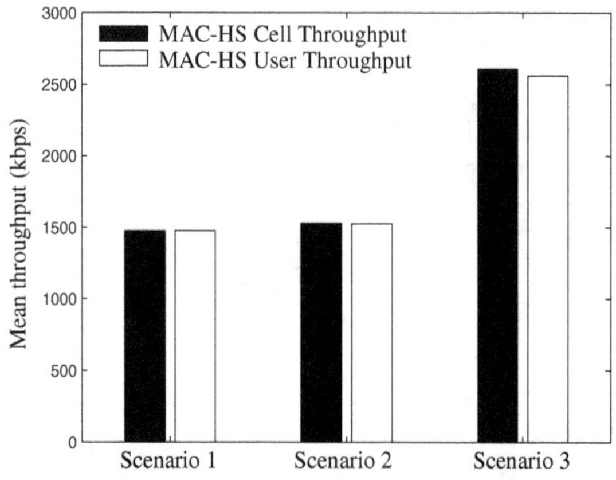

Partial Conclusion

In this first part of the investigation it was possible to conclude that the problem was related neither to radio Access nor to Node B nor to Iub interface. So the troubleshooting required extension across the e2e chain.

Cross-Domain Investigation

Traces were set up to extract the TCP/IP packets within a number of selected interfaces (Iu-PS, Gn, and Gi) as per Fig. 3.35. Moreover, UE traces and FTP server logs were also collected. For this test case a HSDPA category 12 device was used.

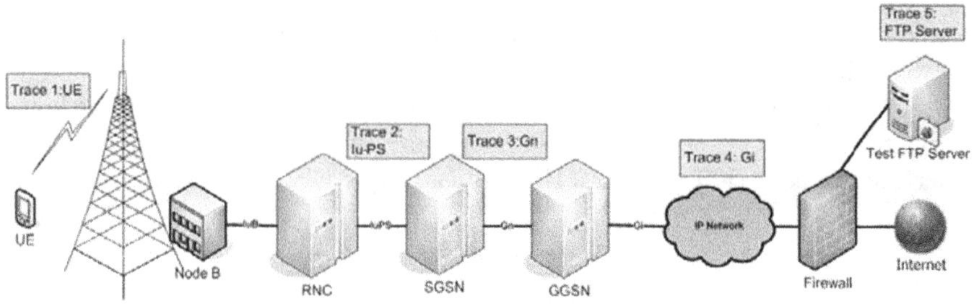

Fig. 3.35 Network interface traces options.

The results showed that a high rate of TCP/IP packets were either delayed or lost in the path, which was causing retransmission or even TCP time out. This would have an impact proportional to the level of throughput to be delivered, i.e., the higher the peak rates supported by the devices, the higher the impact of the packet loss.

For one typical TCP/IP session a packet retransmission ratio of around 1.5% was verified. At this level of retransmission ratio, it is not possible to attain, in a single TCP flow, the expected performance for a HSDPA CAT 12 UE, i.e., an average throughput of approximately 1.4 Mbps. This can be verified using (3.2).

After analysis of all the traces, it was apparent that the most likely problem was that the TCP/IP packets were lost before arriving in the Gn interface. Therefore, the focus of our investigation moved towards the gateway GPRS support node (GGSN) and the IP backbone.

IP Backbone Investigation

This last step was focused on the GGSN and IP backbone. The first action was to trace the path of the IP packet. This can be easily done by using trace route command in MS-DOS. Trace route sends a packet with incremental TTL to get information for all the hops. A diagram of the network was also available as illustrated in Fig. 3.36.

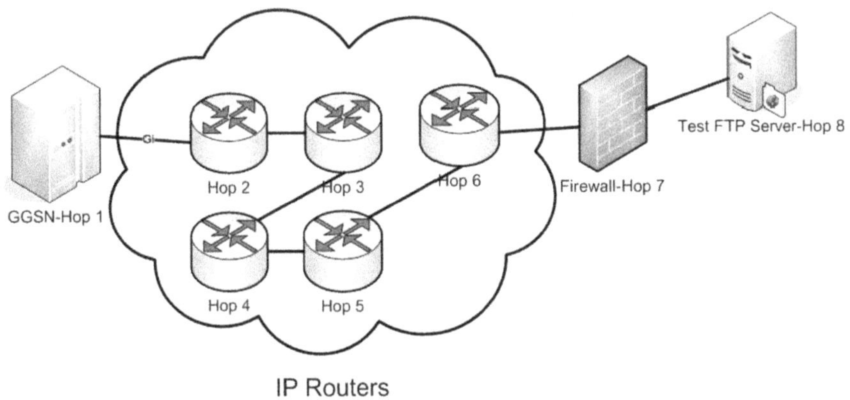

Fig. 3.36 IP network.

Once the hops were identified, it was possible to run a continuous ping to each of the Hops. The ping results showed a high packet loss rate on Hop 6 and nearly no losses on all other Hops. This indicated that the problem might rely between Hop 2 and Hop 6. In order to confirm this, a temporary FTP server was set up at the node immediately after the GGSN. The FTP results are presented in Figs. 3.37 and 3.38 showing a clear improvement over the original problematic situation. The average bit rate is now around 1.4 and 2.8 Mbps for HSDPA category 12 and 6 UE, respectively. So it was possible to conclude that up to the GGSN and IP router (Hop 2) the performance was good. Hence the problem began from the IP backbone onwards (Hop 3 onwards).

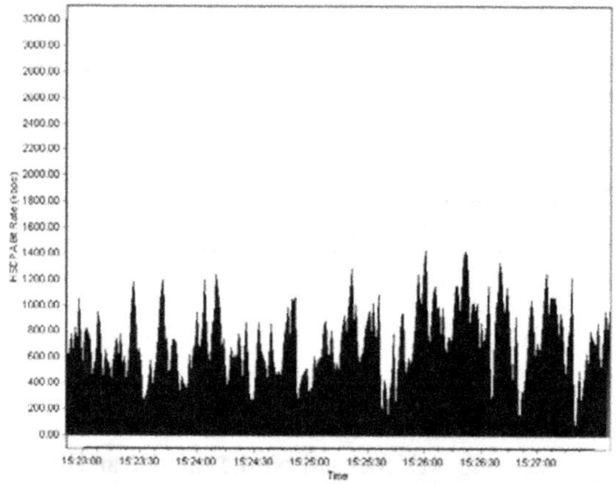

Fig. 3.37 HSDPA bit rate for UE category 6 before e2e troubleshooting.

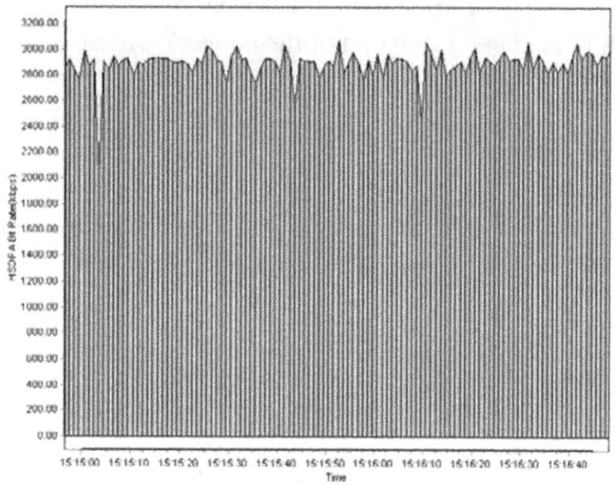

Fig. 3.38 HSDPA bit rate for UE category 6 after e2e troubleshooting.

Final Conclusion

After detailed analysis of the router interfaces, cyclic redundancy check (CRC) errors were presented in the WAN (wide area network) links between Hop 3 and Hop 4. After the router was replaced by the standby router, the problem was resolved with the expected bit rates achieved.

3.7 Capacity Planning

Since the launch of mobile broadband services with flat rates, data traffic has experienced considerable growth. However, this rapid increase in data traffic can result in a capacity bottleneck earlier than expected; therefore careful capacity planning is required. Residential broadband usage also shifts the peak hours from the typical network voice busy hour.

Rapid data traffic growth has been experienced by mobile operators since the launch of HSDPA; the higher data usage could also potentially be attributed to two factors: first the better user experience provided by HSDPA, e.g., the higher bit rates and second the flat subscription rates introduced by some operators. Figure 3.39 shows a typical example of the relative traffic growth for data and voice in a given 3G network. The simple activation of HSDPA resulted in over 75% increase in the 3G data traffic volumes. The launch of the mobile broadband product for the residential market, which reduced the data tariff to flat rates, and the fast uptake of new broadband customers have kept thc continuous growth of data traffic. It can also be seen that the data have increased much more than speech traffic throughout the same period.

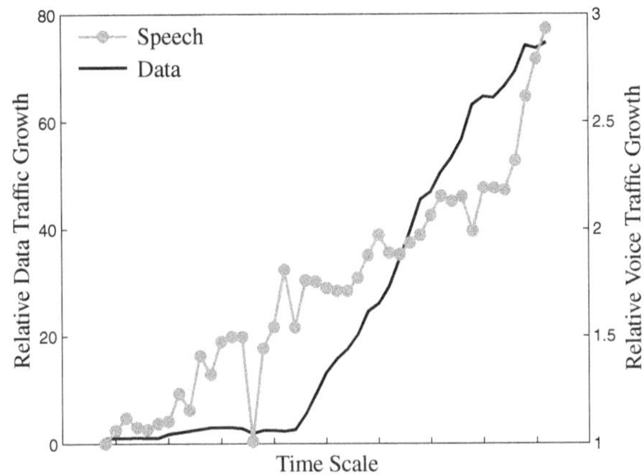

Fig. 3.39 Relative 3G traffic growth for speech and data.

Mobile broadband has affected the usage profile of the mobile network when it comes to Internet usage; most of the traffic is carried out of office hours during the week days and is more spread during weekends and holidays. Figure 3.40 shows typical data and voice usage across the day. It is worth noting that speech and data

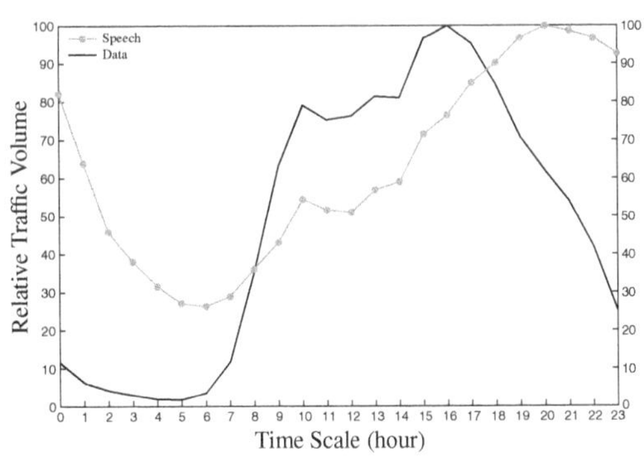

Fig. 3.40 3G speech and data traffic hourly distribution.

busy hours may be different. This can bring benefits in terms of maximizing network resources usage.

The point of interest in terms of capacity planning is that the type of applications used by your end users is likely to have a large impact on the type of network re-sources affected. For example, "keep alive" messages in certain types of applications could hold uplink-dedicated resources in the radio for a long period of time in case of non-optimized channel switching parameters. In summary, it is important to understand the dominant type of application in order to dimensionalize and parameterize your network.

Another key input in dimensioning is how the traffic is shared between downlink and uplink. This also depends on the type of application and the bit rates supported in both directions. The uplink share could expand with the increased usage trend of more peer-to-peer applications and higher upload bit rates supported with HSUPA. Nevertheless, it is expected that a typical 3:1 to 2:1 traffic volume ratio between downlink and uplink is kept for some time.

Finally, Fig. 3.41 shows an example of the monthly data usages per subscriber for a given month. As can be seen in this example, the majority of the subscribers (69%) have a monthly data usage of less than 1 GB, which corresponds to 10% of total traffic. However, 4% of the subscribers with usage greater than 10 GB are carrying 46% of the total traffic. This information can be used by the broadband service providers to define their strategy regarding fair usage policies, traffic shaping, pricing decisions based on usage, and hard limits on data volumes per users.

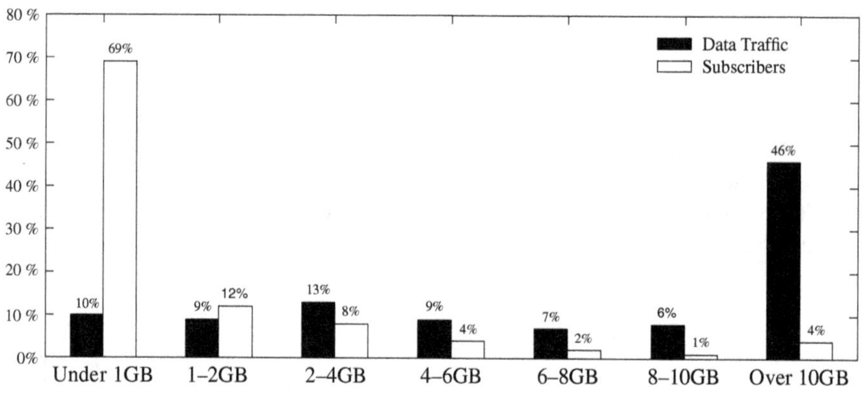

Fig. 3.41 Monthly average usage per subscriber.

3.7.1 Capacity Planning Guidelines for the Residential Market

Capacity planning for broadband services is a complex exercise and involves not only radio, but several other domains such as transmission, core networks, IP backbone, firewalls, and DNS servers. The radio capacity is usually the starting point of the dimensioning; nevertheless the other network nodes need to have the capacity to support the aggregate traffic. Figure 3.42 shows an example of a network diagram.

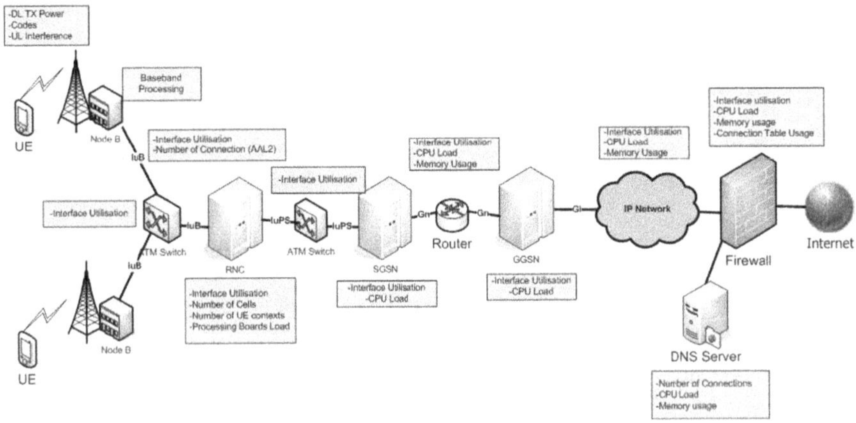

Fig. 3.42 Network diagram example with key capacity limitations.

The network elements and their key components regarding dimensioning and capacity monitoring are as follows:

- **Node B**: The downlink transmit power, the code usage, and the uplink interference consist of the radio capacity of the cell, whereas the baseband processing cards are related Node B equipment capability, and they determine the hard limit number of dedicated connections.
- **Transport**: Iub interface utilization at the Node B and asynchronous transfer mode (ATM) switches, and number of connections.
- **RNC**: In the data-centric traffic, the Iub and Iu-PS interface utilization and processing boards are the key items. However, the number of cells and the number of UE contexts should not be neglected.
- **service GPRS support node (SGSN)**: Iu-PS and Gn interface utilization and CPU load of the interface cards.
- **GGSN**: Gi and Gn interface utilization and CPU load of the interface cards.
- **Routers**: Interface utilization, CPU load, and memory usage are the key components.
- **Firewall**: Interface utilization, CPU load, memory usage, and connection table usage.
- **DNS server**: The number of connections, CPU and memory usage.

Radio Capacity

Radio capacity and coverage are interrelated in UMTS more than in other systems, i.e., the system capacity will depend on the size of the cell. The coverage is typically uplink limited, which determines the signal thresholds used for coverage planning. It is worth illustrating that the HSDPA radio capacity depends on several aspects. Therefore, some of the key factors impacting the HSDPA radio capacity are as follows:

- **Scheduler algorithms**: The scheduler selected can have an impact on the system capacity. Trade-off between capacity and fairness is essential.
- **UE types**: Higher data rates category UE with advanced receiver's techniques such as equalizer and two-way receiver diversity provide higher system capacity. Therefore, it is important that a high percentage of these devices are presented in the network.
- **Code multiplexing**: Enabling two or more HS-SCCH channels allows more than two users to be scheduled in the same TTI. This can increase the system capacity by enabling the scheduler to use more or even all the codes available independent of the UE category profile in the network.
- **HS-SCCH power**: Power control should be switched on for the HS-SCCH channel if more channels are enabled, otherwise the HS-SCCH is likely to require a level of power similar to the pilot channel, thereby reducing the power available for HS-PDSCH.
- **Available codes for HSDPA**: The system capacity will depend on the number of codes available for HSDPA. The maximum number of codes is 15; however, in order to deploy more than 5 codes for HSDPA in a single carrier, a dynamic code allocation is recommended. This means that HSDPA will not statically be reserved for the codes, which would increase the blocking probability of dedicated (DCH) connections (e.g., speech) due to code shortage. Instead, it will use the spare codes available at a given time, optimizing radio resource allocation.
- **Power available for HSDPA**: The percentage of power available for HSDPA is a key input to calculate the HSDPA cell capacity. If mixed services such as speech plus data are present, this will likely reduce the capacity available for the data as less downlink transmit power would be available for HSDPA. In the absence of the simulation results, a rough estimation could be assumed where the data capacity is directly proportional to the available HSDPA power.

However, assuming the coverage plan was already carried out and the network site count and locations are known, the HSDPA cell capacity obtained in simulations or field trials can be used to determine the number of subscribers that can be supported. A simplistic formula is used in [22] to estimate the number of subscribers:

$$V_{dm} = [(N_c \times C_c \times K_S/K_C) \times U_{dbh}/S_{dbh}] \times 30, \tag{3.3}$$

where V_{dm} [GB] is the total downlink traffic volume in a month per site; U_{dbh} is the percentage of network utilization during the busy hour (for example, 80%); S_{dbh} is the percentage of total daily downlink traffic that is carried during the busy hour (for example, 10%); C_c [bps] is the cell capacity (for example 1,116 kbps from simulations in Section 3.3.1); N_c is the number of cells per site (in case of only one carrier, it can be assumed as 3 due to sectorization); K_s [s] is a constant with the number of seconds in the hour, i.e., 3600 s; K_C is a constant for conversion of bps to gigabytes, i.e., $8 \times 1024^3 = 8,589,934,592$.

Applying the values from the paragraph above to 3.3, it can be derived that the total downlink traffic volume in a month is 334 GB per site. However, this corresponds to the downlink only. In order to calculate the total traffic, a typical ratio of 3:1 between downlink and uplink traffic may be assumed. Based on that, the number of subscriptions per site versus the monthly average usage per subscriber can be obtained. For instance, taking an average monthly usage per subscriber of 2 GB (corresponding to the usage pattern of over 80% of the users according to Fig. 3.41), an average capacity of 228 subscriptions per site can be estimated. It is worth noting that this estimation is based on a best effort service. If a minimum or guaranteed QoS is envisaged by the operators, then it could reduce significantly the number of subscribers supported by the system depending on the QoS target value.

3.8 Conclusion and Research Directions

HSPA is a great success as a wireless broadband enabler. Wireless broadband services growth has been significant and according to many forecasts, this trend will continue. High data usage combined with flat rates in the broadband services will require a lower operational cost from the operators and a high-capacity radio access network. Therefore, capacity enhancement will probably be the main focus of interest as far as HSPA networks' evolution and research are concerned. The increased capacity can be achieved through more efficient RRM algorithms, higher performance devices, and introduction of new features to improve spectral efficiency. Radio link improvements such as multiple input and multiple output (MIMO) antennas and 64-QAM modulation will be available to operators as part of 3GPP Release 7. Further specification efforts are aiming at a multicarrier version of HSPA in Release 8.

Finally, another topic that will increase in importance is QoS and service differentiation, driven by the necessity to offer an acceptable level of service to broadband users with different service-level agreements and subscription plans.

References

1. 3GPP Organizational Partners: 3GPP TR 25.848 Physical Layer Aspects of UTRA High Speed Downlink Packet Access, release 4 edn. (2001). Version 4.0.0
2. 3GPP Organizational Partners: 3GPP TS 25.211 Physical Channels and Mapping of Transport Channels onto Physical Channels (FDD), release 5 edn. (2005). Version 5.8.0
3. 3GPP Organizational Partners: 3GPP TS 25.211 Physical Channels and Mapping of Transport Channels onto Physical Channels (FDD), release 6 edn. (2005). Version 6.9.0
4. 3GPP Organizational Partners: 3GPP TS 25.214 Physical Layer Procedures (FDD), release 5 edn. (2005). Version 5.11.0

5. 3GPP Organizational Partners: 3GPP TS 25.215 Physical Layer – Measurements (FDD), release 5 edn. (2005). Version 5.7.0
6. 3GPP Organizational Partners: 3GPP TS 25.309 FDD Enhanced Uplink – Overall Description, release 6 edn. (2006). Version 6.6.0
7. 3GPP Organizational Partners: 3GPP TS 25.211 Physical Channels and Mapping of Transport Channels onto Physical Channels (FDD), release 6 edn. (2007). Version 6.9.0
8. 3GPP Organizational Partners: 3GPP TS 25.306 UE Radio Access Capabilities, release 6 edn. (2007). Version 6.12.0
9. 3GPP Organizational Partners: 3GPP TS 25.308 UTRA High Speed Downlink Packet Access (HSDPA) – Overall Description, release 6 edn. (2007). Version 6.4.0
10. 3GPP Organizational Partners: 3GPP TS 25.321 Access Control (MAC) Protocol Specification, release 6 edn. (2008). Version 6.15.0
11. 3GPP Organizational Partners: 3GPP TS 25.331 Radio Resource Control (RRC), release 6 edn. (2008). Version 6.18.1
12. Analytics, S.: Internal Report (2007). URL `http://www.strategyanalytics.com`
13. Assaad, M., Zeghlache, D.: TCP Performance over UMTS-HSDPA Systems. Auerbach Publications, New York (2006)
14. Chevallier, C., Brunner, C., Garavaglia, A., Murray, K.P., Baker, K.R. (eds): WCDMA (UMTS) Deployment Handbook: Planning and Optimization Aspects. Wiley, Chichester (2006)
15. Dahlman, E., Parkvall, S., Sköld, J., Beming, P.: 3G Evolution. Academic Press, London (2007)
16. Das, A., Khan, F., Sampath, A., Hsuan-Jung, S.: Performance of hybrid ARQ for high speed downlink packet access in umts. IEEE 54th Vehicular Technology Conference **4**(1), 2133–2137 (2001)
17. Elliott, R.C., Krzymien, W.A.: Scheduling algorithms for the cdma2000 packet data evolution. Vehicular Technology Conference, (2002). Proceedings. VTC 2002-Fall. 2002 IEEE 56th. **1**, 304–310 (2002)
18. Forouzan: TCP/IP Protocol Suite. McGraw Hill, New York (2006)
19. Gutierrez, P.J.A.: Packet scheduling and quality of service in HSDPA. PhD, Aalborg University, Denmark (2003)
20. Hai, W., Lei, W., Min, L.: HSDPA link adaptation based on novel quality model. Ericsson Res., Ericsson (China) Co. Ltd., Beijing, China; Vehicular Technology Conference, 2005. VTC 2005-Spring. 2005 IEEE 61st, 30 May–1 June **1**, 334–338 (2005)
21. Helmersson, K., Englund, E., Edvardsson, M., Edholm, C., Parkvall, S., Samuelsson, M., Wang, Y.P., Jung-Fu, C.: Wcdma enhanced uplink – principles and basic operation. IEEE 61st Vehicular Technology Conference **3**, 1427–1431 (2005)
22. Holma, H., Toskala, A. (eds): HSDPA/HSUPA for UMTS: High Speed Radio Access for Mobile Communications. Wiley, Chichester (2006)
23. Holtzman, J.M.: Cdma forward link waterfilling power control. IEEE 51st Vehicular Technology Conference **3**, 1663–1667 (2000)
24. Holtzman, J.M.: Asymptotic analysis of the proportional fair algorithm. 12th IEEE International Symposium Indoor and Mobile Radio Communications (2001)
25. Jalali, A., Padovani, R., Pankaj, R.: Data throughput of CDMA-HDR a high efficiency-high data rate personal communication wireless system Vehicular Technology Conference Proceedings, 2000. VTC 2000-Springer Tokyo. 2000 IEEE 51st **3**, 1854–1858 (2000).
26. Kelly, F.P., Maulloo, A.K., Tan, D.K.H.: Rate control for communication networks: shadow-prices, proportional fairness and stability. Journal of the Operational Research Society **49**, 237–252 (1998)
27. Mathis, M., Semke, J., Mahdavi, J., Ott, T.: The macroscopic behavior of the tcp congestion avoidance algorithm. ACM SIGCOMM Computer Communication Review **27**(3), 67–82 (1997)
28. Peisa, S.P.J., Torsner, J., Sagfors, M., Malm, P.: Wcdma enhanced uplink – principles and basic operation. IEEE 61st Vehicular Technology Conference **3**, 1411–1415 (2005)

Chapter 4
Congestion Control for Wireless Cellular Systems with Applications to UMTS

Emanuel B. Rodrigues, Francisco R. P. Cavalcanti, and Stefan Wänstedt

4.1 Introduction

A wireless cellular network operates normally when the offered traffic load stays around or below a target point defined in the network planning phase. The network is able to work in an optimized way thanks to the operation of radio resource management (RRM) algorithms, which meet the contracted Quality of Service (QoS) requirements for the majority of users causing few service outages, and maintain the planned coverage. However, there can be a mismatch between the network capacity and the users' current demands due to system dynamics, which can cause network overload and service outage (congestion situations). Network congestion can be caused by some factors, such as voice and data traffic dynamics, network utilization pattern during specific periods of the day (busy hours), random behavior of the external interference, subscribers' profiles (commercial and residential areas) and their call distributions, different mobility profiles, and geographical location of mobile terminals. These factors can cause variations on the QoS experienced by the users and the cell load. In order to guarantee the resources necessary for the provision of real-time (RT) services despite load/QoS fluctuations, a congestion control (CC) technique must be employed.

Providing QoS, in particular meeting the data rate and packet delay constraints of RT flows, is one of the main requirements in communication networks. However, it is quite a challenge to provide statistical QoS guarantees for RT services in highly dynamic traffic environments, such as cellular systems. The problem is worse in packet-switched (PS) networks, since they were first conceived for best effort services. In order to do that, efficient RRM algorithms must be employed. In the context of wireless cellular networks, CC technique is a framework that joins several RRM algorithms in order to prevent, avoid, and recover from congestion situations.

Wireless cellular systems were first deployed with circuit-switched (CS) connections over dedicated channels. In order to further enhance the system capacity and lower the operating cost, next generation systems will rely on the PS mode, where

F. Cavalcanti, S. Andersson (eds.), *Optimizing Wireless Communication Systems*,
DOI 10.1007/978-1-4419-0155-2_4,

a single all-IP-based network with flat architecture will be used. The legacy CS cellular networks, which have a remarkable number of deployed infrastructures, and emerging high-speed PS cellular networks will co-exist for a long time. Therefore, it is of utmost importance to employ CC frameworks that are suitable for each of these modes, CS and PS.

The CC technique has been successfully employed in many communication networks. It was originally proposed for CS telephone networks, but nowadays it is an indispensable player in the Internet, asynchronous transfer mode (ATM) networks, and several wireless systems, such as cellular, sensor, and multihop networks.

The application of the CC technique as a QoS management tool is explained in Section 4.2. The description of the RRM algorithms that can compose a CC framework is presented in Section 4.3. A new taxonomy of CC schemes based on the consumption of radio resources and service quality is described in Section 4.4. In this work, two CC frameworks are proposed: a resource-based framework suitable for CS systems and a QoS-based framework suitable for PS systems. The former is described in Section 4.5, while the latter is presented in Section 4.7. Each one of the CC frameworks are evaluated in the 3rd Generation Partnership Project (3GPP) Universal mobile telecommunications system (UMTS). The performance evaluation of the first case study (resource-based CC framework) is presented in Section 4.6, while the performance evaluation of the second case study (QoS-based CC framework) is shown in Section 4.8. Finally the general conclusion and the perspectives of future studies are presented in Section 4.9.

4.2 Congestion Control and QoS Management

Network congestion is a resource-sharing problem that will result if the resources in the network cannot meet all of the users' current demands. In simple terms, if for any time interval, the total sum of users' demands on the network is more than its available capacity, the network is said to be congested for that interval.

Network congestion is caused by the evolution of system dynamics: increase in the offered load, traffic variability, mobility aspects for the case of mobile systems, increase in interference for the case of wireless systems, etc. [35]. An increase in the offered load causes congestion by increasing the demand for the resources in the network. Taking the example of mobile communication system, if several users in a cell suddenly move far from the base station (BS) and it tries to compensate this by means of power control, two things will probably happen: there may not be enough power to satisfy all the links' qualities simultaneously and excessive interference will be generated to neighboring cells. And this may lead to a congestion situation.

Congestion may lead to a network collapse, where the quality of service, error rate, and reception delays reach unacceptable values. This leads to packet discarding, thereby triggering retransmission of packets and causing a severe decrease in total network throughput. In this situation, QoS requirements cannot be guaranteed, network capacity is lower than required, and in the case of wireless systems, the planned coverage area is not provided [28].

In order to prevent, avoid, or recover from network congestion, a CC technique is needed. The CC technique is a very important network management tool that assures network stability and reliability, so that both efficiency and user fairness are provided. The CC scheme is required to match the demand dynamically to the available capacity. Thus, it should ask users to increase the demand when additional capacity becomes available, and to decrease it if the demand exceeds the capacity. The demand curve should follow the capacity curve very closely [26]. Thus, CC algorithms need to monitor the network status continuously in order to correct overload situations. The monitoring will be based on averaged measurements in order to avoid both false congestion detections and congestion non-detection. Furthermore, the CC algorithm needs to exhibit a fast reactivity under overload conditions in order to prevent degradation of the quality of the connections [35]. A conceptual illustration of the system dynamics and the phases of the operation of the CC technique is presented in Fig. 4.1.

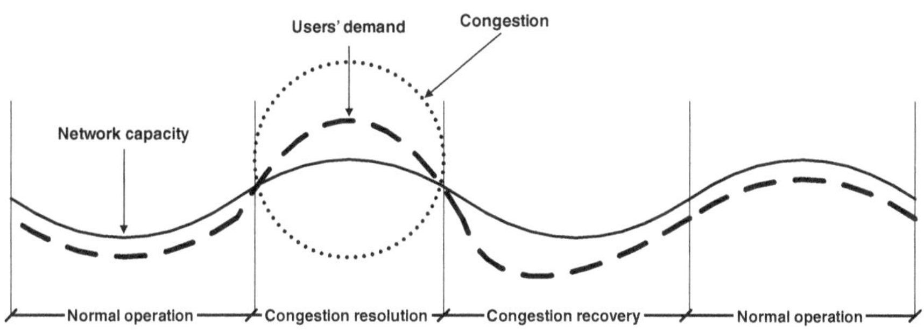

Fig. 4.1 Conceptual illustration of the system dynamics with congestion control.

A well-known CC technique that is widely adopted in the Internet is transmission control protocol (TCP). It is based on four intertwined algorithms: slow start, congestion avoidance, fast retransmit, and fast recovery [24]. The reader can find a detailed description of these algorithms in Chapter 3 of this book. One can make an association between the TCP algorithms and the general phases in a CC operation, which is illustrated in Fig. 4.1. When the network is in normal operation, new connections use the slow start algorithm to increase their transmission windows. Congestion avoidance is responsible for monitoring the system (detection of packet losses) and for loading the system in a gradual manner, because for every useful acknowledgement received, the transmission window is increased linearly. If a packet loss is detected (timeout or duplicate ACK), then TCP enters the congestion resolution phase, dividing the transmission window by 2. Slow start and fast recovery are executed in the congestion recovery phase. If the packet loss was due to timeout, slow start is used, otherwise, if duplicate ACKs were received, fast recovery is executed.

After the explanation of the main phases of the CC operation, we will now address the several kinds of CC techniques that were proposed in the technical literature and/or implemented in practical networks. A taxonomy for CC techniques is presented below based on the terminology employed in several references:

- **Congestion avoidance × Congestion resolution/recovery**: This taxonomy was discussed in [61]. The strategy of congestion avoidance is preventive in nature; it aims to keep the operation of a network at or near the point of maximum capacity, so that congestion will rarely occur. In its turn, the goal of congestion resolution/recovery is to solve the congestion and restore the operation of a network to its normal state after congestion has occurred. Without a congestion resolution/recovery scheme, a network may crash entirely whenever congestion occurs. Therefore, even if a network adopts a strategy of congestion avoidance, congestion resolution/recovery schemes would still be required in the case of abrupt changes in a network that may cause congestion [61].
- **Resource creation × Demand reduction**: This taxonomy was presented in [26]. The resource creation schemes increase the capacity of the network by dynamically reconfiguring them. Some examples are the addition of new communication links only during high usage, power increases on satellite links to increase their bandwidths or path splitting so that extra traffic is sent via alternative routes. Demand reduction schemes try to reduce the demand to the level of the available resources. There are three basic classes of such schemes: service denial, service degradation, and scheduling. Service denial schemes do not allow new sessions to start up during congestion (e.g., admission control). Service degradation schemes ask all users to reduce their loads. There is a technique known as traffic shaping that smoothes out the traffic on the transmitter side, regulating the average rate (and burstiness) of data transmission, thus reducing congestion [53]. Finally, scheduling schemes ask users to schedule their demands so that the total demand is less than the capacity. One can notice that all packet scheduling algorithms are a special case of the service degradation approach.
- **Open loop × Closed loop**: This classification is based on Control Theory and was proposed by Yang and Reddy [61]. As the name indicates, the problem of congestion control is basically a control problem. Open-loop CC algorithms, different from the closed-loop algorithms, are the ones in which the control decisions do not depend on any sort of feedback information from the congested spots in the network and so do not monitor the state of the network dynamically. Due to this fact, the former schemes are not robust enough and cannot guard the network against all traffic patterns [61].

This work also propose a further classification of the CC techniques: **Resource based × QoS based**. The former scheme monitors continuously the consumption of resources in the network so that it can avoid and react on overload situations. In this case, the network overload is defined when the resource utilization is above some predefined threshold. The latter scheme considers that the system is overloaded when the QoS of one or more services is degraded as measured by the network considering some network-level metric (e.g., the average of packet error rates of all users during a specific time window). If the service quality is below a predefined level (service outage), the QoS-based CC will act in order to perform QoS differentiation among services so that the most prioritized service can have its QoS guaranteed in a statistical manner.[4.1] More details on this proposed approach are given in Section 4.4.

As can be concluded from the explanations above, any communication network needs a CC framework in order to provide statistical QoS guarantees. CC techniques are employed successfully in many communication networks, such as CS telephone networks [33, 38], ATM networks [18, 23], Internet [32, 54], multicast networks [51, 62], wireless sensor networks [57, 63], multihop wireless networks [30, 52], and wireless cellular networks [43, 44].

4.3 Congestion Control Framework and Radio Resource Management

This work will focus on the proposal of CC solutions for wireless cellular systems, in particular the 3GPP UMTS. Four RRM algorithms are the fundamental components of a CC framework that is able to provide QoS guarantees for RT services in CS and PS wireless cellular networks. These algorithms are admission control (AC), rate adaptation (RA), packet scheduling (PSC), and load control (LC). Figure 4.2 presents a simplified system architecture and a brief description of how the proposed general CC framework works.

This work assumes two scenarios: CS and PS. In the former it is assumed that all voice and data services are provisioned in dedicated channels, while in the latter it is assumed that the voice and data services use high-speed shared channels. Taking these strict assumptions into consideration, 3GPP systems such as Release 99 (based on wideband code division multiple-access (WCDMA) air interface) and Release 5 (based on high speed packet access (HSPA) air interface) can be thought of as examples of CS and PS systems, respectively.

3GPP Release 99 QoS architecture was designed to provide QoS differentiation [2, 3]. This QoS architecture defines four service classes: conversational, streaming, interactive, and background. The main distinguishing factor between the four traffic classes is how delay-sensitive the traffic is: the conversational class is meant for very delay-sensitive traffic, while the background class is the most delay-insensitive. There are, further, three different priority categories, called allocation/retention priority (ARP) categories, within each QoS class. Interactive has also three traffic-handling priorities. Conversational and streaming class parameters also include the guaranteed bit rate and the transfer delay parameters.

[4.1] There is a difference between statistical and strict QoS assurance schemes. The former provides soft guarantees relying on the statistical behavior of the system to achieve multiplexing gains in the resource utilization. The latter has to assure that the QoS requirements will not be violated at any circumstances, requiring strict and fast QoS control mechanisms, such as dropping, which can decrease the system performance. For further information see [39].

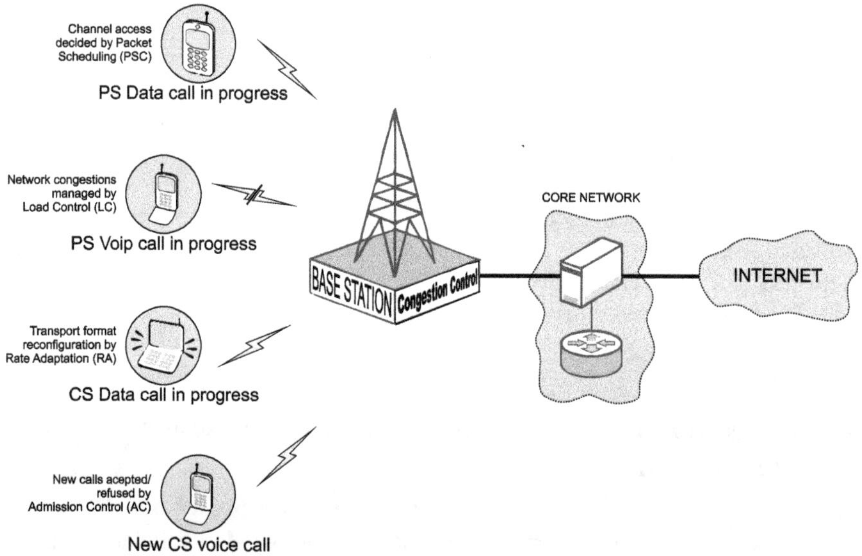

Fig. 4.2 Simplified system architecture and the general CC framework used to provide QoS guarantees for real-time services in wireless cellular networks.

Some RRM algorithms can take advantage of QoS differentiation at the level of radio access network, improving the efficiency of the network utilization when CS services are provided. These algorithms are mainly AC, RA, and LC.

QoS differentiation becomes useful for network efficiency during periods of high load and when there are services with different delay requirements. If the radio network has knowledge about the delay requirements of the different services, it will be able to prioritize the services accordingly and improve the efficiency of the network. Considerable efficiency gains can be obtained just by introducing a few prioritization classes within interactive or background class by using ARPs. Flows within the same QoS and ARP class will share the available capacity. The RA algorithm is a way to enforce this QoS hierarchy by means of bit rate adaptation. Bit rates can be adapted by using transport format (TF) selection [7]. If the number of flows is simply too high, they will all suffer from bad quality. In that case it would be better if the AC algorithm blocks a few users to guarantee the quality of the existing connections. AC can estimate the available radio capacity and block an incoming user if there is no room to provide the required resource without sacrificing the quality of the existing connections. Notice that these actions should be coordinated according to the network load, which is continuously monitored by the LC algorithm.

Regarding enhanced PS services, a new QoS concept is being agreed within the 3GPP [6], which aims at providing means to enhance QoS for services or data flows that require QoS beyond what the default IP mechanism provides. One such example is session initiation protocol (SIP) [25] signaling to handle multimedia telephony services over IMS (MTSI) sessions, where IMS stands for IP Multimedia Subsystem [14]. If MTSI were to replace CS telephony, call-related signaling must be prioritized and should have the potential to pre-empt other traffic to ensure delivery of, e.g., signaling related to an emergency call. Another example is real-time media, such as voice over ip (VoIP) and video. The scheduling prioritization among SIP and VoIP was studied for 3GPP HSPA and long-term evolution (LTE) networks in [48] and [60], respectively. Furthermore, studies conducted in [16, 58] compare the performance between VoIP and CS voice over WCDMA and HSPA, respectively, showing that VoIP can be a CS replacement with the same QoS demands, i.e., the quality, accessibility, and retainability of CS voice calls.

The access bearer is related to an edge-to-edge association between the gateway (GW), which provides the IP connectivity, and the mobile station (MS) [14]. The access bearer is associated with uplink filters in the MS and downlink filters in the GW of the mobile core network. The filters only match certain packets and provide differentiation of packets belonging to different service flows. This allows the scheduling mechanisms in the MS and the BS to use different scheduling strategies for packets belonging to different service flows. Enhanced QoS for certain PS service flows, such as SIP signaling, VoIP, or video, is achieved by mapping the service flows to different radio bearers. The radio bearer is simply the radio part of the access bearer, and each radio bearer is associated with one scheduler configuration based on its QoS demand, e.g., fixed priority, proportional fair (based on fairness and channel quality), or delay threshold (based on a service-dependent delay budget). Different service flows may be mapped to the same scheduler configuration.

The SIP, the voice, and the video flows use different access bearers in order to allow for prioritization. The bearers can be classified primarily into two different types: guaranteed bit rate (GBR) and non-GBR bearers. Depending on the GBR value, the GBR bearer requires dedicated resources that are permanently allocated at the bearer establishment or modification by an admission control function. Sources sending at a rate lower than the GBR capacity may assume that congestion-related packet drops will not occur or at least will be extremely rare. However, transient link outages, which will always occur in a radio access system, are exceptions.

The resources for GBR services in the PS domain can be ensured by three RRM algorithms: AC, PSC, and LC. AC is used to check if there are free resources and if the service quality of the ongoing connections is not threatened by the admission of new users. PSC can assure that GBR services are served frequently enough by using QoS-sensitive utility functions. LC will primarily be used to monitor the system load and/or QoS and, in case of overload or service outage, order AC to block new calls and order PSC to reduce the bit rate for the non-GBR bearers to ensure that the guarantees for GBR services can be sustained. Therefore, services over non-GBR bearers should be prepared to experience congestion-related delays or packet drops. This work assumes that GBR services have the highest short-term priority among all services, and also that other lower priority services agree to have their short-term quality degraded as long as their long-term QoS requirements (e.g., mean session throughput) are guaranteed. Typical examples of GBR and non-GBR services include VoIP and web browsing, respectively.

4.4 Resource-Based and QoS-Based Congestion Control

As pointed out in Section 4.2, in this work is proposed an extra classification for CC techniques: resource based and QoS based. A more detailed explanation will be given in Sections 4.4.1 and 4.4.2.

4.4.1 Resource-Based CC

In Section 4.2, the importance of keeping the air interface load of wireless cellular systems under predefined thresholds is explained, because excessive loading can prevent the network from guaranteeing the needed requirements. Resource-based CC techniques for wireless cellular systems base their decisions on the consumption of radio resources, i.e., the air interface load. Thus, this load must be measured or estimated. It is well known that the uplink (UL) and downlink (DL) can be asymmetrically loaded in code-division multiple–access (CDMA)-based wireless systems. For this reason, the tasks of congestion control have to be done separately for both links. There are two different approaches to measure/estimate the air interface loads in CDMA-based systems (of which UMTS is an example) [22, 28]: power based and throughput based. Both approaches for the uplink and downlink are described in Sections 4.4.1.1 and 4.4.1.2, respectively.

4.4.1.1 Uplink

In the UL power-based CC technique, the BS measures the total received power, which is given by

$$p_{\mathrm{r}} = p_{\mathrm{r}}^{\mathrm{own}} + p_{\mathrm{r}}^{\mathrm{oth}} + p_{\mathrm{noise}}, \tag{4.1}$$

where p_{r} is the total received power of a given BS, $p_{\mathrm{r}}^{\mathrm{own}}$ is the received power from users in the own cell (intra-cell interference), $p_{\mathrm{r}}^{\mathrm{oth}}$ is the received power from users in the surrounding cells (inter-cell interference), and p_{noise} is the total noise power. Using the parameters presented in (4.1), the UL load factor drives the decision of UL power-based CC techniques and is given by [22, 28]

$$\eta_{\mathrm{UL}} = \frac{p_{\mathrm{r}}^{\mathrm{own}} + p_{\mathrm{r}}^{\mathrm{oth}}}{p_{\mathrm{r}}}. \tag{4.2}$$

UL throughput-based CC technique is based on the definition of a load factor for each MS and is defined as [22, 28]

$$\eta_j^{\mathrm{UL}} = \frac{1}{1 + \dfrac{W}{\rho_j \cdot r_j \cdot v_j}}, \tag{4.3}$$

where j is the index of the MS, η_j^{UL} is the UL load factor, W is the chip rate, ρ_j is the E_b/N_0 requirement, r_j is the bit rate, and v_j is the service activity factor. The total UL load factor of the cell is given by

$$\eta_{\mathrm{UL}} = (1+\varphi_{\mathrm{int}}) \cdot \sum_{j=1}^{J} \eta_j^{\mathrm{UL}} = (1+\varphi_{\mathrm{int}}) \cdot \sum_{j=1}^{J} \frac{1}{1+\dfrac{W}{\rho_j \cdot r_j \cdot v_j}}, \tag{4.4}$$

where J is the number of MSs, φ_{int} is the other-to-own cell interference ratio, and η_j^{UL} is the UL load factor of the jth MS given by (4.3).

4.4.1.2 Downlink

DL power-based CC techniques are based on the BS transmission power consumption. A usual metric is the DL load factor defined as [22, 28]

$$\eta_{\mathrm{DL}} = \frac{p_{\mathrm{t}}}{p_{\mathrm{max}}}, \tag{4.5}$$

where p_{t} is the total transmit power of a given BS and p_{max} is the maximum transmit power capability of the cell. Notice that DL power-based CC techniques can also use absolute power thresholds instead of normalized power measures as in (4.5).

There are two ways to estimate the throughput-based DL load factor. The first approach uses the sum of the allocated bit rates as follows [22, 28]:

$$\eta_{\mathrm{DL}} = \frac{\sum_{j=1}^{J} r_j}{r_{max}}, \tag{4.6}$$

where J is the number of MSs, r_j is the bit rate of the jth MS, and r_{max} is the maximum allowed throughput of the cell. The other way to define the throughput-based DL load factor is to weigh the MS bit rates with E_b/N_0 requirements as follows:

$$\eta_{\mathrm{DL}} = \left[(1-\overline{\alpha}_{\mathrm{orth}}) + \overline{\varphi}_{\mathrm{int}}\right] \cdot \sum_{j=1}^{J} \frac{\rho_j \cdot r_j \cdot v_j}{W}, \tag{4.7}$$

where J is the number of MSs, j is the index of the MS, $\overline{\alpha}_{\mathrm{orth}}$ is the average orthogonality factor, $\overline{\varphi}_{\mathrm{int}}$ is the average DL other-to-own cell interference ratio, W is the chip rate, ρ_j is the E_b/N_0 requirement, r_j is the bit rate, and v_j is the service activity factor.

Section 4.4.1.1 and this section described how resources in a CDMA-based cellular system can be estimated in the UL and DL and how this mathematical approach can be used in a resource-based CC framework. However, this reasoning can be extended to other radio access technologies, such as time division multiple access (TDMA), frequency division multiple access (FDMA), and orthogonal frequency division multiple access (OFDMA). Since these systems have radio resources well defined in the time–frequency domain, it is possible to easily measure/estimate the cell load directly from the consumption of these resources. Therefore, a resource-based CC framework can be adapted with slight modifications to most of the cellular systems.

4.4.2 QoS-Based CC

In this section are described QoS-based CC techniques. Unlike resource-based CC techniques which use different criteria depending on which link is being considered (UL or DL), the QoS-based techniques employ a unified framework, since the decisions of the CC algorithm is service quality centric. Furthermore, since the QoS criteria is more general, this new approach is not directly dependent on which wireless cellular system is deployed. It can be used without much modifications in every current and next-generation wireless cellular system. The idea behind this paradigm is to consider a QoS metric of the highest priority service in the system as the network "resource." The CC framework has to monitor the quality of the most important service continuously. If this quality becomes worse than a predefined QoS target, the system can be considered congested, because the QoS guarantees of the service with highest priority cannot be guaranteed in this "overload" situation. On the other hand, if the service quality is good, it means that the system is operating in normal conditions and no congestion resolution actions need to be taken.

Since this QoS-based paradigm is service centric, the monitored QoS metrics are dependent on which service is being prioritized in the network. For example, let us assume that the VoIP service, which is a RT delay-sensitive service, is the one with highest priority in the system. Some QoS metrics that are important to this service and could be used in the CC framework are delay and frame erasure rate (FER). Notice that the FER metric is more general, because packets that arrive in the receiver play-out buffer with delay higher than the VoIP delay budget are considered lost and are accounted in the FER metric. Thus, the FER also takes into consideration packet delay. Another example is the video telephony service. The CC framework aims to guarantee the QoS of this service; some important quality metrics are delay, delay jitter, and guaranteed bit rate.

4.4.3 CS × PS Congestion Control: Discussion

From what was exposed above, it can be observed that the resource-based and QoS-based CC frameworks have some characteristics that made them more suitable for either CS or PS cellular systems. CS cellular systems need to allocate both radio and core network resources for the entire session. Furthermore, the services are provided in dedicated channels that are usually power controlled. These characteristics allow a good estimation of resource consumption, which is of utmost importance for the resource-based CC techniques to work properly. For this reason, resource-based CC frameworks are well suited for CS cellular systems. On the other hand, cellular networks are evolving toward an all-IP architecture where only PS services are provided over high-speed shared channels that use channel-dependent adaptive modulation and coding (AMC) to guarantee block error rate (BLER) performance and maximize capacity. PS systems were conceived for bursty traffic and the BSs operate most of the time at their maximum power allocation. Thus, it is difficult to make accurate and up-to-date estimates/measurements of the air interface load. For this reason, it is more advantageous for the PS networks to employ QoS-based CC frameworks instead of resource-based ones. It is worthy to say that the QoS-based CC techniques could also be used in CS networks, since their focus is on the QoS requirements of multimedia services rather than the access technology.

Section 4.5 presents the proposal of a CC framework that is based on the BS transmit power and is suitable for the downlink of CS cellular networks. Section 4.7 describes a proposed CC framework for the downlink of PS cellular networks that is based on the FER of an RT service.

4.5 Resource-Based Framework for Circuit-Switched Networks

In this section, the proposed resource-based CC framework for CS cellular networks is described. It was originally proposed in [44]. According to the taxonomies presented in Section 4.2, this framework can be classified as a scheme that performs both congestion avoidance and resolution/recovery, reduces the demand in overload situations, performs a closed-loop control, and is based on radio resources. This CC framework is composed of three RRM algorithms: AC, RA, and LC. Figure 4.3 presents how these RRM techniques interact in order to avoid, resolve, and recover from congestion situations.

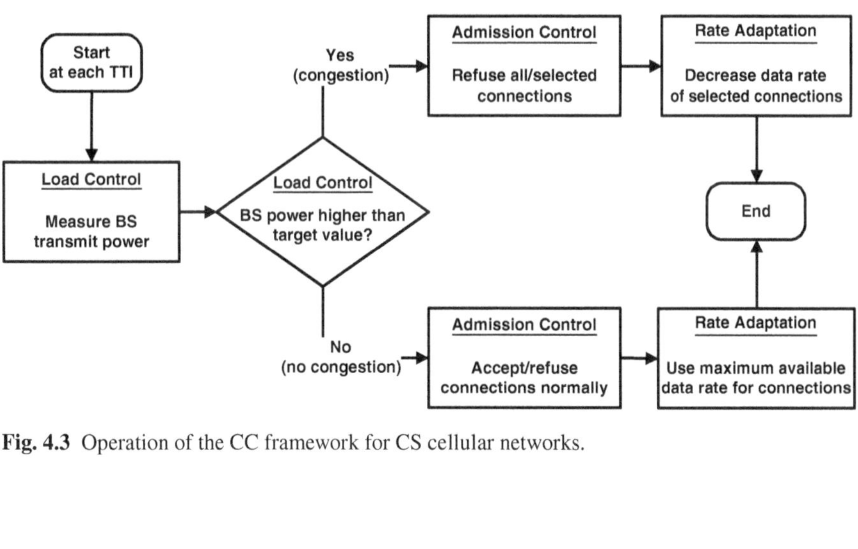

Fig. 4.3 Operation of the CC framework for CS cellular networks.

The CC is based on the usage of the BS transmission power (air interface load in the DL). In normal operation, the AC algorithm is able to keep the DL air interface load in an acceptable level by means of service denial, thus avoiding the system from entering in congestion most of the time. However, due to the dynamics of the system, the LC algorithm must monitor the BS transmission power constantly in order to detect an overload. If this occurs, it will interact with the AC and RA algorithms in order to resolve the congestion. The LC will order the AC to block any incoming call during the congestion resolution phase so that the air interface load is not increased anymore. Moreover, the LC will order the RA to change the TFs of some flows aiming the reduction of the air interface load. When the BS transmission power achieves an acceptable level again, the LC can start the congestion recovery phase, where the admission of new calls is allowed and the RA can reconfigure the TF of the degraded flows to the previous values in a conservative manner. In this way, it is expected that the system comes back to stability after the end of the recovery phase. The proposed CC framework has a completely automated and adaptive framework. The framework adjusts itself in order to prevent the BS from working with an unacceptable transmission power, and so avoiding a possible network collapse.

In the congestion resolution phase, prioritization is a very important issue. In general, ongoing calls have priority in comparison with new calls, because the latter are not allowed to enter in the system when the system is congested. In a multiservice scenario, the AC can perform service prioritization, since in a first stage only low-priority calls (e.g., non-real time (NRT) service) are blocked, and only if the system continues in congestion, the high-priority calls (e.g., RT service) would be blocked. RA can perform prioritization on CS cellular networks, both in single-service and multiservice scenarios. In a single service scenario, different priority levels can be determined, for example different adaptive multirate (AMR) modes in a voice service (12.2, 7.95, and 4.75 kbps) or different rate requirements in a data service (64, 128, and 384 kbps). Adaptive TF selection can be performed within the same service depending on the chosen priority policy. Regarding the multiservice scenario, QoS differentiation can be performed by means of TF selection in the RA algorithm. For example, in the congestion resolution phase the RA algorithm can first degrade the QoS of the best-effort service by means of data rate decrease. As a final option, if the congestion is not resolved, the TFs of the voice service would be downgraded by means of AMR mode reconfiguration. The inverse reasoning is applied in the congestion recovery phase. The high-priority RT service will be the first to have their calls readmitted by the AC algorithm and also the first to have their TFs upgraded by the RA algorithm.

Compared to other resource-based CC techniques for wireless cellular systems available in the literature, the proposed DL power-based CC framework has a clear contribution. Some works studied the UL, such as [20, 36, 49], which proposed throughput-based CC techniques, and the work [10] that evaluated a power-based CC framework. The DL is contemplated in [34], but the proposed CC technique is based on throughput rather than transmission power. A DL power-based CC is evaluated in [37], but the study was focused on the interactive service class only.

Furthermore, none of those works propose a unified CC framework composed of the AC, RA, and LC algorithms, as the present work does.

The adaptive proposed CC framework is scalable to several services. However, telephony is, and will be for some time, by far the most important of all personal mobile communication services. Therefore, a particular version of the resource-based CC framework adjusted to the traditional CS voice service is described in the following sections. Sections 4.5.1, 4.5.2, and 4.5.3 explain the operation of the AC, RA, and LC algorithms, respectively. The algorithms and approaches presented below can be readily extended to other CS services besides voice.

4.5.1 Admission Control

The AC algorithm used in the CC framework for the downlink of CS cellular networks is the link admission control (LAC), which is a standard AC technique that monitors only the radio resources in order to decide whether to accept or not a new radio link in the network, in case of new call admission, or in the cell, in case of a handover call.

In interference-limited systems, transmission power will typically be the limiting factor for capacity before other resources are exhausted, for example spreading codes in CDMA-based systems. The BS power usage and the DL interference are directly related. In normal operation, if the LAC algorithm roughly controls the BS power usage by means of a power admission threshold that must be respected by the new and handover calls, the DL interference can be kept in acceptable levels most of the time. For these reasons, the LAC algorithm used in the resource-based CC framework will be based on the BS transmission power measurements and reports in each cell and on a power demand estimation of a new desired link (new or handover call). If the addition of the new link does not violate the transmission power limitations of the cell, the new link is created, otherwise it is denied.

The used LAC algorithm is closely related to the initial power allocation strategy in the selected cell. An open-loop power control must be used to provide a coarse initial power setting for new and handover connections using downlink measurement reports from the MS. A possible open-loop power control algorithm for CDMA-based systems can be found in [28].

When the jth MS tries to set up a new connection or make a handover to the ith BS, the LAC algorithm admits or denies this attempt in accordance with

$$p_i^{\mathrm{tot}} + p_{i,j} \leq p_i^{\mathrm{th}}, \tag{4.8}$$

where p_i^{tot} is the total transmission power of the ith BS, $p_{i,j}$ is the initial transmission power of the ith BS toward the jth MS estimated by the open-loop power control, and p_i^{th} is the BS power threshold for the admission decision determined in the radio access planning phase.

The LAC used in the proposed CC framework is one among many other available schemes. The reader can get more information on advanced LAC schemes for CDMA-based CS networks in [45].

4.5.2 Rate Adaptation

Video and voice codecs with variable bit rates are widely used in communication networks. Some examples are H.264 and MPEG-4 Visual [40], for video, and 3GPP's AMR [1], for voice.

The H.264 and the MPEG-4 Visual standards allow the use of a rate control algorithm [40]. Typically, an encoder with constant parameters will produce fewer bits when there is low motion and/or detail and more bits when there is high motion and/or detail in the input sequence. For many practical transmission mechanisms this variation in bit rate can be a problem. For example, a CS channel, which is a constant bit rate channel, cannot transport a variable bit rate data stream. A PS network can support varying throughput rates but the mean throughput at any point in time is limited by factors such as link rates and congestion. In these cases it is necessary to adapt or control the encoding parameters in order to maintain a target output bit rate and match the available bit rate of the transmission mechanism.

The 3GPP's AMR voice codec has eight operation modes (source rates): 12.2, 10.2, 7.95, 7.40, 6.70, 5.90, 5.15, and 4.75 kbps [1]. The introduction of the AMR voice codec extends the quality range perceived by the conversational service class compared with a fixed rate voice codec. The quality achieved by the AMR codec varies from very high with AMR-12.2 mode to acceptable with AMR-4.75 mode. The AMR codec is capable of changing its transmission rate dynamically at each voice frame of 20 ms by means of commands of the radio access network, which takes this decision depending on the air interface load (total cell transmission power), the radio link quality, and the transmission power required by the traffic-dedicated channels.

A CC framework can take advantage of the flexibility of variable rate video and voice codecs in order to avoid and resolve congestion situations. In the present work, the RA algorithm used in the proposed CC framework for CS cellular networks use AMR, where the TF of the voice flows can be reconfigured dynamically.

The transmission power is one of the most important radio resources in the forward link of interference-limited systems, such as CDMA-based systems. Different AMR modes require distinct power levels. The lower the AMR mode, the lower the data rate, and therefore, the higher the processing gain. The processing gain increase is more pronounced than the E_b/N_0 requirement increase for lower data rates (see Table 4.1 and [21, 22]). Thus, there is an improvement in the signal-to-interference-plus-noise ratio (SINR) of the MS, causing transmission power consumption saving.

It is exactly this advantageous relation between AMR mode selection and power consumption saving that is explored by the proposed CC framework. The CC framework uses this rate flexibility of the RA algorithm in order to reduce the air interface

Table 4.1 E_b/N_0 requirements for different AMR modes.

Transmission rate (kbps)	E_b/N_0 requirement (dB)
12.2	7.0
7.95	7.5
4.75	8.0

load (BS transmission power) during overload situations. For simplicity, the RA algorithm used in the CC framework works with the AMR modes 12.2, 7.95, and 4.75 kbps.

4.5.3 Load Control

The soft overload is a very important aspect of cellular systems that provides a compromise between system stability, number of voice calls being served by the system, and the received voice quality (conversation intelligibility). An increase of the blocking rate and a decrease of the voice intelligibility quality for a short time are assumed to be an acceptable behavior for a system in overload situations. In other words, the system degrades the service quality in a gradual and controlled fashion in congestion situations, so that the ongoing calls and the system stability are guaranteed.

The main task of the LC algorithm is to monitor, detect, and manage situations where the system reached or is close to reach an overload state. It means that a region of the coverage area possibly has run out of resources. LC must take decisions and make the system return to its stable state in a gradual and controlled manner.

The LC algorithm described in this section was proposed in [42]. In general, the LC algorithm functionality is divided into two phases, congestion resolution and congestion recovery, whose start and end are determined by a congestion detection procedure. The LC procedure is based on the transmission power of the active traffic channels of a generic BS i. The algorithm formulation is depicted in Fig. 4.4 and is described below.

1. **Congestion detection**: The decision criterion for the initiation of the congestion resolution/recovery phase is based on the BS active transmission power. Let us define the thresholds ϕ_{resol} and ϕ_{recov}, whose values are relative to the maximum transmission power available for traffic channels and are defined in the network planning phase. If the BS transmission power stays above ϕ_{resol} during a time window of ΔT_{resol}, the LC algorithm supposes that the cell is congested and must initiate a phase to resolve the overload. After the actions needed to reduce the cell load, if the sector transmission power stays below ϕ_{recov} during a time window of ΔT_{recov}, the algorithm initiates a new phase in order to recover from the congestion situation.

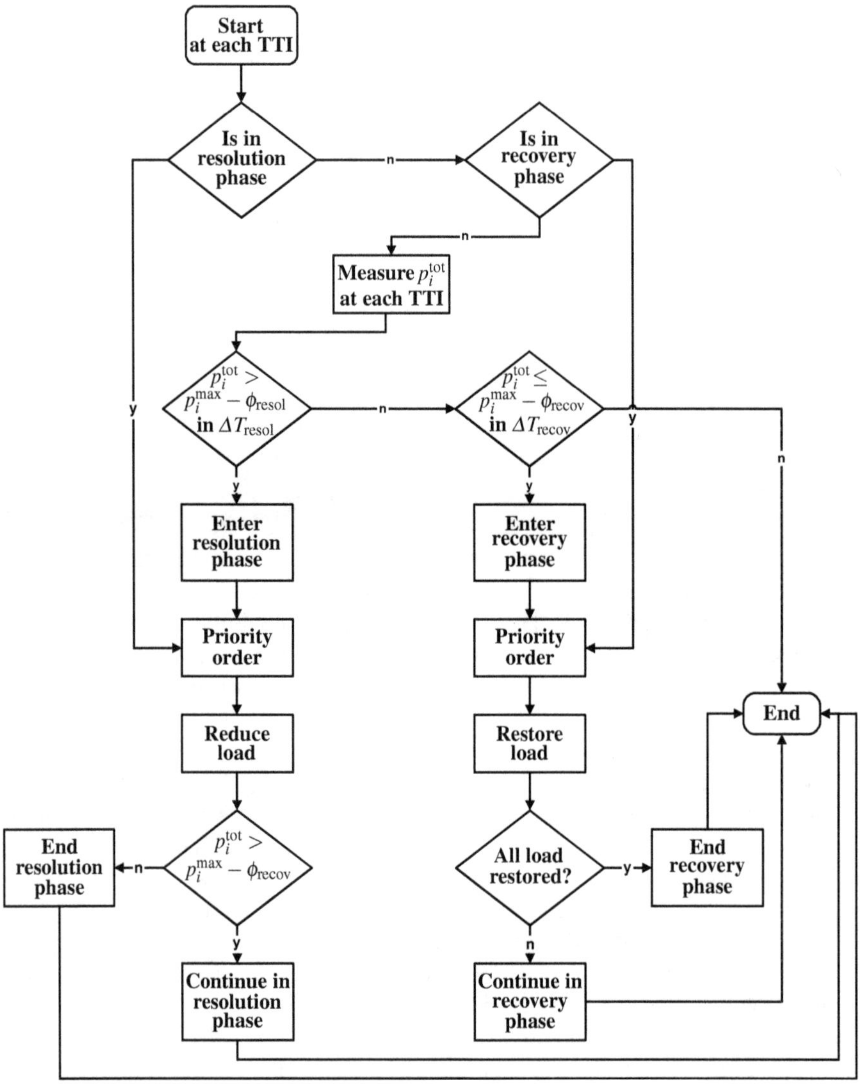

Fig. 4.4 Operation of the LC algorithm of the CC framework for CS cellular networks.

2. **Congestion resolution**: An algorithm based on the following three steps can be used in order to guarantee the network stability:

   a. **Priority order**: A criterion based on random choice or radio link quality can be used to order the different users from the lowest priority to the highest.
   b. **Load reduction**: Some actions taken from the interaction between the LC, LAC, and RA algorithms are performed in order to reduce the cell load in congestion periods.

c. **Load check**: The actions to reduce the load must be performed until the BS transmission power is lower than ϕ_{recov}, the same threshold used by the congestion detection procedure to decide the beginning of the recovery stage. If the congestion persists, the algorithm has to return to step (b) and contemplate the next users in the priority list.

3. **Congestion recovery**: In this phase, the inverse order of the priority list adopted in the congestion resolution phase is followed in order to restore the previous transmission parameters of the flows and the standard configurations of the RRM algorithms.

This sophisticated control requires stability all the time. Thus, some time constants must be defined so that the network is able to react rapidly and firmly. The time windows used to detect overload situations are sliding windows with duration ΔT_{resol} and ΔT_{recov}, relative to the decision of the resolution and recovery actions, respectively. For instance, consider an observation window of 200 ms. Considering a transmission time interval (TTI) of 20 ms, 10 samples of the mean BS transmission power will be collected. A percentile of Υ_{cong} (%) is considered in the distribution of BS transmission power samples collected every TTI of 20 ms within the observation window. Assuming that Υ_{cong}=80%, at least eight of the samples must satisfy the criterion defined to trigger the congestion resolution and recovery processes.

It is relevant to mention that the decision thresholds ϕ_{resol} and ϕ_{recov} are defined in logarithmic scale and they represent percentages of the maximum BS transmission power available to traffic channels.

In order to re-establish the cell stability by means of load reduction in the congestion resolution phase, LC commands LAC to reject any admission of new calls or connections coming from other cells (handover). Furthermore, the RA algorithm is used to reduce the transmission powers of the traffic channels by means of mode selection of the AMR voice codec. It was mentioned in Section 4.5.2 that inferior AMR modes (AMR-7.95 and AMR-4.75) require less transmission power than the AMR-12.2 mode. In this way, the total BS transmission power can be decreased so that the congestion problem is solved. Afterward, in the congestion recovery phase, the AMR modes used by the MSs before the congestion detection will be restored and the call admission will be liberated.

The dynamics of the LC operation is illustrated in Fig. 4.5. Figure 4.5(a) depicts a typical variation of the BS transmission power. Sometimes the BS transmission power stays above the allowed threshold (superior threshold of Fig. 4.5(a)) during a given period of time. In this moment, the congestion resolution phase is triggered (status 1 of Fig. 4.5(b)) and load reduction actions are performed, decreasing the transmission powers of the traffic channels, and consequently the interference generated. When it is detected that the cell is out of the congestion situation, in other words when the BS transmission power stays below a specified threshold (inferior threshold of Fig. 4.5(a)), the congestion recovery stage is initiated (status -1 of Fig. 4.5(b)). When all the users recover their initial transmission parameters, the cell returns to its normal state and is assumed not congested (status 0 of Fig. 4.5(b)).

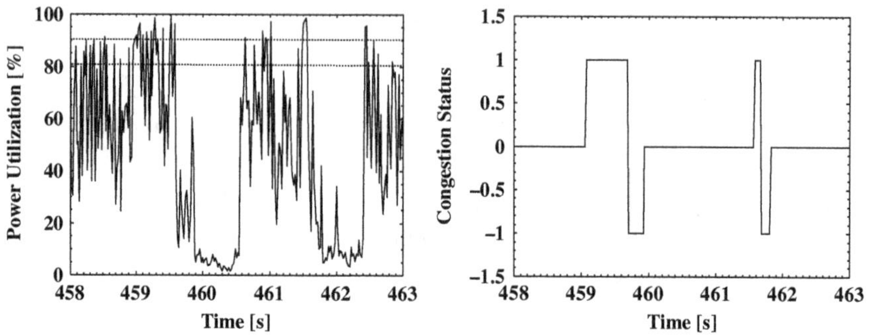

(a) Normalized active transmission power utilization of a sample BS (maximum value = 16.5 W); the detection thresholds for congestion resolution (upper) and recovery (lower) are indicated in dotted lines.

(b) Congestion status: 0 (without congestion); 1 (congestion resolution phase); –1 (congestion recovery phase).

Fig. 4.5 Load control algorithm functionality illustrating the congestion resolution and recovery phases.

4.6 Case Study: WCDMA Performance with Circuit-Switched Voice

In Section 4.5, a resource-based CC framework for CS cellular networks was presented. Now, the performance of the aforementioned framework will be evaluated in a case study where the voice service is provisioned in dedicated channels of the UMTS WCDMA system.

The simulation models will be presented in Section 4.6.1, while the simulation results will be shown in Sections 4.6.2 and 4.6.3. The performance evaluation of the CC framework will be classified according to its main procedures. In Section 4.6.2, the congestion detection function will be studied with the variation of decision thresholds and observation windows. The congestion resolution and recovery stages are evaluated in Section 4.6.3, in particular the user priority selection criterion in the phases when the cell load is reduced or restored. Finally, the conclusions for this particular case study are presented in Section 4.6.4.

The values of quality and capacity metrics presented in the results section should not be regarded as absolute performance indicators. In fact, the reader should focus on the relative comparisons presented.

4.6.1 System Modeling

A discrete time system-level dynamic simulator that models the forward link of the UMTS WCDMA Release 99 system was used. This section comprises the most important computational models used in this software tool.

The simulation environment is a hexagonal multi-cell deployment with uniformly distributed MSs. A wrap-around technique is used to avoid border effects in interference calculations. The macrocell test environment that was considered is the vehicular test environment with low-speed users at 3 km/h [55]. The propagation effects considered are mean path loss, slow variation in the signal due to shadowing, and rapid variation due to multipath effects and scattering. The impacts of multipath propagation on channel fading, downlink orthogonality loss, and RAKE receiver performance are included in the simulations.

Voice call requisitions are generated according to a Poisson process and call durations are exponentially distributed. An ON–OFF traffic pattern is modeled, with activity and silent periods also being generated by an exponential distribution [55].

When a medium access control (MAC) transport block is transmitted on a dedicated channel (DCH) in a 20 ms TTI, the corresponding block error probability (BLEP) is read from the average value interface (AVI) look-up tables that depend on the channel quality and the channel profile [41].

Our power control (PC) and soft handover (SHO) strategies are strongly based on the 3GPP standards [4, 8, 9, 22].

The system is assessed considering the following performance metrics:

- call blocking probability ($\mathscr{P}_{\text{block}}$);
- call dropping probability due to handover ($\mathscr{P}^{\text{HO}}_{\text{drop}}$);
- grade of service ($GoS = \mathscr{P}_{\text{block}} + 2 \cdot \mathscr{P}^{\text{HO}}_{\text{drop}}$). It is assumed that a dropping is more annoying for the user than a blocking;
- frame erasure rate (FER) due to errors in the wireless channel;
- rejection rate of PC commands, i.e., the percentage of PC commands to increase the power that are rejected by the BS due to power unavailability;
- congestion rate, i.e., the percentage of time that the system remains in the congestion resolution phase;
- user satisfaction, i.e., a user is considered satisfied if he perceives a FER lower than a maximum allowed value at the end of his connection.

The system capacity will be represented by the theoretical voice traffic load in Erl/Cell, which comes from the call arrival process analysis and the call mean duration, and the spectral efficiency in the QoS limit point in Erl/MHz/Cell, which is a simulation output result. The QoS limit assumed is $FER = 2\%$ (see Table 4.2).

In Section 4.6.3, the proposed CC framework will be compared with a **Reference Scenario**. This scenario is characterized by

- use of a LAC algorithm, as described in Section 4.5.1;
- Non-adoption of any RA functionality (fixed AMR data rate of 12.2 kbps);
- Non-adoption of any LC functionality.

The main general simulation parameters considered in the performance evaluation presented in this section are pointed out in Table 4.2.

Table 4.2 General simulation parameters.

Parameter	Value	Unit
Deployment		
Number of cells (torus grid)	48	–
Maximum BS transmission power	20	W
Percentage of reserved power for common channels	17.5	%
Cell radius	0.89	km
MS speed	3	km/h
Traffic		
Voice E_b/N_0 requirement	7	dB
Activity factor of the voice service	0.5	–
Mean voice activity period	1	s
Propagation		
Path loss [55]	$128 + 37.6 \cdot log_{10}(d)$	dB
Lognormal shadowing standard deviation	7	dB
Shadowing decorrelation distance	50	m
Shadowing correlation between BSs	0.5	–
Small-scale fading	Single-path Rayleigh	–
RRM		
PC step	1	dB
SHO threshold	2	dB
SHO 1A, 1B, and 1C events hysteresis	1, 1, and 2	dB
SHO measurements reporting time	20	ms
SHO filtering window duration	80	ms
SHO time-to-trigger	320	ms
SHO active set size	1	–
Satisfaction		
Voice FER threshold (satisfaction)	2	%
System satisfaction threshold (capacity limit)	95	%

4.6.2 Congestion Detection

The study of the variation of the congestion detection thresholds ϕ_{resol} and ϕ_{recov}, related to the indication of the resolution and recovery phases, has demanded dynamic system-level simulations whose parameters associated to the LC algorithm are presented in Table 4.3. The values of ϕ_{resol} and ϕ_{recov} considered in the simulations (−0.5, −1, and −1.5) are presented in unit dB, once they are assumed to be relative to the maximum BS transmission power. The percentages equivalent to those values are 89.13% (−0.5 dB), 79.43% (−1 dB), and 70.79% (−1.5 dB). For this specific performance evaluation, the values of the time observation windows ΔT_{resol} and ΔT_{recov} are considered equal to 100 ms.

First, the system congestion rate, which represents the percentage of time that the cell spent in the congestion resolution phase, is depicted in Fig. 4.6(a). The investigation of the figure shows that the lower the decision thresholds, the higher the probability that the CC framework will consider the system congested.

Table 4.3 Simulation parameters for the analysis of the congestion detection thresholds and observation windows.

Parameter	Value	Unit	Notation
Measurement window of the BS transmission power	667	μs	–
Filtering window of the BS transmission power measurements	20	ms	–

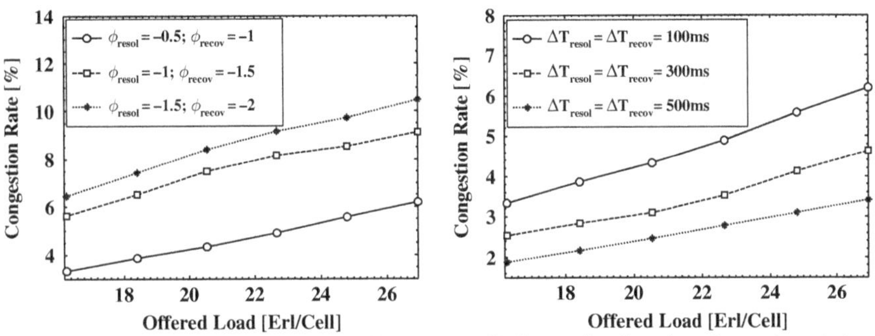

(a) Congestion detection thresholds (b) Congestion detection observation windows

Fig. 4.6 System congestion rate varying the congestion detection thresholds and observation windows (resolution and recovery).

One of the main tasks of the LC algorithm is to disallow the admission of new calls and reject handover attempts when the cell is in the congestion resolution phase. Thus, the call blocking probability $\mathscr{P}_{\text{block}}$ presents the same behavior verified in Fig. 4.6(a). The lower the decision thresholds, the higher the call blocking probability. It was also observed in the simulations that the more frequent is the LC action (lower decision thresholds), less power is consumed in the network, and therefore lower handover drops due to power unavailability occur (low $\mathscr{P}_{\text{drop}}^{\text{HO}}$). However, the effect of the blocking of new calls was more preponderant than the dropping of handover connections. Therefore, once the system has passed longer periods in the congestion resolution phase for the case of lower decision thresholds, a performance loss in terms of *GoS*, which is a combination of the metrics $\mathscr{P}_{\text{block}}$ and $\mathscr{P}_{\text{drop}}^{\text{HO}}$, was verified.

Regarding the FER of all users in the system, it was observed in the simulations that prolonged actions of the LC algorithm (lower decision thresholds), which uses rate adaptation via AMR voice codec in order to reduce the system load in congestion periods, resulted in a better reception of voice frames. Analyzing the simulation results, it was verified that the variation of the detection thresholds regarding congestion resolution and recovery introduces a compromise between the performance metrics *GoS* and FER.

The time observation windows ΔT_{resol} and ΔT_{recov}, which are related to the decision of when to start the congestion resolution and recovery phases, are sliding windows where the mean transmission power samples (over a TTI) are collected. In this work it is assumed that $\Upsilon_{\text{cong}} = 80\%$ (80th percentile) of the samples must satisfy the algorithm criterion in order to initiate or terminate the congestion resolution and recovery stages. The parameters presented in Table 4.3 were used to obtain the simulation results related with the variation of the time observation windows. For this specific performance evaluation, in this work, the values of the congestion detection thresholds ϕ_{resol} and ϕ_{recov} are considered equal to -0.5 and -1 dB, respectively.

The investigation of the variation of the observation windows starts with the presentation of the system congestion rate in Fig. 4.6(b). It is verified that the CC framework considers that the system is less congested when a time window of 500 ms is utilized. This was expected, once the transmission power of the BSs would have to remain for a longer time above the congestion resolution detection threshold ϕ_{resol} so that the load reduction procedure could be initiated.

As said before, the percentage of blocked calls follows the same behavior of the percentage of time in congestion. The shorter the observation window, the higher the probability of the cell to enter in the congestion resolution phase, and so, higher call blocking rates were verified in the simulations. The ongoing calls are benefited with a higher actuation frequency of the LC algorithm. This fact could be concluded from the observation of the simulation results regarding the dropping probability due to handover. In this case, the benefits achieved by the dropping probability reduction was more pronounced than the drawback of the blocking rate increase. Thus, a shorter congestion detection observation window provided better performance in terms of *GoS*.

The quality perceived by the MSs and the transmission power allocation of the BSs can be visualized indirectly by means of the rejection rate of the transmit power control (TPC) commands. When the sector is very congested and there is not any transmission power available, the BS rejects power control commands from the MS. It was noticed in the simulation campaigns that an observation window of 100 ms presented higher power availability due to the fact that the LC algorithm performed more frequently in the system. This resource availability caused a lower rejection rate of the inner-loop power control commands.

Assessing the results of congestion rate and rejection rate of the TPC commands, one should consider the fact that longer congestion observation windows can cause a system inertness when it has to react rapidly in overload situations. This probable short sight of the CC framework impacts system performance dramatically. The CC framework assumes that the system is not congested, when in fact, the contrary applies.

4.6.3 Congestion Resolution and Recovery

Section 4.6.2 evaluated the parameters related to the detection process to decide the start of the congestion resolution and recovery phases. In the present section, these phases will be studied in more detail, mainly the utilization of a priority ordination to reduce and restore the transmission parameters of the MSs. These criteria will be compared with the reference scenario, which was described in Section 4.6.1. The parameters of the simulations conducted for this section were the same presented in Table 4.3. For this specific performance evaluation, the values of the congestion detection thresholds ϕ_{resol} and ϕ_{recov} are considered equal to -0.5 and -1 dB, respectively, and the time observation windows ΔT_{resol} and ΔT_{recov} equal to 100 ms.

As indicated in Section 4.5.3, three procedures compose the congestion resolution phase: priority order definition, load reduction, and load check. The transmission parameters (AMR modes) that the MSs were using before the load reduction in the congestion resolution phase are restored in the congestion recovery phase. The reduction and restoring of the transmission parameters follow a priority order, which is based on a criterion that will define which MSs will be the first to have their transmission data rate decreased (resolution) or increased (recovery). In the resolution stage, the priority ordination is performed from the lowest to the highest priority. In the recovery phase, the inverse order is adopted. The following priority definition criteria are proposed:

1. **Best** E_c/N_0: The lowest priority is allocated to those MSs who present the best common pilot channel (CPICH) E_c/N_0, which is the received chip energy relative to the total power spectral density of the CPICH channel on the downlink. They will be the first ones to have their AMR mode decreased in the congestion resolution phase and the last ones to have their AMR mode restored during the congestion recovery phase.
2. **Worst** E_c/N_0: The MSs with the worst propagation channel quality have the lowest priority.
3. **First-in-first-out (FIFO)**: In this criterion, the first MSs that were admitted in the system have lower priority. In comparison to the other two criteria, the FIFO strategy can be regarded as a random choice, since the radio link quality of the oldest users in the system is not known.

One of the main performance indicators of the CC framework is the system congestion rate, which is the percentage of time that the cells are considered congested based on the criterion used to determine the start of the resolution phase. This metric is presented in Fig. 4.7(a). As expected, the FIFO criterion turned out to be a compromise between the two other criteria (best and worst CPICH E_c/N_0). Furthermore, the three cases that used the CC framework presented lower congestion levels compared to the reference scenario, showing that the load reduction procedure of the resolution phase worked properly.

The CC framework makes sure that a congested cell will deny any new admission requisition or handover connections coming from other cells. For this reason, the blocking probability is strongly influenced by the portion of time that the system remains in congestion situations. The longer the duration of the congestion resolution phase, the higher the number of blocked calls. However, this is necessary in order to protect the users already admitted to the system (ongoing calls), whose QoS requirements are threatened by the overload situation. This fact can be verified by the

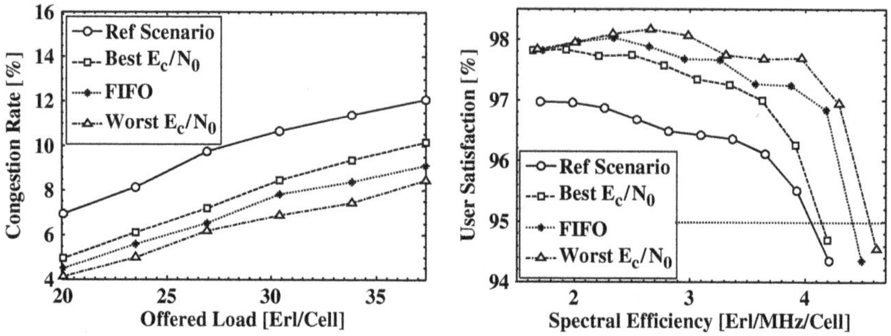

(a) System congestion (b) Percentage of satisfied users $(FER \leq 2\%)$

Fig. 4.7 Performance evaluation of the resource-based CC framework considering different priority selection criteria.

reduction of the dropping probability due to handover. Even when a congested cell denies access to a handover connection, the user remains connected to the same cell. It is assumed that in this special case (overload situation), the user would accept to possibly experiment a degraded channel quality instead of being dropped. The CC framework is able to provide better QoS in terms of *GoS* compared to the reference scenario, which does not use any technique to control the congestion.

In order to conclude the analysis, Fig. 4.7(b) depicts the percentage of satisfied users considering a voice FER limit of 2%. It can be observed that the proposed CC framework is capable of guaranteeing the QoS requirements of the MSs even for high offered loads. Furthermore, as indicated by the other performance indicators, the selection criterion that gives lower priority to those users with worst radio link quality obtained the best results. Therefore, during the load reduction procedure of the congestion resolution phase, it is advantageous to decrease the data rate (lower AMR mode) of those MSs that experiment low CPICH E_c/N_0. This priority order is inverted in the congestion recovery stage, where the original AMR modes must be restored to each MS that had been selected in the resolution phase.

4.6.4 Conclusions

The power thresholds for congestion detection ϕ_{resol} and ϕ_{recov}, relative to the resolution and recovery phases, have a direct impact on the system performance because they determine the actuation frequency of the LC algorithm. Regarding the sliding observation windows for congestion detection ΔT_{resol} and ΔT_{recov}, relative to the resolution and recovery phases, it was observed that when these parameters are configured to high values, the CC framework tends not to consider the sector congested. The correct determination of the parameters related to congestion detection (power thresholds and observation windows) ought to be based on the network operator experience; the QoS perceived by the MSs must be monitored in the network deployment phase, considering a large range of offered loads.

The reference scenario, which does not use any technique to deal with overload situations, presented higher rejection rate of TPC commands, higher congestion rate, and higher *GoS* than the scenarios where the CC framework was used. Finally, the selection criterion that gives lower priority to those users that experience the worst propagation conditions (worst CPICH E_c/N_0) presented the highest user satisfaction. The CC framework based on this criterion provided a capacity gain of 12.6% in comparison with the reference scenario, considering a satisfaction threshold of 95%. Thus, the actions of reduction/restoration of the transmission parameters (AMR mode) in the resolution/recovery phases should follow a priority order based on this criterion to adapt the data rate of the voice calls via AMR mode selection. However, this criterion yields a high amount of uplink signaling load. The network operator must evaluate whether this higher complexity is worthy. The FIFO criterion seems to be a trade-off, with low complexity and performance close to that presented by the worst CPICH E_c/N_0 criterion.

The general conclusion of Sections 4.5 and 4.6 is that the proposed resource-based CC framework is able to improve the efficiency of any CS cellular network. The concepts presented are general enough to be particularized to any CS services and any CS cellular networks, where the framework parameters needed to be chosen according to particular network experiments/measures or the operator experience. The chosen case study presented a performance evaluation of the CC framework in a UMTS WCDMA network with CS voice service. The proposed CC framework was able to provide statistical QoS guarantees to the voice service, while increasing the system capacity in comparison to a reference scenario without congestion control.

4.7 QoS-Based Framework for Packet-Switched Networks

In this section, the proposal of a QoS-based CC framework suitable for PS cellular networks is presented. According to the taxonomy presented in Section 4.2, this framework can be classified as a scheme that performs both congestion avoidance and resolution/recovery, reduces the demand in overload situations, performs a closed-loop control, and is based on service quality. The framework is composed of three RRM algorithms: AC, PSC, and LC, which are described in more detail in Sections 4.7.1, 4.7.2, and 4.7.3, respectively. Figure 4.8 describes the operation of the aforementioned CC framework, whose main ideas were originally proposed in [43].

The QoS-based CC framework does not take the network radio resources into account directly because it is based on service quality. In this framework, the service outage is thought as a congestion. In other words, a congestion occurs when the quality of the most prioritized service, for example an RT service, reaches unacceptable levels. The LC functionality adjusts parameters of the AC and PSC algorithms

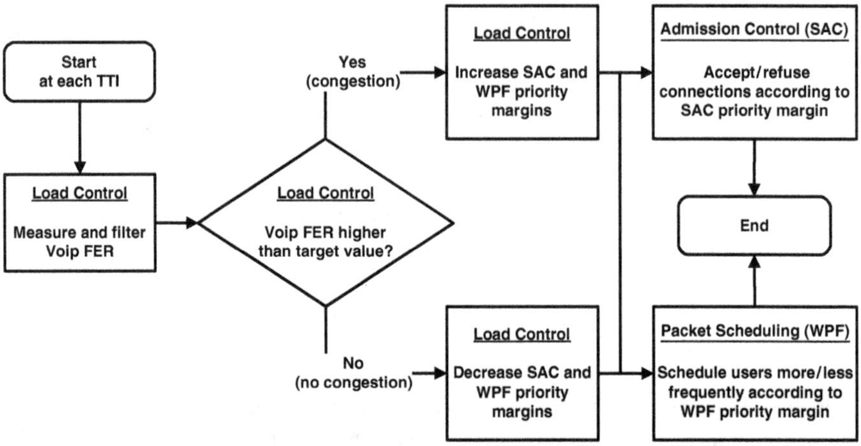

Fig. 4.8 Operation of the CC framework for PS cellular networks.

depending on whether the QoS of this RT service is acceptable or not. This is done in order to define service prioritization levels among this service and other lower priority services, e.g., NRT services. The means to perform service prioritization is through statistical resource reservation at call (AC) and packet (PSC) levels: depending on the priority levels defined by LC, the RT flows already admitted in the system will have more or less chances to access the channel (free resources). The proposed general CC framework is scalable to several services, i.e., the quality of the most prioritized service can be maintained, no matter how many other lower priority services are provided in the network.

This soft QoS balancing performed by the CC framework, which aims to improve the quality of the services with stricter requirements, is able to find a trade-off between QoS guarantees and the efficient network resource usage in mixed services scenarios. In this way, it is capable of maximizing the overall system capacity as long as the service with stricter requirements limits the system capacity.

Without loss of generality, VoIP service, which is an RT delay-sensitive service, is assumed as the most prioritized service in our framework and the World Wide Web (WWW) service, as an example of best-effort NRT service.

The proposed CC framework adjusts its parametric structure adaptively in order to follow the temporal behavior of a VoIP quality measure and assures that it will be kept around a planned value. Since it is commonly agreed in the literature that the VoIP capacity is most impacted by the FER metric, the framework considers the VoIP FER as the quality measure to be controlled. All the time-variant aspects of the system that can affect VoIP capacity, such as user speed, propagation environment, physical layer performance, and interference are taken into account by the FER metric. This is true because the FER takes into consideration two causes of packet loss: packet discard at the receiver play-out buffer due to unacceptable delay (higher than the VoIP delay budget) and errors caused by the wireless channel.

Since the proposed CC framework is based directly on the VoIP FER, it provides statistical QoS guarantees for the VoIP service by means of autonomous adaptation of its regulating parameters to the different wireless scenarios.

There are few works in the literature that dealt with CC frameworks based on service quality. In this direction, some works have proposed adaptive CC mechanisms for RT services based on both resource consumption and service quality [27, 31]. Reference [17] proposed a CC framework based only on QoS, but the framework was composed only of the LC and PSC algorithms, and it was suitable only for NRT traffic. Therefore, one can clearly identify the contribution provided by the CC framework proposed in the present work. Not only is it fully based on service quality, but it is also scalable for any number of services. Furthermore, the proposed framework is the only one based on the functionalities of the AC, PSC, and LC, which will be described in more detail in Sections 4.7.1, 4.7.2, and 4.7.3.

4.7.1 Session Admission Control

The AC algorithm used in the proposed CC framework is the session admission control (SAC) scheme [13]. The SAC is a tool which aims to provide QoS guarantees to flows with high priority, through the degradation of lower priority flows. This degradation consists in denying the admission of flows when a system overload situation in the high priority flows is detected. The overload condition in high priority flows can be detected through regular observation of performance metrics related to those flows. Some admission thresholds have to be defined for those metrics, and when the monitored metric surpasses the threshold, the admission controller starts to block the access to the system. When the monitored metric returns to an acceptable value (below the specified threshold), the admission controller gives access to flows normally. Note that SAC also has the flexibility of different admission thresholds for flows with different priorities.

In our work, the SAC scheme is employed to guarantee VoIP QoS. This scheme considers the VoIP delay as the resource to be shared among users in the system. Therefore, the VoIP packet delays are regularly measured and filtered at the BS and then possibly reported to a network controller entity. At each session admission event, this filtered measurement is added to the estimated resource demand of the incoming flow, and the result of this summation is compared to a given admission threshold, which is dependent on the type of service of the flow. If the summation is higher than the threshold, the access of the incoming session is blocked, otherwise the incoming session is submitted to a LAC algorithm which would verify radio resources availability, e.g., power, codes, bandwidth, and sub-carriers. Examples of LAC schemes for mixed traffic scenarios in a UMTS high speed downlink packet access (HSDPA) network are presented in [47].

In the proposed CC framework, the ratios between the VoIP admission threshold and the thresholds of other lower priority services are called SAC priority margins. From now on, it is assumed that the admission threshold of the VoIP service is fixed to a constant value. For example, consider that the VoIP and a lower priority service (e.g., WWW) are provided in the cellular network. Let us define α_{SAC} in dB scale as the priority degradation margin of the WWW service compared to VoIP in the SAC algorithm. Consider the SAC admission threshold for the VoIP service is $D_{VoIP}^{th} = 150$ ms. If $\alpha_{SAC} = -3$ dB, then the SAC admission threshold for the WWW service is given by $D_{WWW}^{th} = D_{VoIP}^{th} \cdot 10^{\frac{\alpha_{SAC}}{10}} \simeq 75$ ms. In this way, the SAC priority margin measures the level of priority degradation of a given service compared to VoIP, i.e., the more negative in dB the SAC priority margin associated to a given service, the higher the VoIP priority over the flows of this service.

4.7.2 Packet Scheduling

The PSC algorithm used in the proposed CC framework is the weighted proportional fair (WPF) [12], which is a variation of the well-known proportional fair (PF) scheduler. The WPF scheduler provides a fixed priority for each service. The priority value of the jth service flow is given by

$$p_j = W_j(s) \cdot \left(\frac{r_j}{T_j}\right), \tag{4.9}$$

where $W_j(s)$ is a multiplicative weight, which is dependent on the service class s of the jth flow and is used for QoS differentiation among services; r_j is the estimated instantaneous bit rate of the jth flow for the next transmission attempt; and T_j is the throughput experienced by the jth flow in a given time window, which is filtered by an exponential low-pass filter.

Besides WPF, there are other PSC algorithms that are able to provide QoS differentiation between VoIP and other flows, such as a scheduler that gives strict priority to the VoIP flows or a delay-based scheduler. Although the former is the simplest way to provide service prioritization, it can cause a starvation problem for the lower priority services. The latter is aware of the delay requirements of each specific user and provides good performance results in a mixed traffic scenario [12, 15], but its use in an adaptive framework is not straightforward. On the other hand, the WPF scheduler does not present the starvation problem and has a simple parametric structure that can be controlled easily by the LC algorithm in an automatic manner. These characteristics make the WPF scheduler specially suited for the proposed CC framework.

Assuming that the network resource is the shared channel access, the WPF scheduler can improve the VoIP capacity through the assignment of a higher weight factor $W_j(s)$ for VoIP compared to other services. The VoIP weight factor is assumed to be fixed to a constant value and the weight factors of other lower priority services are given by the multiplication between the VoIP weight factor and the respective WPF priority margins. For example, consider that the VoIP and a lower priority service (e.g., WWW) are provided in the cellular network. Let us define β_{WPF} in dB scale as the priority degradation margin of the WWW service compared to VoIP in the WPF algorithm. Consider the WPF priority weight for the VoIP service is $W_{VoIP}^{prio} = 1$. If

$\beta_{WPF} = -3$ dB, then the WPF priority weight for the WWW service is given by $W^{prio}_{WWW} = W^{prio}_{VoIP} \cdot 10^{\frac{\beta_{WPF}}{10}} \simeq 0.5$. In this way, by using negative values in dB for the priority margin, the WPF algorithm is able to provide service prioritization for the VoIP service, decreasing the overall VoIP delay at the expense of the degradation of WWW session throughput.

4.7.3 Load Control

As could be seen in Sections 4.7.1 and 4.7.2, the strategy of assigning non-zero SAC and WPF priority margins gives precedence to VoIP flows in the access to the shared radio resources and can improve the VoIP quality through the QoS degradation of the existing lower priority services. The adaptation of the priority margins by the proposed LC algorithm provides dynamic service prioritization allowing efficient resource reservation for the VoIP service.

Let us assume again α_{SAC} and β_{WPF} as the priority degradation margins of the WWW service compared to VoIP in the SAC and WPF strategies, respectively. The proposed LC algorithm is composed of two loops, an outer and an inner loop. This framework is similar to the framework of the WCDMA outer-loop power control (OLPC) algorithm. Table 4.4 shows the similarities between the OLPC and the adaptive LC algorithm, which can facilitate the understanding of its operation.

Table 4.4 Comparison between outer-loop power control and the load control frameworks.

	Outer-loop power control	**Load control**
Resource in inner loop	Transmission power	VoIP delay
Resource in outer loop	SINR	Service prioritization
Output	Dynamic SINR target	Dynamic SAC priority margin (α_{SAC}) Dynamic WPF priority margin (β_{WPF})
Desired quality	Block error rate (BLER)	VoIP frame erasure rate (FER)

In the following, a more detailed description of the proposed LC algorithm within the QoS-based CC framework is given (see Fig. 4.8). The objective of the LC algorithm is to make sure that the FER of the VoIP users connected to a given BS is kept at the planned value. LC is composed of an outer and an inner loop. The former monitors the VoIP quality in the cell regularly, checking whether the VoIP QoS requirement is being fulfilled or not, and changing the target values of the outer-loop resource accordingly, which are the SAC (α_{SAC}) and WPF (β_{WPF}) priority margins between the VoIP and WWW services. If the VoIP quality is excessively good (VoIP quality better than a desired target), more priority can be given to the WWW service, so that the radio resources are used more efficiently. Otherwise, α_{SAC} and β_{WPF} will be updated in order to degrade the WWW quality and direct more resources to the VoIP service. The inner loop of LC is characterized by the actuation of the SAC and WPF algorithms: SAC accepts/refuses WWW connections and WPF schedules WWW flows more/less frequently. By means of these actions, these algorithms have indirect control of the inner-loop resource, which is the delay experienced by the VoIP service flows. This inner-loop control tries to fulfill the service prioritization target that was chosen by the outer loop. One should keep in mind that the SAC and WPF algorithms have also direct control over the VoIP service flows, which must consider their fixed admission threshold ($D^{\mathrm{th}}_{\mathrm{VoIP}}$) and the fixed priority weight ($W^{\mathrm{prio}}_{\mathrm{VoIP}}$).

The proposed CC framework is flexible enough to enable only one or both of the parameters α_{SAC} and β_{WPF} to be adjusted. For example, if only parameter α_{SAC} is updated, it means that the LC algorithm relies only on the SAC scheme to guarantee the QoS of the VoIP service. On the other hand, if only parameter β_{WPF} is updated, it means that the LC algorithm relies only on the WPF scheme to guarantee the QoS of the VoIP service. This characteristic of the CC framework is very important because it allows the cellular network operators to decide what is more suitable to their interests.

There are different ways to adapt the α_{SAC} and β_{WPF} parameters of the LC algorithm in order to achieve a target VoIP FER. Two possible algorithms were proposed in [46] and are described in detail in the following.

4.7.3.1 Jump-Based Load Control

The jump-based load control (JLC) algorithm was proposed in [46]. It was inspired by [20], which studied an adaptive uplink load control for CDMA systems based on uplink load (noise rise) thresholds.

The update of the α_{SAC} and β_{WPF} parameters by the JLC algorithm is done by the well-known jump algorithm, which was proposed by [50] in the WCDMA OLPC framework.

On one hand, the trigger of the OLPC jump algorithm is the cyclic redundancy check (CRC) at every frame reception (success/failure), which is necessary for the BLER calculation. On the other hand, in the proposed LC algorithm, the trigger will be a VoIP QoS outage event Ev_{METH}. An outage event occurs when the VoIP packet delay averaged over all users connected to a given cell is higher than the delay budget of the VoIP service. More information about the VoIP delay budget can be found in [11].

The algorithm is described in Algorithm 4.1, where Δ is the fixed step size of the JLC algorithm in dB. In the OLPC jump algorithm, a typical range of values of Δ is [0.3, 0.5] dB. Ev_{SAC} and Ev_{WPF} are variables that indicate if there was a VoIP QoS outage event during the last time window of the SAC and WPF algorithms, respectively. $K \geq 1$ is an integer that is related to the jump in the target values of α_{SAC} and β_{WPF} when there was not an outage event in the last time window. According to [50], the jump algorithm tries to maintain the desired quality always lower or equal to $1/(K+1)$. Therefore, we must have $K = \frac{1}{FER^{\mathrm{target}}_{\mathrm{VoIP}}} - 1$. The α_{SAC} parameter is constrained to the range $[\alpha^{\min}_{\mathrm{SAC}}, \alpha^{\max}_{\mathrm{SAC}}]$ in dB, while the β_{WPF} parameter

Algorithm 4.1 Proposed jump-based load control (JLC) Algorithm.

for all t **do**
  if Ev_{SAC} = TRUE **then**
    $\alpha_{\mathrm{SAC}}(t) = \alpha_{\mathrm{SAC}}(t-1) - \Delta$
  else
    $\alpha_{\mathrm{SAC}}(t) = \alpha_{\mathrm{SAC}}(t-1) + \Delta/K$
  end if
  if Ev_{WPF} = TRUE **then**
    $\beta_{\mathrm{WPF}}(t) = \beta_{\mathrm{WPF}}(t-1) - \Delta$
  else
    $\beta_{\mathrm{WPF}}(t) = \beta_{\mathrm{WPF}}(t-1) + \Delta/K$
  end if
end for

is constrained to the range $[\beta_{\mathrm{WPF}}^{\min}, \beta_{\mathrm{WPF}}^{\max}]$ in dB. Normally, $\alpha_{\mathrm{SAC}}^{\max}$ and $\beta_{\mathrm{WPF}}^{\max}$ are equal to 0 dB, so that the WWW priority in the SAC and WPF strategies cannot be higher than the VoIP priority. Furthermore, $\alpha_{\mathrm{SAC}}^{\min}$ and $\beta_{\mathrm{WPF}}^{\min}$ must have a negative value in dB. Assuming they are equal to -10 dB, in the lowest degradation case allowed by the LC algorithm, the WWW's priority will be ten times lower than the VoIP's.

4.7.3.2 Error Feedback-Based Load Control

The error feedback-based load control (EFLC) algorithm was proposed in [46]. It was inspired by [27], which studied the performance of admission control, diversity control, and router control in a best-effort all-IP CDMA cellular network.

Since all downlink VoIP traffic will be scheduled at the BS, each cell can calculate the FER averaged over all the VoIP flows served by it. The priority margins α_{SAC} and β_{WPF} are calculated periodically in each cell by comparing the monitored VoIP FER with a target VoIP FER value. The monitored VoIP FER is measured and filtered in every control interval by means of a sliding time window. The time basis (duration of the sliding window) for the calculation of the α_{SAC} parameter may be different of the time basis for the calculation of the β_{WPF} parameter. This filtered VoIP FER is represented as $FER_{\mathrm{VoIP}}^{\mathrm{filt}}(t)$ and the target VoIP FER value is represented as $FER_{\mathrm{VoIP}}^{\mathrm{target}}$. In this way, the new LC parameters α_{SAC} and β_{WPF} are calculated as

$$\alpha_{\mathrm{SAC}}(t) = \alpha_{\mathrm{SAC}}(t-1) - \sigma_{\alpha}^{\mathrm{SAC}} \cdot \left(FER_{\mathrm{VoIP}}^{\mathrm{filt}}(t) - FER_{\mathrm{VoIP}}^{\mathrm{target}}\right), \tag{4.10}$$

$$\beta_{\mathrm{WPF}}(t) = \beta_{\mathrm{WPF}}(t-1) - \sigma_{\beta}^{\mathrm{WPF}} \cdot \left(FER_{\mathrm{VoIP}}^{\mathrm{filt}}(t) - FER_{\mathrm{VoIP}}^{\mathrm{target}}\right), \tag{4.11}$$

where the parameters $\sigma_{\alpha}^{\mathrm{SAC}}$ and $\sigma_{\beta}^{\mathrm{WPF}}$ control the adaptation speed of the LC parameters α_{SAC} and β_{WPF}, respectively. The α_{SAC} and β_{WPF} parameters are constrained to the ranges $[\alpha_{\mathrm{SAC}}^{\min}, \alpha_{\mathrm{SAC}}^{\max}]$ and $[\beta_{\mathrm{WPF}}^{\min}, \beta_{\mathrm{WPF}}^{\max}]$ in dB, respectively.

For a constrained integral controller, higher values of $\sigma_\alpha^{\mathrm{SAC}}$ and $\sigma_\beta^{\mathrm{WPF}}$ lead to a faster response. However, such higher values can cause oscillations or even instabilities. One could ask, how big can $\sigma_\alpha^{\mathrm{SAC}}$ and $\sigma_\beta^{\mathrm{WPF}}$ be without encountering states $\alpha_{\mathrm{SAC}}^{\min}$, $\alpha_{\mathrm{SAC}}^{\max}$, $\beta_{\mathrm{WPF}}^{\min}$, and $\beta_{\mathrm{WPF}}^{\max}$? Let $\varepsilon_\alpha^{\mathrm{SAC}}$ and $\varepsilon_\beta^{\mathrm{WPF}}$ be the largest errors that occur once the closed-loop system is in operation. Then, if $\sigma_\alpha^{\mathrm{SAC}} \leq \frac{\alpha_{\mathrm{SAC}}^{\max}-\alpha_{\mathrm{SAC}}^{\min}}{\varepsilon_\alpha^{\mathrm{SAC}}}$ and $\sigma_\beta^{\mathrm{WPF}} \leq \frac{\beta_{\mathrm{WPF}}^{\max}-\beta_{\mathrm{WPF}}^{\min}}{\varepsilon_\beta^{\mathrm{WPF}}}$, then there will be no transitions into the extreme states [27]. Let us consider $\alpha_{\mathrm{SAC}}^{\max} - \alpha_{\mathrm{SAC}}^{\min} = \beta_{\mathrm{WPF}}^{\max} - \beta_{\mathrm{WPF}}^{\min} = 10$ and also $\varepsilon_\alpha^{\mathrm{SAC}} = \varepsilon_\beta^{\mathrm{WPF}} = 1 - FER_{\mathrm{VoIP}}^{\mathrm{target}} = 0.99$, for the case where $FER_{\mathrm{VoIP}}^{\mathrm{target}}$ is 1%. Considering these values, the gains $\sigma_\alpha^{\mathrm{SAC}}$ and $\sigma_\beta^{\mathrm{WPF}}$ of the integral controllers must be lower than 10. In order to make a comparison as fair as possible between the JLC and EFLC algorithms and to guarantee the algorithm convergence, $\sigma_\alpha^{\mathrm{SAC}} = \sigma_\beta^{\mathrm{WPF}} = \frac{\Delta}{1-FER_{\mathrm{VoIP}}^{\mathrm{target}}}$ is used, where Δ is the fixed step size of the JLC algorithm in dB.

Reference [27] claims that if one adjusts the controlled parameter in a nonlinear (e.g., exponential) fashion, it is possible to obtain fast reactivity of the integral controller with small variance. Furthermore, [29] presents a study where it is concluded that an exponential filter-based OLPC controller is a feasible alternative to the jump algorithm-based OLPC for WCDMA EUL systems. Based on that, it was decided to use a filtered value of the VoIP FER ($FER_{\mathrm{VoIP}}^{\mathrm{filt}}(t)$) and use it in (4.10) and (4.11). A simple exponential smoothing (SES) filter [19], which is a first-order infinite impulse response (IIR) filter suitable for time series with slowly varying trends, was used to suppresses short-run fluctuations and smooth the time series $FER_{\mathrm{VoIP}}(t)$.

The following equation is a recursive relation and defines the simple exponential smoothing filter:

$$FER_{\mathrm{VoIP}}^{\mathrm{filt}}(t) = \eta_{\mathrm{f}} \cdot FER_{\mathrm{VoIP}}(t) + (1-\eta_{\mathrm{f}}) \cdot FER_{\mathrm{VoIP}}^{\mathrm{filt}}(t-1), \tag{4.12}$$

where $0 \leq \eta_{\mathrm{f}} \leq 1$ is the filter smoothing constant, $FER_{\mathrm{VoIP}}(t)$ is the time series to be processed by the filter, and $FER_{\mathrm{VoIP}}^{\mathrm{filt}}(t)$ is the processed result at time instant t. Notice that $FER_{\mathrm{VoIP}}(t)$ is the average value of the VoIP FER considering all the VoIP flows served by a given BS during the last control interval, whose duration can be different for the α_{SAC} and β_{WPF} parameters, as explained before.

The SES method also has a forecasting property since it learns from the past errors: the estimate for period $t+1$ is increased if the actual value for period t is greater than what was estimated to be and decreased otherwise. The relative influence of recent and older data is regulated by the smoothing constant.

The main differences between JLC and EFLC are the way these algorithms decide if the VoIP QoS requirement was met or not, and the step size for the adaptation of the α_{SAC} and β_{WPF} parameters. On the one hand, JLC uses fixed step sizes (up and down) depending on the occurrence of a VoIP QoS outage based on delay. On the other hand, EFLC uses dynamic step sizes that update α_{SAC} and β_{WPF} and are calculated as the multiplication of the control error ($FER_{\mathrm{VoIP}}^{\mathrm{filt}}(t) - FER_{\mathrm{VoIP}}^{\mathrm{target}}$) by

$\sigma_{\alpha}^{\mathrm{SAC}}$ and $\sigma_{\beta}^{\mathrm{WPF}}$, respectively. Due to this fact, a more fine-tuned control of the VoIP FER toward the desired value is expected with EFLC.

The way the EFLC algorithm monitors the VoIP FER and adapts the WPF priority margin (β_{WPF} parameter) over time is presented in Fig. 4.9 (the α_{SAC} parameter behaves in a similar fashion). In this example, $\sigma_{\beta}^{\mathrm{WPF}} = 0.5$ dB, $FER_{\mathrm{VoIP}}^{\mathrm{target}} = 1\%$, $\beta_{\mathrm{WPF}}^{\min} = -10$ dB, $\beta_{\mathrm{WPF}}^{\max} = 0$ dB, and a mixed services scenario with VoIP and WWW. Table 4.5 presents more information on traffic and other simulation models. The β_{WPF} parameter behaves in accordance with the general operation of the CC framework described in Fig. 4.8. When $FER_{\mathrm{VoIP}}^{\mathrm{filt}}(t)$ is zero, β_{WPF} remains in its maximum value. When $FER_{\mathrm{VoIP}}^{\mathrm{filt}}(t)$ is above $FER_{\mathrm{VoIP}}^{\mathrm{target}}$, the EFLC algorithm decreases β_{WPF}, which degrades the WWW quality and brings the VoIP FER to acceptable levels (congestion resolution). A congestion recovery phase can take place when $FER_{\mathrm{VoIP}}^{\mathrm{filt}}(t)$ becomes lower than $FER_{\mathrm{VoIP}}^{\mathrm{target}}$. In this recovery, EFLC increases β_{WPF} so that more priority is given to the WWW service.

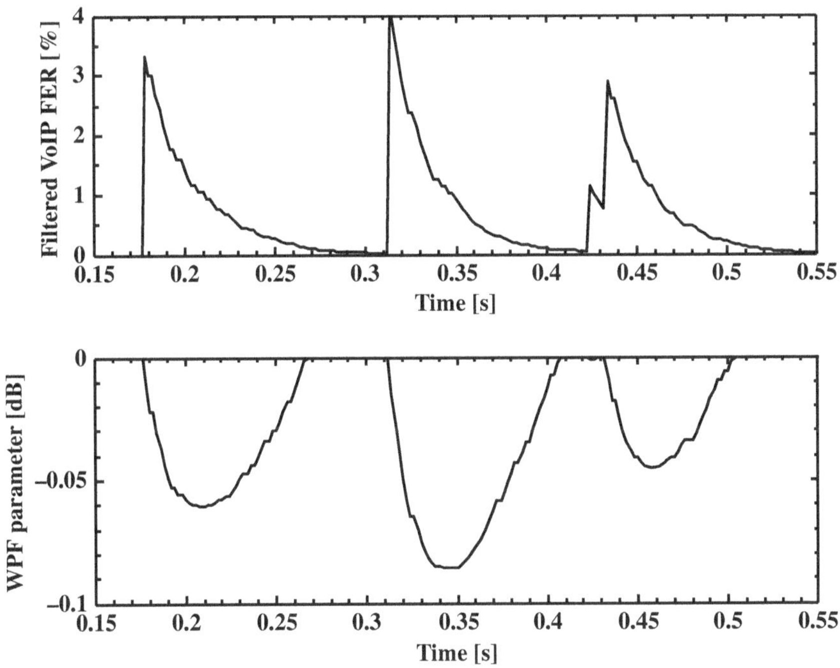

Fig. 4.9 Adaptation of the WPF priority margin (β) over time depending on the filtered VoIP FER.

A complete performance evaluation and comparison between the two proposed LC algorithms and a reference scenario considering various mixed traffic scenarios was conducted and the simulation results are presented in the following section.

4.8 Case Study: HSDPA Performance with VoIP and WWW Services

In Section 4.7, a QoS-based CC framework for PS cellular networks was presented. Now, the performance of the aforementioned framework will be evaluated in a case-study where the VoIP and WWW services are provisioned in the high-speed downlink shared channel (HS-DSCH) of the UMTS HSDPA system.

In this case study, many important assumptions and restrictions of a real HSDPA network were taken into account. For instance, the values for the actuation period of the CC framework were chosen in accordance with the network architecture restrictions. It is expected that the shorter the time period for the adaptation of the α_{SAC} and β_{WPF} parameters, the higher the efficiency in the CC framework operation and the better the system performance. However, the UMTS terrestrial radio access network (UTRAN) architecture imposes constraints on these adaptation periods. Since the SAC algorithm is run at the radio network controller (RNC), the α_{SAC} parameter must be calculated at the BS and reported to the RNC using the Iub interface and the node b application part (NBAP) signaling. Taking the example of other measurements in the NBAP signaling, the time period for the calculation and reporting of the α_{SAC} parameter is in the order of hundreds of milliseconds. Since the WPF algorithm is performed at the BS itself, there is no need to report the β_{WPF} parameter to the RNC and it can be calculated at each HSDPA TTI.

The simulation models will be presented in Section 4.8.1, while the simulation results will be shown in Section 4.8.2. The performance results regarding the comparison between different traffic mixes and the joint capacity regions will be presented in Sections 4.8.2.1 and 4.8.2.2, respectively. Finally, the conclusions concerned with the present case study will be drawn in Section 4.8.3.

4.8.1 System Modeling

A discrete time system-level dynamic simulator that models the downlink of the 3GPP WCDMA Release 5 system (HSDPA) was used in this case study.

Important aspects related to HSDPA were modeled, such as code multiplexing where the available BS transmission power for HSDPA is equally shared among all the channelization codes (physical channels) of the multiplexed users; hybrid automatic repeat request (H-ARQ) chase combining; H-ARQ stop-and-wait (SAW) processes; AMC based on link conditions and on the amount of data available in the MAC-hs buffer (no feedback error assumed).

An AVI link-to-system interface was considered. When a MAC-hs transport block is transmitted on the HS-DSCH in a 2 ms TTI, the corresponding BLEP is read from the look-up tables that depend on the channel profile, the modulation and coding scheme, and the channel quality. The SINR-BLER mapping curves used in the simulations are presented in Fig. 7.5.

Detailed radio propagation models are incorporated in the simulator, such as distance attenuation [55], spatial correlated shadow fading, and single-path Rayleigh small-scale fading.

LAC for associated dedicated physical channel (A-DPCH) is based on the power not used by HSDPA (dedicated and common channels), also called the non-HS power. For more information on LAC algorithms for HSDPA systems, see [47]. The A-DPCH is power controlled and can be in soft handover mode.

The WWW and VoIP traffic models were taken from [55]. These services use radio link control (RLC) acknowledged mode (AM) and unacknowledged mode (UM), respectively. A voice frame is generated every 20 ms during activity periods by the AMR 12.2 kbps vocoder. A MAC-d service data unit (SDU) payload of 39 bytes was used, which is appropriate for VoIP service with header compression. Thus, it is assumed that the total average protocol overhead including all the protocol layers is composed of 7 bytes. Discontinuous transmission (DTX) packets are not considered.

According to [56], in order to achieve an acceptable quality for the VoIP call, the one-way mouth-to-ear delay should be less than 250–300 ms. This total delay should account for all the nodes in the communication path. The present research is interested in the delay budget inside UTRAN. This delay budget should be enough for all the BS functionalities and the user reception of VoIP packets. Delay budgets inside UTRAN in the range of 80–150 ms were considered in [12, 56, 59]. This range should be sufficient for scenarios where the VoIP call is between two mobiles or between a landline and a mobile user. This work considered a fixed delay budget of 150 ms. To compensate for variations in delay, the receiving terminal employs a play-out buffer. This buffer might discard packets that arrive later than a deadline, which is the upper bound of the tolerable delay budget.

A WWW data user is regarded as satisfied if its average session throughput is higher or equal to a given threshold and it is not blocked. A VoIP user is assumed as satisfied if it has a FER lower than or equal to a given threshold and it is not blocked. The thresholds values are given in Table 4.5. Notice that the satisfaction definition takes into account the most important QoS metrics related to WWW and VoIP: blocking rate, channel error probability, WWW session throughput, and VoIP packet delay.

The system offered load will be represented by the estimated total number of users of all service classes in each cell. This estimate considers the mean session duration of each service class and Poisson arrival rate. The system capacity regions are defined as the set of expected number of users for which acceptable system-level quality is sustained for all service classes simultaneously. The capacity region is constructed varying the traffic mix among the considered service classes, including single-service evaluations.

The most important simulation parameters are presented in Table 4.5, whose values are typical of a macrocellular HSDPA network.

Table 4.5 General simulation parameters.

Parameter	Value	Unit
Deployment		
Number of cells (torus grid)	27	–
Cell radius	500	m
Maximum BS power	20	W
Power reserved for common channels	3	W
User speed	3	km/h
HSDPA		
Number of codes reserved for HSDPA	5	–
Number of H-ARQ parallel processes	6	–
Max. H-ARQ retransmissions	5	–
Average power per HS-SCCH	0.5	W
User equipment class	5	–
Link-level performance	BLER curves for CQI 1-22	–
Propagation		
Path loss [55]	$128 + 37.6 \cdot log_{10}(d)$	dB
Lognormal shadowing standard deviation	8	dB
Shadowing decorrelation distance	50	m
Shadowing correlation between BSs	0.5	–
Small-scale fading	Single-path Rayleigh	–
RRM		
RRM for A-DPCHs	PC, SHO, and LAC	–
Traffic		
VoIP and WWW traffic models	According to [55]	–
AMR codec rate	12.2	kbps
Voice activity factor	0.5	–
VoIP delay budget	150	ms
Congestion control		
VoIP SAC delay threshold ($D_{\text{VoIP}}^{\text{th}}$)	150	ms
VoIP WPF priority weight ($W_{\text{VoIP}}^{\text{prio}}$)	1	–
VoIP FER target ($FER_{\text{VoIP}}^{\text{target}}$)	1	%
Time basis for adaptation of α_{SAC}	100	ms
Time basis for adaptation of β_{WPF}	2	ms
Maximum value of α_{SAC} and β_{WPF} ($\alpha_{\text{SAC}}^{\text{max}}$, $\beta_{\text{WPF}}^{\text{max}}$)	0	dB
Minimum value of α_{SAC} and β_{WPF} ($\alpha_{\text{SAC}}^{\text{min}}$, $\beta_{\text{WPF}}^{\text{min}}$)	-10	dB
SAC step size ($\sigma_{\alpha}^{\text{SAC}}$)	0.5	dB
WPF step size ($\sigma_{\beta}^{\text{WPF}}$)	0.5	dB
Satisfaction		
VoIP and WWW satisfaction threshold	90	%
VoIP FER threshold (satisfaction)	1	%
WWW throughput threshold (satisfaction)	64	kbps
Simulation		
Simulation stop criterion	At least 5,000 finished sessions of each service	–

4.8.2 Performance Evaluation Results

The CC framework is compared with a **reference scenario**, where no LC algorithm was employed, and the SAC and WPF algorithms were configured with a fixed priority margin of 0 dB. The objective in this scenario is to remove service prioritization between VoIP and WWW given in the admission and scheduling procedures.

Three service mixes are considered: a mix where the WWW dominates in number of users (25% VoIP/75% WWW), a scenario where both services have equal proportions (50% VoIP/50% WWW), and a scenario that comprises a domination of the VoIP service (75% VoIP/25% WWW). The satisfaction ratio of both services considering the three mixes are shown in Section 4.8.2.1, while Section 4.8.2.2 summarizes the capacity limits of all traffic mixes, including single-service scenarios.

4.8.2.1 Traffic Mixes Comparison

The satisfaction ratio of both VoIP and WWW services for the reference scenario and the two proposed LC algorithms are presented in Fig. 4.10(a)–(c). Regarding the VoIP performance, no matter which LC algorithms are used or the traffic mixes considered, the figures clearly show that the proposed CC framework is efficient at maximizing the VoIP satisfaction compared to the reference scenario. Furthermore, as the proportion of VoIP users in the traffic mix increases, the gain of EFLC over JLC regarding the VoIP satisfaction is higher. The advantages of EFLC compared to JLC explain the difference in performance: faster response when leading to congestion situations due to the forecasting property of the exponential filtering; variable LC step size and, consequently, fine-tuning control of the LC parameters α_{SAC} and β_{WPF}; and better synchronized action of the SAC and WPF schemes. At low and moderate offered loads, both LC algorithms mostly rely on the adaptive WPF scheme to control the delay experienced by the VoIP packets. However, at high offered loads, the SAC scheme in EFLC was stricter than in JLC. The former started earlier to prevent a huge number of WWW users from entering the system in order to guarantee the QoS of the ongoing VoIP flows. Looking at the performance curves of JLC and EFLC in Fig. 4.10(a)–(c), it can be observed that the satisfaction of both service classes decreases when the proportion of VoIP users in the traffic mix increases. This majority of VoIP users translates to a more challenging network scenario because the CC framework needs to guarantee the QoS of a larger number of highly demanding VoIP users and softly pre-empt network resources of fewer WWW users.

A capacity study based on Fig. 4.10(a)–(c) is now described. The joint system capacity is defined as the minimum capacity between the service classes, so that acceptable system-level quality is sustained for all service classes simultaneously. Considering the reference scenario and a satisfaction threshold of 90% for both service classes, one can see that the joint system capacity was strongly limited by the VoIP service (approximately 32, 31, and 37 users for mixes v25w75, v50w50, and v75w25, respectively), while the WWW QoS was excessively good. The CC

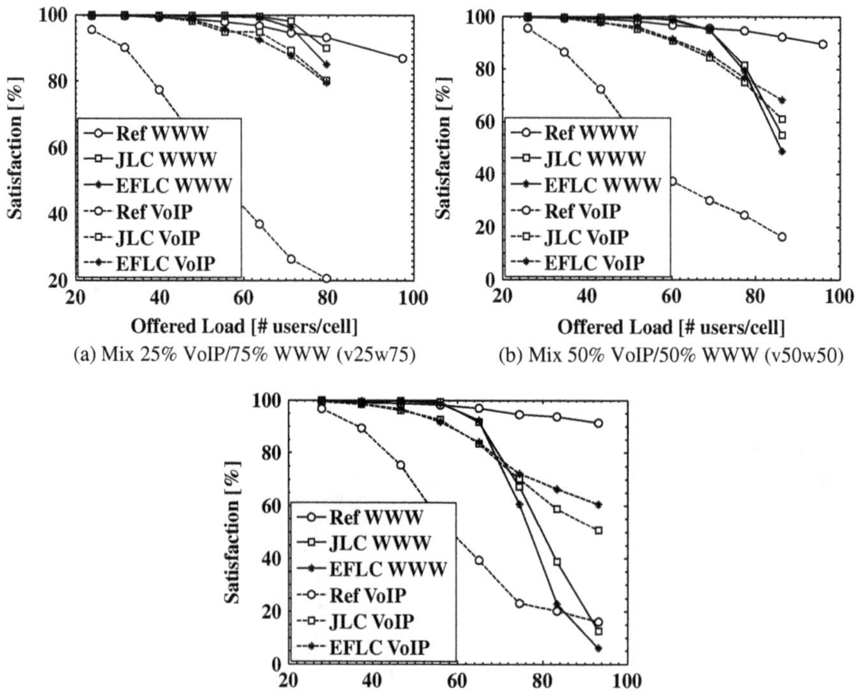

(a) Mix 25% VoIP/75% WWW (v25w75)
(b) Mix 50% VoIP/50% WWW (v50w50)
(c) Mix 75% VoIP/25% WWW (v75w25)

Fig. 4.10 User satisfaction ratio for different traffic mixes and different LC algorithms.

framework performed a smooth and controlled degradation of the WWW quality (see the WWW curves in the figures) in order to free network resources and maintain the VoIP FER around the planned value, providing a considerable increase in the VoIP satisfaction (see the VoIP curves in the figures). Although the VoIP continues to be the limiting service, this QoS balancing provided an increase in the joint system capacity. Taking the EFLC algorithm as example, the joint system capacity was increased to approximately 68, 63, and 58 users for the mixes v25w75, v50w50, and v75w25, respectively.

4.8.2.2 Joint Capacity Regions

The system capacity regions for the CC framework, which was evaluated with each of the LC algorithms, and the reference scenario are shown in Fig. 4.11. The satisfaction curves presented in Fig. 4.10(a)–(c) and single-service performance evaluations are used to build the capacity regions. Notice that the single-service performance results are depicted in the extreme points of the capacity curves in Fig. 4.11. The system capacity curve for the reference scenario is built in the following manner: for

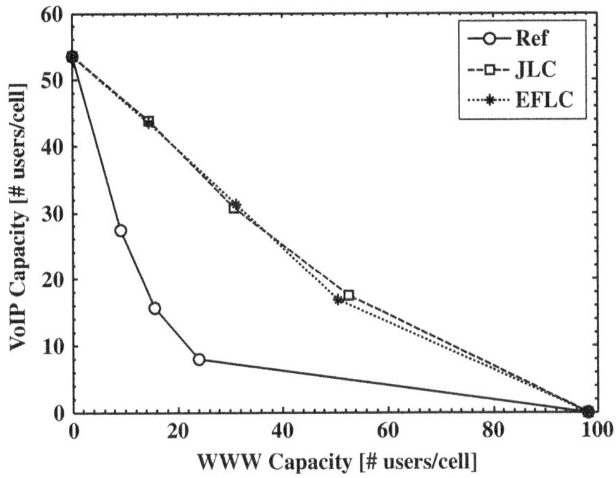

Fig. 4.11 System capacity regions for different traffic mixes.

each traffic mix (Fig. 4.10(a)–(c)), go to the reference scenario satisfaction curves and take the total offered load (number of users) where the satisfaction limit of 90% was first crossed by any of the services. Each point in the reference scenario capacity curve (ordered pair) is the multiplication of this offered load limit by the corresponding traffic mix. The capacity curves for the CC framework with JLC and EFLC are built in a similar manner.

The main result that Fig. 4.11 presents is that the CC framework was able to increase the overall system capacity in comparison with the reference scenario, while respecting a minimum satisfaction limit of 90%. The larger the area below the capacity curve, the higher the number of satisfied users. Although capacity loss is observed for the WWW service when analyzed individually (see WWW satisfaction curves in Fig. 4.10(a)–(c) and Table 4.6), the system capacity regions show a remarkable joint capacity gain obtained with the CC framework.

Table 4.6 Global and individual capacity gains of the EFLC algorithm over the reference scenario for the three traffic mixes: 25% VoIP/75% WWW (v25w75), 50% VoIP/50% WWW (v50w50) and 75% VoIP/25% WWW (v75w25).

	v25w75	v50w50	v75w25
VoIP gain (%)	110.99	100.45	59.06
WWW gain (%)	−14.57	−24.36	−29.47
Global gain (%)	110.99	100.45	59.06

Table 4.6 shows the relative capacity gains (global and for each service class) achieved by the EFLC algorithm over the reference scenario for all traffic mixes. It can be observed that not only did the EFLC algorithm protect the QoS of the VoIP service, but it maximized its capacity as well. This was possible due to the soft and controlled QoS degradation of the WWW service. The more VoIP users that exist in the system, the lower the VoIP capacity gain and the higher the WWW capacity loss.

It is important to notice that since the VoIP was the capacity-limiting service for all the traffic mixes, the overall system capacity gain was equal to the VoIP capacity gain. Depending on the traffic mix considered, the EFLC algorithm presented an approximate global capacity gain from 59 to 111%.

4.8.3 Conclusions

In a specific case study where VoIP and WWW service flows compete for shared access in an HSDPA wireless cellular network, the proposed CC framework was able to increase the overall system capacity from 59 to 111% depending on the traffic mix, while keeping the system operating optimally in its target QoS profile.

Regarding the LC algorithms, EFLC presented equal or better VoIP satisfaction than JLC for all the ranges of traffic loads and traffic mixes considered in the simulations. This was due to the advantages of the EFLC over the JLC, such as variable load control step size and, consequently, fine-tuning control of the LC parameters α_{SAC} and β_{WPF}; more quickness when leading to congestion situations due to the forecasting property of the exponential filtering; and better synchronized action of the SAC and WPF schemes.

The conclusions drawn from this particular case study serve as a proof-of-concept of the general concepts of the QoS-based CC framework proposed in Section 4.7. Since this general CC framework is service quality centric, its parameters can be easily adjusted to any present and future cellular wireless system and to provide statistical QoS guarantees to any RT service.

4.9 Conclusions and Research Directions

In this work, two adaptive CC frameworks for wireless cellular systems were proposed. These frameworks have the objective of avoiding network collapse and providing statistical QoS guarantees for high-priority services (e.g., real-time) even in overload/outage situations (congestion). They were classified in a proposed taxonomy: resource based and QoS based. The first one is based on the consumption of radio resources and is suitable for CS networks, like WCDMA with dedicated channels. The second one is based on service quality and is suitable for PS networks, like HSDPA with a high-speed shared channel. Both frameworks were evaluated by means of dynamic system-level simulations considering the UMTS system. Although the performance evaluation focused on a CDMA-based system, both parametric CC frameworks can be easily adjusted to wireless cellular systems based on other multiplexing schemes, such as TDMA, FDMA, and OFDMA.

From the CC frameworks' structure and performance, the following general conclusions can be drawn:

Resource-based CC framework

- This framework needs a good estimation of the resource consumption. Thus, it is well suited for CS cellular systems, in which both radio and core network resources are allocated for the entire sessions and the services are provided in power-controlled dedicated channels.
- The framework is scalable to several services and has a completely automated and adaptive framework. It adjusts itself in order to prevent the BS from working with an unacceptable transmission power, and so avoiding a possible network collapse.
- The resource-based CC technique, which is composed of the AC, RA, and LC algorithms, was able to increase the user satisfaction, and therefore system capacity, compared to a reference scenario, where no CC technique was used. This was achieved while protecting the QoS of the voice users even in overload situations.

QoS-based CC framework

- The decisions of this CC framework is centered on the service quality. Since the QoS criteria is more general, this new approach can be adapted to any current and next-generation wireless cellular systems.
- The proposed framework is scalable to several services, i.e., the quality of the most prioritized service can be maintained no matter how many other lower priority services are provided in the network. This can be achieved because the framework adjusts its parametric structure adaptively in order to follow the temporal behavior of the QoS measure of the most prioritized service and assures that it will be kept around a planned value.
- The QoS-based CC framework, which is composed of the AC, PSC, and LC algorithms, was able to guarantee the QoS of the VoIP service by means of a soft-controlled QoS degradation of the WWW service. This provided an overall system capacity gain from 59 to 111%, depending on the considered traffic mix.

As future research directions, self-configuration and management can be highlighted, which was addressed in a simpler scenario in the present work.

Radio access networks (RANs) are evolving to a scenario where service quality has to be guaranteed simultaneously for a variety of service classes (conversational, streaming, interactive, and background) and a number of user classes (gold, silver, premium) and where the network deployment follows an irregular and uncontrolled spatial pattern, e.g., the "Home Node-B" (femtocell) concept [5].

This work believes that the key solution for this problem is the utilization of intelligent, adaptive, automatic, and pro-active RRM techniques in a self-configuration and self-management paradigm. These RRM algorithms must be able to measure and/or predict the changes of the radio mobile environment, optimize a high-dimensional resource allocation problem, control/adapt their own configuration parameters in different time scales, and guarantee the service quality of flows from several service classes and several user classes, providing an hierarchical level of prioritization.

The CC frameworks proposed in the present work are self-configured frameworks composed of RRM algorithms. The perspective for future research is to use a generalized form of this self-configured RRM framework to solve the complex problem stated before. This work believes that a mix of simple heuristic algorithms coupled with some more advanced machine learning algorithms (e.g., neural networks, genetic algorithms) could entail a feasible solution, depending also on the time scale of interest. Solutions based on Control Theory (feedback control and Kalman filters) are also envisaged. The ultimate result of such an approach would be minimum human administration of the network and an almost organic capability of network configuration and management.

References

1. 3GPP: Adaptive multi-rate (AMR) speech codec – General description. Tech. Rep. TS 26.071 v7.0.1 – Release 7, 3rd Generation Partnership Project, Sophia Antipolis, France (2007)
2. 3GPP: End-to-end quality of service (QoS) concept and architecture. Tech. Rep. TS 23.207 v7.0.0 – Release 7, 3rd Generation Partnership Project, Sophia Antipolis, France (2007)
3. 3GPP: Quality of service (QoS) concept and architecture. Tech. Rep. TS 23.107 v7.1.0 – Release 7, 3rd Generation Partnership Project, Sophia Antipolis, France (2007)
4. 3GPP: Radio resource management strategies. Tech. Rep. TR 25.922 v7.1.0 – Release 7, 3rd Generation Partnership Project, Sophia Antipolis, France (2007)
5. 3GPP: 3G Home NodeB study item technical report. Tech. Rep. TR 25.820 v8.2.0 – Release 8, 3rd Generation Partnership Project, Sophia Antipolis, France (2008)
6. 3GPP: 3GPP system architecture evolution (SAE): Report on technical options and conclusions. Tech. Rep. TR 23.882 v1.15.1 – Release 8, 3rd Generation Partnership Project, Sophia Antipolis, France (2008)
7. 3GPP: Medium access control (MAC) protocol specification. Tech. Rep. TS 25.321 v8.2.0 – Release 8, 3rd Generation Partnership Project, Sophia Antipolis, France (2008)
8. 3GPP: Physical layer – Measurements (FDD). Tech. Rep. TS 25.215 v8.1.0 – Release 8, 3rd Generation Partnership Project, Sophia Antipolis, France (2008)
9. 3GPP: Physical layer procedures (FDD). Tech. Rep. TS 25.214 v8.2.0 – Release 8, 3rd Generation Partnership Project, Sophia Antipolis, France (2008)
10. Abrardo, A., Giambene, G., Sennati, D.: Capacity evaluation of a mixed-traffic WCDMA system in the presence of load control. IEEE Transactions on Vehicular Technology **52**(3), 490–501 (2003)
11. Braga, A.R., Rodrigues, E.B., Cavalcanti, F.R.P.: Packet scheduling for voice over IP over HSDPA in mixed traffic scenarios with different end-to-end delay budgets. In: Proc. VI International Telecommunications Symposium – ITS2006, pp. 83–88 (2006)
12. Braga, A.R., Rodrigues, E.B., Cavalcanti, F.R.P.: Packet scheduling for VoIP over HSDPA in mixed traffic scenarios. In: Proc. 17th Annual IEEE International Symposium on Personal, Indoor and Mobile Radio Communications – PIMRC, pp. 1–5 (2006)
13. Braga, A.R., Wanstedt, S., Ericson, M.: Admission control for VoIP over HSDPA in a mixed traffic scenario. In: Proc. VI International Telecommunications Symposium – ITS2006, pp. 97–402 (2006)
14. Chakraborty, S., Peisa, J., Frankkila, T., Synnergren, P.: IMS Multimedia Telephony over Cellular Systems: VoIP Evolution in a Converged Telecommunication World. John Wiley & Sons (2007)

15. Ericson, M., Wänstedt, S.: Mixed traffic HSDPA scheduling – impact on VoIP capacity. In: Proc. 65th Vehicular Technology Conference – VTC Spring, pp. 1282–1286 (2007)
16. Ericson, M., Wänstedt, S., Pettersson, J.: Effects of simultaneous circuit and packet switched voice traffic on total capacity. In: Proc. 63rd Vehicular Technology Conference – VTC Spring, vol. 2, pp. 538–542 (2006)
17. Eryilmaz, A., Srikant, R.: Fair resource allocation in wireless networks using queue-length-based scheduling and congestion control. IEEE/ACM Transactions on Networking **15**(6), 1333–1344 (2007)
18. Gaiti, D., Boukhatem, N.: Cooperative congestion control schemes in ATM networks. IEEE Communications Magazine **34**(11), 102–110 (1996)
19. Goot, R.E., Mahalab, U., Cohen, R.: Nonlinear exponential smoothing (NLES) algorithm for noise filtering and edge preservation. HAIT Journal of Science and Engineering **2**, 2–10 (2005)
20. Gu, J., Che, X.: Adaptive uplink load control in CDMA systems. In: Proc. 62nd IEEE Vehicular Technology Conference – VTC Fall, vol. 2, pp. 1193–1196 (2005)
21. Holma, H., Melero, J., Vainio, J., Halonen, T., Mäkinen, J.: Performance of adaptive multirate (AMR) voice in GSM and WCDMA. In: Proc. 57th IEEE Vehicular Technology Conference – VTC Spring, vol. 4, pp. 2177–2181 (2003)
22. Holma, H., Toskala, A. (eds.): WCDMA for UMTS: Radio Access for Third Generation Mobile Communications, 3rd edn. John Wiley & Sons (2004)
23. Hong, D., Suda, T.: Congestion control and prevention in ATM networks. IEEE Network **5**(4), 10–16 (1991)
24. IETF: TCP slow start, congestion avoidance, fast retransmit. Tech. Rep. RFC 2001, Internet Engineering Task Force (1997)
25. IETF: SIP: Session initiation protocol. Tech. Rep. RFC 3261, Internet Engineering Task Force (2002)
26. Jain, R.: Congestion control in computer networks: Issues and trends. IEEE Network **4**(3), 24–30 (1990)
27. Kasera, S.K., Ramjee, R., Thuel, S.R., Wang, X.: Congestion control policies for IP-based CDMA radio access networks. IEEE Transactions on Mobile Computing **4**(4), 349–362 (2005)
28. Laiho, J., Wacker, A., Novosad, T. (eds.): Radio Network Planning and Optimisation for UMTS, 2nd edn. John Wiley & Sons (2006)
29. de Lima, C.H.M., Larsson, K., Ökvist, P.: RNC-located filtering-based OLPC solution for E-UL. In: Proc. 17th Annual IEEE International Symposium on Personal, Indoor and Mobile Radio Communications – PIMRC, pp. 1–5 (2006)
30. Lin, X., Shroff, N.B., Srikant, R.: A tutorial on cross-layer optimization in wireless networks. IEEE Journal on Selected Areas in Communications **24**(8), 1452–1463 (2006)
31. Liu, T.K., Silvester, J.A.: Joint admission/congestion control for wireless CDMA systems supporting integrated services. IEEE Journal on Selected Areas in Communications **16**(6), 845–857 (1998)
32. Low, S.H., Paganini, F., Doyle, J.C.: Internet congestion control. IEEE Control Systems Magazine **22**(1), 28–43 (2002)
33. Mase, K., Yamamoto, H.: Advanced traffic control methods for network management. IEEE Communications Magazine **28**(10), 82–88 (1990)
34. Mückenheim, J., Bernhard, U.: A framework for load control in 3rd generation CDMA networks. In: Proc. IEEE Global Telecommunications Conference – GLOBECOM, vol. 6, pp. 3738–3742 (2001)
35. Pérez-Romero, J., Sallent, O., Agustí, R., Díaz-Guerra, M.A.: Radio Resource Management Strategies in UMTS. John Wiley & Sons (2005)
36. Pérez-Romero, J., Sallent, O., Agustí, R., Sánchez, J.: Managing radio network congestion in UTRA-FDD. Electronics Letters **38**(22), 1384–1386 (2002)

37. Rave, W., Köhler, T., Voigt, J., Fettweis, G.: Evaluation of load control strategies in an UTRA/FDD network. In: Proc. 53rd IEEE Vehicular Technology Conference – VTC Spring, vol. 4, pp. 2710–2714 (2001)
38. Regnier, J., Cameron, W.H.: State-dependent dynamic traffic management for telephone networks. IEEE Communications Magazine **28**(10), 42–53 (1990)
39. Reisslein, M., Ross, K.W., Rajagopal, S.: A framework for guaranteeing statistical QoS. IEEE/ACM Transactions on Networking **10**(1), 27–42 (2002)
40. Richardson, I.E.G.: H.264 and MPEG-4 Video Compression: Video Coding for Next Generation Multimedia. John Wiley & Sons (2003)
41. Rodrigues, E.B.: Proposal of a congestion control framework for WCDMA cellular networks and performance evaluation. Master's thesis, Federal University of Ceará – UFC, Fortaleza, Brazil (2004) (in Portuguese)
42. Rodrigues, E.B., Cavalcanti, F.R.P., de Lima, C.H.M., de Souza Jr., V.A.: (Re)Active load control based on radio link quality for the UMTS/WCDMA forward link. In: Proc. IEEE 61st Vehicular Technology Conference – VTC Spring, vol. 1, pp. 363–367 (2005)
43. Rodrigues, E.B., Cavalcanti, F.R.P., Wänstedt, S.: QoS-driven adaptive congestion control for voice over IP in multiservice wireless cellular networks. IEEE Communications Magazine **46**(1), 100–107 (2008)
44. Rodrigues, E.B., de Lima, C.H.M., de Souza Jr., V.A., Cavalcanti, F.R.P.: Power-based congestion control framework for downlink WCDMA systems. In: Proc. IEEE 61st Vehicular Technology Conference – VTC Spring, vol. 3, pp. 1870–1874 (2005)
45. Rodrigues, E.B., de Lima, C.H.M., de Souza Jr., V.A., Cavalcanti, F.R.P., Braga, A.R., Maciel, T.F.: QoS and load management via admission control in UMTS forward link. In: Proc. IEEE Vehicular Technology Conference – VTC Spring, vol. 4, pp. 2425–2429 (2004)
46. Rodrigues, E.B., Lima, F.R.M., Cavalcanti, F.R.P.: Load control for VoIP over HSDPA in mixed traffic scenarios. In: Proc. 18th Annual IEEE International Symposium on Personal, Indoor and Mobile Radio Communications – PIMRC, pp. 1–5 (2007)
47. Rodrigues, E.B., Olsson, J.: Admission control for streaming services over HSDPA. In: Proc. Advanced Industrial Conference on Telecommunications/Service Assurance with Partial and Intermittent Resources Conference/E-Learning on Telecommunications Workshop – AICT/SAPIR/ELETE, pp. 255–260 (2005)
48. Rui, F., Min, W., Ericson, M., Wänstedt, S.: Evaluation and analysis of SIP and VoIP performance with presence traffic over HSPA. In: Proc. 18th Annual IEEE International Symposium on Personal, Indoor and Mobile Radio Communications – PIMRC, pp. 1–5 (2007)
49. Sallent, O., Pérez-Romero, J., Agustí, R., Casadevall, F.: Provisioning multimedia wireless networks for better QoS: RRM strategies for 3G W-CDMA. IEEE Communications Magazine **41**(2), 100–106 (2003)
50. Sampath, A., Kumar, P.S., Holtzman, J.M.: On setting reverse link target SIR in a CDMA system. In: Proc. 47th IEEE Vehicular Technology Conference – VTC Spring, vol. 2, pp. 929–933 (1997)
51. Shapiro, J.K., Towsley, D., Kurose, J.: Optimization-based congestion control for multicast communications. IEEE Communications Magazine **40**(9), 90–95 (2002)
52. Tan, K., Jiang, F., Zhang, Q., Shen, X.S.: Congestion control in multihop wireless networks. IEEE Transactions on Vehicular Technology **56**(2), 863–873 (2007)
53. Tanenbaum, A.S.: Computer Networks. Prentice Hall (2003)
54. Tsao, S.C., Lai, Y.C., Lin, Y.D.: Taxonomy and evaluation of TCP-friendly congestion-control schemes on fairness, aggressiveness, and responsiveness. IEEE Network **21**(6), 6–15 (2007)
55. UMTS: Selection procedures for the choice of radio transmission technologies of the UMTS. Tech. Rep. TR 101 112 V3.2.0 - UMTS 30.03, Universal Mobile Telecommunications System, Sophia Antipolis, France (1998)
56. Wang, B., Pedersen, K.I., Kolding, T.E., Mogesen, P.E.: Performance of VoIP on HSDPA. In: Proc. 61st IEEE Vehicular Technology Conference – VTC Spring, vol. 4, pp. 2335–2339 (2005)

57. Wang, C., Sohraby, K., Li, B., Daneshmand, M., Hu, Y.: A survey of transport protocols for wireless sensor networks. IEEE Network **20**(3), 34–40 (2006)
58. Wänstedt, S., Ericson, M., Sandlund, K., Nordberg, M., Frankkila, T.: Realization and performance evaluation of IMS multimedia telephony for HSPA. In: Proc. 17th Annual IEEE International Symposium on Personal, Indoor and Mobile Radio Communications – PIMRC, pp. 1–5 (2006)
59. Wänstedt, S., Rui, F., Ericson, M., Nordberg, M.: Providing reliable and efficient VoIP over cellular networks. In: Proc. Future Telecommunications Conference (2005)
60. Wernersson, M., Wänstedt, S., Synnergren, P.: Effects of QoS scheduling strategies on performance of mixed services over LTE. In: Proc. 18th Annual IEEE International Symposium on Personal, Indoor and Mobile Radio Communications – PIMRC, pp. 1–5 (2007)
61. Yang, C.Q., Reddy, A.V.S.: A taxonomy for congestion control algorithms in packet switching networks. IEEE Network **9**(4), 34–45 (1995)
62. Yang, Y.R., Lam, S.S.: Internet multicast congestion control: A survey. In: Proc. ICT (2000)
63. Zawodniok, M., Jagannathan, S.: Predictive congestion control protocol for wireless sensor networks. IEEE Transactions on Wireless Communications **6**(11), 3955–3963 (2007)

Chapter 5
Resource Allocation in Multiuser Multicarrier Wireless Systems with Applications to LTE

Walter Freitas Jr., F. Rafael M. Lima, Ricardo B. Santos
and Francisco R. P. Cavalcanti

5.1 Introduction

International mobile telecommunications (IMT)-advanced systems are mobile systems that include the new capabilities that go beyond those of IMT-2000 as specified by the International Telecommunication Union (ITU). Such systems provide access to a wide range of telecommunication services including advanced mobile services, supported by mobile and fixed networks, which are increasingly packet based. Key features of IMT-advanced systems are [23]

- a high degree of commonality while retaining the flexibility to support a wide range of services and applications in a cost-efficient manner;
- compatibility of services within IMT and with fixed networks;
- capability of interworking with other radio access systems;
- high-quality mobile services;
- user equipment suitable for worldwide use;
- user-friendly applications, services, and equipment;
- worldwide roaming capability;
- enhanced peak data rates to support advanced services and applications (100 Mbit/s for high and 1 Gbit/s for low mobility were established as initial targets).

To achieve those challenging targets, multicarrier Orthogonal Frequency Division Multiple Access (OFDMA[5.1]) was chosen consensually as the radio interface technology. OFDMA, as the name implies, is based on orthogonal frequency division multiplexing (OFDM) as modulation technique. OFDM enables the transmission of multiple parallel low data rate narrowband channels by sub-dividing a

[5.1] Hereafter, multicarrier and OFDMA will be used indistinctly, even though some of the approaches presented here for multicarrier OFDMA could also be extended to other multicarrier systems, e.g., multicarrier code division multiple access (MC-CDMA).

F. Cavalcanti, S. Andersson (eds.), *Optimizing Wireless Communication Systems*,
DOI 10.1007/978-1-4419-0155-2_5, © Springer Science+Business Media, LLC 2009

wider bandwidth into so-called subcarriers. As a consequence, different diversity dimensions can be exploited in an OFDMA system, such as frequency and multiuser diversities. In the frequency diversity, as the different subcarriers tend to present different channel fading states if separated by one or more coherence bandwidths, the subcarriers scheduled to a given user equipment (UE) may be chosen so that only subcarriers in good channel state are used by that UE. In multiuser diversity each UE will be in a different location and consequently will experiment different channels fading states. A subcarrier with a low SNR for a given UE may be in better condition with respect to other UEs. These channel diversities make possible the use of radio resource allocation (RRA) schemes that, for example, distribute the radio resources fairly among UEs.

A suitable RRA scheme can determine the use of some radio resources in order to provide a specific goal respecting some system conditions. Constrained optimization techniques are used to seek solutions that minimize or maximize a cost function while a set of constraints or restrictions are satisfied. Based on these methods, a large range of RRA strategies are possible, such as time-frequency assignment, where basic resource units can be allocated dynamically to different UEs in the frequency–time plane (see Fig. 5.1 for an illustration of this method). RRA schemes adaptively assign the system radio resources (subcarrier, power, and bit rate) as a function of traffic load, channel condition, channel information availability, and QoS requirements. These schemes provide a greater improvement in the system performance if compared to static schemes, which do not take advantage of frequency, time, and multiuser diversities. The RRA in OFDMA systems can be divided into two main problems:

- Subcarrier allocation – the subset of subcarriers on which each UE will transmit is determined;
- Power allocation – the transmit power for each subcarrier is determined.

In this work it is assumed that, once these two steps are performed, the bit rate is determined as a consequence by an adaptive modulation and coding scheme according to the channel quality.

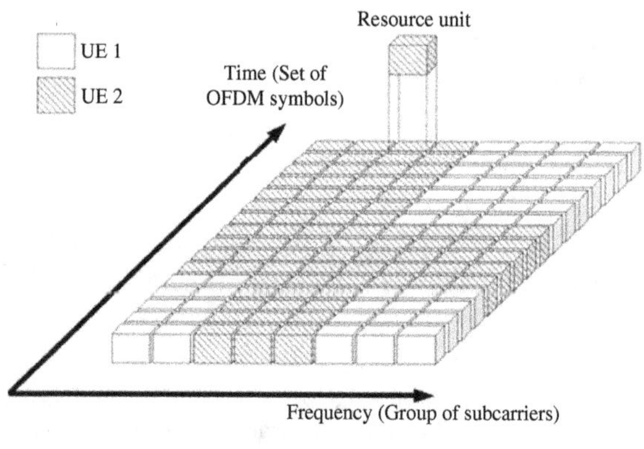

Fig. 5.1 Frequency–time resource grid in OFDMA.

One notable aspect of these emerging systems is the plurality of services supported. Classifying the services and applications provided by wireless networks is not a trivial task because they are continuously evolving to integrated and complex applications. Besides that, the services can be classified in terms of time dependency (time or non-time based), delivery requirements (real-time (RT) or non-real-time (NRT)), directionality (unidirectional or bi-directional), symmetry of the communications (symmetric or asymmetric), interactivity, and number of parties [40]. Time-based services are the ones where the information should be presented at specific instants so as to have a meaning because time is an integral part of the information to be communicated, e.g., video and audio. Examples of non-time-based services are images and text. Bi-directional communication can be either symmetric or asymmetric. Web browsing is a classical example of an asymmetric application where only commands are transmitted in one link direction.

Note that the classification according to the delivery requirements is different from the one regarding intrinsic time dependency. As an example, imagine an online game where uncorrelated images are displayed to the users, and the number of points a user scores is dependent on how quickly the user reacts to the images (performs some actions). In this case, the images have to be displayed to the users with tight delay requirements so as to assure interactive response to the users (RT application). However, the images do not need any synchronization in order to make sense to the user (non-time-based application). In this work the classification according to the delivery requirements is emphasized.

The rest of this chapter is organized as follows: after the introduction of possible scenarios to RRA in multicarrier OFDMA systems, fundamental problems in such scenarios and key performance metrics are described, followed by a discussion of the optimization approaches capable of solving the problems optimally; after that some algorithms in RRA are reviewed and it is proposed one focused in the user satisfaction metric considering the 3GPP LTE system. Finally, trends and directions for further evolution of resource allocation in multicarrier OFDMA systems are discussed in Section 5.9.

5.2 Scenarios for Radio Resource Allocation

To understand the complexity of multicarrier OFDMA systems, its challenges, and characteristics, three scenarios of increasing complexity will be presented. Thus, the previous scenarios will always be a particular case of the next one.

5.2.1 Single Link

In this scenario, a single link between one UE and its respective BS is studied. This is essentially a link optimization problem as only one UE is considered.

Figure 5.2 presents a generic transmission chain of a single OFDM link. At the transmitter side, the data is coded and modulated according to the channel state. The power of each symbol of the modulated data is then adjusted by the power allocation block. Finally, the output from the power allocation is converted to the time dimension by an inverse fast fourier transform (IFFT) block, receives a cyclic prefix (CP), and is transmitted through the channel.

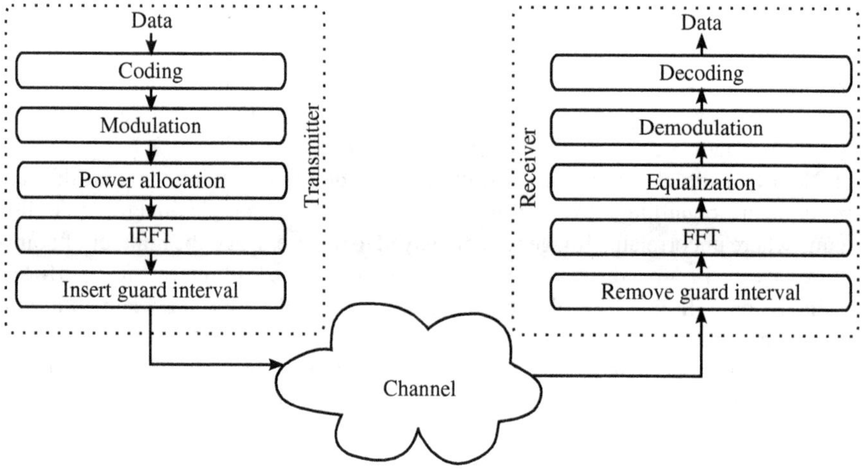

Fig. 5.2 Single link scenario.

After passing through the channel, the signal has its CP removed and is converted back to the frequency dimension by a fast fourier transform (FFT) block. The equalization block removes the effects of channel and power allocation. The signal is then demodulated and decoded to provide the transmitted data. A more detailed description about OFDM transceivers can be found in Chapter 9 of this book.

In this scenario, N contiguous subcarriers will be used to transmit data from the BS to the UE in the downlink direction. Each subcarrier n will present a different channel gain g_n that will be correlated to the channel gains of the adjacent subcarriers. This correlation depends on the inter-subcarrier spacing and the level of frequency selectivity of the channel, which can be measured by the coherence bandwidth.

The data will be transmitted using the power p_n allocated to subcarrier n by an appropriate algorithm. The sum of the transmitted powers will be limited to $p_{\max}$. Each subcarrier power p_n leads to an SNR γ_n defined as

$$\gamma_n = \frac{p_n \cdot g_n}{\sigma^2}, \tag{5.1}$$

where σ^2 is the receiver's additive white Gaussian noise (AWGN) power per subcarrier, assumed equal on all subcarriers.

The data rate r_n achieved at subcarrier n is a function of γ_n, given by

$$r_n = F(\gamma_n), \tag{5.2}$$

where $F(\cdot)$ is a link adaptation function. In general, the link adaptation function is a monotonically increasing function of γ_n. Link adaptation is usually implemented in practice by means of adaptive modulation and coding schemes (MCSs). The most obvious objective in this scenario is to maximize the sum of the subcarrier data rates, also known as sum-rate capacity.

5.2.2 Multiple Links Without Co-channel Interference

In this scenario a single BS serving J UEs is considered. Each UE j has a channel gain $g_{j,n}$ on subcarrier n that is independent of other UEs. This scenario is illustrated in Fig. 5.3. Frequency diversity is caused by the different channel states on each subcarrier. Data rates $r_{j,n}$ are proportional to each subcarrier channel quality, according to, e.g., a water-filling solution (see Section 5.4.1 for the formulation of the optimization problem and Section 5.5.1.1 for the solution). In this way a higher UE total data rate r_j may be achieved without increasing the probability of transmission error.

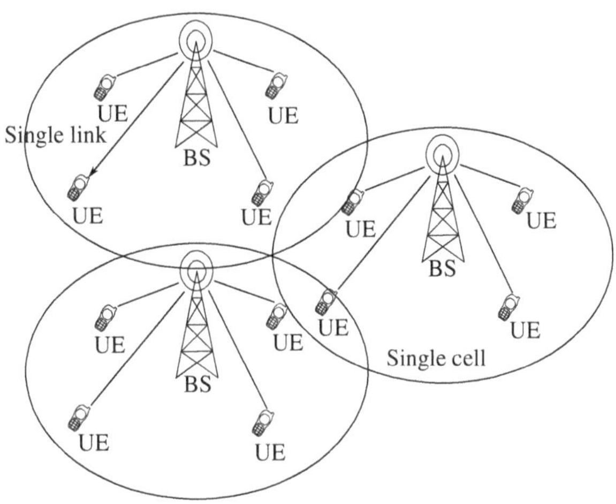

Fig. 5.3 Multiple cells with co-channel interference scenario.

In this scenario, besides the maximization of the sum-rate capacity, other objectives can be formulated, for example, minimization of the total power consumption, maximization of the minimum data rate among all UEs, and maximization of the QoS satisfaction. These objectives will be addressed in more detail in Section 5.4.

5.2.3 Multiple Cells with Co-channel Interference

In this scenario, multiple cells exist in the system and transmission in one cell interferes with the other cell transmissions due to frequency reuse. In this case the SINR is a more meaningful channel quality indicator. Thus, the variable $\gamma_{j,n}$ will from now on represent the SINR of the UE j on subcarrier n. This scenario is illustrated in Fig. 5.3.

At least three different RRA approaches are possible in this scenario. The first is the centralized approach. In this approach, a central controller receives channel quality measurements from all cells and all links and then reports the RRA decisions back to the cells. The main advantage of this approach is the potential global optimality of the RRA solution. Disadvantages of this approach include high complexity for RRA decision making, high signaling load and feedback bandwidth required, and a potential communication delay from measurements to actuation. This approach is illustrated in Fig. 5.4(a).

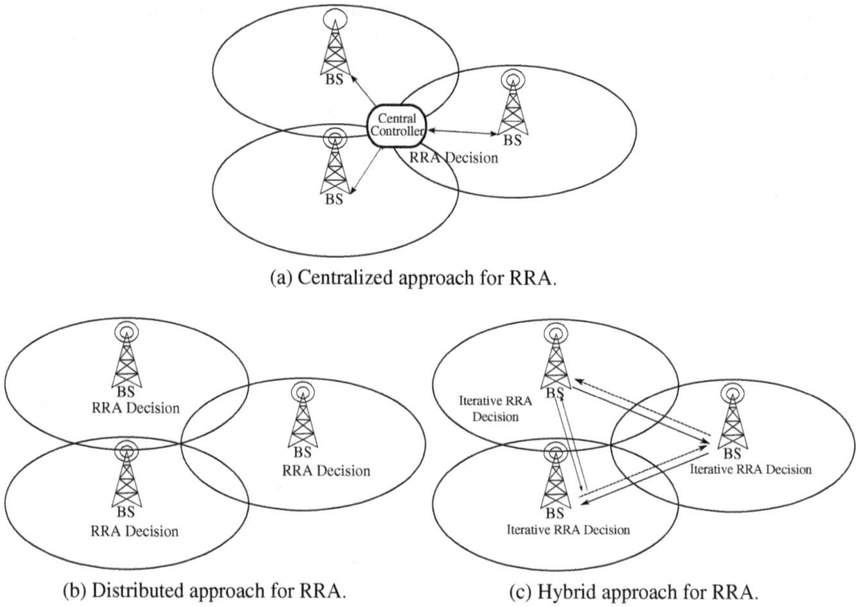

(a) Centralized approach for RRA.

(b) Distributed approach for RRA.

(c) Hybrid approach for RRA.

Fig. 5.4 RRA approaches.

The other extreme approach is the fully distributed case. In this approach, each BS optimizes its transmission parameters independently and the impact of the decision on the interfering links may only be estimated by measuring the interference from other cells. The advantages of this approach are the exact opposites of the centralized one: lower complexity, signaling load, and communication delays. The drawback is, as expected, that the resulting RRA solution may be far from optimal and interference is not actually managed, but reduced to a certain level. This approach is illustrated in Fig. 5.4(b).

Another approach is based on features of both the fully centralized and distributed ones. It is usually known as hybrid or locally distributed approach for RRA. There are many possible configurations of hybrid RRA approaches, some more like a centralized one, some more like a distributed one. For instance, consider that each cell takes on RRA decisions by itself, as in the distributed approach. Then these decisions are communicated to neighboring cells and a refinement of each cell's solution is performed. Or, alternatively, a central controller can be introduced again for collecting the RRA decisions and refining them altogether. There are other possibilities. Figure 5.4(c) illustrates one case where each cell decides its transmission parameters but communicates it to other cells to increase the solution efficiency.

5.3 Radio Resource Allocation Fundamental Problems

To evaluate radio resource allocation in multicarrier wireless systems there are several possible criteria to be considered when designing solutions to solve the optimization problems of resource allocation. The possible solutions are classified in accordance to their efficiency, applicability, guarantee of QoS, and fairness.

5.3.1 Maximization of Total Data Throughput

A very common criteria to measure the efficiency of the RRA is the sum-rate capacity. The optimal solution to this objective is the allocation of resources to UEs with high channel quality followed by power allocation, e.g., using water-filling, which captures multiuser diversity and maximizes the sum-rate capacity. In spite of achieving the optimal sum-rate capacity, full applicability of channel-aware resource allocation is limited since complete and perfect channel quality information per UE in each resource allocation unit would be necessary. Therefore, actual applicability of any resource allocation scheme must be analyzed against the available feedback bandwidth for control measurements.

5.3.2 Fairness

Fairness is used, in wireless systems, to assure that all served UEs will receive a fair share of the system resources. However, fairness is a subjective concept and what is a fair resource allocation depends strongly on the studied scenario.

The most straightforward kind of fairness concept comes from the idea that a fair distribution happens when every UE receives the same share of radio resources aiming at a data rate balancing. A resource allocation is considered fair in this approach if all UEs of set $\mathscr{J}$ achieve the same data rate, that is,

$$r_i = r_j; \quad \forall i, j \in \mathscr{J}, i \neq j. \tag{5.3}$$

A well-known fairness definition based on this concept is the Jain fairness index [25], which is based on the concept of generalized processor sharing for wireline systems, and defined as

$$JF(x_1, \ldots, x_N) = \frac{\left(\sum_1^N x_i\right)^2}{N \cdot \left(\sum_1^N x_i^2\right)}, \tag{5.4}$$

where x_i could be, for example, data rate or delay. Regarding data rate fairness, the Jain fairness index JF is 1 in the fairest case in which all UEs have the same average data rate and decreases as a subset of UEs begins to receive a higher data rate in disfavor of other UEs.

Another approach seeking to guarantee a certain degree of fairness among UEs is to maximize the minimum achievable data rate among all UEs (max–min fairness [24]). A straightforward manner to measure the max–min fairness is the data rate of the UE with the lowest data allocation. The main advantage of the max–min concept with respect to the balancing among the data rate of all the UEs is that the max–min does not punish allocations in which some UEs achieve high data rate.

Kelly et al. in [30] propose the proportional fairness criteria to evaluate rate control for communication networks. The proportional fairness criteria applied to wireless systems state that a fair distribution implies in resources proportional to the UE channel condition. A resource allocation solution $\mathbf{r}^P$ is considered proportionally fair when compared with other $\mathbf{r}^S$ if

$$\Delta r = \sum_{j \in \mathscr{J}} \frac{r_j^S - r_j^P}{r_j^P} \leq 0, \tag{5.5}$$

where r_j^P is the data rate of UE j using resource allocation solution P, for any feasible resource allocation solution S. Still in [30] the authors proved that a proportional fair algorithm should maximize the sum of the logarithm of r_j formulated as

$$\max \sum_{j \in \mathscr{J}} \log(r_j). \tag{5.6}$$

The proportional fair solution in a multiple link OFDMA scenario was proposed by [43] as

$$j^* = \arg \max_{j \in \mathscr{J}} \frac{\mathrm{F}(\gamma_n)}{\bar{r}_j / r_j^{\mathrm{req}}}, \tag{5.7}$$

where $\bar{r}_j$ is the average data rate received by UE j which is normalized by the respective data rate requirement r_j^{req} due to the different QoS requirements.

Another metric capable of capturing the fairness of different RRA schemes is the average worst-delay metric as proposed in [29] that measures the average time until all UEs are served at least once. Thus, the more fair is the scheme, the lower is this time.

All consideration about fairness is not very important when the network faces congestion due to overload. Those abnormal scenarios should only happen on extreme situations when a complete congestion control framework is incapable of keeping the system in manageable loads [37].

Most of the time, a wireless cellular network operates normally with an offered traffic load below or around a target point defined in the network planning phase. Congestion (overload and/or outage) situations can be caused by a random behavior of external interference, different mobility profiles and geographical location of mobile terminals, network utilization patterns during specific periods of the day (busy hours), voice and data traffic dynamics, and subscribers' profiles (commercial and residential areas) and their call distributions. In these cases users will face a degradation of the QoS experienced.

In such situation, traditional RRM functionalities like subcarrier and power allocations do not work well in OFDMA systems. Thus, it is necessary for a set of functionalities that manipulate how network resources are allocated through time to different service flows to assure QoS requirements. Examples of these functionalities include call admission control (CAC), packet scheduling and load control (LC) algorithms in order to avoid and counteract congestion situations. For the rest of this chapter, it will be assumed that the load in the network can be controlled to be under normal conditions.

5.3.3 QoS Satisfaction

In RT services, there is the requirement of a short time response between the communicating parts. In general, RT services impose strict requirements regarding packet delay and jitter. Examples of this kind of service are online games that require quick responses from the users and VoIP. Specifically, VoIP has been extensively studied with radio resource management so as to provide good solutions to replace CS speech [8, 15]. The main challenge is to provide the same or improved QoS to VoIP compared to the conventional CS speech when the radio resources are no more dedicated but shared among other services.

In contrast, NRT services do not have tight requirements concerning packet delay. In fact, when transmitting NRT services, the major constraint is the information integrity, i.e., information loss is not tolerable. Therefore, applications of this type must have error-correction or recovery mechanisms. Examples of NRT services are Web browsing and FTP. QoS requirements represent the minimum resources necessary to maintain the user satisfied with the service. Failing to achieve this minimum requirement means waste of resources as the user will not be satisfied with the service provided anyway.

QoS will be a very important factor in IMT-Advanced networks. Therefore, the proposal of RRA schemes that take into account QoS explicitly is necessary. Fairness should still be considered but in conjunction with QoS. QoS can be taken into account by means of the user satisfaction ratio which is the fraction of UEs achieving their QoS targets in a given time period. Network capacity, in turn, can be defined according to a minimum level of user satisfaction. Later on, in this chapter, this issue will be focused.

5.4 Optimization Problems in Multicarrier Resource Allocation

It is a basic premise that radio resources in wireless systems should be utilized in an efficient and optimum way. Therefore, RRA problems in general are formulated as optimization problems.

An optimization problem is basically composed of an objective function, constraints, and decision variables [9]. In RRA problems, the objective function represents the goal to be searched by the system designer. The objective function depends on the considered scenario and it can be, for example, the maximization of sum-rate capacity in a single link or the minimization of the inter-cell interference in multiple co-channel links in a co-channel interference scenario. The constraints have the function to limit the feasible solutions of the optimization problems and can be practical system restrictions such as maximum power and available bandwidth or performance requirements, e.g., a minimum data rate. Finally, the decision variables are the resources that the system designer can adjust or control in order to find the best solution(s) regarding a given objective. In RRA for multicarrier systems the decision variables can be the assignment of frequency resources to terminals and power distribution among frequency resources.

RRA problems in multicarrier systems can assume several forms depending on the system characteristics and scenarios. Furthermore, they can easily become too complex depending on the number of decision variables, nature of the objective and constraint functions. So, to introduce these problems to the reader, the following sections describe some basic optimization problems applicable to the context of the downlink of a multicarrier system corresponding to Scenario 2 defined in Section 5.2.2 (single BS and multiple links). A discussion about this topic is also found in [20].

5.4.1 Rate Maximization

The rate maximization in a multiple link scenario consists of a classic problem whose objective is to utilize the system bandwidth in an efficient way in the sense that the BS can transmit, in a given transmission opportunity, with the maximum aggregated data rate to the served UEs [27]. This problemis also known as sum-rate capacity maximization and is formulated for the k -th transmission time interval (TTI) as

$$\begin{aligned} &\max_{\mathbf{p},\mathbf{X}[k]} \sum_j \sum_n \mathrm{F}\left(\frac{p_n[k]\cdot g_{j,n}[k]}{\sigma^2}\right)\cdot x_{j,n}[k] \quad \text{subject to} \\ &\sum_j x_{j,n}[k] \leqslant 1, \quad \forall n \\ &\sum_n p_n[k] \leqslant p_{\max}. \end{aligned} \tag{5.8}$$

As the decision variable $x_{j,n}[k]$ comprising the assignment matrix among $\boldsymbol{J}$ users and $\boldsymbol{N}$ subcarriers $\mathbf{X}[k]$ in the problem is binary and, in real networks, the power levels are quantized, problem (5.8) is a combinatorial one. This problem has two constraints. The first one represents the fact that a subcarrier cannot be shared by UEs served by the same BS, which means no intra-cell interference. The last constraint concerns the limitation in the total available power in the BS.

Although the solution of problem (5.8) leads to a high spectral efficiency, it is not suitable for current systems that intend to provide QoS to users. More specifically, in order to increase the bit rate, the BS tends to assign more subcarriers and power to the UEs with good channel conditions that, in general, are located near to the transmit antenna. Therefore, UEs at the border of the cell would starve for transmission opportunities and consequently experience a poor QoS.

As well-known water-filling approach is the most efficient method to power allocation in a single link scenario. But, the performance gain of the usage of water-filling is negligible when the average SNR is maintained high enough when compared with equal allocation of power among all the resource units. This condition could be verified in a multiple links scenario (see Fig. 5.3). Thus, in this chapter the problem of power allocation will be treated after and independently of the subcarrier allocation problem.

5.4.2 Margin Adaptive

In the margin-adaptive problem, the main objective is to guarantee a minimum QoS requirement for all UEs with the lowest possible utilized power [32, 42]. The problem is presented in the following:

$$\begin{aligned} &\min_{\mathbf{p},\mathbf{X}[k]} \sum_j \sum_n p_n[k]\cdot x_{j,n}[k] \quad \text{subject to} \\ &\sum_j x_{j,n}[k] \leqslant 1, \quad \forall n \\ &\sum_n \mathrm{F}\left(\frac{p_n[k]\cdot g_{j,n}[k]}{\sigma^2}\right)\cdot x_{j,n}[k] \geqslant r_j^{\mathrm{req}}, \quad \forall j. \end{aligned} \tag{5.9}$$

Similar to problem (5.8), problem (5.9) is a combinatorial one and has the same decision variables. However, as commented before, the objective function is to minimize the BS transmit power while fulfilling the minimum QoS requirements of the flows r_j^{req} that is here represented by the allocated rate.

Besides the advantage of saving power, the solution to this problem can be suitable for a multiple link scenario with co-channel interference where the available bandwidth is reused throughout the cells. In this scenario, the transmitted power in adjacent cells is received by UEs as interference turning the correct reception more difficult. In this way, low transmit power can improve the channel quality by reducing the overall system interference.

5.4.3 Rate Adaptive

In the rate-adaptive problem, the objective is to assign subcarriers and allocate power so as to improve the fairness among flows. The problem formulation follows [36, 45]:

$$\begin{aligned} \max_{\mathbf{p},\mathbf{X}[k]} \quad & \varepsilon \quad \text{subject to} \\ & \sum_j x_{j,n}[k] \le 1, \quad \forall n \\ & \sum_n p_n[k] \le p_{\max} \\ & \sum_n \mathrm{F}\left(\frac{p_n[k]\cdot g_{j,n}[k]}{\sigma^2}\right)\cdot x_{j,n}[k] \ge \varepsilon, \qquad \forall j. \end{aligned} \tag{5.10}$$

The first two constraints are similar to the ones discussed in the previous problems. Furthermore, the decision variables in this problem are subcarrier assignment and power which characterize problem (5.10) as a combinatorial one, as commented before.

The fairness criterion utilized in this problem is the max–min fairness, discussed before in Section 5.3.2 which states that the lower UE data rate, in this case floor data rate ε, is maximized. Consequently, the solution of this optimization problem leads to a rate balancing in the cell with similar UE data rates. However, depending on the UE distribution in the cell, there can be a resource waste by allocating too many resources to poor channel UEs in order to increase the floor rate.

5.4.4 User Satisfaction Ratio Maximization

This problem is based on the QoS satisfaction criteria and the objective is to maximize the number of satisfied users [38]. User satisfaction involves several aspects not only on the technical scope such as experienced data rate in a file download, but also on economical aspects like the costs that the subscriber has to afford in order to utilize the service. However, in this chapter only technical aspects are considered in the satisfaction formulation. Specifically, a user is satisfied if a minimum QoS requirement is fulfilled. The user satisfaction maximization problem is presented as follows:

$$\begin{aligned} \max_{\mathbf{p},\mathbf{X}[k]} \quad & \sum_j \mathrm{U}(j,k) \quad \text{subject to} \\ & \sum_j x_{j,n}[k] \leqslant 1, \quad \forall n \\ & \sum_n p_n[k] \leq p_{\max}. \end{aligned} \tag{5.11}$$

The problem constraints are similar to those presented in the previous problems and concern the power availability in the cell and subcarrier assignment restrictions to assure no intra-cell interference. $\mathrm{U}(j,k)$ is a step function in the problem objective representing the satisfaction state of flow j at TTI k. In other words, this function assumes 1 when flow j is satisfied at TTI k and 0 otherwise.

5.5 Optimization Tools for Multicarrier Resource Allocation Problems

A great variety of optimization tools exist deriving from both mathematics and computer science. These tools are based on a variety of methods – ranging from linear to evolutionary programming techniques. In this section are listed some optimization problems in the context of multicarrier systems, as well as the suitable mathematical tools used to solve them focusing on its applicability.

These tools can be divided into two approaches: exact and approximate search approaches.

5.5.1 Exact Search Approaches

Exact search approaches offer an optimal solution at the expense of the computational complexity involved in such search, becoming the problem dimension unpractical to solve.

5.5.1.1 Lagrangian Multipliers

The Lagrangian's method of multipliers is a classical non-linear tool for optimization problems with constraints. It can be used to find the optimum power allocation of a single link scenario of an OFDM system with constraint of maximum power, where the objective is to maximize the total throughput T of the connection. This problem is a single-user version of the rate maximization problem. This classic problem is known as finite tones water-pouring (or water-filling) and can be expressed by

$$\begin{aligned} &\max_{\mathbf{p}} \quad T = \Delta f \sum_{n=1}^{N} \log_2 \left(1 + \frac{p_n \cdot g_n}{\sigma^2}\right) \quad \text{subject to} \\ &\sum_{n=1}^{N} p_n \leq p_{\max}, \end{aligned} \tag{5.12}$$

where Δf is the bandwidth of each subcarrier.

The Lagrangian's method of multipliers consists of determining the Lagrangian function $L(\mathbf{p}, \lambda_L)$ of the problem, making its gradient equal to zero and using the fact that the total power budget $p_{\max}$ is completely allocated. With the N partial derivatives of the gradient equal to zero and the last assumption about the power constraint, the following equation system of $N+1$ equations results:

$$\begin{aligned} &\begin{pmatrix} \frac{\partial L}{\partial p_1} \\ \vdots \\ \frac{\partial L}{\partial p_N} \end{pmatrix} = \mathbf{0}, \\ &\lambda_L \left(\sum_{n=1}^{N} p_n - p_{\max} \right) = 0. \end{aligned} \tag{5.13}$$

This equation system has $N+1$ variables: $p_n, n = 1, 2, \ldots, N$ and the variable λ_L, which is the Lagrange multiplier [28]. Solving this equation system, one obtains [20]

$$p_n = \frac{1}{N} \left(\sum_{i=1}^{N} \frac{\sigma^2}{g_i} + p_{\max} \right) - \frac{\sigma^2}{g_n}. \tag{5.14}$$

This is the water-filling solution to the multicarrier case. The expression (5.14) may yield negative powers for some subcarriers. In this case, a possible solution would be to exclude these subcarriers from the set of valid subcarriers and solve again the problem for the remaining ones.

Following this approach, this solution can be interpreted as pouring water to a vessel having unevenly shaped base and the term $\frac{1}{N}\left(\sum_{i=1}^{N} \frac{\sigma^2}{g_i} + p_{\max}\right)$ represents the water level while the total volume of water corresponds to the available transmit power $p_{\max}$. However, this technique is only feasible when continuous transmission power is assumed, as well as continuous multilevel modulation and infinitely small subcarriers. It can, however, be adapted to more practical conditions, e.g., an actual link adaptation mapping with practical modulations, with good results [12].

5.5.1.2 Feasible Sequential Quadratic Programming

Sequential quadratic programming (SQP) method is a class of efficient algorithms for solving non-linearly constrained optimization problems. It has received much attention and its application in OFDMA multiple links scenario can be found in [44]. This work considers a centralized power allocation algorithm that maximizes the throughput T of a set of M co-channel cells under the bit error rate (BER) and maximum transmit power constraints, similar to the rate maximization problem in a multicell version.

Mathematically, this problem can be formulated as

$$\begin{aligned} \max_{\mathbf{p}} \quad & T = \sum_{n=1}^{N} \sum_{m=1}^{M} \log_2(1+\xi\,\gamma_{n,m}(\mathbf{p})) \quad \text{subject to} \\ & \sum_{n=1}^{N} p_{n,m} \le p_{\max,m}, \ \text{for all } m \\ & p_{n,m} \ge 0, \ \text{for all } n \text{ and } m, \end{aligned} \tag{5.15}$$

where $\gamma_{n,m}$ and $p_{n,m}$ are the SINR and the transmit power in the subcarrier n in the cell m, respectively; $p_{\max,m}$ is the maximum transmit power of cell m and ξ is a constant which depends on the target BER of the system. The solution of this problem is given by the power vector $\mathbf{p} = \left[p_{1,1}\ p_{2,1}\ \cdots\ p_{N,M}\right]$ composed of NM elements. Notice that the solution of this problem provides only the power allocation in each subcarrier, but not the subcarrier assignment. Such an assignment has to be previously defined through another method.

Since problem (5.15) is a constrained non-linear programming problem, SQP is well suited for its solution. SQP is an iterative algorithm and works as follows. At each iteration i, SQP method approximates the Hessian matrix of the Lagrangian function of the problem using a quasi-Newton updating method that guarantees super-linear convergence by exploiting the second-order information. The approximated Hessian matrix is then used to formulate a quadratic programming (QP) subproblem whose solution is used to compute the search direction $\mathbf{d}(i)$ in this iteration:

$$\begin{aligned} \min_{\mathbf{d}} \quad & \frac{1}{2}\mathbf{d}^{\mathrm{T}}\mathbf{H}(i)\mathbf{d} + \nabla^{\mathrm{T}} T(\mathbf{p}(i))\mathbf{d} \quad \text{subject to} \\ & \sum_{n=1}^{N} p(i)_{n,m} - p_{\max,m} + \nabla^{\mathrm{T}}\left(\sum_{n=1}^{N} p(i)_{n,m} - p_{\max,m}\right)\mathbf{d} = 0, \ \text{for all } m, \end{aligned} \tag{5.16}$$

where $\mathbf{H}(i)$ is the positive definite approximation of the Hessian matrix of the Lagrangian function of problem (5.15) in the iteration i. The operator $(\cdot)^{\mathrm{T}}$ indicates the transposition of vector.

Then, a line search procedure (which is a one-dimensional minimization problem) is performed in order to determine a step length $t(i)$ and finally, the next solution is $\mathbf{p}(i+1) = \mathbf{p}(i) + t(i)\mathbf{d}(i)$.

However, in the traditional SQP method, the subproblem (5.16) may be inconsistent, that is, its feasible set may be empty. To overcome this shortcoming, a variant called feasible sequential quadratic programming (FSQP) algorithm was proposed to generate solutions for the subproblem (5.16) in the feasible region, along the iterations. Further study on FSQP algorithms can be found in [17].

5.5.1.3 Branch-and-Bound

Branch-and-bound (BB) is an approach developed for solving combinatorial optimization problems. This optimization tool combines enumeration of all possible solutions by means of "branches" and "pruning" of some of them. Each branch contains a set of nodes where each one corresponds to a stage of decision of the complete solution. The strategy consists of covering the branches of the enumeration tree, one by one. When an unpromising or infeasible node is reached, the correspondent branch is pruned without covering it completely, and there is no need to consider their descendent nodes. If enough branches can be pruned off, the processing time may be reduced to a computationally manageable value. Notice that the algorithm does not ignore those solutions in the leaves of the branches that were pruned. Instead of this, it has left them out of consideration after it has made sure that the optimal solution cannot be at any one of these branches. Thus, the BB approach is not an approximating procedure, but it is an exact optimization procedure that finds an optimal solution.

Three questions are fundamental in the implementation of a BB algorithm:

- How to do the enumeration of the solutions: What kind of partitions will be used in the ramifications?
- Which branches should be covered first?
- The efficiency of the pruning: how to evaluate whether a branch will not lead to worse solutions?

In [35], a BB method was applied in an multicarrier OFDMA system in order to solve the classic RRA problems, margin-adaptive (MA) and rate-adaptive (RA), previously presented in Sections 5.4.2 and 5.4.3, respectively. The algorithm enumerates the solutions in a tree of N levels of nodes, where each one corresponds to a subcarrier, associated in an arbitrary way. With this node structure, a UE j and modulation scheme m are allocated in each node. Hence, there are $J \cdot M$ possibilities for each node, where J and M are the number of UEs and modulation schemes, respectively.

In order to cover each branch, the UEs are ordered in an arbitrary way. For the first node, the first UE of the ordering is assigned using the modulation of higher order. Now it is necessary to evaluate the cost of this assignment. As the descendent nodes are not already assigned, the rest of the solution (rest of the branch) is obtained by solving the relaxed version of an integer linear programming subproblem with the first node already set. If this subproblem is infeasible, the modulation order of the first node is decreased and the process is repeated. If all modulations for this UE leads to infeasible solutions, the UE associated to this node is changed to the second in the ordering and so on. On the other hand, if the subproblem is feasible, the cost value is calculated and it is set as the bound. From now, solutions with higher cost are immediately pruned. If a better solution is found, the cost associated to this new solution is used as the new bound to prune new branches. This process is repeated for the other $(N-1)$ nodes.

5.5.2 Approximate Search Approaches

Approximated approaches avoid searching in all the solutions space, e.g., as the exact search approaches, saving computational complexity.

5.5.2.1 Genetic Algorithm

Genetic algorithm (GA) is an established stochastic search method based on the theory of natural selection. The theory of natural selection, proposed by Charles Darwin [19], assumes that individuals adapted to the natural environment have more chances of survival and consequently transfer their characteristics to their offspring. An individual is said to be well adapted if its genetic characteristics (genotype) demonstrate in favor of its existence in the environment in which it is inserted.

GA is robust and effective [13, 19] in combinatorial optimization problems. They are effective because of their ability to exploit favorable characteristics of previous solutions and successively produce better solutions. Another advantage of GA is the fact that it is not necessary to know if the objective function is continuous or differentiable. Moreover, genetic algorithms are easy to implement.

The application of GA in a multicarrier OFDMA system was proposed in [41]. This work considers the margin-adaptive problem for an OFDM symbol, subject to QoS restrictions. The QoS restrictions refer to a maximum target BER and the number of bits of each UE that needs to be transmitted in an OFDM symbol. Mathematically, this problem can be formulated as

$$\begin{aligned} &\min_{c_{j,n}} \quad P_{total} = \sum_{n=1}^{N} \sum_{j=1}^{J} \frac{f(c_{j,n})}{g_{j,n}} \quad \text{subject to} \\ &BER_n \leq BER_{target}, \text{for all } n \\ &\sum_{n=1}^{N} c_{j,n} = r_j, \text{for all } j \\ &\text{where} \\ &f(c_{j,n}) = \frac{\sigma^2}{3}\left[Q^{-1}\left(\frac{BER_n}{4}\right)\right]^2 (2^{c_{j,n}}-1), \end{aligned} \tag{5.17}$$

where $c_{j,n}$, $f(c_{j,n})$, $Q(\cdot)$, and $g_{j,n}$ are, respectively, the number of bits allocated to user j on subcarrier n, the received power necessary to transmit $c_{j,n}$ bits with a bit error rate BER_n, the numeric Q-function and the channel gain.

Genetic algorithms (GA) work with a set of P encoded solutions, called population. An initial population of size P is generated randomly. It is necessary to represent each solution through a string, denominated chromosome. An intelligent codification scheme improves the quality of the solution as well as decreases the processing time. The search from a set of solutions imposes an implicit parallelism in the search (thus accelerating the time of search) and avoids the final solution from being a local optimum point.

One of the objectives of GA in the problem above is to decide the best assignment of subcarriers to the UEs. Therefore, the chromosome structure is composed of N elements, where each element represents a subcarrier in the OFDMA system. The value of each element in the array is confined to the set $1, 2, \ldots, J$, which represents the UEs, illustrated in Fig. 5.5.

Subcarrier1	Subcarrier2	Subcarrier3	Subcarrier4		Subcarrier N–2	Subcarrier N–1	Subcarrier N
UE4	UE7	UE1	UE J	...	UE7	UE2	UE4

Fig. 5.5 Structure of the chromosome.

Therefore, the codification of the chromosome gives only the subcarrier allocation. In order to achieve the bit loading and calculate the overall transmit power, the water-filling method is employed for each chromosome in the population. The fitness function used to evaluate the chromosome is the overall power, which is the objective function. Chromosomes with higher fitness (lower overall power) have higher priority of being selected for mating. The mating is carried out through crossover. In each crossover, two chromosomes generate two offspring. Before applying crossover, the C (a predefined number) best solutions are selected to compose the new generation (elitism). At the same time, the C worst solutions are directly discarded. The remaining chromosomes together with the best C chromosomes are selected to carry out crossover. Therefore, the crossovers will generate $P-C$ offspring. Two-point crossover is used in the algorithm. After crossover, mutation is applied to the offspring with a given probability. In order to prevent the algorithm from converging on a local optimal solution, the probability of mutation is increased when no better solution is found in 5, 10, and 15 consecutive generations.

The $P-C$ new offspring are combined with the best C chromosomes in the last generation to form the new population of the next generation. These steps are repeated until the predefined number of generations is reached or no better solution is found in q consecutive generations.

5.5.2.2 Simulated Annealing

Simulated annealing (SA) is a random-search technique which exploits an analogy between the way in which a metal cools and freezes into a minimum energy crystalline structure (the annealing process) and the search for a minimum in a more general system.

One typical feature of SA is that, besides accepting solutions with improved cost, it might also accept solutions with deteriorated cost, with a given probability. This feature gives to the algorithm the "hill climbing" capability and, consequently, the ability to avoid becoming trapped in local minima. Different from GA, SA works with only one solution instead of a set of solutions.

The work in [34] proposes the application of SA in an OFDMA system. This work considers the problem of the maximization of the sum of individual utility functions $\mathrm{U}(j)$ of the J UEs at the system. This individual utility function takes into account the rate $r_{j,n,m}$ in each subcarrier n assigned to the UE j using a modulation of order m, the long-term QoS requirements Q_j^{req}, and the priority q_j, which depends on the type of service of the UE j. Therefore, the utility $\mathrm{U}(j)$ is also a function of the power allocation.

The rate $r_{j,n,m}$ in the subcarrier n for the UE j using a modulation of order m depends on the allocated power $p_{j,n,m}$, for a specific required bit error rate BER_{target}. The function Q_j^{req} depends on the delay and the packet dropping ratio for RT services. For NRT services, it is a function of the rate. The priority function $P(j)$ can assume only two constants values: ρ_{RT} and ρ_{NRT}, so that $\rho_{RT} + \rho_{NRT} = 1$. In order to give higher priority to RT services, ρ_{RT} must be greater than ρ_{NRT}.

Restrictions about the total power p_{total} and the maximum power $p_{\max}$ per subcarrier are also considered. The proposed problem consists of deciding the UE, the power level, and the modulation in each subcarrier in order to maximize U. Mathematically, the problem is formulated as

$$\begin{aligned} \max_{\mathbf{p},x_{j,n,m}} \quad & \sum_{n=1}^{N}\sum_{j=1}^{J}\sum_{m\in M} \mathrm{U}(\mathbf{p})\cdot x_{j,n,m} \quad \text{subject to} \\ & \sum_{n=1}^{N}\sum_{j=1}^{J}\sum_{m\in M} p_{j,n,m}\cdot x_{j,n,m} \leq p_{total} \\ & \sum_{j=1}^{J}\sum_{m\in M} p_{j,n,m}\cdot x_{j,n,m} \leq p_{\max} \text{ for all } n \\ & \sum_{n=1}^{N}\sum_{m\in M} r_{j,n,m}\cdot x_{j,n,m} \leq r_j^* \text{ if } j \text{ is an RT user,} \end{aligned} \tag{5.18}$$

where

$$x_{j,n,m} = \begin{cases} 1, & \text{if the subcarrier } n \text{ is allocated to UE } j \text{ using modulation order } m \\ 0, & \text{otherwise} \end{cases}. \tag{5.19}$$

One fundamental aspect is the codification of the solution in this problem. The solution is represented by a power vector $\widehat{\mathbf{p}} = [\widehat{p}_1\ \widehat{p}_2\ \cdots\ \widehat{p}_N]$ composed of N elements. Beyond the power level, each element of the solution vector determines the UE allocated in the correspondent subcarrier and the modulation order. The element $\widehat{p}_n$ belongs to a set of power levels of size $(J \cdot M)$:

$$\widehat{p}_n \in \{p_{1,n,1}, p_{1,n,2}, \cdots, p_{1,n,M}, \cdots, p_{J,n,1}, p_{J,n,2} \cdots, p_{J,n,M}\}, \qquad (5.20)$$

where the values $p_{j,n,m}$ are pre-calculated according to j, n, and m. If $\widehat{p}_n = p_{j,n,m}$, it means that the subcarrier n will be allocated to the UE j and the modulation order is m. With this codification, the solution vector has $(J \cdot M)^N$ possible combinations.

In order to execute simulated annealing (SA), it is necessary to define the structure of the neighborhood. In this problem, two solution vectors are considered neighbors when they differ from each other in only one element of the vector. With this neighborhood structure, the algorithm works as follows: SA starts at an initial solution (generated randomly), and then randomly generates a new neighbor solution. The process compares the new neighbor solution with the initial solution to see if it is a better design. If it is better, it will accept this new solution and this one becomes the current solution; otherwise, it will accept it with a certain probability; if not accepted, it will go to the previous solution and start the process again, and this iterative process will continue until a solution close to the optimal one is found.

The probability of accepting a worse solution changes as the temperature changes. It starts at a high probability, meaning that the process will accept worse solutions many times in the beginning to make sure that the current solution is not at a local optimal solution. As temperature decreases (annealing process), this probability also decreases. Then, SA will accept little or no worse solutions in the end of the process in order to get closer to the optimal solution.

5.5.3 Comparison Among Optimization Tools

Naturally, each of the listed tools present its particular strong and weak points. Usually, the choice of which tool will be employed is a captious question. A general comparison for every situation is impractical, and an application-driven choice is recommended. Thus, an interesting challenge consists of pondering all important conveniences and drawbacks in the sense to answer a single question: *which tool is the most suitable under specific scenario, purpose, and implementation limitations?* Thereupon, applicability, method power, and computational complexity queries must be considered.

Applicability is a very important query because it can dismiss some tools for choice. For a few tools, there are known outcome accounts for RRA applications. Such RRA problems can be mathematically formulated and, as a consequence, it can highlight peculiar characteristics including linearity and allowed decision variable values. Aside from identical characteristics, any comparison among such tools may be unfair.

The solving method query can show too much about the power of the listed tools. An optimization tool can (or cannot) be able to do a complete scanning of the feasible region in order to evaluate the objective function. All the listed tools have the potential skill of locating the global maximum or minimum value of the objective function. Furthermore, some problems cannot be solved in an exact way, rather a suitable heuristic method can be adopted, which can yield an approximate solution, albeit a good one.

The evaluation of the computational complexity query has high significance especially in the practical implementation field. Unrealistic demands of computational swiftness and storing can hide the qualities of a method, which in principle could be suitable and powerful. For the sake of comparison, the well-known asymptotic worst-case time complexity is usually assumed as an estimation of the computational complexity. Note that, as a worst-case measure, the evaluated computational complexity may mean that a unique instance will require this time, while the majority of problem instances might actually require considerably less time than that. There are some exponential time algorithms that have been quite useful in practice. This is, for instance, the case of branch-and-bound (BB) under certain conditions and suppositions [18].

Table 5.1 hints a general characterization frame of the listed optimization tools as well as some specific applications in which these tools were used. It is important to mention that the column "functional constraints type" refers to the specific application discussed in this section, indicating that the respective optimization tool can also be applied in other configurations.

Table 5.1 Tools' characteristics

Tool	Example of RRA problem		Solution method		Computational complexity
	Application	**Functional constraints type**	**Search result**	**Stopping criterion**	
Lagrangian multipliers	Maximize sum-rate capacity s.t. power restrictions in a single link scenario	Linear	Exact	None	$O\left(N^2\right)$
FSQP	Maximize sum-rate capacity s.t. power restrictions in multiple links scenario	Linear	Exact	Number of iterations q	$> O\left(q(KN)^3\right)$
Genetic algorithm	Minimize total power s.t. QoS restrictions in multiple links without interference scenario	Non-linear	Approximate	Number of generations q	$O\left(1.65 \cdot 2^{0.21N} \cdot N^2\right)$
Simulated annealing	Maximize utility function s.t. power and QoS restrictions in multiple links without interference scenario	Non-linear	Approximate	Number of iterations q	$O\left(qJ^3M^3N^3\right)$
Branch-and-bound	Margin- or rate-adaptive problems	Non-linear	Exact	None	$\sqrt{(2)}^{JMN}$

5.6 Algorithms for Frequency Resource Assignment

One of the advantages of multicarrier OFDMA-based system is the opportunity to take benefit from frequency and multiuser diversities. A mechanism for taking advantage of the frequency and multiuser diversities is the employment of scheduling algorithms. Scheduling algorithms are responsible for selecting which UEs will have access to the system resources and with which configuration. In this way, scheduling algorithms have a great impact on system performance.

The most basic resource for a multicarrier OFDMA system is bandwidth. As the bandwidth is divided into subcarriers, the use of different parts of the system bandwidths by different UEs becomes simple. Because the frequency assignment is directly related to the channel-aware packet scheduling functionalities, the two functionalities will be unified and be referred to as a scheduler. The minimal assignable frequency resource can comprise one or a group of subcarriers and will be referred to as resource unit (RU) for the rest of this section.

Scheduling algorithms are an important functionality to perform QoS control among users utilizing different packet-based services. Some general requirements of a good scheduler are:

- **Efficient link utilization**: Scheduler must be opportunistic in the sense of taking advantage of multiuser diversity so as to utilize the channel efficiently.
- **Delay bound**: The scheduler must guarantee delay bounds for individual flows in order to support delay-sensitive applications.
- **Fairness**: A certain level of fairness should be assured in the system in order to avoid flows with QoS overprovision.
- **Implementation complexity**: A low-complexity algorithm is a necessity in high-speed networks in which scheduling decisions have to be made very rapidly.
- **Isolation**: The algorithm should isolate a session from the ill effects of misbehaving sessions. The QoS guarantees for a session should be maintained even in the presence of sessions whose demands are in excess of their reserved values.
- **Delay/bandwidth decoupling**: For most schedulers, the delay is tightly coupled to the reserved rate; that is, a higher reserved rate provides a lower delay. However, some high-bandwidth applications, such as Web browsing, can tolerate relatively large delays.
- **Scalability**: The algorithm should operate efficiently as the number of sessions sharing the channel increases. Moreover, the scheduler must be flexible enough to work well in different scenarios, e.g., different traffic mix proportions. Schedulers can be classified according to the information that it utilizes to support its decision. In this way, depending on whether the scheduler utilizes channel state information of the UEs it can be classified into either channel aware or channel unaware. Another possible classification is the ability to deal with multiple services. Hence, QoS-differentiated schedulers are capable of prioritizing flows according to the QoS demands and service. Otherwise, the scheduler is considered non-QoS-differentiated.

On all presented algorithms, each UE j is implicitly removed from the UE set $\mathscr{J}$ when the total UE rate r_j is sufficient to transmit all data in its respective transmission buffer.

5.6.1 Maximum Rate

The maximum rate (MR) algorithm was proposed in [27] with the objective of maximizing the system data rate of OFDMA systems as shown in Section 5.4.1.

The solution of the MR problem is quite simple. The algorithm assigns each RU to the UE where the highest channel gain is verified for that RU. The algorithm continues to assign the best channel RU to the respective UEs until all RUs have been assigned. The MR scheduler is presented in Algorithm 5.1.

Algorithm 5.1 Maximum Rate Algorithm.

while $\mathscr{N} \neq \emptyset$ and $\mathscr{J} \neq \emptyset$ **do**
  $(j^*, n^*) \leftarrow \max_{j,n}\{\gamma_{j,n} | j \in \mathscr{J} \text{ and } n \in \mathscr{N}\}$
  $\mathscr{N} \leftarrow \mathscr{N} - n^*$
  $r_{j^*} \leftarrow r_{j^*} + \mathrm{F}(\gamma_{j^*,n})$
end while

5.6.2 Round Robin

The idea of the RR scheduler is to be fair by assigning the same number of RUs to all active UEs. The scheduler operates by generating a randomly ordered list with all active UEs and assigning random RUs to each UE following the list order. The process starts again from the beginning of the list once all UEs received RUs. But the previous order of the list stays fixed.

Note that an equal number of RUs does not result in equal data rate. The RUs of the UEs will be in different channel states resulting in different data rates. The RR scheduler is presented in Algorithm 5.2.

Algorithm 5.2 Round Robin Algorithm.

$j^* \leftarrow$ first element in $\mathscr{J}$
for all $n \in \mathscr{N}$ **do**
  $r_{j^*} \leftarrow r_{j^*} + \mathrm{F}(\gamma_{j^*,n})$
  if $\mathscr{J} = \emptyset$ **then**
    end algorithm
  else
    $j^* \leftarrow$ next element in $\mathscr{J}$
  end if
end for

5.6.3 Multicarrier Proportional Fair

The multicarrier proportional fair (MPF) scheduler [31] is an extension of the classical single-carrier proportional fair algorithm [26] to multicarrier systems. As in the single-carrier version, the algorithm uses a filtered measurement of the average data rate of each UE to provide information about the past data rates and support its assignment decisions.

The MPF assigns the RUs by searching the combination of UE j^* and RU n^* that maximizes the equation

$$\frac{r_{j,n}}{T_j[k]}, \tag{5.21}$$

where $r_{j,n}$ is the achievable data rate of UE j on RUs n and $T_j[k]$ is the filtered data rate of UE j at each TTI k that is given by

$$T_j[k] = \left(1 - \frac{1}{t_a}\right) \cdot T_j[k-1] + \left(\frac{1}{t_a}\right) \cdot r_{j,k}, \tag{5.22}$$

where t_a is a filtering time constant used to configure the time window to define the filtered rate. The MPF scheduler is presented in Algorithm 5.3.

Algorithm 5.3 Multicarrier Proportional Fair Algorithm.

while $\mathscr{N} \neq \emptyset$ and $\mathscr{J} \neq \emptyset$ **do**
  $(j^*, n^*) \leftarrow \max_{j,n} \left\{ \frac{r_{j,n}}{T_j[k]} | j \in \mathscr{J} \text{ and } n \in \mathscr{N} \right\}$
  $\mathscr{N} \leftarrow \mathscr{N} - n^*$
  $r_{j^*} \leftarrow r_{j^*} + F(\gamma_{j^*,n})$
end while

5.6.4 Satisfaction-Oriented Resource Allocation (SORA) Algorithm

As stated in Section 5.4, several different objectives may be pursued by a resource allocator. In this section is described an example of scheduler to maximize the user satisfaction of the system, the satisfaction-oriented resource allocation (SORA) algorithm.

One of the interesting aspects of the SORA algorithm is its flexibility to deal with flows from different services. Consider that the system has ρ_{total} data flows and ρ_s is the number of data flows of a service $s \in \Psi$, where Ψ is the set of available services.

The objective of the scheduler is to decide which flows will transmit on which RUs. But when multiple services share the channel, the problem from Section 5.4.4 needs to be adapted.

The objective now implies not only in maximizing the number of satisfied flows, but also to balance the rate of satisfaction between services. This results in the new user satisfaction maximization problem in

$$\max_{\mathbf{X}} \min_{s} \left(Q_s[k] | s \in \Psi\right) \quad \text{subject to} \\ \sum_j x_{j,n}[k] \leqslant 1, \quad \forall n, \tag{5.23}$$

where Q_s is satisfaction rate of the flows of service s at TTI k, given by

$$Q_s = \frac{\rho_s^{sats}}{\rho_s}, \tag{5.24}$$

where ρ_s^{sats} is the number of satisfied UEs of the service $s \in \Psi$.

The SORA algorithm is divided into two parts:

- **The resource allocation part**: This part uses the current state of each UE to determine the number of RUs required by each UE, the relative priority of the UEs among themselves, and the number of UEs to be scheduled from each service. This part is different for each service because it depends on service-specific parameters to measure the degree of satisfaction of the UEs.
- **The resource assignment part**: This part does the actual mapping between RUs and UEs. This part is performed by ordering the chosen UEs by their channel states and assigning the RUs with the best channel state to each UE. This step does not depend on the service type.

5.6.4.1 SORA: Resource Allocation Part

The first step of the resource allocation part is to calculate the data rate required by each flow. Using this data rate, the algorithm generates the priority list $\mathbf{p}_s$ for each service $s \in \Psi$. The relative priority of the flows of the service s is given by the order of the flows in the list $\mathbf{p}_s$.

After the priority lists for all services are generated, the algorithm generates the allocation list **a**. The allocation list contains the flows that will transmit at the current TTI. The allocation list **a** is generated keeping the proportion of flows of each service. Thus, if a service has twice the service-active flows of an other, it will also have twice the number of flows in **a**. The number of flows in **a** is restricted by the maximum number of transmitting UEs allowed in the BS.

In the following, the specific implementation of the resource allocation part for each service is presented. Here, the specific implementations for the NRT and RT services will be presented.

5.6.4.2 Resource Allocation for Non-real-Time Services

This is the resource allocation step for NRT services such as TCP-based traffic, Web browsing, and FTP. NRT services do not have strict packet delay requirements. In fact, subscribers utilizing this service type only expect that their average data rate be maintained above a given target. Fluctuations of the average data rate around the target are tolerable.

The QoS indicator for NRT services is the time-averaged data rate until the TTI k, $\bar{r}_j[k]$ defined as

$$\bar{r}_j[k] = \frac{s_j[k]}{t_j[k]}, \tag{5.25}$$

where s_j and t_j are the total transmitted data in bits and the total active time of flow j until TTI k, respectively.

An NRT flow is considered satisfied if its time-averaged data rate $\bar{r}_j$ is higher than or equal to its averaged data rate requirement $\bar{r}_j^{\text{req}}$. Using the current state of the flow, the required data rate Δr_j to make the flow satisfied at the next TTIs is given by

$$\Delta r_j[k] = \bar{r}_j^{\text{req}} \cdot (t_j[k] + z) - \bar{r}_j[k-1] \cdot t_j[k], \tag{5.26}$$

where z is a constant. Note that if the flow is already satisfied, Δr_j will be non-positive.

The Δr_j is then used to calculate the number of RUs m_j required by the flow j, calculated as

$$m_j = \max\left(1, \frac{\Delta r_j[k]}{F\left(\bar{\gamma}_j[k]\right)}\right), \tag{5.27}$$

where $\bar{\gamma}_j[k]$ is the mean SINR among all available RUs of the flow j at TTI k.

The values of m_j are then used to construct the priority list $\mathbf{p}$. Firstly, the unsatisfied flows are ordered by increasing m_j and added to the list $\mathbf{p}$. Then the satisfied flows are also ordered by increasing m_j and added to the end of the list $\mathbf{p}$.

5.6.4.3 Resource Allocation for Real-Time Services

This section presents the resource allocation part utilized for RT services. This service type is delay sensitive and imposes strict requirements in the packet loss rate and delay variation.

The FER is considered as the QoS indicator for the RT service. The FER is defined as

$$FER_j[k] = \frac{n_j^{\text{lost}}[k]}{n_j^{\text{lost}}[k] + n_j^{\text{succ}}[k]} \quad \forall j \in \mathcal{J}, \tag{5.28}$$

where $n_j^{\text{succ}}[k]$ is the number of successfully transmitted RT packets from flow j until TTI k and $n_j^{\text{lost}}[k]$ is the number of lost RT packets from flow j until TTI k.

The parameter w_j is the equivalent to the m_j for the RT services. The w_j represents the distance, in packets, to the target FER FER_j^{req} of the flow j. For a flow below the target FER (satisfied flow), it is the number of packets that should be lost to achieve the target FER. For a flow above the target FER (unsatisfied flow), it is the number of packets that should be successfully transmitted to achieve the target

FER. The specific calculation of w_j is given by

$$w_j[k] = \begin{cases} \left\lfloor \dfrac{\left(n_j^{\text{succ}}[k] + n_j^{\text{lost}}[k]\right) \cdot FER_j^{\text{req}} - n_j^{\text{lost}}[k]}{1 - FER_j^{\text{req}}} \right\rfloor, & \text{if } FER_j[k] \leq FER_j^{\text{req}} \\ \left\lceil \dfrac{n_j^{\text{lost}}[k] - \left(n_j^{\text{succ}}[k] + n_j^{\text{lost}}[k]\right) \cdot FER_j^{\text{req}}}{FER_j^{\text{req}}} \right\rceil, & \text{otherwise,} \end{cases} \tag{5.29}$$

where $\lfloor v \rfloor$ represents the first integer equal to or lower than v and $\lceil v \rceil$ is the first integer equal to or greater than v.

The required data rate Δr_j of a RT flow is simply the rate necessary to transmit the oldest packet of the flow j.

Different from the NRT services, the generation of the priority list **p** for the RT services gives priority to the satisfied flows. This procedure is adopted because the RT services are very sensitive to fluctuations in the experienced QoS. Therefore, the idea is to keep the highest number of flows with acceptable QoS instead of trying to recover flows from dissatisfaction as is done for NRT services.

The satisfied flows are ordered by decreasing i_j using

$$i_j = \left((D^{\text{discard}} - D_j^{\text{oldest}}[k]) \cdot (w_j[k] + 1)\right)^{-1}, \tag{5.30}$$

where $D_j^{\text{oldest}}[k]$ is the delay of the oldest packet of flow j at TTI k and D^{discard} is the maximum allowable packet delay before discard. Thus, the flows with higher delays (first term) and the ones closer to the unsatisfied state (second term) are prioritized. In the group of unsatisfied flows, the ones with higher i_j are also prioritized, which are the flows with higher delays (first term) and can become satisfied more easily (second term).

5.6.4.4 SORA: Resource Assignment Part

The resource assignment part is based on assignment phases. On each phase, the allocation list **a** is ordered according to the channel quality of the best RU of the flow. Thus, the flow with the best channel quality RU will be the first to receive an RU (its best one), followed by the flow with the second best RU and so on. In this way, each flow will have one RU at the end of the first phase.

If, after receiving an RU, a flow j achieves its required data rate $\Delta r_j[k]$, this flow stops receiving RUs from the scheduler.

The phases continue, with the flows being ordered again at the beginning of each phase, until all RUs are assigned or there is no active flow. In case of all flows achieving the required rate $\Delta r_j[k]$ and there are still unused RUs, all flows that comprised list **a** and with remaining buffered data will compete for resources again in the same fashion.

5.7 Subcarrier Assignment in 3GPP's Long-Term Evolution (LTE)

This section presents a case study where scheduling algorithms are employed in the 3GPP's LTE system. LTE or evolved UMTS terrestrial radio access network (E-UTRAN) is an evolution of 3GPP radio access technology in order to improve the performance of current systems in terms of bit rate, latency, and packet-based services. The following section provides a short description of LTE radio access, followed by some performance results through the application of RRA algorithms in this system.

5.7.1 LTE System Overview

LTE radio access network (RAN) is connected to a core network that has been called evolved packet core (EPC) that provides a packet-optimized framework to connect multiple RANs. EPC is out of the scope of this chapter and further details can be found in [7].

The RAN is shown in Fig. 5.6. The RAN is connected to the core network through the S1 interface that provides access to the RAN radio resources for the transport of user plane and control plane traffic. The RAN part comprises several enhanced Node Bs (eNBs) that are the standardized name to the base stations. The eNBs are responsible, among other tasks, for the RRA in downlink and uplink. eNB are also in charge of controlling one or more cells and are connected among each other through the X2 interface. The X2 interface allows eNB to exchange control and user plane information such as handover messages and buffered data.

In order to standardize the data representation and establish rules for signaling, data transmission, and error recovery, the radio interface is organized in protocol layers that performs closely related subtasks and communicates with each other. These protocol layers are disposed in protocol stacks that are based on the open systems interconnection (OSI) model [14]. The protocols layers present in the LTE

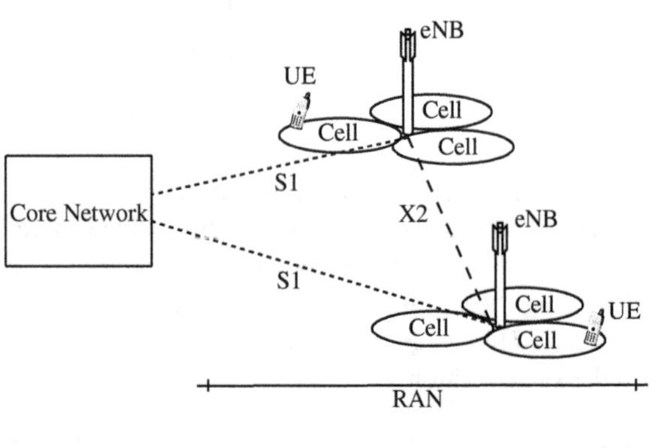

Fig. 5.6 Overview of LTE including the main nodes and interfaces.

radio interface are depicted in Fig. 5.7 split in the control and user plane protocol layers. User plane protocols are in charge of carrying user data through the access stratum (AS), i.e., within the RAN. On the other hand, control plane protocols are responsible for controlling the connection between the UE and the network. Non-access stratum (NAS) messages, i.e., messages exchanged between the core network and UEs, are also transmitted utilizing control plane protocols.

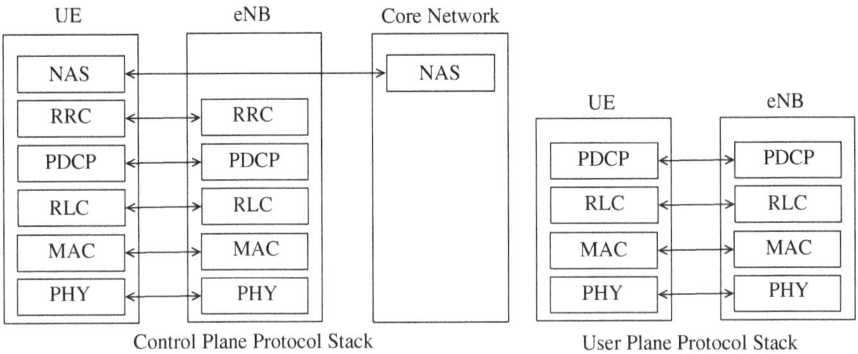

Fig. 5.7 Radio interface protocols in the control and user planes.

The radio resource control (RRC) is a pure control plane protocol layer, which is established between the UE and the enhanced node b (eNB). This protocol takes care of control aspects and higher layer signaling of RRM functions [6]. The packet data convergence protocol (PDCP) layer is responsible among other tasks for header compression and decompression, ciphering, and integrity protection of user plane and control plane data [4]. The RLC protocol sublayer is essentially related to the transfer of packets. Its main responsibility is to provide reliable data transfer to the upper layers [5]. Another important task is the packet discard mechanism. In the following section the MAC and physical (PHY) layers are described in more details.

5.7.1.1 Medium Access Control

The MAC sublayer is responsible for performing the following tasks: mapping between logical and transport channels, multiplexing of RLC protocol data units (PDUs) into transport block (TB) and demultiplexing of TB into RLC PDUs, traffic volume measurement reporting, error correction through hybrid automatic repeat request (H-ARQ), priority handling between logical channels of one UE, priority handling between UEs by means of dynamic scheduling, and transport format selection [3].

The following techniques, which are either implemented at the MAC level or have a certain interaction with the MAC layer, are briefly discussed here:

- **Scheduling**: The eNB scheduler has the function of determining which UEs have access to which radio resources at a given time. This can be interpreted as a mix of classical channel allocation and time scheduling algorithms.
- **Link adaptation**: It adapts the modulation and coding schemes according to reported channel quality measurements.
- **H-ARQ**: According to 3GPP the H-ARQ should be based on incremental redundancy (IR). In the downlink, the H-ARQ is asynchronous and adaptive. In the asynchronous operation the H-ARQ retransmissions can take place at any time. When adaptive H-ARQ is utilized the scheduler has the flexibility of modifying the MCS of previous (re)transmissions.

5.7.1.2 Physical Layer

A basic configuration of LTE may consist of a system having a downlink bandwidth ranging from 1.4 to 20 MHz which corresponds to 72 and 1,200 subcarriers effectively available for transmission [2]. The subcarrier spacing is 15 kHz. FDD frames have a time duration of 10 ms and are divided into 20 slots of 0.5 ms each. A short CP of approximately 4.7 μs will be considered in each OFDM symbol [1]. With short CP a slot has seven OFDM symbols. The basic modulation schemes are quadrature phase shift keying (QPSK), 16 and 64 quadrature amplitude modulation (QAM).

The radio bearer (RB) is defined as a block of M subcarriers and N subsequent OFDM symbols. In this way, a reduced signaling effort in the system is obtained compared to, e.g., an allocation on a subcarrier basis.

The subcarrier distribution in the RB is localized, i.e., RBs are composed of adjacent subcarriers. By adaptively allocating RBs to UEs in good channel conditions, considerable multiuser diversity gains can be obtained. Nevertheless, frequency diversity gains might also be obtained by allocating multiple RBs spread over the system bandwidth to the same UE [1].

For example, consider $M = 12$ and $N = 7$ which means that an RB is defined as 12 adjacent subcarriers in a 7-symbol slot in the time–frequency grid. The minimum allocable resource block or RU is defined as two consecutive RBs in the time domain, i.e., 14 OFDM symbols with 12 subcarriers considering the short CP.

The data destined to the UE will be adequately modulated, interleaved, and coded. The channel coding scheme in LTE is the turbo code with a coding rate of 1/3, two 8-state constituent encoders and a contention-free quadratic permutation polynomial (QPP) turbo code internal interleaver [1].

5.7.2 Radio Resource Allocation in LTE

Before presenting the simulation results, it is important to define some performance metrics necessary to understand the results and the schedulers used for comparison and also show the main simulation parameters.

5.7.2.1 Definitions and Simulation Parameters

This section shows the performance results of RRA algorithms that are applied to LTE system. Specifically, simulations were performed with the following schedulers: SORA, delay scheduler (DS), RR, and MR. These schedulers have different characteristics and input parameters. Therefore, through the analysis of these schedulers, the importance of different aspects can be assessed when designing scheduling algorithms such as channel quality state and packet-related information.

RR schedules the flow with the longest starving period, i.e., it gives opportunity to the flow whose last scheduling time is the oldest. This scheduling algorithm has some similarities to the frequency resource assignment algorithm presented in Section 5.6.2. The scheduled flow receives the number of RUs necessary to transmit all awaiting data. In case after scheduling a flow there are unused resources, the next flow with the longest starving period is chosen. When the UE is selected to receive data, the scheduler chooses its best RU, which means that this scheduler is not a pure RR scheduler.

DS prioritizes flows with the greatest packet delay in the transmit buffer [21]. The flow receives resources until all the existing data in the transmit buffer can be transmitted. The flow with the second highest packet delay is scheduled when there are unused resources. The RUs are selected in an opportunistic way, i.e., the UE gets assigned its best RUs in terms of channel quality. So, to differentiate VoIP and Web flows, the packet delay of VoIP flows is multiplied by a constant factor. The objective of this is to balance the QoS of both services since VoIP flows are time sensitive and have smaller packets than Web flows. This constant factor can be made equal to the ratio between the average required rate for Web service (128 kbps) and the data rate generated by a VoIP application (12.2 kbps) which, in this particular case, amounts to 10.5.

MR is a complete opportunistic scheduler, i.e., it schedules the UE that is in better channel condition. This scheduler is based on the frequency resource assignment algorithm presented in Section 5.6.1. In this way, this scheduler is supposed to maximize the aggregated data rate in the downlink by utilizing high data-rate MCSs.

The following simulation results show the user satisfaction ratio, cell throughput and capacity. The main parameters utilized in the simulations are shown in Table 5.2. A Web flow is considered satisfied if the average Web throughput is greater than an average required throughput while a VoIP flow is considered satisfied if its VoIP FER is lower than the required FER.

The average Web throughput is the ratio between the number of correctly received bits at the TCP layer of the UE and the total session active time. Total session active time is the total time in which the Web flow was active. The Web flow is considered active in the period between the transmission of the hypertext transfer protocol (HTTP) request from the client (UE) to the server and the complete reception of the requested Web page at the UE.

On the other hand, the VoIP packet delay is the time that a VoIP frame takes from the transmitter behind the Internet to the receiver at the UE. The VoIP FER in this study is defined as the ratio between the number of lost packets and the total expected packets. There are two cases in which a packet is considered lost: if it does

Table 5.2 Simulation parameters.

LTE network		
Parameter	**Value**	**Unit**
Bandwidth	3	MHz
Carrier frequency	2	GHz
Duplexing scheme	FDD	–
Number of RUs	15	–
Total cell power	20	W
Number of scheduled UE per TTI (PDCCH limit)	5	–
Number of eNBs	3	–
Number of cells per eNB	3	–
Number of UE antennas	2	–
Number of cell antennas	1	–
Cell radius	500	m
Frequency reuse	1/3	–

Web service		
Parameter	**Value**	**Unit**
Web page size (fixed)	10,000	bytes
Mean reading time	1.5	s
Average required throughput	128	kbps
VoIP service		
Mean talk period time	5	s
Voice activity	0.5	–
Frame size	264	bits
Frame period	20	ms
Maximum VoIP frame delay	140	ms
Required FER	1	%

not arrive at the receiver or if its reception is performed with a delay greater than the maximum VoIP packet delay. The first case can be caused by either an RLC SDU discard or an H-ARQ failure, e.g., by reaching the maximum number of H-ARQ retransmissions allowed.

In case of a single service scenario, the system capacity is defined as the maximum offered load (e.g., number of flows in the cell) in which the user satisfaction ratio is greater than the user satisfaction threshold. In case of mixed service scenarios, the system capacity is the maximum offered load in which all services have a user satisfaction ratio greater than their respective satisfaction thresholds. In this case study the satisfaction threshold for Web and VoIP are considered equal to 95% and 90%, respectively.

Finally, the cell throughput is calculated at PDCP above the RLC layer for both Web and VoIP flows. This metric presents an insight of how well the system resources have been utilized.

5.7.2.2 Results

First we present some results in the mixed traffic in which 75% of the flows are from the VoIP service and the other 25% are from the Web service. This is a likely scenario in the future when the circuit-switched voice service would have been replaced by the packet-switched VoIP service. In Fig. 5.8 we illustrate the cell throughput for the scheduling algorithms. In general, the cell throughput increases with the system load due to the higher resource utilization (RU and power) and multiuser diversity gain [33].

The cell throughput provided by the MR scheduler is not the one expected of an opportunistic scheduler that prioritizes the flows in better channel conditions and, therefore, is supposed to maximize the cell throughput. The reasons for this

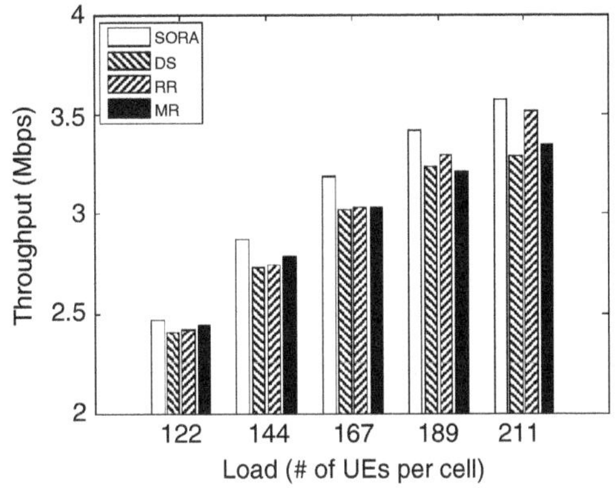

Fig. 5.8 Average cell throughput in the mixed scenario of 75% VoIP and 25% Web.

degraded performance are the presence of VoIP traffic and limitation in the number of scheduled UEs. As the scheduled flows have good channel conditions and can quickly transmit the buffered data, the scheduled flows in general do not have much awaiting data to transmit. Consequently, the scheduling process does not end with the limitation in the number of RUs, but with the limitation in the number of scheduled UEs leaving unused resources in the system.

DS takes into account the packet delay when scheduling flows. This scheduling criterion works quite well with low traffic rate and delay-sensitive services such as VoIP. However, with NRT services such as Web that has large packets and burst traffic, the performance is not so good. The problem is due to the fact that the Web packets are very large compared to VoIP frames. Furthermore, the flows with higher packet delays are in general the ones in poor channel conditions. So, the Web flows with higher packet delays need many system resources before transmitting completely the buffered data. In this way, DS presents a poor performance in the cell throughput.

The selection criterion of RR gives equal transmission opportunities to all active flows in the system and does not take channel quality state into account. However, when a flow is selected it gets assigned the resources in better channel state. Therefore, RR has an intermediate position in cell throughput.

SORA scheduler has the best performance in cell throughput. Despite this is not the key objective of SORA, this is mainly consequence of its Resource Assignment part where the flows get assigned system resources exploiting the frequency and multiuser diversities.

In Fig. 5.9, the user satisfaction ratio is shown for the Web service. The schedulers DS and RR perform poorly for Web traffic, while the schedulers SORA and MR present quite good results. The inferior performance of DS is explained by the inadequacy of DS scheduler for Web traffic. When Web traffic is concerned, it is important to exploit the channel quality state and the burst traffic nature to achieve high data rates. This explains the performance difference between RR and MR.

The SORA scheduler achieves the best user satisfaction ratio for Web traffic. This comes from the smart QoS control that considers the current satisfaction state of each user so as to increase the number of satisfied users. With this strategy, the

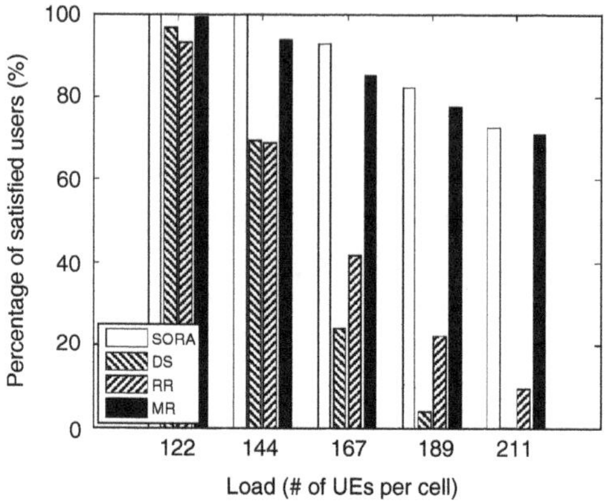

Fig. 5.9 Web user satisfaction ratio in the mixed scenario of 75% VoIP and 25% Web.

flows that are in good QoS conditions are preempted by the ones that need more resources.

With regard to VoIP service we can see in Fig. 5.10 that, some schedulers perform differently compared to the Web service. The channel opportunistic behavior of MR that does not consider packet delay in its formulation has not presented good results in this case. Although DS takes into account packet delay, the resource waste with Web flows compromises the overall performance. The RR scheduler performs well with VoIP flows even though it does not take into account the packet delay directly. However, the starvation period that is the selection criterion of RR has some similarities with packet delay.

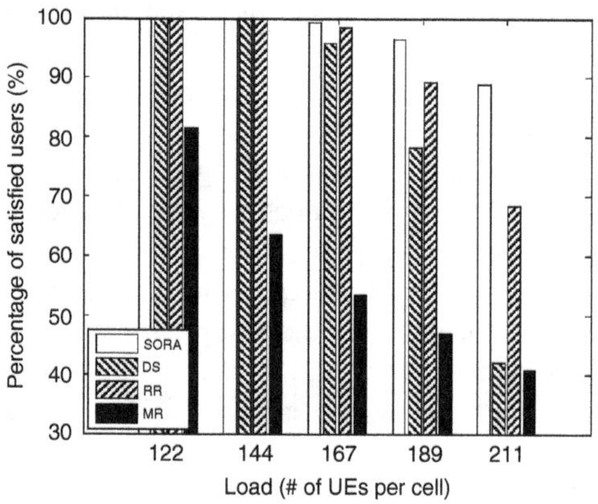

Fig. 5.10 VoIP user satisfaction ratio in the mixed scenario of 75% VoIP and 25% Web.

The SORA scheduler repeats for VoIP service the good performance achieved with the Web service. This is a consequence of the better resource sharing accomplished by SORA. First, the flows of each service are sorted according to a priority that intends to avoid QoS overprovision. Then, the most important flows of each service are selected to be scheduled considering the number of active flows from each

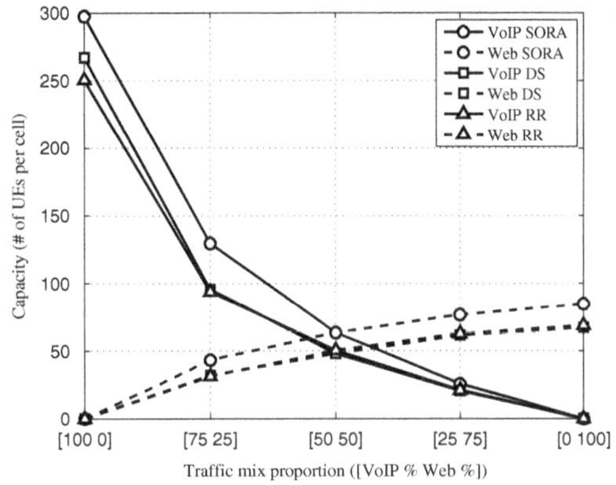

Fig. 5.11 Overall capacity of the simulated schedulers in several traffic mixed scenarios.

service. As a consequence, the resources are better distributed among the services leading to a user satisfaction balancing.

Finally, this section presents the downlink capacity in Fig. 5.11 so as to provide a complete picture of the performance of the schedulers in other service mixes. In order to improve the system capacity in mixed traffic scenarios, the system resources should be fairly distributed among the service classes. With this in mind, the SORA scheduler achieves gains in system capacity in the presented traffic mixes that ranges from 11 to 35% over DS and from 18 to 38% compared with RR.

5.8 Power Allocation Algorithms and Performance in OFDMA

Another important functionality in multicarrier OFDMA systems is power allocation (PA) among the subcarriers. The motivation to employ PA algorithms in OFDMA systems is the fact that several subcarrier assignment algorithms assume that the power in each subcarrier is constant and equally distributed. Of course, this approach is not optimum when the objective is to maximize the sum-rate throughput, because a possible power reallocation could improve the transmission rates for some subcarriers. For instance, the power transferred of some subcarrier to another could improve the transmission mode used on the last, without degrading the transmission mode of the donor subcarriers. A transmission mode is a combination of the transmission parameters such as modulation order and channel coding rate referred to here as MCS. Furthermore, one MCS can be considered better than another if it can transmit more bits with the same transmission resources.

Different solutions have been proposed in the literature for the power allocation problem. One of the most known is the Hughes-Hartogs algorithm [22]. In this algorithm, for each subcarrier, the amount of power required to transmit data with the worst MCS is calculated. Then, the subcarrier which requires the least amount of power is selected; this amount of power is allocated to it, decreasing the available total power, and the required additional power for applying the next better MCS is calculated. This process is repeated until all subcarriers reach the best MCS or the available power for allocation is completely utilized.

Chow and Bingham proposed in [10] an algorithm faster than Hughes-Hartogs algorithm. The objective of this algorithm is to minimize the transmit power while maintaining a target data rate. This algorithm starts with equal power in each subcarrier and then adjusts these powers in order to reach the target rate. After getting the bit rate in each subcarrier using constant power, the algorithm iteratively increases or decreases the transmit power margin in each subcarrier, depending on the difference between the currently achieved data rate and the target data rate. After these adjustments have been carried out for a finite amount of time, the power allocation is finalized with a last refinement.

An extension of the algorithm proposed in [32] was presented in [16]. In this extension, the objective is to minimize the bit error probability while achieving a target data rate. Another interesting approach is that utilized in [39], where instead of iteratively adding bits to subcarriers, the algorithm removes bits iteratively.

The following section revises the Hughes-Hartogs algorithm and presents a proposal called multiuser residual power allocation (MURPA), which involves a mechanism of power reallocation so that the assumptions took into account by the scheduling algorithm is respected.

5.8.1 Hughes-Hartogs Algorithm

The Hughes-Hartogs algorithm was proposed in [22]. This algorithm can be performed after any subcarrier assignment algorithm and can be adapted for the multiuser case in the following way. The total available power p_j^{tot}, assumed for each UE j, is directly proportional to the number of subcarriers N_j assigned to it, which is given by the subcarrier assignment algorithms, assuming that the power in each subcarrier is constant and equally distributed. Once the p_j^{tot} is determined for each UE j, the algorithm is run individually for each UE.

In the algorithm initialization, it is necessary to calculate the amount of power $pow_{m,n}$ necessary to achieve each MCS m in the subcarrier n which is already assigned to a specific UE. Of course, these values depend on the channel gain in each subcarrier and the link adaptation mapping. After that, the power values $\Delta pow_{m,n}$ necessary to improve from the MCS $m-1$ to the m in subcarrier n are calculated. Note that this power can be different according to the channel gain and modulation level. After that, the algorithm increases, one by one, the MCS of the subcarriers that require the lower amount of power to achieve the next better MCS, according to the link adaptation. This process is performed until all subcarriers reach the best MCS level or the available power for the corresponding UE, p_j^{tot}, is completely utilized. At the end, this algorithm achieves the maximum data rate, according to the power constraints and subcarriers channel states. This process is summarized in Algorithm 5.4, where $\mathcal{J}$ is the set of active UEs in the system, $\mathcal{S}_j$ and N_j are the set of subcarriers and the number of subcarriers assigned to UE j, respectively; $\mathcal{D}$ is the set of possible MCSs, p_{cell} is the cell power, and μ_n and p_n represent the MCS and the power in the subcarrier n, respectively.

Algorithm 5.4 Hughes-Hartogs Algorithm.

```
𝒥 ← {1,...,J}
for all j ∈ 𝒥 do
    𝒮_j ← {1,...,N_j}
    𝒟 ← {0,...,M}
    Calculate p_j^tot = (N_j / N) · p_cell
    for all n ∈ 𝒮_j do
        for all m ∈ 𝒟 do
            Calculate pow_{m,n}
            Calculate Δpow_{m,n} = pow_{m,n} − pow_{m−1,n}
        end for
    end for
    P_used ← 0
    for all n ∈ 𝒮_j do
        μ_n ← 0
    end for
    p_n ← 0 ∀n ∈ 𝒮_j
    while (P_used < p_j^tot) and ( Σ_{n∈𝒮} μ_n ≠ N_j · M ) do
        n* ← arg min_{n∈𝒮_j} Δpow_{1,n}
        if Δpow_{1,n*} > p_j^tot − P_used then
            break
        end if
        μ_{n*} ← μ_{n*} + 1
        p_{n*} ← pow_{μ_{n*},n*}
        P_used ← P_used + pow_{μ_{n*},n*}
        Δpow_{m,n*} ← Δpow_{m+1,n*}   ∀m ∈ 𝒟
    end while
end for
```

In terms of computational complexity, this algorithm demands a significant processing time. Supposing that there are N subcarriers and M available modulation levels, it is necessary to calculate two matrices of elements $pow_{m,n}$ and $\Delta pow_{m,n}$, respectively, both having dimensions $M \times N$. As will be shown in the next section, the MURPA algorithm demands less computational calculations.

5.8.2 Multiuser Residual Power Allocation Algorithm

The optimization problem to be solved by the MURPA algorithm is to maximize the bit rate of the system, keeping or improving the MCSs of all subcarriers (in relation to the scheme with uniform power) limited to a restriction of total power in system. The problem formulation follows:

$$\max_{\mathbf{p}} \quad \sum_{n} r_n \quad \text{subject to} \\ \mu_n(p_n) \geq m_n^{\text{EP}} \quad \forall n, \tag{5.31}$$

where $\mathbf{p} = [p_1, p_2, \dots, p_N]$ is the power vector whose element p_n represents the power allocated to the subcarrier n, r_n is the bit rate in the subcarrier n which depends on the allocated power p_n to it, μ_n is the MCS in the subcarrier n, which also depends on p_n, and m_n^{EP} is the MCS given by the subcarrier assignment algorithm supposing equal power in the subcarrier n. The problem constraint means that the MCS after power allocation must be better than or equal to the achieved MCS using equal power allocation on subcarrier n.

Unlike Hughes-Hartogs algorithm, in MURPA, the power is not divided proportionally among the UEs, according to the number of subcarriers assigned to it. That is, the power belonging to a subcarrier of a given UE can be reallocated to another subcarrier of a different UE.

The proposed algorithmic solution for problem (5.31) is as follows. In each TTI, the subcarrier assignment is carried out with constant and equally distributed powers in the subcarriers. Then, the MURPA algorithm calculates, for each subcarrier, the power p_n^{ded} which can be deducted from the subcarrier without degrading its current MCS. These deducted powers are accumulated in a pool of power P^{reall} and then the step of power reallocation starts. This step is similar to the one of Hughes-Hartogs algorithm, that is, the algorithm improves, one by one, the MCS of the subcarriers that require the lowest amount of power to reach the next better MCS, according to the link adaptation curve. This process is performed until all subcarriers reach the best MCS or until the amount of power of the pool is insufficient to improve the MCS of any subcarrier. This process is summarized in Algorithm 5.5, where $\mathscr{S}$, $\mathscr{D}$, and N are the set of subcarriers, the set of possible MCSs, and the number of subcarriers in the system, respectively.

Regarding the computational complexity, notice that differently from the Hughes-Hartogs algorithm, the MURPA algorithm does not need to calculate all the components $pow_{m,n}$ and $\Delta pow_{m,n}$. Only the corresponding elements which are equal or superior to the MCS yielded by the subcarrier assignment algorithm (with constant and equally distributed power) are computed, i.e., the elements of the sets $PowerSet_n$ and $\Delta PowerSet_n$ in Algorithm 5.5. For example, if the subcarrier assignment algorithm results in a MCS μ_n for the subcarrier n, only the elements $\mu_n, \mu_n + 1, \dots, M$ of the column n need to be calculated. Therefore, this algorithm is less computationally complex than the Hughes-Hartogs algorithm.

Algorithm 5.5 Multiuser Residual Power Allocation Algorithm.

$\mathscr{S} \leftarrow \{1,\ldots,N\}$
$\mathscr{D} \leftarrow \{0,\ldots,M\}$
for all $n \in \mathscr{S}$ **do**
  Determine m_n^{EP} of the subcarrier n using constant and equally distributed power
  $\mu_n \leftarrow m_n^{\mathrm{EP}}$
  $\mathscr{D}_n \leftarrow \{m_n^{\mathrm{EP}},\ldots,M\}$
  Calculate p_n^{ded} without decreasing μ_n
  $p_n \leftarrow pow_{\mu_n,n}$
  $PowerSet_n \leftarrow \emptyset$
  $\Delta PowerSet_n \leftarrow \emptyset$
  for all $m \in \mathscr{D}_n - \{\mu_n\}$ **do**
    Calculate $pow_{m,n}$
    $PowerSet_n \leftarrow PowerSet_n \bigcup \{pow_{m,n}\}$
    Calculate $\Delta pow_{m,n} = pow_{m,n} - pow_{m-1,n}$
    $\Delta PowerSet_n \leftarrow \Delta PowerSet_n \bigcup \{\Delta pow_{m,n}\}$
  end for
end for
Calculate $P^{reall} = \sum_n p_n^{ded}$
while $(P^{reall} > 0)$ and $(\Delta PowerSet_n \neq \emptyset \quad \forall n)$ **do**
  $n^* \leftarrow \arg\min_{n \in \mathscr{S}} \Delta pow_{\mu_n,n}$
  if $\Delta pow_{\mu_{n^*},n^*} > P^{reall}$ **then**
    break
  end if
  $\mu_{n^*} \leftarrow \mu_{n^*} + 1$
  $p_{n^*} \leftarrow pow_{\mu_{n^*},n^*}$
  $P^{reall} \leftarrow P^{reall} - p_{n^*}$
  $\Delta PowerSet_{n^*} \leftarrow \Delta PowerSet_{n^*} - \{\Delta pow_{\mu_{n^*},n^*}\}$
end while

5.8.3 Performance of Power Allocation Algorithms

The performance of MURPA and Hughes-Hartogs algorithms are evaluated in a system consisting of a single cell in which all UEs are uniformly distributed in the cell area at the simulation start. These UEs are static during the whole simulation, but their channel gains are variable due to fast fading modelling according to the Jakes Model. The simulation parameters are as follows. The system model is LTE-like with 100 subcarriers and the total power in the cell 5 W. The cell radius is 500 m. The bandwidth of each subcarrier is 15 kHz and the system operating frequency is 2 GHz. The shadowing standard deviation is 8 dB and the noise power is -123.24 dBm. In each TTI of time length 0.5 ms the subcarriers and power are allocated for the active UEs. For this system, it is assumed that the service utilized by the users is World Wide Web (www).

In the simulations, the adaptive modulation and coding scheme are carried out in each subcarrier based on a SNR-to-rate table, where the SNR values indicate the switching levels among consecutive MCSs, corresponding to the M-QAM modulations ($M = 2^m$; $m = 1,2,3,4,5,$ and 6). Each MCS corresponds to the total of bits that can be transmitted in a TTI, with a given BER, and therefore, it is equivalent to maximum achievable rate in that subcarrier. In the computation of these SNR values, the SNR gap formula is used with a BER = 10^{-6} [11]. The SNR switching levels are shown in Table 5.3.

Table 5.3 SNR switching levels.

SNR (dB)	**9.5**	**14.4**	**18.2**	**21.7**	**25.1**	**28.3**
MCSs	1	2	3	4	5	6
Rate (kbps)	2	4	6	8	10	12

In order to compare the algorithms, the maximum rate (MR) is utilized as subcarrier assignment algorithm (Section 5.6.1) for the WWW service; and as performance metrics the user satisfaction and the mean UE rate are employed. The user satisfaction is defined as the percentage of UEs that are satisfied at the end of the simulation. A UE is satisfied if its mean throughput is higher than or equal to the UE average minimum rate requirement which is considered as 50 kbps.

First, in Fig. 5.12 is illustrated the pattern of power distribution in each subcarrier as well as their respective MCSs for both PA algorithms in the same channel realization. The channel realization is shown in Fig. 5.12(a). The MCS and power allocation for Hughes-Hartogs algorithm are presented in Fig. 5.12(b) and (c), respectively. For the MURPA algorithm, the resulting MCS and power allocation are shown in Fig. 5.12(d) and (e), respectively. In this realization, MR assigns the subcarriers in the ranges 1–30 and 73–100 to one UE (UE A) and the other subcarriers, 31–72, to another one (UE B). In these figures can be observed some differences between the allocation pattern of the algorithms. Higher channel gains make possible better MCSs; however, for the same channel realization, MURPA reaches better MCSs for the UE A than Hughes-Hartogs algorithm, in the neighborhood of the subcarrier 80, even using the same total power in the cell. This happens because the MURPA algorithm transfers power of the UE B, which would be used by Hughes-Hartogs algorithm, to the UE A, since MURPA does not have the restriction of power division among the UEs. Therefore, MURPA allocates the total power more efficiently than Hughes-Hartogs algorithm in this example.

In terms of user satisfaction, it can be observed in Fig. 5.13(a) that the MURPA algorithm outperforms the Hughes-Hartogs algorithm, labelled as H-H in the figure. The user satisfaction can be translated into capacity, when associated with the capacity as the maximum number of users which the system can support, so that a minimum percentage of users are satisfied with their average rates. For Hughes-Hartogs, the absolute and relative capacity gains in relation to the scheme using equal power distribution are 22 UEs and 34.9%, respectively. Now, comparing MURPA, the capacity gains are equivalent to 32 UEs and 50.8%, respectively.

Concerning the average UE rate, the performances of both algorithms are also superior to the scheme with uniform power allocation, as can be seen in Fig. 5.13(a).

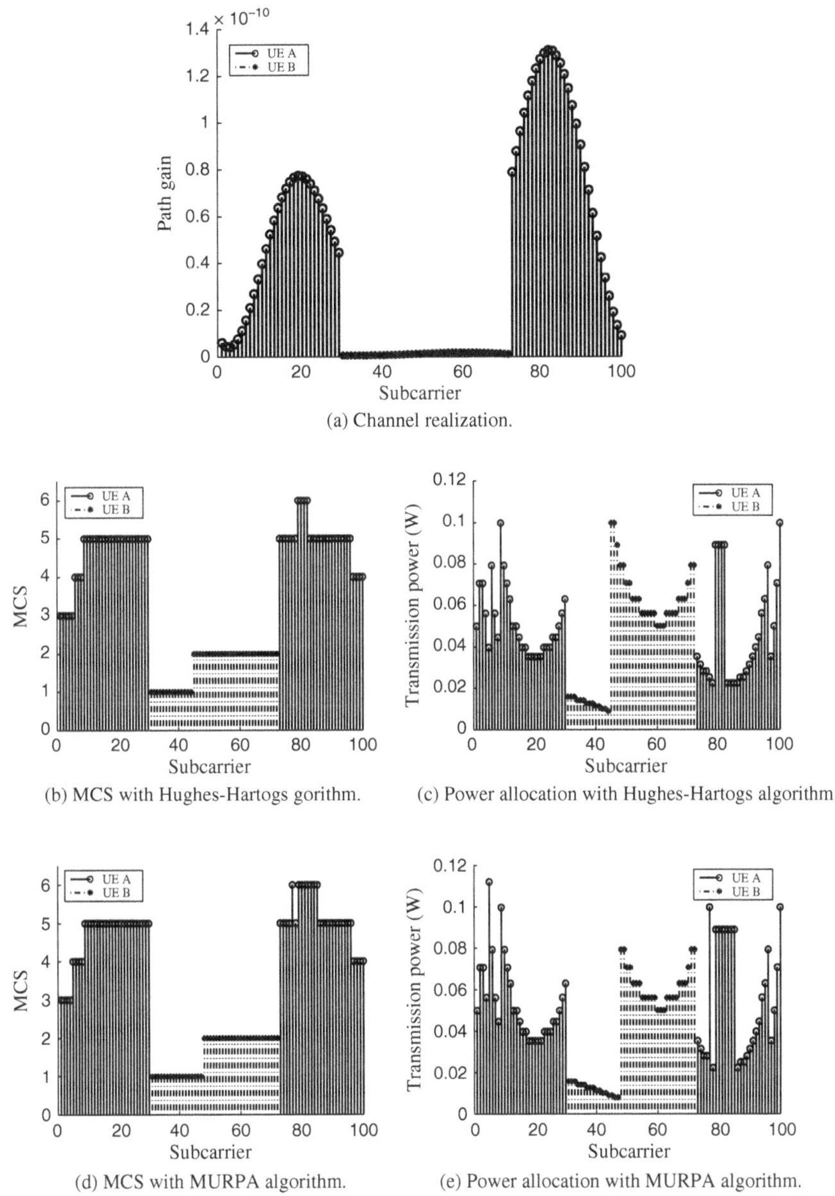

(a) Channel realization.

(b) MCS with Hughes-Hartogs gorithm.

(c) Power allocation with Hughes-Hartogs algorithm.

(d) MCS with MURPA algorithm.

(e) Power allocation with MURPA algorithm.

Fig. 5.12 MCS and power allocation for Hughes-Hartogs and MURPA algorithms.

Comparing the two PA algorithms, it can be observed that MURPA outperforms Hughes-Hartogs algorithm in all simulated loads.

Therefore, MURPA is superior in terms of both mean UE rate and UE satisfaction. MURPA's degree of freedom in allocating power without restrictions permits

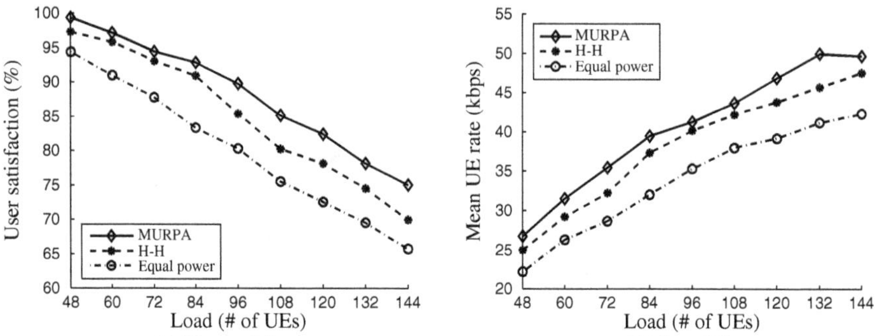

(a) User satisfaction (b) Mean UE rate

Fig. 5.13 User satisfaction and mean UE rate for the schemes with and without power allocation using RM algorithm with average rate requirements of 50 kbps for WWW service.

a better distribution of power among the subcarriers and consequently a better performance.

Another advantage of MURPA in relation to Hughes-Hartogs algorithm lies in its computational complexity. For example, in a simulation of 60 s for a load of 100 UEs, where each TTI provides a sample of the number of hops, the average number of hops for MURPA is 30 and for H-H is 321. A hop is equivalent to a change in the MCS to the immediately superior level in a subcarrier.

5.9 Conclusions and Research Directions

This chapter presented the main problems of RRA in multicarrier OFDMA systems. To take advantages of the diversities presented in such systems, schedulers are proposed in the literature in accordance to objectives, such as maximization of sum-rate capacity, minimization of the transmit power, guarantee of fairness, and maximization of user satisfaction.

Furthermore, in this chapter, some scheduling algorithms were presented that have different characteristics and objectives. As it was presented in a case study on long-term evolution (LTE) system, these different approaches lead to different performance results. Among the presented schedulers, this work highlights the satisfaction-oriented resource allocation (SORA) scheduler that aims at maximizing the number of satisfied users in the system. As presented in the results, this scheduler is able to increase the system capacity in multiservice scenarios where there are unbalanced traffic mixes.

In the literature most of the schedulers consider the subcarrier assignment (SA) and power allocation (PA) problems separately, so, to the subcarrier assignment function the power is uniform among subcarriers. But, after the SA stage a PA algorithm can be used. In this chapter was proposed the MURPA as PA algorithm, which improves the total throughput and user satisfaction of an OFDMA cell with a lower complexity when compared with the traditional Hughes-Hartogs.

There are still some open issues in RRA in multicarrier OFDMA systems as follows:

- Most schedulers designed to multicarrier OFDMA systems need the knowledge of the complete states of the channel per subcarrier, and due to limitations in the control channel in the uplink direction, this knowledge becomes prohibitive. The impact of the limited feedback measurements in RRA in multicarrier OFDMA systems is an open issue.
- Another open issue is the consideration of the spatial dimension in RRA problems in multicarrier OFDMA systems. Multiple-input multiple-output (MIMO) transceivers have been intensively studied and the consideration of joint MIMO and RRA in OFDMA is a promising research topic.

References

1. 3GPP: Physical Layer Aspects for Evolved Universal Terrestrial Radio Access (UTRA). Tech. Rep. TR 25.814 V7.1.0 – Release 7, 3rd Generation Partnership Project (2006). URL `http://www.3gpp.org`
2. 3GPP: Evolved Universal Terrestrial Radio Access (E-UTRA); Base Station (BS) Radio Transmission and Reception. Tech. Rep. TS 36.104 V8.1.0 – Release 8, 3rd Generation Partnership Project (2008)
3. 3GPP: Evolved Universal Terrestrial Radio Access (E-UTRA); Medium Access Control (MAC) Protocol Specification. Tech. Rep. TS 36.321 V8.1.0 – Release 8, 3rd Generation Partnership Project (2008)
4. 3GPP: Evolved Universal Terrestrial Radio Access (E-UTRA); Packet Data Convergence Protocol (PDCP) Specification. Tech. Rep. TS 36.323 V8.1.0 – Release 8, 3rd Generation Partnership Project (2008)
5. 3GPP: Evolved Universal Terrestrial Radio Access (E-UTRA); Radio Link Control (RLC) Protocol Specification. Tech. Rep. TS 36.322 V8.1.0 – Release 8, 3rd Generation Partnership Project (2008)
6. 3GPP: Evolved Universal Terrestrial Radio Access (E-UTRA) Radio Resource Control (RRC); Protocol Specification. Tech. Rep. TS 36.331 V8.2.0 – Release 8, 3rd Generation Partnership Project (2008)
7. 3GPP: General Packet Radio Service (GPRS) Enhancements for Evolved Universal Terrestrial Radio Access Network (E-UTRAN) Access. Tech. Rep. TS 23.401 V8.2.0 – Release 8, 3rd Generation Partnership Project (2008)
8. Choi, Y.J., Bahk, S.: Scheduling for VoIP Service in CDMA2000 1x EV-DO. In: Communications, 2004 IEEE International Conference, Vol 3, pp. 1495–1499 (2004). DOI 10.1109/ICC.2004.1312760
9. Chong, E.K.P., Zak, S.H.: An Introduction to Optimization, 3rd edn. John Wiley & Sons (2008)
10. Chow, P.S., Cioffi, J.M., Bingham, J.A.C.: A practical discrete multitone transceiver loading algorithm for data transmission over spectrally shaped channels. IEEE Transactions on Communications **43**(234), 773–775, (Feb/Mar/Apr 1995). DOI 10.1109/26.380108
11. Chung, S.T., Goldsmith, A.J.: Degrees of freedom in adaptive modulation: A unified view. IEEE Transactions on Communications **49**(9), 1561–1571 (2001). DOI 10.1109/26.950343
12. Cover, T., Thomas, J.: Elements of Information Theory. John Wiley & Sons (1991)
13. Davis, L.: Handbook of Genetic Algorithms. Van Nostrand Reinhold (1991)

14. Day, J.D., Zimmermann, H.: The OSI Reference Model. Proceedings of the IEEE **71**(12), 1334–1340 (1983)
15. Ericson, M., Wanstedt, S.: Mixed Traffic HSDPA scheduling – Impact on VoIP Capacity. In: Vehicular Technology Conference, 2007. VTC2007-Spring. IEEE 65th, Ericsson Res., Lulea pp. 1282–1286. Dublin (2007). DOI 10.1109/VETECS.2007.269
16. Fischer, R., Huber, J.: A new loading algorithm for discrete multitone transmission. IEEE Proc. Globecom (1996)
17. Fletcher, R.: Practical Methods of Optimization, Vol 2. John Wiley & Sons (1980)
18. Garey, M.R., Johnson, D.S.: Computers and Intractability: A Guide to the Theory of NP-Completeness. W. H. Freeman & Co (2003)
19. Goldberg, D.E.: Genetic Algorithms in Search, Optimization and Machine Learning. Addison-Wesley (1989)
20. Gross, J., Bohge, M.: Dynamic mechanisms in OFDM wireless systems: A survey on mathematical and system engineering contributions. Tech. Rep. TKN-06-001, Technical University Berlin – Telecommunication Networks Group (2006). URL `http://www.tkn.tu-berlin.de/publications/papers/TKN_Report_06_001.pdf`
21. Hosein, P.: Scheduling of VoIP traffic over a time-shared wireless packet data channel. In: Personal Wireless Communications, 2005. ICPWC 2005. 2005 IEEE International Conference, pp. 38–41 (2005). DOI 10.1109/ICPWC.2005.1431297
22. Hughes-Hartogs, D.: Ensemble Modem Structure for Imperfect Transmission Media. United States Patent (4.679.227) (1987)
23. ITU-R: Framework and overall objectives of the future development of IMT-2000 and systems beyond IMT-2000. Tech. Rep. Recommendation ITU-R M.1645 (2006)
24. Jaffe, J.: Bottleneck flow control. IEEE Transactions on Communications [legacy, pre-1988] **29**(7), 954–962 (1981)
25. Jain, R., Chiu, D., Hawe, W.: A quantitative measure of fairness and discrimination for resource allocation in shared computer systems. Tech. Rep. TR-301, DEC Research Report TR-301 (1984)
26. Jalali, A., Padovani, R., Pankaj, R.: Data throughput of CDMA-HDR a high efficiency-high data rate personal communication wireless system. IEEE Vehicular Technology Conference Proceedings **3**, 1854–1858 (2000)
27. Jang, J., Lee, K.B.: Transmit power adaptation for multiuser OFDM systems. IEEE Journal on Selected Areas in Communications **21**(2), 171–178 (2003). DOI 10.1109/JSAC.2002.807348
28. Jeffery, A.: Mathematics for Engineers and Scientists, 5th edn. Chapman & Hall, London (1996).
29. Jorswieck, E.A., Sezgin, A., Zhang, X.: Framework for analysis of opportunistic schedulers: average sum rate vs. average fairness. Modeling and Optimization in Mobile, Ad Hoc, and Wireless Networks and Workshops, 2008. WiOPT 2008. 6th International Symposium, pp. 100–105 (2008). DOI 10.1109/WIOPT.2008.4586049
30. Kelly, F., Maulloo, A., Tan, D.: Rate control in communication networks: shadow prices, proportional fairness and stability. Journal of the Operational Research Society **49**(3), 237–252 (1998)
31. Kim, H., Kim, K., Han, Y., Yun, S.: A proportional fair scheduling for multicarrier transmission systems. In: Vehicular Technology Conference, 2004. VTC2004-Fall. 2004 IEEE 60th, Vol 1, pp. 409–413 (2004). DOI 10.1109/VETECF.2004.1400034
32. Kivanc, D., Liu, H.: Subcarrier allocation and power control for OFDMA. In: Signals, Systems and Computers, 2000. Conference Record of the Thirty-Fourth Asilomar Conference, Vol 1, pp. 147–151 (2000). DOI 10.1109/ACSSC.2000.910933
33. Knopp, R., Humblet, P.A.: Information capacity and power control in single-cell multiuser communications. In: Communications, 1995. ICC '95 Seattle, 'Gateway to Globalization', 1995 IEEE International Conference on, Vol 1, pp. 331–335 (1995). DOI 10.1109/ICC.1995. DOI 525188

34. Lee, L.T.H., Chang, C.J., Chen, Y.S., Shen, S.: A utility-approached radio resource allocation algorithm for downlink in OFDMA cellular systems. Proceedings of IEEE Vehicular Technology Conference, pp. 1798–1802 (2005)
35. Mao, Z., Wang, X.M., Lin, J.: Fast optimal radio resource allocation in OFDMA system based on branch-and-bound method. Proceedings of IEEE Pacific Rim Conference on Communications, Computers and Signal Processing, pp. 348–351 (2005)
36. Rhee, W., Cioffi, J.M.: Increase in capacity of multiuser OFDM system using dynamic subchannel allocation. In: Vehicular Technology Conference Proceedings, 2000. VTC 2000-Spring. 2000 IEEE 51st, Vol 2, pp. 1085–1089. Tokyo (2000). DOI 10.1109/VETECS.2000.851292
37. Rodrigues, E.B., Cavalcanti, F.R.P., Wanstedt, S.: Qos-driven adaptive congestion control for voice over ip in multiservice wireless cellular networks. Communications Magazine, IEEE **46**(1), 100–107 (2008). DOI 10.1109/MCOM.2008.4427237
38. Santos, R.B., Lima, F.R.M., Freitas, W.C., Cavalcanti, F.R.P.: QoS based radio resource allocation and scheduling with different user data rate requirements for OFDMA systems. In: Proceedings of 18th PIMRC 2007, pp. 1–5 (2007). DOI 10.1109/PIMRC.2007.4394699
39. Sonalkar, R.V., Shively, R.R.: An efficient bit-loading algorithm for DMT applications. IEEE Communications Letters **4**(3) 80–82 (Mar 2000). DOI 10.1109/4234.831031
40. Velez, F.J., Correia, L.M.: Classification and characterisation of mobile broadband services. In: Vehicular Technology Conference, 2000. IEEE VTS-Fall VTC 2000. 52nd, Vol 3, pp. 1417–1423 (2000). DOI 10.1109/VETECF.2000.886329
41. Wang, Y., Chen, F., Wei, G.: Adaptive subcarrier and bit allocation for multiuser OFDM system based on genetic algorithm. Proceedings of International Conference on Communications, Circuits and Systems, pp. 242–246 (2005)
42. Wong, C.Y., Cheng, R.S., Lataief, K.B., D.Murch, R.: Multiuser OFDM with adaptive subcarrier, bit, and power allocation. IEEE Journal on Selected Areas in Communications **17**(10), 1747–1758 (1999). DOI 10.1109/49.793310
43. Yanhui, L., Chunming, W., Changchuan, Y., Guangxin, T.: Downlink scheduling and radio resource allocation in adaptive OFDMA wireless communication systems for user-individual QoS. Proceedings of World Academy of Science, Engineering and Technology, pp. 221–225 (2006)
44. Yih, C.H., Geranotis, E.: Centralized power control algorithms for OFDM cellular networks. Proceedings of IEEE Military Communications Conference, pp. 1250–1255 (2003)
45. Yin, H., Liu, H.: An efficient multiuser loading algorithm for OFDM-based broadband wireless systems. In: Global Telecommunications Conference, 2000. GLOBECOM '00. IEEE, Vol 1, pp. 103–107 (2000). DOI 10.1109/GLOCOM.2000.891705

Chapter 6
Common Radio Resource Management for Multiaccess Wireless Networks

Alex P. da Silva, Leonardo S. Cardoso, Vicente A. de Sousa Jr., and Francisco R. P. Cavalcanti

6.1 Introduction

Multiaccess (MA) networks are systems created by the integration of two or more radio access technologies (RATs) which cooperate, in order to provide a better use of their complementary features, aiming at a higher performance as a whole. In order to achieve this integration, a logical entity called common radio resource management (CRRM) is adopted, whose role is to perform a higher level management. CRRM allows the communication and coordination among different RATs to accomplish overall performance goals.

Even though it seems quite obvious to combine radio access technologys (RATs) that have complementary features, the cost–benefit tradeoff behind this aggregation is uncertain, as it may demand modifications to the standards and addition of extra network nodes.

Radio access networks consisting of multiple integrated RATs are called here as MA networks.

Historically, the initial motivation for creating MA networks was the fact that people roaming from one place to the other could not use their mobile terminals, especially in the case where technologies were different. As such, since the beginning of the third generation (3G) standardization efforts, the possibility to integrate a broad range of technologies, ranging from cellular networks to low earth orbit satellites [42], was considered. This integration was initially proposed with the creation of the international mobile telecommunications (IMT)-2000 by the International Telecommunication Union (ITU) in the 1990s.

Real interoperability between access technologies came some years later, with the standards coordinated by the 3rd. Generation Partnership Project (3GPP). 3GPP's Universal Mobile Telecommunication System (UMTS) was created to support the integration of the wideband code division multiple-access (WCDMA) and global system for mobile communication (GSM)/enhanced data rate for GSM evolution (EDGE), so as to enhance the lifetime of the GSM equipment and provide a smooth upgrade transition to 3G [27].

F. Cavalcanti, S. Andersson (eds.), *Optimizing Wireless Communication Systems*,
DOI 10.1007/978-1-4419-0155-2_6, © Springer Science+Business Media, LLC 2009

More recently, MA networks are motivated by the harmonization of wireless networks towards a unified infrastructure able to provide seamless communications at high rates, with extended coverage and with a wider range of services.

An illustration of an MA network is as follows. A specific set of RATs can be distributed in public areas such as restaurants, shopping centers, and airports, either allowing users to select the desired network or seamlessly maintaining their connectivity while roaming through them. Relevant technical aspects in this scenario include identification of parameters that influence the MA network performance such as cost, security, billing, mobility, quality-of-service (QoS), capacity, and coverage.

In light of this context, the deployment of CRRM is critical to the success of MA networks. CRRM is the key functionality integrating the multiple RATs at the access layer and, therefore, providing the complementarity that may exist among them in terms of link capacity, latency, and coverage.

The remaining sections of this chapter are organized as follows. Section 6.2 presents a detailed vision of MA networks, its main features, architectures, and interfaces. Section 6.3 presents a general introduction to CRRM, its functionalities, the access selection (AS) problem, and algorithms. Section 6.4 presents results for the two defined generalized access selection problem (GASP) and strict version of the access selection problem (SASP) formulations. Section 6.5 presents performance results for active set (AS) in practical systems for real and non-real-time services. Section 6.6 presents performance results for the joint operation of AS and vertical handover (VHO) in practical systems. Section 6.7 presents performance results for AS in a UMTS terrestrial radio access network (UTRAN)-wireless local area network (WLAN). Section 6.8 presents a summary of the chapter and presents research directions on MA networks.

6.2 Multiaccess Networks

This section describes the MA networks to allow a better understanding of their characteristics and unveils the exploitable degrees of freedom of resource management. It starts with a review of the current research work, those developed or proposed by industry, academy, and standardization entities. Section 6.2.2 briefly depicts some aspects of the integration and interoperability of different wireless networks. Finally, the implementation of architectures and protocols is described.

6.2.1 State of the Art

MA networks have caught the attention of standardization entities for quite a while. For 3GPP, the possibility of integrating their technologies with other systems meant a greater diversity in the access possibilities, the ability to better serve traffic hots pots, and also to augment the lifespan of their currently standardized systems. The 3GPP foresees the integration of its systems with Institute of Electrical and Electronics Engineers' (IEEE) 802.11 WLAN ones, but does not exclude the possibility to integrate other WLAN radio technologies if they comply with a basic set of features [7]. From their point of view, the 802.11 networks are to be connected to entities within the 3GPP core network and shall be able to offer 3GPP services as well as Internet access.

3GPP describes the 3GPP-WLAN interworking in [7, 10]. These documents discuss the extension of 3GPP functionalities to WLAN. Such functionalities include accounting, authentication, authorization, services, security, deployment scenarios, terminal modes, WLAN ownership, and the requirements for this interworking. They also specify how these procedures occur and the required interaction between the 3GPP core network (CN) and WLAN. In [1], a description of the architecture required to connect WLAN to the 3GPP system, their interfaces, and procedures, is given.

Within the context of the 802 series of standards, IEEE has also devoted some attention to this concept. Their point of view focuses on VHO. The IEEE 802.21 standard "media-independent handover (MIH)" focuses on the handover procedure across standards. The main purpose of 802.21 is to determine the steps required for the handover initiation and preparation, rather than its execution. The fundamental ideas supporting this set of MIH protocols are the cooperative decision making and the use of centralized information of the network topology. IEEE has spun off new standards within each of its RATs with the 802.11u and 802.16g being responsible for implementing the 802.21 modifications into the 802.11x and 802.16x worldwide interoperability for microwave access (WIMAX) technologies, respectively. The internet engineering task force (IETF) [29] and 3GPP [26] have started their own efforts to integrate their technologies to 802.21.

It is also worthwhile to mention a number of R&D projects that have focused on MA networks. The Monasidre project [23] proposes a framework for the cooperation of the RATs, providing a management software for that purpose. It also describes the operation and the procedures required for this multi-RAT integration. A different approach was considered by the WINNER project [41], whose main goal was to introduce a new RAT, while providing an integration framework to the legacy RATs. Their integration framework includes such aspects as QoS, seamless connectivity, architecture, and service types. The EVEREST project [20, 30] studied the integration of RATs such as WCDMA, GSM/EDGE, and WLANs, defining end-to-end QoS architectures and radio resource management (RRM) procedures for both the individual RATs and for the MA network. Another project working on this theme is Ambient Networks [38], in which, various RATs can be integrated in a decentralized manner to compose an MA network. It also defines the inclusion of control entities (e.g., CRRM) for several network tiers. The correct functioning of this MA network is guaranteed by a generic link layer that enables the interfacing to the various RATs.

6.2.2 Multiaccess Features and Procedures

One of the most important aspect of MA networks is the interworking. It is the actual integration and interoperability of different wireless networks to promote the exploitation of diverse characteristics in favor of an enriched user experience when using wireless services. Interworking involves common mobility and radio resource management among multiple RATs.

Mobility management is a subset of procedures that enable seamless mobility among heterogeneous systems of an MA network. It includes procedures such as handover and location. In MA networks, a common mobility management may include vertical or inter-system handovers.

CRRM extends the traditional RRM techniques for individual RATs to such procedures as AS, inter-system scheduling, and multi-RAT congestion control.

AS and VHO are fundamental procedures that allow the execution of the RAT selection and inter-system mobility. They will be exposed in more details in Section 6.3.1.

6.2.3 Multiaccess Architectures and Interfaces

MA network architectures can be classified according to the degree of coupling among composing RATs. Depending on aspects such as how data and signaling traffic is handled, if billing is commonly managed, and if radio resources are shared at the access layer, MA networks can be classified as loosely or tightly coupled. A range of characterizations of this kind have been proposed by 3GPP [10], European Telecommunications Standards Institute (ETSI) [22] and others [48, 49]. As a general rule, the tighter the coupling, the more flexible the possibilities of jointly exploiting the complementary features of composing RATs. In this chapter, a tightly coupled MA network, employing the AS and VHO functionalities, is assumed.

3GPP has adopted internet protocol (IP) multimedia subsystem (IMS) to provide interworking-able core networks with other IP-based networks [11, 13–15]. For instance, interworking between 3GPP and WIMAX can be provided by the WLAN access gateways (WAG) located at 3GPP core network domain. Interworking between 3GPP and IEEE 802.11 WLANs is also well described in 3GPP standards [1, 3–6, 8, 9, 12].

6.3 Common Radio Resource Management

As mentioned in Section 6.2.2, a successful interworking requires knowledge of RATs characteristics to manage the resources and to optimize the overall MA network performance. A well-accepted solution is the adoption of an entity called CRRM [2], whose main functionalities will be detailed in this section.

6.3.1 Access Selection and Vertical Handover

CRRM manages the RRM entities of each individual RAT through measurement reports and decisions. There are two fundamental CRRM functionalities: AS and VHO. These decisions must be made taking into account the following possibilities:

- Different RATs co-exist in the same area, potentially with overlaping coverage.
- Different RATs operate on different frequency bands.
- Users employ multi-mode mobile terminals to connect to different RATs.
- Each RAT may also belong to a different network operator with integration via roaming agreements.

6.3.1.1 Access Selection

It corresponds to the initial access of a terminal to the MA network. At call setup, AS intends not only a better resource utilization of the MA network, but also it aims at providing the users with better service times by only selecting an appropriate RAT. The choice of which RAT to serve a connection can be based on different aspects: QoS and service requirements, user preferences and policies, link quality, system load and cost.

6.3.1.2 Vertical Handover

It is a procedure in which the terminals can switch from their current RAT to another one. It differs from the traditional handover (horizontal handover) because the latter promotes the reassignment of terminals in the context of a unique RAT, mainly based on signal strength and interference criteria. In addition, VHO decisions can be based on aspects like service requirements and link quality. Opposed to AS, VHO is performed during an ongoing call, thus being a more complex operation and requiring fast signaling within the MA network.

6.3.2 Inter-system Scheduling

Generally speaking, the packet flow to and from a mobile terminal is served by a single RAT. In MA networks, it is possible to assign different packets to be served by different RATs, according to load conditions, packet deadlines, and throughput requirements. This scheduling across RATs is known as inter-system scheduling, whose objective is to exploit the inter-system diversity at the packet level. In Fig. 6.1, an example of inter-system scheduling procedure, where a terminal connected to both RATs transmits either through the macrocell or the microcell according to packet scheduling policies, is illustrated.

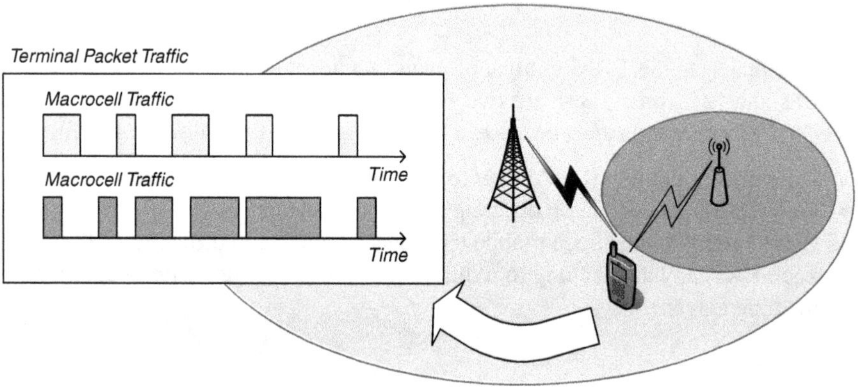

Fig. 6.1 Inter-system scheduling functionality.

Flow control is a particular case of the inter-system scheduling. It comprehends the assignment of the mobile terminal data at the flow level (as opposed to the packet level) to the most suitable RAT. As an example, the flow of a file download can be sent through a high-rate RAT, while a voice flow could be sent through a low-latency one. Despite the potential performance gains, inter-system packet scheduling presupposes a high degree of coordination and fast signaling among composing RATs.

6.3.3 Congestion Control

Congestion control takes place when the CRRM entity identifies a congestion situation. It works with other CRRM procedures to re-establish MA network stability. When a RAT is overloaded, congestion control can trigger a VHO in order to alleviate the overloaded RAT. With the same goal, subsequent new connections can be assigned to the less loaded RAT. The congestion control functionality for an MA network composed of two RATs is illustrated in Fig. 6.2, covering a macro and a microcell. The illustration shows a case of load balancing that can be achieved by forcing some connections to handover from one RAT to another.

In spite of the apparent simplicity shown in Fig. 6.2, congestion control is a challenging engineering problem. Load measurement does not lend an easy definition when multiple and diverse RATs need to be compared. Not less important is the definition of an overload threshold for each RAT. Finally, signaling requirements could be significant when forcing several inter-system handovers in a short period of time.

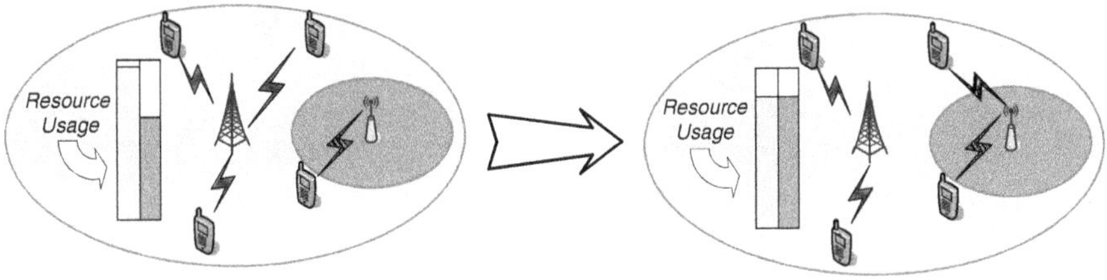

Fig. 6.2 Congestion control functionality.

6.3.4 Access Selection Problem Formulation

AS is one of the simplest CRRM procedures in MA networks. It can be understood as the basic and the first CRRM functionality above which other CRRM procedures can be built.

In terms of decision making, AS can be network-centric or user-centric. In the user-centric case, each mobile terminal is able to independently select which RAT to connect. Aspects such as signal strength, access cost, and link capacity may drive this decision, that may or may not reflect a direct choice of the user. As for the network-centric AS, the goal is to maximize the overall MA network capacity considering the type of demanded services, capabilities of the RATs, as well as radio access conditions at call setup.

This section is dedicated to present formulations for the AS problem. First, bounds for the AS problem are investigated by means of a mathematical optimization problem, named generalized access selection problem (GASP). GASP is formulated considering that the CRRM entity can assign new connections by AS and reallocate ongoing connections by means of VHO.

This problem can be mathematically formulated as a generalized assignment problem (GAP) [24, 36]. GAP seeks the maximum profit assignment of n tasks to m agents subject to capacity restrictions on the agents and that one task is assigned to one and only one agent, requiring a certain amount of the resources from the agent.

Translating to the AS perspective, let us consider a wireless network composed of M_R RATs (numbered $m = 1,\ldots,M_R$) and the set of connections (numbered $j = 1,\ldots,J$) to be assigned to these RATs. GASP consists in maximizing the objective function $f(x_{m,j}, r_{m,j}, w_{m,j}, G_m)$ subject to some restrictions. Each RAT has a limited radio resource quantity G_m. When a connection j is allocated to a RAT m, it consumes a radio resource quantity $r_{m,j}$ and generates a revenue $w_{m,j}$. The revenue generated corresponds to a function $H_{m,j}(r_{m,1},\ldots,r_{m,n},\ldots,r_{m,J})$ and it is characterized by the specific radio capabilities of the mth RAT, which depends on the resource consumption of the new and ongoing connections. $x_{m,j}$ is a binary variable assuming 1 if connection j is allocated to RAT m, or 0 otherwise. Mathematically, GASP can be expressed as

$$\begin{aligned} \max \quad & f(x_{m,j}, r_{m,j}, w_{m,j}, G_m) \quad \text{subject to} \\ & \sum_{1 \leq m \leq M_R} x_{m,j} \leq 1, \text{for } j = 1,2,\ldots,J \\ & \sum_{1 \leq j \leq J} r_{m,j} \cdot x_{m,j} \leq G_m, \text{for } m = 1,\ldots,M_R, \end{aligned} \tag{6.1}$$

where

$$\begin{aligned} & w_{m,j} = H_{m,j}(r_{m,1},\ldots,r_{m,j},\ldots,r_{m,J}) \\ & x_{m,j} \in \{0,1\}, \text{ for } m = 1,\ldots,M_R \text{ and } j = 1,2,\ldots,J. \end{aligned} \tag{6.2}$$

The formulation (6.1) aims to maximize a specific objective function taking into account two constraints: each connection must be allocated to one or no RAT (in this case the connection is blocked) and each RAT has a limited amount of resources available.

Considering the real operation of an MA system, when its resource consumption ($\sum r_{m,j} \cdot x_{m,j}$) reaches a maximum value (G_m), the MA network might decide to allow or deny new connection requests. This depends on the wireless provider's strategy and defines two important admission strategies:

- **User blocking admission strategy:** it is a traditional admission control strategy considering those cellular systems capable of providing voice services only. The system using this strategy blocks new connection requisitions when the system resource consumption reaches its limit;
- **Bandwidth sharing admission strategy:** it is an admission policy able to take advantage of the burst transmission behavior of some data services to increase the bandwidth usage efficiency. Instead of blocking connections, they are still active, but sharing transmission resources.

Formulation (6.1) considers the user blocking admission strategy, i.e., AS excludes connections when the required resource is unavailable. However, the problem for the bandwidth sharing admission strategy (without blocking) can be formulated mathematically by

$$\begin{aligned} \max \quad & f(x_{m,j}, r_{m,j}, w_{m,j}, G_m) \quad \text{subject to} \\ & \sum_{1 \leq m \leq M_R} x_{m,j} = 1, \text{for } j = 1,2,\ldots,J, \end{aligned} \tag{6.3}$$

where

$$\begin{aligned} & w_{m,j} = H_{m,j}(r_{m,1},\ldots,r_{m,j},\ldots,r_{m,J}) \\ & x_{m,j} \in \{0,1\}, \text{ for } m = 1,\ldots,M_R \text{ and } j = 1,2,\ldots,J. \\ & r_{m,j} = \begin{cases} r^{Req}{}_{m,j}, \text{if } \sum_{1 \leq j \leq J} r_{m,j} \cdot x_{m,j} \leq G_m \\ G_{m,j}(\mathbf{v}_m, G_m, J_m), \text{if } \sum_{1 \leq j \leq J} r_{m,j} \cdot x_{m,j} > G_m \end{cases}, \\ & \mathbf{v}_m = [r^{Req}_{m,1},\ldots,r^{Req}_{m,j},\ldots,r^{Req}_{m,J}], \text{for } m = 1,\ldots,M_R \end{aligned} \tag{6.4}$$

where ${r^{Req}}_{m,j}$ is the minimum resource consumption that meets the connection's QoS requirement[6.1] of connection j in RAT m and J_m is the number of connections assigned to RAT m.

Formulation (6.4) states that when a RAT has its resources completely consumed, they are shared among all connections (without blocking). This resource distribution depends on a scheduling policy (represented by $G_{m,j}(\cdot)$) which can be a function of the total RAT resources (G_m), number of ongoing connections (J_m), and the amount of resources ($\mathbf{v}_m$) required to meet the user's QoS.

In a second front, the AS problem is reformulated as the assignment of one connection at a time to a RAT, as decided by an algorithm (based on simple test of objective function and constraints), with no complementary VHO. This second formulation is called SASP.

In fact, SASP may be seen as a version (or subproblem) of GASP with suitable modifications. The main modification is related to the number of controlled variables. In SASP, only the user requesting a new connection can be managed, i.e., ongoing connections are kept unchanged. Basically, when a connection j is requested to an MA network, it will be served by the RAT m that maximizes the objective function $f(x_{m,j}, r_{m,j}, w_{m,j}, G_m)$ instantaneously (see formulation (6.5)):

$$m = \arg\max\left[f(x_{m,j}, r_{m,j}, w_{m,j}, G_m)\right]. \tag{6.5}$$

One additional problem can be defined when it is permitted to manage both the user requesting a new connection and a small fraction of already connected users, i.e., those users which are potential candidates for a VHO. This solution is an intermediate case between GASP and SASP. This case will be investigated in Section 6.6.

6.3.5 Criteria and Algorithms for Access Selection

This section covers general criteria and algorithms for the problems defined in formulations (6.1) and (6.5).

The problem of maximizing $f(x_{m,j}, r_{m,j}, w_{m,j}, G_m)$ in formulation (6.1) can be seen as a multi-variable decision problem. The revenue ($w_{m,j}$) generated by assigning a connection j to a RAT m depends on the resources consumed by the connections already served. Additionally, the resource consumption $r_{m,j}$ is composed of several parameters which depends on the candidate RAT. Then, the variable $r_{m,j}$ might be a composition of power, code, and/or frequency resources and the variable $w_{m,j}$ might be a key performance indicator (KPI) such as data throughput value, signal-to-interference-plus-noise ratio (SINR), blocking/dropping probabil-

[6.1] Even having no guaranteed QoS, best-effort services may require a minimum resource consumption that corresponds to a satisfaction criterion. The main difference is that, for best-effort services, the $r_{m,j}$ can be higher than ${r^{Req}}_{m,j}$ even if $\sum_{1\leq j\leq J} r_{m,j}\cdot x_{m,j} \leq G_m$.

ities, among others. The objective functions in (6.1) and (6.5) can also be seen as utility functions implementing specific system goals.

From the utility function perspective, AS solutions can be categorized according to their criteria. They can be divided into the following three groups.

6.3.5.1 Utility Function for Balancing

It follows a formulation that aims at balancing instead of maximizing a specific KPI among RATs [39, 46, 52]. One possible utility function for balancing is given by

$$f(x_{m,j}, r_{m,j}, w_{m,j}, G_m) = \frac{1}{\sum_{m=1}^{M_R} (c_m - \overline{c})^2}, \tag{6.6}$$

where M_R is the number of RATs of the MA network, c_m is the KPI to be balanced, and $\overline{c}$ its average. One well-known algorithm is called load balancing, where the normalized load is used as the balance indicator. In this way, c_m is defined as

$$c_m = \frac{\sum_{j=1}^{J} r_{m,j} \cdot x_{m,j}}{G_m}. \tag{6.7}$$

Note that to define a load balancing criterion, the utility function (f) depends on $r_{m,j}$ (resource to manage, in this case, the normalized load), G_m (total available resource of RAT m, i.e., the maximum tolerated load), and the binary selection variable ($x_{m,j}$).

6.3.5.2 Utility Function for Revenue Maximization

It provides a decision-making criterion based on the maximization of the function that maps the consumed resource in a numerical profit [17, 33]. This profit depends on the amount of resources granted to a connection and how each RAT converts such resource quantity into QoS. For the case of two RATs, a general utility function for revenue maximization can be formulated as

$$f(x_{m,j}, r_{m,j}, w_{m,j}, G_m) = \sum_{j=1}^{J} (w_{1,j} \cdot x_{1,j} + w_{2,j} \cdot x_{2,j}). \tag{6.8}$$

This time f is a function of $w_{m,j}$ (generated revenue) and $x_{m,j}$. As $w_{m,j}$ is a function of $r_{m,j}$ (consumed resource), f depends also on $r_{m,j}$. The profit is usually mapped into KPIs such as consumed power, perceived throughput and delay, SINR, among others.

6.3.5.3 Policy-Based Utility Function

AS decisions may also be governed by policies which are incorporated in the design of utility functions. This rule can be based on radio access capabilities and conditions, user's profile, bandwidth requirements, and terminal capabilities [31, 40]. For instance, one AS policy-based algorithm can establish that data traffic of low-speed users will be served by a low-coverage RAT and the data traffic of high-speed users by a high-coverage one. This rule can also be associated with a customer's individual profile so that high-priority users are always assigned to a high-capacity RAT and with a pre-defined guaranteed bandwidth. The coverage-oriented AS in [51] uses a policy-based utility function which depends on radio access condition (coverage test on the perceived signal level) and an explicit policy (connection will be served by the highest capacity RAT when the terminal senses its coverage).

6.4 Performance of Access Selection

This section presents studies concerning the two already defined GASP and SASP formulations (refer to Section 6.3.4). In the GASP evaluation, the upper bound performance is determined by assuming that the CRRM entity can reallocate ongoing connections by a VHO procedure at will. For this purpose, a heuristic strategy based on a genetic algorithm (GA) guided by specific objective functions which will be defined afterward[6.2] is used. In the SASP evaluation, the performance of less complex algorithms based on simple tests of objective functions and constraints is the focus.

The approach presented here assumes a network-centric AS procedure and an MA network modeled by a single-cell including a hot spot and two generic RATs, as shown in Fig. 6.3. It is also assumed that the terminals are multi-mode, i.e., they are able to support more than one RAT. Then, terminals within the coverage range

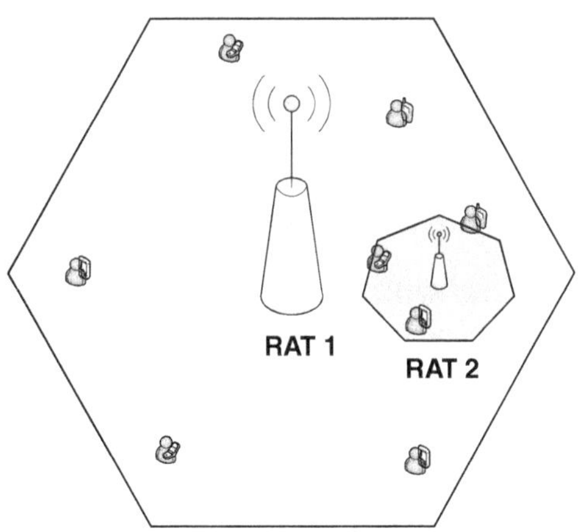

Fig. 6.3 Multiaccess deployment model.

[6.2] See more details about genetic algorithms in [47].

of RAT 2 can also connect to RAT 1 if instructed to do so. In this section a network-centric AS procedure is assumed.

The MA network performance is assessed based on a system model which represents the access of the shared channel by the link utilization concept, originally presented in [50]. Connections are assumed to generate an average traffic z_j (kbps). To obtain the occupation of the transmission medium by a connection j in RAT m, a desired link utilization factor is defined as the ratio of generated traffic z_j and the mapped radio link bit rate $R_{m,j}$:

$$\rho_{m,j}^{Des} = z_j / R_{m,j}. \tag{6.9}$$

A scheduling rule is assumed to be proportional to the radio link rate in which, for high loads, the remaining capacity is shared among connections with worst link quality following a maximum rate scheduling policy. Then, the best connections in terms of link quality are first allocated with their desired $\rho_{m,j}^{Des}$ until system capacity is reached. In this manner, the highest possible link utilization $\left(\rho_m^{Res}\right)$ for the worst connections is given by

$$\sum_{j=1}^{J} \min\left\{\rho_{m,j}^{Des}, \rho_m^{Res}\right\} = 1. \tag{6.10}$$

Then, the effective link utilization factor of the connection j in RAT m is

$$\rho_{m,j} = \min\left\{\rho_{m,j}^{Des}, \rho_m^{Res}\right\}. \tag{6.11}$$

Finally, the wireless radio link performance is evaluated by means of the circuit-switched equivalent (CSE) bit rate which is calculated as follows [50]:

$$CSE_{m,j} = R_{m,j} \cdot \left(1 - \sum_{i \neq j} \rho_{m,i}\right). \tag{6.12}$$

Performance results are based on a Monte Carlo simulation approach in which the CSE of all connections are averaged over a determined number of snapshots. Instead of modeling all wireless radio link and protocol stack aspects, the signal-to-noise ratio (SNR) distribution in the cell is expressed by a Gaussian distributed random variable due to path loss and long-term fading. The link quality of a RAT is then simply represented by the mean and the standard deviation of the SNR experienced by users connected to it. In order to evaluate the connection's bit rate, an idealized link adaptation model is considered. It models the specific link capacity as a linear and upper-limited mapping of SNR into bit rate, as shown in Fig. 6.4. SNR_{knee}, SNR_{sat}, R_{min}, and R_{max} delineate this idealized model. While R_{min} and R_{max} define the achieved RAT capability in terms of bit rate, SNR_{knee} and SNR_{sat} express RAT capability in terms of physical layer aspects (e.g., modulation, transmission power, and transceiver sensitivity).

Specializing the general formulation of GASP (6.3), solutions are formulated based on the three different utility functions as defined in Section 6.3.5: policy-based, utility function for revenue maximization, and utility function for balancing. Such solutions are exposed in more details in the next sections.

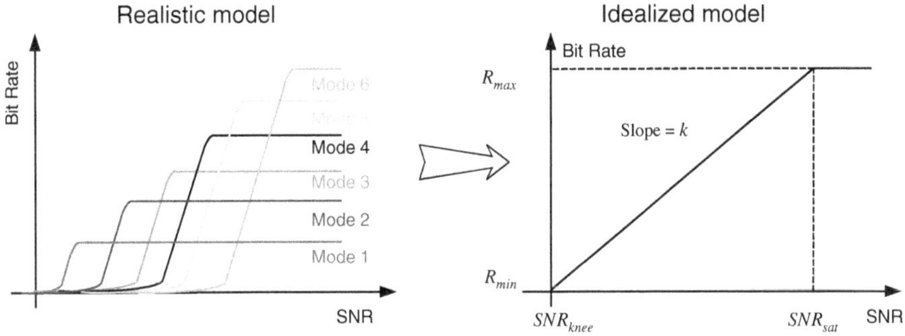

Fig. 6.4 Ideal link adaptation model.

6.4.1 Coverage Threshold Algorithm (CTA)

Coverage threshold algorithm (CTA) assigns connections to the RAT 2 if there is coverage, i.e., if the SNR experienced in RAT 2 is higher than a specified threshold SNR_{min}, regardless of the situation in RAT 1. If there is no coverage in RAT 2, i.e., SNR $< SNR_{min}$, the algorithm tries to assign connections to RAT 1 even if the coverage in this RAT is not guaranteed. It is a policy-based AS algorithm and taken as performance reference because of its simplicity. This strategy is the simplest one because it requires only information of the connection's SNR relative to the RAT covering the hot spot area.

6.4.2 Load Balancing Algorithm (LBA)

Load balancing algorithm (LBA) represents a utility function for balancing as defined in Section 6.3.5. Its GASP version intend to balance the offered load in both RATs:

$$f(x_{m,j}, r_{m,j}, w_{m,j}, G_m) = f(x_{m,j}, z_j, R_{max_m}) = \frac{1}{\left| \sum_{j=1}^{N} \left(\frac{z_j \cdot x_{1,j}}{R_{max_1}} - \frac{z_j \cdot x_{2,j}}{R_{max_2}} \right) \right|}, \quad (6.13)$$

where f is function of z_j (offered traffic load), R_{max_m} (maximum bit rate capacity of RAT m), and the binary decision variables $x_{m,j}$.

The SASP version of LBA assigns a new connection to RAT 1 or to RAT 2 so that the normalized load in both RATs is kept similar:

$$\frac{\text{Consumed Capacity at RAT}_1}{\text{Total Capacity of RAT}_1} \approx \frac{\text{Consumed Capacity at RAT}_2}{\text{Total Capacity of RAT}_2} \quad (6.14)$$

6.4.3 Link Utilization Balacing Algorithm (LUBA)

Link utilization balancing algorithm (LUBA) represents a utility function in which the target is to balance the sum of link utilization factors of RATs. Its behavior is similar to LBA, however, LUBA also takes into account the link quality, unlike LBA which considers only the traffic generated by users (see definition of link utilization factor in (6.9)). The utility function that implements this strategy is given by

$$f(x_{m,j}, r_{m,j}, w_{m,j}, G_m) = f(x_{m,j}, \rho_{m,j}) = \frac{1}{|\sum\limits_{j=1}^{N} (\rho_{1,j} \cdot x_{1,j} - \rho_{2,j} \cdot x_{2,j})|}. \tag{6.15}$$

The main objective of this strategy is to promote the equivalence of the resource consumption between both RATs in the sense of the offered load and bit rate capacity of each individual link.

In the SASP version, the status of the connections in each RAT is monitored continuously. When a new connection arrives, its desired link utilization factor in each RAT is computed. According to these values, it is assigned to the RAT so that the difference between the sums of the desired link utilization factors of the RATs is minimized.

6.4.4 Rate Maximization Algorithm (RMA)

Rate maximization algorithm (RMA) is an algorithm based on a utility function for revenue maximization. The GASP version of RMA is given by

$$f(x_{m,j}, r_{m,j}, w_{m,j}, G_m) = f(x_{m,j}, R_{m,j}) = \sum_{j=1}^{J} (R_{1,j} \cdot x_{1,j} + R_{2,j} \cdot x_{2,j}), \tag{6.16}$$

where $R_{m,j}$ is the radio link transmission rate of connection j in RAT m.

The SASP version is based on an estimate of link rate achieved in both RATs. The connection is admitted to that RAT in which the link experiences the highest instantaneous transmission rate.

6.4.5 CSE Maximization Algorithm (MCSE)

Cse maximization algorithm (MCSE) is an AS algorithm based on the maximization of the estimated CSE for new connections. In order to perform this estimate, the link utilization of all connections are recalculated yielding a criterion that considers the impact to admit a new connection to each RAT. Its utility function for GASP is defined as

$$f(x_{m,j}, r_{m,j}, w_{m,j}, G_m) = f(x_{m,j}, \rho_{m,j}) = \sum_{j=1}^{J} (CSE_{1,j} \cdot x_{1,j} + CSE_{2,j} \cdot x_{2,j}), \quad (6.17)$$

where $CSE_{m,j}$ is the CSE bit rate of connection j relative to RAT m.

In the SASP version, the average CSE is estimated for each RAT considering the admission of the new connection. The definitive admission is accomplished to that RAT which experiences the highest average CSE.

6.4.6 Comparison

A set of scenarios is defined in Table 6.1 and used for displaying the performance of solutions for GASP and SASP formulations. SNR of all scenarios is a Gaussian-distributed random variable as mentioned before. The mean (SNR_{mean_m}, regarding a RAT m) is given in the table and the standard deviation is 4 dB for links inside the hot spot connected to RAT 1 and 8 dB elsewhere.[6.3] The proportion of users indicates the load difference between areas inside and outside the hot spot and it is represented by ζ. For instance, for $\zeta = \frac{1}{4}$ and a load of 20 users per cell, there are 16 users inside and 4 users outside the hot spot. As stated in (6.9), connections are assumed as continuous traffic which is represented by an average traffic z_j.

Table 6.1 MA scenarios.

Scenario	Expected offered load (ζ)	Bit rate capacity [Mbps]	Link quality [dB]
Scenario 1	4/1	$R_{max_1} = 6$, $R_{max_2} = 6$	$SNR_{mean_1} = 10$, $SNR_{mean_2} = 16$
Scenario 2	1/4	$R_{max_1} = 6$, $R_{max_2} = 6$	$SNR_{mean_1} = 10$, $SNR_{mean_2} = 16$
Scenario 3	1/1	$R_{max_1} = 54$, $R_{max_2} = 6$	$SNR_{mean_1} = 10$, $SNR_{mean_2} = 26$

Results focus on two main cases: (i) evaluation of the proportion of users inside–outside the hot spot (Scenarios 1 and 2) and (ii) bit rate capacity evaluation (Scenario 3). Case (i) illustrates the influence of expected offered load distribution (ζ) in the performance of GASP strategies while Case (ii) the effect of link quality and link adaptation models in a scenario composed of RATs with different bit rate capacities.

Figures 6.5 and 6.6 show the performance of AS in terms of CSE versus a range of offered loads for Scenarios 1 and 2. The AS is attractive for similar RATs when the proportion of users in the hot spot is higher than the corresponding one in the macrocell RAT, as shown in Fig. 6.5. The opposite happens in Fig. 6.6 suggesting that the proportion of users inside the hot spot is a fundamental parameter to measure the potential gain available from AS in MA network setups. Another important

[6.3] This Gaussian distribution for the cell SNR and the proposed standard deviations were attested in an actual system simulator of an MA network setup.

observation is that the MCSE algorithm has superior performance compared to the other ones, mainly for high loads. This is expected as CSE is the metric used for measuring the performance.

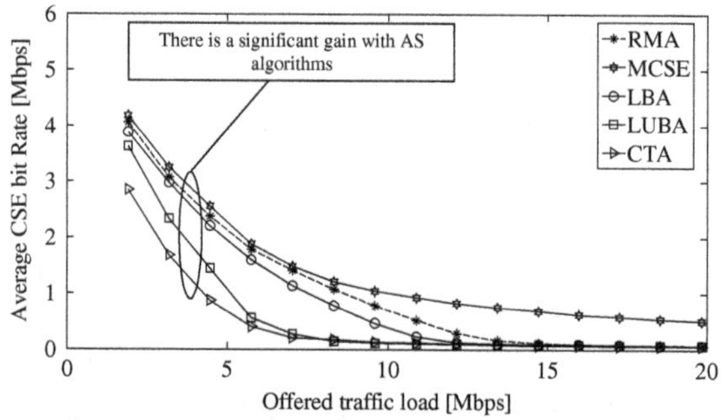

Fig. 6.5 Scenario 1 – average CSE bit rate

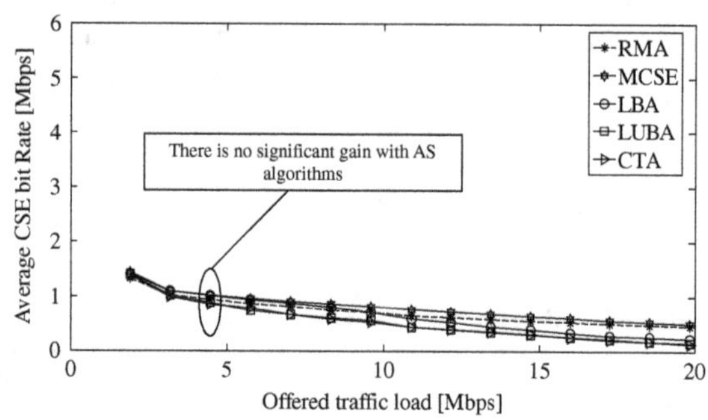

Fig. 6.6 Scenario 2 – average CSE bit rate

In Scenario 3, the interesting situation where RAT 1 has higher radio link capacity but at the same time worse average SNR as compared to RAT 2 is analyzed.

As expected, the possibility of rate improvement with AS is strongly dependent on the maximum system capacities and has significant influence on the AS performance. Then, although the link quality is not favorable in RAT 1, AS provides gains when the maximum rate capacity of the RAT 1 is much higher than that of the RAT 2, as can be seen in Fig. 6.7.

Now, a comparison between SASP and the upper bound AS solution provided by GASP is presented. Considering its optimization nature, the GASP is an NP-hard problem, claiming for an evolutionary computation approach, while SASP has low complexity, enabling its implementation in an on-line fashion.

The performance evaluation is conducted in Scenario 1 as defined in Table 6.1. Figure 6.8 summarizes the performance comparison between AS strategies by GASP and SASP approaches. As pointed out before, the SASP solution is a specific case of the GASP approach in the sense that there is no connection reassignment at each call arrival, but the optimization criterion is performed at each admission request

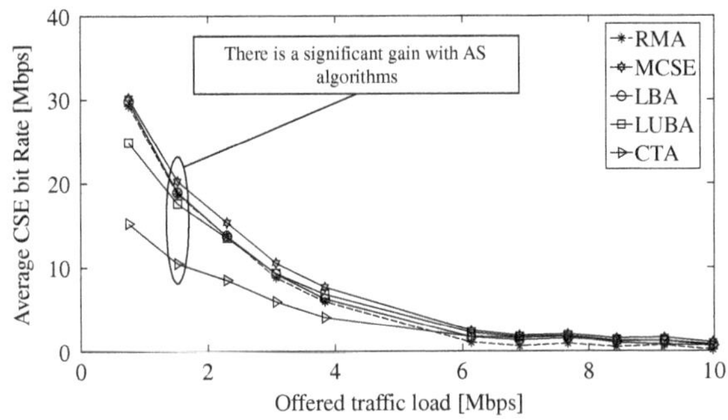

Fig. 6.7 Scenario 3 – average CSE bit rate regarding the maximum rate capacity of RAT 1 higher than RAT 2.

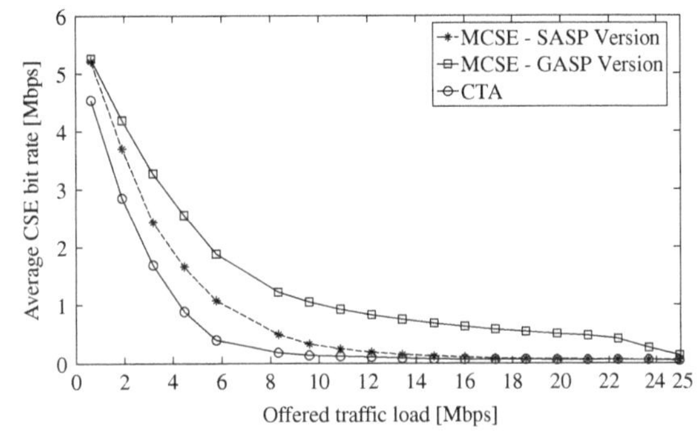

Fig. 6.8 Scenario 1 – average CSE bit rate regarding GASP and SASP solutions.

only for the arriving connections. The results achieved by GASP indicate that a meaningful gain is obtained when the AS algorithm performs the reassignment of ongoing connections. Then, VHO becomes a promising approach to be initialized together with the AS at call setup. The other algorithms (LBA and RMA) performed similarly, although the gain of GASP over SASP was observed over a shorter load range (up to 10–15 Mbps).

For all cases, MCSE has superior performance compared to the other algorithms. This fact stimulates the investigation of practical issues (e.g., availability of measurements) in order to properly implement the AS algorithm based on the maximization of throughput. These practical issues will be considered afterward.

6.5 Access Selection Solutions Performance in Practical Scenarios

The previous section focused on the limiting performance gains of AS when considering mainly the GASP formulation. In this section, the analysis is extended to a more practical SASP approach by deriving the performance of AS in more realistic scenarios including actual traffic models for non-real-time and real-time wireless services.

The first corresponds to a type of service whose QoS level is designed for applications that do not require transmission to take place in real time. Examples of non-real-time services are World Wide Web (WWW) and e-mail applications. The latter corresponds to a type of service that requires more strict restrictions for QoS, such as delay and packet loss requirements. Voice over IP (VoIP) is a well-known real-time service.

6.5.1 Results for Non-real-Time Service Class

In this section, the impact of AS in a MA network, whose only service is WWW best-effort (modeled according to [32]), is evaluated.

An algorithm, namely, **estimated RAT throughput algorithm (ERT),** based on the already successfully transmitted data is presented. If a user j connected to a RAT m has some data transmitted successfully, its current throughput is given by

$$T_{m,j} = \frac{b_{m,j}}{D_{m,j}}, \tag{6.18}$$

where $b_{m,j}$ and $D_{m,j}$ denote, respectively, the total amount of successful transmitted data of the jth connection and its corresponding transmission delay (queueing and service times). Then, the average throughput at RAT m is defined as

$$T_m = \frac{1}{J_m} \sum_{j=1}^{J_m} T_{m,j}, \tag{6.19}$$

where J_m corresponds to the number of ongoing calls (with successfully transmitted data) in RAT m. Then, a user is connected to RAT 2 if $(T_2 \geq T_1)$ and to RAT 1 otherwise.

A second approach is to use information from both the arriving call and the network. Two algorithms were conceived following this approach. The **estimated user and RAT throughput algorithm (URT)** performs a combined version of RMA and ERT. The call is admitted to the RAT in which the sum of the estimated RAT average throughput (T_m) plus the incoming call link rate $(R_{m,j})$ is the highest. Then, a user is connected to RAT 2 if $\left(T_2 + R_{2,j} \geq T_1 + R_{1,j}\right)$ and to RAT 1 otherwise.

As it can be observed, ERT and URT are only based on throughput. However, another important performance indicator is the delay. This fact motivates the proposition of an algorithm which contemplates both throughput and delay. Assuming a user j requesting a connection, a utility function U^m for each RAT m is defined as follows:

$$U_{m,j} = \frac{R_{m,j}}{D_m}, \tag{6.20}$$

where $R_{m,j}$ is the mapped link bit rate of the arriving call j in the RAT m and D_m corresponds to average delay seen in that RAT.

The proposed utility function represents the satisfaction caused by low delay and high bit rate. It attempts to provide a tradeoff between link quality (bit rate) and system load (RAT delay). The **utility-based algorithm (UTA)** assigns the arriving call to RAT 2 if $\left(U_{2,j} \geq U_{1,j}\right)$ and RAT 1 otherwise.

Table 6.2 shows the set of parameters of three case studies (MA network scenarios) evaluated in the following.

Table 6.2 Studied scenarios.

Parameter	Scenario 1	2	3
Maximum capacity in RAT 1 (Mbps)	2	2	6
Maximum capacity in RAT 2 (Mbps)	6	2	2

Common Configurations	
Parameter	**Value**
SNR_{max} in RAT 1	10 dB
SNR_{max} in RAT 2	10 dB
Proportion of users ζ	4/1

As mentioned before, in order to create AS algorithms which offer performance gains relative to CTA, the research community is assuming to explore the freedom to assign connections inside the hot spot to RAT 1. This action can provide two gains: (i) link bit rate enhancement gain, due to an increase in the user's link quality and (ii) the statistical multiplexing gain, relative to the load management between RATs. The three studied scenarios are designed in order to explore, in different ways, these two kinds of gains. The first scenario is more restrictive in terms of bit rate enhancements. This way, it is expected to have a better performance for those AS algorithms that explore the statistical multiplexing gain. On the other hand, Scenario 3 gives a great opportunity for link bit rate enhancement gains. An intermediate case is represented by Scenario 2 in which the degree of dominance between the two kinds of gains and their relation with the proposed AS algorithms can be observed.

Figure 6.9 presents the gain of average session throughput for three different loads (3.33 requests/s (Req/s) for Scenario 1; 2 requests/s for Scenario 2; and 6.67 requests/s for Scenario 3). These loads are chosen in order to represent offered loads near to the capacity limit (90% of users having throughput higher than 128 kpbs).

Comparing the algorithms with high performance, it can be concluded that rate maximization algorithm (RMA) provides a considerable performance gain in all tested scenarios. Nevertheless, the UTA has better performance among tested algorithms considering Scenario 1. As a consequence, the UTA algorithm provides a higher capacity gain in this scenario (see Fig. 6.10). As expected, in Scenario 2, the RMA provides the best capacity gain, because the overall improvement from

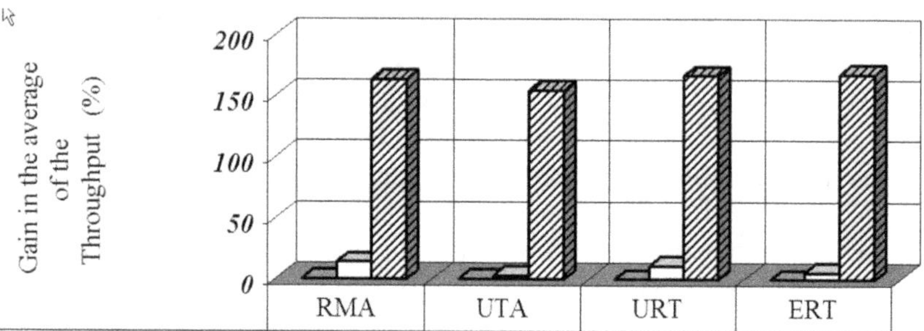

	RMA	UTA	URT	ERT
Scenario 1 (3.33 Req/s)	0.11	0.30	-0.70	-0.67
Scenario 2 (2 Req/s)	14.04	3.13	11.13	5.71
Scenario 3 (6.67 Req/s)	163.49	154.70	166.30	167.61

Fig. 6.9 Gains in the average session throughput of AS algorithms (relative to CTA).

the bit rate enhancement is higher than that from the queue load management in a scenario composed of similar RATs.

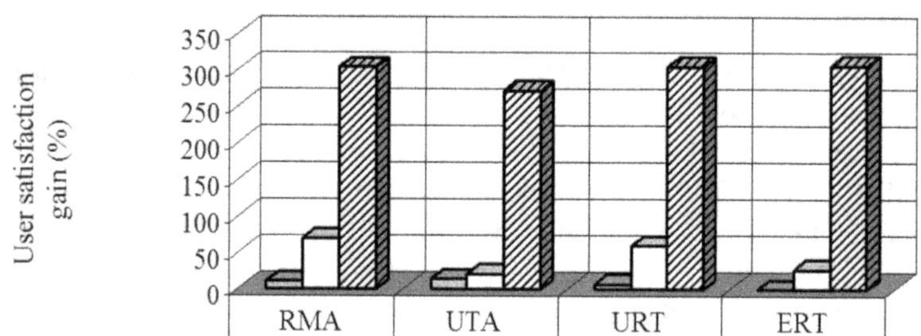

	RMA	UTA	URT	ERT
Scenario 1 (3.33 Req/s)	10.82	14.16	7.00	-0.18
Scenario 2 (2 Req/s)	69.78	21.27	59.50	27.05
Scenario 3 (6.67 Req/s)	303.57	271.47	303.95	305.23

Fig. 6.10 Gains in the capacity (regarding the user satisfaction) relative to CTA.

In Scenario 3, the RMA, URT, and the ERT have similar performance, with the latter performing slightly better. This is due to the low utilization of RAT 2 considering the RMA and URT criteria, which assign calls to the RAT with the highest bit rate (in this case, RAT 1) regardless of the load. These results indicate that the gain from the load management between the RATs (more evident in Scenario 1) is less significant and the great source of the gains with the AS algorithms is the bit rate improvement (represented by the RMA algorithm).

6.5.2 Results for Real-Time Service Class

The purpose of this section is to evaluate four AS algorithms for a VoIP service in an MA network. The VoIP traffic is modeled according to a traditional approach, where the arriving calls follow a Poisson process. In this model, a VoIP call is composed of active and silent periods exponentially distributed and the frames are generated with constant bit rate. More details of VoIP traffic modeling can be found in [45].

CTA, LBA, and RMA have already been defined in Section 6.4. Additionally, the satisfaction balancing algorithm (SBA) which is based on the proportion of satisfied ongoing calls in the RAT is proposed. A call is assumed as satisfied if the lost packet rate is lower than 2%. Otherwise, the call is unsatisfied.

SBA assigns terminals to RAT 2 if $\left(\frac{s_2}{J_2} \geq \frac{s_1}{J_1}\right)$ and to RAT 1 otherwise, where s_i and J_i are the total of satisfied users and the total number of connections, respectively, in RAT i. This is a coherent rule because a low percentage of satisfied users in a given RAT indicates a bad performance in this RAT. Similar to LBA, satisfaction balancing algorithm (SBA) aims at balancing the momentary satisfaction of users in RATs. More details about SBA can be found in [45].

Once more an MA network composed of two RATs covering a macro and an embedded microcell (or hot spot) is assumed.

The performance comparison of AS algorithms is presented in a specific scenario: $\zeta = 1/1$, where ζ is the proportion of users inside–outside the hot spot. A reference scenario (Ref) is adopted in order to make coherent comparisons. This corresponds to a non-multiaccess scenario, where the network capacity is the sum of the individual capacities of the RATs operating separately. Table 6.3 presents the simulation parameters.

Table 6.3 Simulation parameters.

VoIP model		**Transmission model**	
Parameter	**Value**	**Parameter**	**Value**
Call duration	90 s	Scheduling strategy	Round robin
Active/silent period	3/3 s	TTI of RAT 1	2 ms
Frame generation rate of the codec	12.2 kbps	TTI of RAT 2	1 ms
MA network model		**QoS parameters**	
Macro/microcell radius	500/100 m	Maximum admissible delay	100 ms
Maximum capacity in RAT 1	4 Mbps	Admissible rate of lost packets	2%
Maximum capacity in RAT 2	54 Mbps	Satisfaction threshold	90%

Figure 6.11 presents results for the evaluated scenario ($\zeta = 1/1$). Note that LBA and SBA present meaningful gains compared to the reference scenario while CTA and RMA provide poor performance.

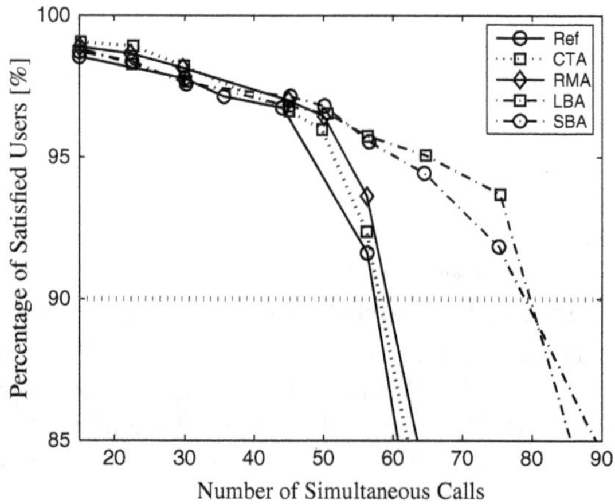

Fig. 6.11 Evaluation of the Scenario $\zeta = 1/1$, where there is one user inside the hot spot area for each user outside.

For the reference scenario, an approximate capacity of 58 simultaneous calls (the capacity is established from the satisfaction threshold) was found. The CTA, RMA, LBA, and SBA algorithms presented a capacity of 58, 60, 80, and 80 calls, respectively. This corresponds to a capacity gain of 0, 3.4, 37.9, and 37.9% for these algorithms with respect to the reference scenario.

CTA and RMA tend to overload RAT 2 while RAT 1 is lightly loaded, since the first RAT presents link bit rates higher than those available in the second RAT.

Thus, LBA and SBA are better AS algorithms for VoIP in the evaluated scenario. In spite of their similar performance, LBA and SBA have differences in terms of practical implementation. While LBA needs to know previously the total RAT capacities, SBA requires the more complex and frequent measurement of user satisfaction.

6.6 Performance of Access Selection and Vertical Handover

Results of this section assume the same general methodology presented in Section 6.5.1. However, the investigations presented here are focused on the joint operation of AS and VHO procedures.

In the GASP formulation, connections are chosen for reassignment according to an optimization criterion, without restrictions on the number of calls that will be involved in the process. The VHO procedure can increase the performance of the overall MA network, as shown in Fig. 6.8. However, it also incurs signaling costs. Then, only a reduced number of connections should be chosen to perform a VHO.

In this section, three criteria to select connections for VHO are evaluated. The simulations are done for a best-effort service class modeled according to [32]. Only users in the coverage area of both RATs (inside the hot spot) can be selected and VHO can take place in both directions. The investigations concentrate in the case in which the VHO algorithm is triggered when AS takes place. Then, this specific VHO process is named as **access selection algorithm with anticipatory vertical handover**. The criteria for the selection of connections are listed below:

- **Users with lowest packet throughput (ULPT)**: the selected users are those inside the hot spot whose packet throughputs are the lowest.
- **Users with highest packet delay (UHPD)**: the selected users are those inside the hot spot whose packet delays are the highest.
- **Users with lowest transmission rate (ULTR)**: the selected users are those inside the hot spot whose link bit rates are the lowest. Additionally, VHO is only performed for that users experiencing better link bit rate in the new RAT.

These are completely heuristic criteria and other possibilities may be investigated in the future. For instance, the choice of VHO criterion can be applied independently of the AS algorithm. The assignment of a new connection to a RAT does not impact the reassignment of the selected ongoing calls. It is also assumed that the calls selected for VHO are reassigned simultaneously, i.e., a call arrival starts the anticipatory VHO process and it is accomplished considering the network status at the time of the call arrival.

Now, the performance gain of the joint use of AS and the anticipatory VHO technique is presented. The algorithm used for AS is RMA. The proposed algorithms are investigated in the two scenarios previously defined for the AS algorithms only (Scenarios 1 and 3 in Table 6.2). As explained in Section 6.5.1, these scenarios represent two opposite cases in terms of the performance increase possibility.

In Scenario 1, the performance gains provided by the different VHO criteria are not very significant. In fact, only gains between 10 and 20% with respect to the AS-only case are possible. This result is explained by both the AS algorithm, which assigns users for the most bit rate effective RAT, and the characteristic of the scenario in test, which limits a meaningful improvement on the aggregate throughput.

UHPD is the best VHO criterion in Scenario 3, as shown in Figs. 6.12 and 6.13. The capacity gain due to one or two call reassignments is significant, over a 100%. This is explained by the handover of the worst users in terms of delay from RAT 2 to RAT 1. Although there is no incentive considering the bit rate, those connections experience better delay in RAT 1 which impact significantly in the 10th percentile of the throughput. On the other hand, with a third reassignment some degradation in performance can be noticed because the gain caused by lower delay does not overcome a likely bit rate degradation.

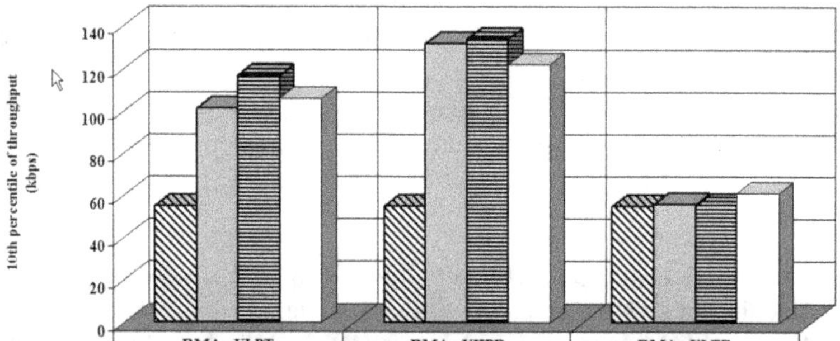

	RMA - ULPT	RMA - UHPD	RMA - ULTR
AS Only	55.11	55.11	55.11
AS+ VHO (1 Reassignment)	101.28	132.00	56.29
AS+ VHO (2 Reassignments)	116.75	134.15	55.78
AS+ VHO (3 Reassignments)	106.16	122.34	61.62

Fig. 6.12 Scenario 3 – performance of the RMA as AS algorithm and VHO criteria.

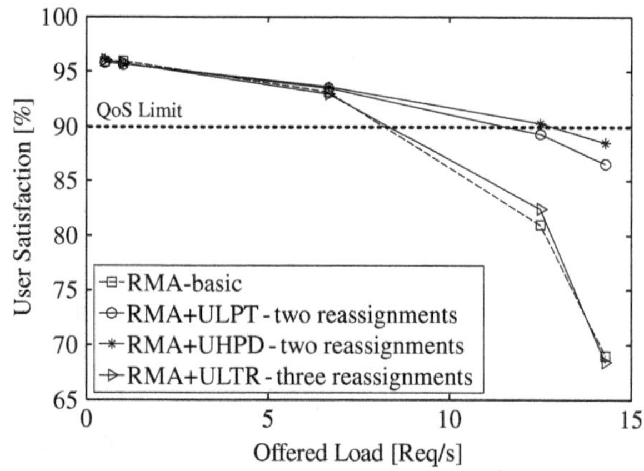

Fig. 6.13 Scenario 3 – user satisfaction performance with RMA as AS algorithm and VHO criterion.

The poor performance with users with lowest transmission rate (ULTR) criterion in Scenario 3 is due to its interaction with RMA and the specific configuration of the tested MA network scenario. Herein, RAT 1 has higher capacity than RAT 2, and as consequence, the RMA algorithm tends to assign the users inside the hot spot to RAT 2. Additionally, there is no incentive (in terms of bit rate) to reassign those users to RAT 1. Then, the number of users managed by VHO with ULTR criterion is low, yielding no significant gains.

As expected, the higher the possibility of bit rate enhancement (Scenario 3), the higher the incremental performance gain with the anticipatory VHO. Considering the performance results of Scenario 3, RMA with UHPD provides the best performance. This is due to the reassignments of users to RAT 2 when the high-capacity RAT 1 is overloaded. As stated before, the RMA algorithm tends to assign the users to RAT 1 (due to its higher bit rate capacity). With the adoption of the anticipatory VHO, some users are reassigned to the lower loaded RAT 2 regardless of its bit rate. This is the main source of gain considering the joint use of RMA and UHPD as the anticipatory VHO criterion.

6.7 Case Study: Access Selection in an UTRAN and WLAN

In the current section, MA network performance is analyzed in a specific case study. The wireless service in focus is the best-effort WWW sessions modeled according to [18, 19]. This part aims to answer two questions:

- **Does it pay off to adopt MA network and CRRM in a 3GPP system?** Here, the focus is on exploring what the feasible gains are in including WLAN under a multiaccess setting as a network capacity enhancing technology to a 3GPP one. It will be identified in which scenarios this adoption can result in higher gains and become interesting for a 3GPP network operator;
- **If a WLAN system is adopted for network capacity enhancement of a 3GPP system, how can the gains be further enhanced?** Here, the focus is at providing an AS scheme that is feasible, simple, and augments the network capacity even further. The AS algorithms presented in the previous sections will be evaluated and also a new one will be proposed, adapted to the specific 3GPP-WLAN scenario.

For the performance and capacity evaluations, a dynamic system-level MA network simulator was used. The simulator implements a detailed, standard-compliant MA network comprised by 3GPP's high-speed downlink packet access (HSDPA) and Institute of Electrical and Electronics Engineers (IEEE) 802.11a WLAN interworking on a tight coupling basis. For further details on the modeling employed in this simulator, refer to the works in [18, 19]. The main configuration parameters for the simulations are shown in Table 6.4.

6.7.1 Impact of the WLAN Adoption Using the Multiaccess Framework

The gains provided by the WLAN access point (AP) addition can be separated in two contributions: the release of resources in 3GPP, which enhance QoS for its remaining users, and the high bit rates provided to the users that connect to WLAN.

Table 6.4 UTRAN(HSDPA) and WLAN main parameters.

UTRAN parameters		**WLAN parameters**	
Parameter	**Value**	**Parameter**	**Value**
Data transmission	HS-DSCH only	Physical layer	802.11a
RLC mode and RLC ACK window	AM with 100 ms	Beacon frame interval	100 ms
MAC-hs retransmissions	3 with 4 parallel processes		
Number of CQIs available	22	Propagation delay	0.33 μs
Grid layout	Tri-sectored with one interference tier	Hot spot radius	60 m
Cell radius	500 m	Path loss model	Breakpoint model
Closed loop power control	A-DCH only		
Open loop PC initial power	0.20989 W		
DTX factor for the A-DCH	20%	Shadowing (standard deviation)	4 dB
2D correlated shadowing (standard deviation)	8 dB with 50 m of correlation distance		
Maximum transmission power	20 W		

In the present case study it is assumed that 80% of the data calls originate from within the hot spot area, creating a scenario where the addition of a WLAN hot spot provides significant gains.

A simple AS algorithm is employed based on the CTA. As expected, the WLAN addition yields considerable gains both in the 3GPP congestion relief and WLAN higher rate regions, which can be attested by Fig. 6.14. Herein, a QoS gain of about 9% in the number of satisfied users allows for a 380% load capacity increase. Thus, the addition of a WLAN AP aiming to improve the performance of the network with a hot spot is a good option, enabling a cellular operator either to use their current WLAN installations to alleviate a 3GPP congestion or to install a new WLAN in areas where known hot spots are formed.

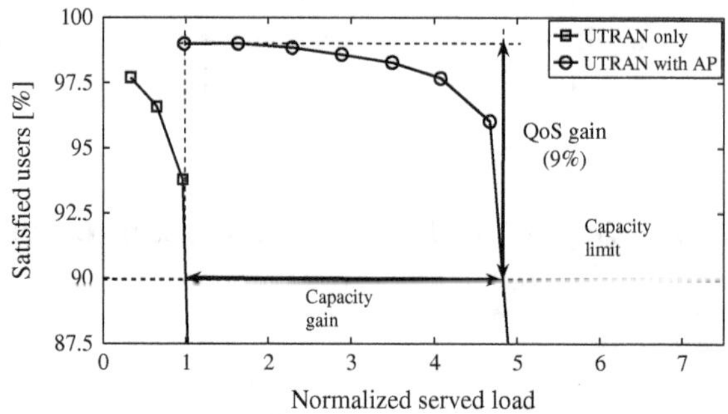

Fig. 6.14 User satisfaction before and after the addition of a WLAN AP to a UMTS cell.

6.7.2 Access Selection Evaluation in the UTRAN-WLAN Multiaccess Network

In this section, more sophisticated AS algorithms are analyzed in the same UMTS and WLAN MA network from the previous section. For the performance comparisons of this section, besides the already defined CTA, RMA, ERT, and UTA algorithms for AS, a new one, called hereafter as the queue-based algorithm (QBA), is employed. Instead of focusing on the packet delay as UTA does, QBA uses the RAT's buffer size to estimate the load [18].

As seen in the previous section, it is indeed interesting to adopt WLAN AP under a multiaccess setting to increase the total load capacity of a 3GPP system. This section will show that by adopting a proper AS scheme, those gains can be further improved.

The AS evaluation is based on the relative proportion of the hot spot call arrival rate in comparison to the macrocellular one. In order to illustrate the potential gains arising from AS, a particular situation where 90% of the data calls come from the hot spot is considered. This situation may be representative of a temporary overload due, e.g., to a sports event in which the total load grows rapidly in a particular location.

To understand the AS algorithm's performance, the algorithm's behavior must be evaluated under increasing offered loads. Regarding the worst case users (10th percentile of session throughput), RMA stands out as the best algorithm under low loads, followed by QBA, UTA, ERT, and CTA, as seen in Fig. 6.15. But, as the load increases, the QBA and UTA algorithms converge to the RMA performance, slightly surpassing it. The better performance for UTA and QBA in relation to RMA is due to the consideration of the load information, agreeing to the conclusions of Yilmaz in [50].

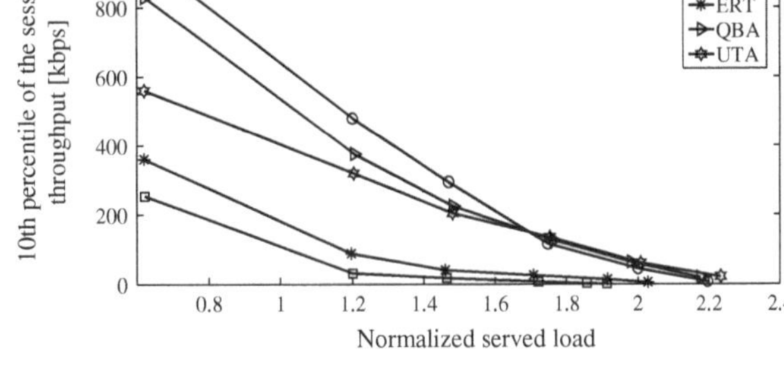

Fig. 6.15 10th percentile of the average session throughput for all AS algorithms.

Comparing to the CTA and ERT algorithms, UTA, QBA, and RMA have a much better performance as these algorithms are based on the bit rates, which happens to be the criterion that provides the best performances.

Although ERT performs better than CTA, it is far behind the other algorithms. This is due to the fact that the RAT choice based solely on RAT's estimated throughput is worse than the ones based on the other criteria that depend on data rates as discussed previously. Another explanation for the good performance of the data bit rate-based approaches is that by assigning users to the RAT in which it experiments the best rates, these bit rate algorithms also improve the RAT's throughput, effectively maximizing it.

All the behaviors stated before can be summarized as in Fig. 6.16, where the user satisfaction is presented. It is interesting to see that QBA and UTA do not only offer gains with the maximum load, but also continue to do so with even higher loads, surpassing the RMA algorithm. ERT offers some gains relative to CTA but stays far behind the rate-based algorithms. The QoS gains, seen in that figure, are of about 9, 8, 8, and 5% for the RMA, QBA, UTA, and ERT, respectively. As for the capacity, UTA offers the best gain of about 117.5% and is closely followed by QBA with 114.5% and RMA with 106%. ERT achieved the worst performance with a capacity gain of 40.5%.

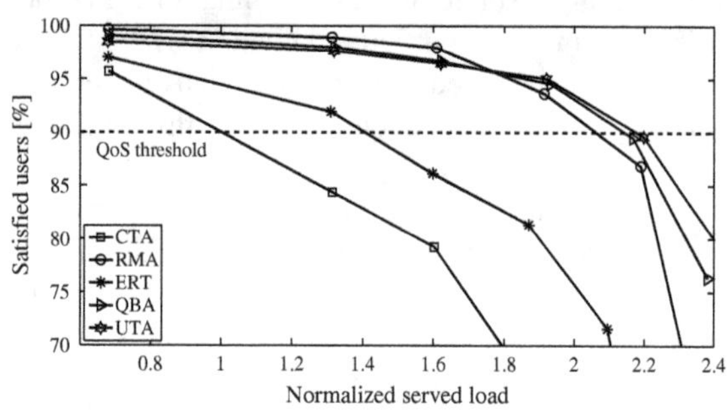

Fig. 6.16 Overall user satisfaction for all AS algorithms.

AS also influences the separate network performance, as seen in Figs. 6.17 and 6.18 which picture the 10th percentile throughput for UTRAN and WLAN, respectively. In Fig. 6.17, all algorithms start with similar performances but, as the load increases, RMA and UTA tend to keep a slightly better session throughput than CTA. This happens because they rely on good AS criteria and only assign users to UTRAN that will experience good QoS in this RAT. As for QBA and ERT, degradation occurs with load increase. Figure 6.18 depicts a different situation for WLAN. In this case, QBA shows a good performance at the low loads, near to the one presented by the RMA. With the load increase, the QBA and UTA converge to RMA, even surpassing it. This behavior is very similar to the one presented in Fig. 6.15 for the overall multiaccess case, confirming the WLAN as the dominant RAT for the MA network performance.

The rather small performance gains for the QBA and UTA algorithms can be explained by the huge difference in the available rates of UTRAN and WLAN. The reason is that, due to the rate scales, a poor rate in WLAN may still be better than a fair equivalent in UTRAN. Thus, the choice of two RATs with comparable rates should provide much better results. In this situation, QBA and UTA should provide a better performance as the load information can be better exploited. Nevertheless, the coverage proportion should also have a similar influence, since with a greater

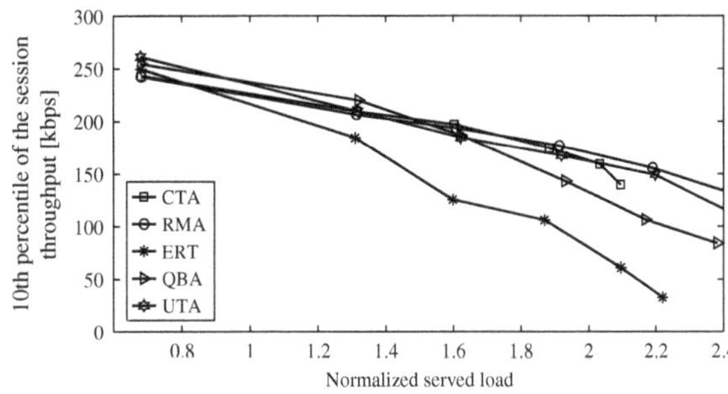

Fig. 6.17 The 10th percentile of the session throughput from the UTRAN perspective.

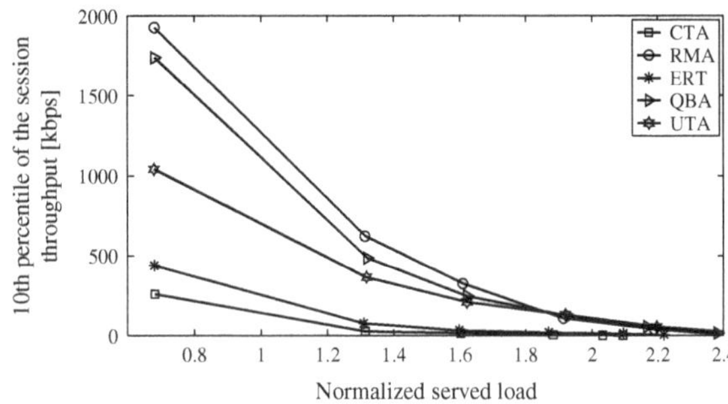

Fig. 6.18 The 10th percentile of the session throughput from the WLAN perspective.

portion of overlapping coverage, more users could be candidates for the AS procedure, making a better use of the multi-user diversity.

6.8 Conclusions and Research Directions

MA networks are a new paradigm for the future wireless communication systems. Although the integration of heterogeneous networks arises as a solution for increasing the system performance as a whole, some challenges take place with respect to implementations of common management strategies.

In this chapter, the MA concept was first introduced and its main features, procedures, architectures, and interfaces were briefly indicated. The CRRM concept was also presented as well as its main functionalities. The focus was mainly on AS and VHO, two fundamental procedures from which the majority of the gains coming from CRRM at a reasonable cost were expected. Several algorithms were proposed and evaluated, considering a typical situation where a macrocell embeds a microcell that can be served by two RATs. Several scenarios and one detailed case study of a UMTS and WLAN MA network were studied, including non-real-time and real-time services. These results serve as an illustration of the potential gains offered by CRRM in MA networks. It is worth noting that the proposed algorithms are scenario independent from a functional point of view. However, the gains extracted from CRRM will certainly vary depending on the specific systemic conditions available. Therefore one important future work is to continue the characterization of CRRM performance according to the different RAT capacities, loading scenarios, and common coverage. These are key aspects at the access layer. There are many other challenges to enable a full-scale employment of MA networks, including aspects related to interworking architecture and security. The following research directions are considered as relevant.

The development of new technologies and applications appears as a range of options to be employed in the MA network context. Nowadays, some efforts have been directed towards the conception and improvement of algorithms, architectures, MA network management, and security, aiming at the provision of a flexible and pervasive MA network platform, where the users can enjoy multiservices anytime and anywhere with any device. In the following, the future perspectives on the MA networks are presented:

- **Architectures for integration of 3G networks**: frameworks for interworking between UMTS and WLAN must be tested in real implementations. The always best connected (ABC) concept arises as a generic architecture model that combines efficiently these systems [25]. It is an alternative to provide ubiquitous access to the users. Another architecture solution aiming at providing session mobility over UMTS-WLAN networks includes the ip multimedia subsystem (IMS) platform [37]. Advantages in the usage of an integrated architecture based on IMS are evident because this framework plays an important role in the provision of IP multimedia services in an MA network due to the unified session control. Another architecture kept in perspective is the integration of mobile ad hoc networks (MANETs) into IP-based systems, aiming at more flexibility of the involved networks in the multiaccess environment [21];
- **VHO solutions**: in order to provide session mobility, efficient VHO strategies must be implemented. With this goal, a lot of research works in the conception of VHO solutions have been proposed. VHO optimization can be performed in UMTS-WLAN network in order to avoid degradation in network utilization due to redundancy retransmissions in handover process [35]. Still for UMTS-WLAN, the performance of VHO can be improved by adopting a fuzzy adaptive handover strategy, since it takes into account multiple criteria and rules based on prior knowledge of the network behavior [28]. For new technologies, secure seamless and soft handover is already envisaged for **WIMAX**-3G networks [16]. Other studies focus on seamless VHO for a wireless broadband (WIBRO)-WLAN network [44];
- **QoS provision**: in the next generation of wireless communication, the growth of service demands will drive the increase of the market competition among companies. Thus, satisfaction of clients is a priority that must be taken into consideration, therefore, QoS requirements of a given service are taken as satisfaction parameters. In an MA network, the QoS provision is more challenging. In fact, some efforts aiming at provisioning real-time services with quality obeying the QoS constraints in the multiaccess context are performed [34];
- **Interworking between emerging technologies**: **WIMAX** and 3G-LTE arise as primary alternatives for fourth generation (4G) networks. Directions for inter-

working strategies between these two systems are under discussion for future communication systems [43]. However, the role of CRRM in the integration of emerging wireless technologies is an open research topic.

References

1. 3GPP: 3GPP System to Wireless Local Area Network (WLAN) Interworking; Functional and Architectural Definition. TR 23.934, 3rd Generation Partnership Project (3GPP) (2002). URL http://www.3gpp.org/ftp/Specs/html-info/23934.htm
2. 3GPP: Improvement of Radio Resource Management (RRM) Across RNS and RNS/BSS. TR 25.881, 3rd Generation Partnership Project (3GPP) (2002). URL http://www.3gpp.org/ftp/Specs/html-info/25881.htm
3. 3GPP: Feasibility Study on Location Services (LCS) for Wireless Local Area Network (WLAN) Interworking. TR 22.935, 3rd Generation Partnership Project (3GPP) (2005). URL http://www.3gpp.org/ftp/Specs/html-info/22935.htm
4. 3GPP: Quality of Service (QoS) and Policy Aspects of 3GPP – Wireless Local Area Network (WLAN) Interworking. TR 23.836, 3rd Generation Partnership Project (3GPP) (2005). URL http://www.3gpp.org/ftp/Specs/html-info/23836.htm
5. 3GPP: Location Services (LCS) Architecture for 3GPP System – Wireless Local Area Network (WLAN) Interworking. TR 23.837, 3rd Generation Partnership Project (3GPP) (2006). URL http://www.3gpp.org/ftp/Specs/html-info/23837.htm
6. 3GPP: 3G Security; Wireless Local Area Network (WLAN) Interworking Security. TS 33.234, 3rd Generation Partnership Project (3GPP) (2007). URL http://www.3gpp.org/ftp/Specs/html-info/33234.htm
7. 3GPP: 3GPP System to Wireless Local Area Network (WLAN) Interworking; System Description. TS 23.234, 3rd Generation Partnership Project (3GPP) (2007). URL http://www.3gpp.org/ftp/Specs/html-info/23234.htm
8. 3GPP: 3GPP System to Wireless Local Area Network (WLAN) Interworking; Stage 3. TS 29.234, 3rd Generation Partnership Project (3GPP) (2007). URL http://www.3gpp.org/ftp/Specs/html-info/29234.htm
9. 3GPP: 3GPP System to Wireless Local Area Network (WLAN) Interworking; WLAN User Equipment (WLAN UE) to Network Protocols; Stage 3. TS 24.234, 3rd Generation Partnership Project (3GPP) (2007). URL http://www.3gpp.org/ftp/Specs/html-info/24234.htm
10. 3GPP: Feasibility Study on 3GPP System to Wireless Local Area Network (WLAN) Interworking. TR 22.934, 3rd Generation Partnership Project (3GPP) (2007). URL http://www.3gpp.org/ftp/Specs/html-info/22934.htm
11. 3GPP: IP Multimedia Subsystem (IMS); Stage 2. TS 23.228, 3rd Generation Partnership Project (3GPP) (2007). URL http://www.3gpp.org/ftp/Specs/html-info/23228.htm
12. 3GPP: Requirements on 3GPP System to Wireless Local Area Network (WLAN) Interworking. TS 22.234, 3rd Generation Partnership Project (3GPP) (2007). URL http://www.3gpp.org/ftp/Specs/html-info/22234.htm
13. 3GPP: Service Requirements for the Internet Protocol (IP) Multimedia Core Network Subsystem (IMS); Stage 1. TS 22.228, 3rd Generation Partnership Project (3GPP) (2007). URL http://www.3gpp.org/ftp/Specs/html-info/22228.htm
14. 3GPP2: 3GPP2-WLAN Interworking – Stage 1 Requirements. Tech. Rep. X.s0028-200-A, 3rd Generation Partnership Project 2 (3GPP2) (2004). URL http://www.3gpp2.org/Public_html/specs/tsgs.cfm

15. 3GPP2: CDMA2000 Packet Data Service; Wireless Local Area Network (WLAN) Interworking – Access to Operator Service and Mobility for WLAN Interworking. Tech. Rep. X.s0028-200-A, 3rd Generation Partnership Project 2 (3GPP2) (2008). URL http://www.3gpp2.org/Public_html/Specs/tsgx.cfm
16. Altaf, A., Iqbal, F.: S3H: A Secure Seamless and Soft Handover Between WIMAX and 3G Networks. In: International Conference on Convergence and Hybrid Information Technology (ICHIT) (2008)
17. Blomgren, M., Hultell, J., Cai, R., Cai, T.: Distributed Demand-Aware Access Selection in Wireless Multi-cell Data Networks. In: Proceedings of the IEEE Symposium on Personal, Indoor and Mobile Radio Communications (PIMRC) (2007)
18. Cardoso, L.: Performance Assessment of a Multi-access Network Composed by 3G Cellular and Wireless LAN. Master's thesis, Universidade Federal do Ceará, Brazil, Fortaleza (2006)
19. Cardoso, L.S., de Sousa Jr., V.A., Pimentel, J.F., Cavalcanti, F.R.P.: On the Adoption of WLAN for Capacity Improvement of 3G Networks Using the Multi-access. In: International Telecommunications Symposium, 2006 (2006)
20. Casadeval, F.: Final Report. Tech. rep., EVEREST IST-2002-001858 (2006)
21. Ding, S.: A Survey on Integrating MANETs with the Internet: Challenges and Designs. Comput. Commun. **1**, 3537–3551 (2008)
22. ETSI: ETSI 101 957: Requirements and Architectures for Interworking Between HIPERLAN/2 and 3rd Generation Cellular Systems. Standard, ETSI (2001). URL www.etsi.org
23. de la Fuente, E.S. et al.: UMTS, MBS and DBS Network and Service Management System Architecture: Technical Requirements and Functionality Description. Tech. rep., MONASIDRE IST-2000-26144 (2001)
24. Garey, M.R., Johnson, D.S.: Computers and Intractability, A Guide to the Theory of NP-Completeness. W. H. Freeman (1979)
25. Gazis, V., Alonistioti, N., Merakos, L.: A Generic Architecture for 'Always Best Connected' UMTS/WLAN Mobile Networks. Int. J. Wireless Mob. Comput. **2**, 248–262 (2007)
26. Gupta, V.: IEEE 802.21 Media Independent Handover – 3GPP Update (2008). URL https://mentor.ieee.org/802.21/file/08/21-08-0144-00-0000-3gpp-update.ppt
27. Holma, H., Toskala, A.: WCDMA for UMTS: Radio Access for Third Generation Mobile Communications, 3rd edn. Wiley (2004)
28. Horrich, S., Jamaa, S.B.: Neural Networks for Adaptive Vertical Handover Decision. In: 5th International Symposium on Modeling Optimization in Mobile, Ad Hoc, and Wireless Networks (2008)
29. IETF: Mobility for IP: Performance, Signaling and Handoff Optimization (mipshop) (2008). URL http://www.ietf.org/html.charters/mipshop-charter.html
30. IST-2002-001858: Everest Project (2002). URL http://www.everest-ist.upc.es/
31. Jin, F., Choi, H.A., Kim, J.H., Sohn, J., Choi, H.I.: Common Radio Resource Management for Access Selection in Multi-access Networks. In: IEEE Radio and Wireless Symposium (2008)
32. Johansson, C., Verdier, L.D., Khan, F.: Performance of Different Scheduling Strategies in a Packet Radio System. IEEE Int. Conf. Universal Pers. Commun. **1**, 267–271 (1998)
33. Jorguseski, L., Litjens, R., Zhiyi, C., Nikookar, H.: Radio Access Selection in Multi-radio Access Systems. In: IEEE Symposium on Communications and Vehicular Technology (2007)
34. Lee, I.: Wireless Video Streaming over Integrated 3G and WLAN Networks. Int. J. Wireless Mob. Comput. (2007)
35. Lin, I.C., Shieh, C.S.: Avoidance of Redundant Retransmission in Vertical Handover by Modified Stream Control Transmission Protocol. In: 3rd International Conference on Innovative Computing Information and Control (ICICIC) (2008)
36. Luenberger, D.G.: Linear and Nonlinear Programming. Addison-Wesley (1989)
37. Munasinghe, K.S., Jamalipour, A.: Interworking of WLAN-UMTS Networks: An IMS-Based Platform for Session Mobility. IEEE Commun. Mag. 46, 184–191 (2008)

38. Niebert, N., Flinck, H., Hancock, R., Karl, H., Prehofer, C.: Ambient Networks – Research for Communication Networks Beyond 3G. In: 13th IST Mobile and Wireless Communications Summit (2004)
39. Ning, G., Zhu, G., Peng, L., Lu, X.: Load Balancing Based on Traffic Selection in Heterogeneous Overlapping Cellular Networks. MINIMICRO SYSTEMS **27**, 2036–2041 (2006)
40. Perez-Romero, J., Sallent, O., Agusti, R.: Policy-Based Initial RAT Selection Algorithms in Heterogeneous Networks. In: Mobile and Wireless Communications Networks (2005)
41. Project, W.I.: Wireless World Initiative New Radio. URL `https://www.ist-winner.org/`
42. Rappaport, T.S.: Wireless Communications: Principles and Practice, 2 edn. Prentice Hall Communications Engineering and Emerging Technologies Series. Prentice Hall PTR (2002)
43. Seol, J.H., Chung, J.M.: IEEE 802.11 MIH Based Handover for Next Generation Mobile Communication Systems. In: 4th International Conference on Innovations in Information Technology (2008)
44. Shin, C., Kim, S., Cho, J.: A Low-Latency L2 Handoff Between WIBRO and CDMA2000 Mobile Networks. In: International Conference on Convergence and Hybrid Information Technology (ICHIT) (2008)
45. da Silva, A.P., Cavalcanti, F.R.P., de O. Neto, R.A.: VoIP Capacity Analysis of Wireless Multi-Access Networks Using Access Selection Schemes. In: Proceedings of the IEEE Symposium on Personal, Indoor and Mobile Radio Communications (PIMRC) (2007)
46. Song, W., Zhuang, W., Cheng, Y.: Load Balancing for Cellular/WLAN Integrated Networks. IEEE Netw. **21**(1), 27–33 (2007)
47. de Sousa Jr., V.A., de O. Neto, R.A., de S. Chaves, F., Cardoso, L.S., Pimentel, J.F., Cavalcanti, F.R.P.: Performance of Access Selection Strategies in Cooperative Wireless Networks Using Genetic Algorithms. In: Wireless World Research Forum (2005)
48. Tsao, S.L., Lin, C.C.: Design and Evaluation of UMTS/WLAN Interworking Strategies. In: Proceedings of the IEEE Vehicular Technology Conference (VTC) (2002)
49. Varma, V., Ramesh, S., Wong, K.D., Barton, M., Hayward, G., Friedhoffer, J.: Mobility Management in Integrated UMTS/WLAN Networks. In: Proceedings of ICC (2003)
50. Yilmaz, O.: Access Selection in Multi-Access Cellular and WLAN Networks. Master's thesis, Royal Institute of Technology, Sweden, Stockholm (2005)
51. Yilmaz, O., Furuskar, A., Pettersson, J., Simonsson, A.: Access Selection in WCDMA and WLAN Multi-access Networks. In: Proceedings of the IEEE Vehicular Technology Conference (VTC), vol. 4, pp. 2220–2224 (2005)
52. Zhou, Y., Rong, Y., Choi, H.A., Kim, J.H., Sohn, J., Choi, H.I.: Utility-Based Load Balancing in WLAN/UMTS Internetworking Systems. In: IEEE Radio and Wireless Symposium (2008)

Part II
Transceiver Architectures

Chapter 7
Strategies for Link-Level Performance Assessment in the Simulation of Wireless Systems

Elvis M. G. Stancanelli, Carlos H. M. de Lima, and Darlan C. Moreira

7.1 Introduction

Performance assessment of wireless communication systems by computer simulations is a valuable and widely adopted tool for research and development, as well as in planning and deployment phases of these systems. It allows the numerical evaluation of a model of the wireless communication system of interest, while featuring speed, inexpensiveness, and flexibility to control the experiments. However, the high number of functionalities to be modeled in typical wireless setups makes a single simulator unfeasible. It is usual to break the system into modules, or layers, of reduced dimensionality. A simulation of the entire system can then be achieved by constructing all necessary modules and inter-connecting them via appropriate interfaces. This chapter describes a two-level organization of the system's functionalities as the simulator is split into system- and link-level modules. These two parts correspond to modeling multiple parallel links and a particular link, respectively. This approach is very popular in wireless simulation studies [11, 17, 20, 23, 30, 35, 43].

More specifically, in this chapter the link-level (LL) simulator part of a wireless network simulator is studied, which attempts to imitate the processes established with the communication between a base station (BS) and a user equipment (UE). LL simulations aim at estimating the quality of radio link during a connection. The chapter has two objectives: (i) an efficient software engineering approach to building simulation tools for link-level evaluation and (ii) proper design of a link-to-system level interface, which fulfills the goal of supplying appropriate metrics to systemic evaluation. While we do not cover system-level (SL) simulations in this chapter, we direct the interested reader to Part I of this book. While not specifically focused on teaching how to perform such simulations, chapters in Part I show plenty of examples where system simulations have been implemented as a tool for radio resource management (RRM) studies.

The rest of this chapter is organized as follows. Section 7.2 concentrates on the explanations of the various fundamental concepts dealt with herein. In Section 7.3, a brief overview of the general aspects regarding LL modeling, identifying, and

F. Cavalcanti, S. Andersson (eds.), *Optimizing Wireless Communication Systems*,
DOI 10.1007/978-1-4419-0155-2_7,

describing the most common functional blocks of a typical cellular LL simulator is given. Afterward, the Link-Level Software Development Framework (LSDF) is introduced in Section 7.4. Finally, the design issue of link-to-system-level (L2S) interfaces is discussed in Section 7.5.

7.2 Rationale for Link-Level Performance Evaluation

The performance of a communication system may be assessed in different ways and probed in several points of the communication chain. Computer simulations are a very convenient method as the experiments are repeatable and allow the researcher to isolate a specific parameter and easily log peculiar effects. In fact, it is much simpler to analyze, test, and evaluate a model of a communication system rather than the real system, given that the model is accurate enough.

Simulating a mobile communication system involves dynamically imitating the individual and mutual behavior of UEs and BSs, and comprising a number of random variables describing demand, users' location, radio channel, and so on. Cell grid configuration, equipments' placement, radio link conditions, RRM, and layered communication procedures are some of the set of features to be comprised. In Fig. 7.1 some of the typical functionalities and phenomena to be modeled in cellular network simulators are presented.

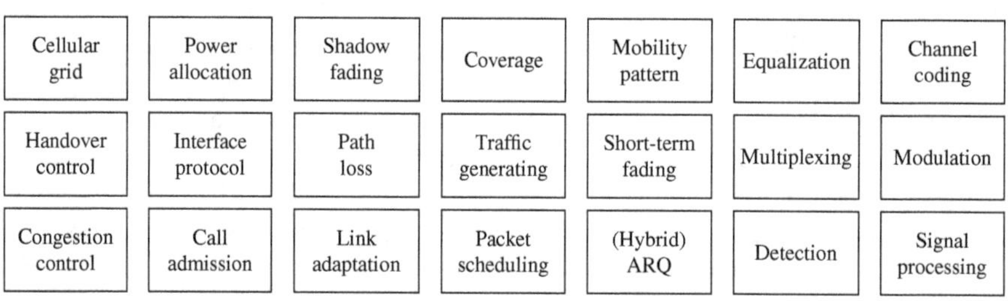

Fig. 7.1 Typical functionalities of cellular network simulators.

However, due to that high number of components in a cellular network, a single and full simulator of the whole communication chain is not able to represent reliably these networks without incurring high complexity. Typically, the simulations of cellular networks are performed separately for link-level (LL) and system-level (SL). While the SL simulator time-scale relates to the life span of connections (or communication sessions), the LL simulator has its time resolution at symbol, bit, or chip (or fraction of those) level. The LL simulator assesses the performance of a specific link established between a given transmitter and receiver, under controlled conditions, which in turn is representative for links in similar conditions throughout the network. The LL outcomes can then be used as inputs to SL simulations, which concern the interactions among multiple simultaneous connections. In Fig. 7.2, a specific link onto a specific cell is highlighted.

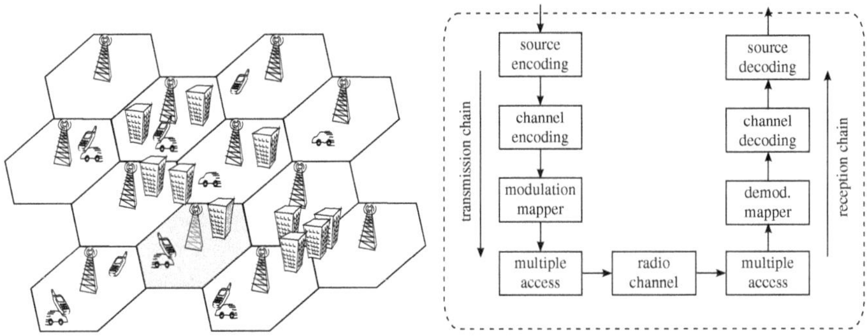

(a) System-level focus for simulations (b) Link-level focus for simulations

Fig. 7.2 The role of LL part on the network context.

The LL simulator evaluates the radio link operation, performing typically the coding, modulation, and radio channel propagation. The SL part aims at performing the systemic characteristics of the networks, including mobility, traffic generation, and radio resource management (RRM). A method for interfacing both simulation levels must then be employed, enabling both result accuracy and computational feasibility. This is the so-called link-to-system-level (L2S) interface.

Bit error rate (BER), block error rate (BLER), and throughput are some of the quality measures that can be obtained at the output of LL simulations with given model parameters, such as carrier frequency, user's velocity, average signal-to-noise ratio (SNR), and other characteristics of the propagation environment. Such measures are present to the SL simulator as well, since the same set of parameters is expected to be matched. Succinctly, it is necessary to supply the SL simulator with LL simulation results from several parameter sets representing the typical range of scenarios found in real networks.

The LL simulator carries out extensive simulation campaigns and summarizes their results by means of averages. Then, average figures of merit – e.g., BLER and signal-to-interference-plus-noise ratio (SINR) – are arranged and stored onto look-up table (LuTs). Whenever the SL needs to assess the LL performance results, it simply consults the look-up table (LuT) at the current channel quality – e.g., SINR – rather than be bothered with in-depth simulations onto the physical layer. Frequently, the parameters presented by the SL must be interpolated over those available in the LuTs.

The border between LL and SL is not well defined, allowing for some freedom on designing the L2S interface. For instance, the hybrid automatic repeat request (H-ARQ) mechanism can be inserted in the LL by making it easier to shape soft combination of transmission and retransmissions, resulting in a LuT with average values of performance, which omits the occurrence of retransmissions to the SL. Alternatively, this insertion can be made in the SL, when one is able to specialize the scheduling for the retransmission events, but unable to combine several retransmission versions.

Yet another possibility is to replicate functionalities on both LL and SL. For instance, the short-term fading could be properly addressed on either level solely, or in both simultaneously as shown in Fig. 7.3. This choice depends on factors such as complexity, accuracy, and flexibility, as it will become clear later on.

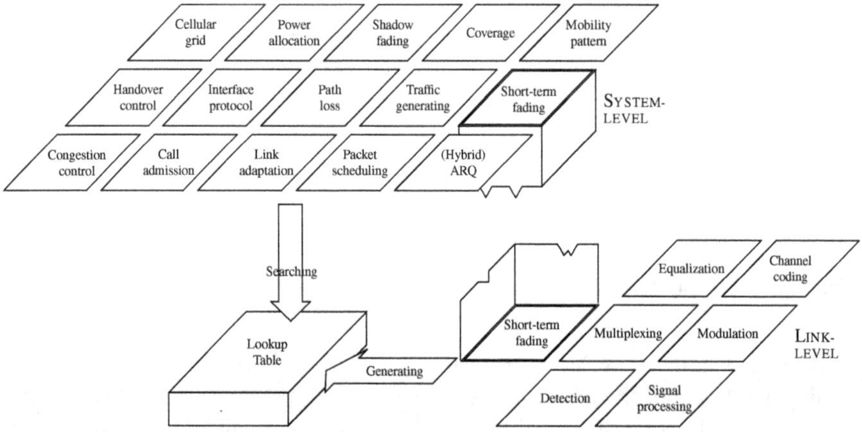

Fig. 7.3 Example with short-term fading inserted in both LL and SL parts.

Before concluding this section, let us review two common error sources in computer simulations [26]: processing error and modeling error. The former is due to computing limitations, such as computation speed, memory, and numerical precision. The latter is associated with mathematical approximation and simplification of models.

Still, it is important to emphasize that usually LL simulators follow the Monte Carlo method, where random processes are implemented using random number generators. In this way, the underlying random processes can be assumed as ergodic. Furthermore, the lower the desired estimated error rate, the higher the number of samples to obtain from the simulation. For instance, consider the estimation of the bit error probability p in a typical additive white Gaussian noise (AWGN) channel. Let $\widehat{p}$ be this estimate according to the Monte Carlo method and N the number of samples used in this estimation. The distribution of $\widehat{p}$ tends to a normal one as N grows, with mean p and variance $p(1-p)/N$ [26]. A practical rule to set N value is to choose a number ranging from $10/p$ to $100/p$ [25], obtaining a 95% confidence interval narrowing respectively from $(1.8\widehat{p};\ 0.55\widehat{p})$ to $(1.25\widehat{p};\ 0.8\widehat{p})$ [26]. As an example, for a bit error rate (BER) of $p = 10^{-3}$ the transmission of at least $10,000$ bits must be simulated, assuming an approximated confidence interval from around half to twice the BER estimate $\widehat{p}$.

7.3 Link-Level Modeling

The conception of the physical layer of cellular networks follows the functional diagram of a generic digital communication system as presented in Fig. 7.2(b). This system is composed of three main parts: transmission chain, radio channel, and reception chain. These constituent parts are addressed in the following sections. As a starting point, consider that the data are delivered by higher layers in so-called transport blocks.

7.3.1 Transmission Chain

A detailed explanation of digital communications is out of the scope of this book, but there exists many books that discuss its foundations such as [40, 48]. In turn, the implementation details of actual functional blocks depend on the system specifications that can be obtained from specific standardization documents.

Generally speaking, the first step to be performed by the physical layer is to calculate cyclic redundancy check (CRC) bits over the transport block and attach them as a redundancy for error detection. Similarly, forward error correction (FEC) schemes are used for producing an output of n bits for an input of k bits at a time, where $n > k$, the code rate given by $r = k/n$. The FEC decoders are able to exploit the redundancy for correcting some errors.

As an example, UMTS terrestrial radio access network (UTRAN) systems [3] usually employ FEC schemes, such as convolutional and turbo codes [9, 10], with available code rates of $1/2$ and $1/3$. These turbo encoders are composed of two eight-state parallel recursive systematic convolutional encoders, one of which is preceded by an interleaver.

Even though the transport block size could vary with time in accordance to the traffic intensity, there is a pre-defined amount of radio capacity at the physical layer available. Eventually this capacity is not enough to match the transport block size. Rate matching is an essential functional block that matches the number of bits arriving from higher layers with the capacity provided by the physical layer. Therefore, either repetition or puncturing (periodical deletion of coded bits [12]) technique can be applied to the bits in such a way that such matching is achieved. Bit streams provided by distinct services subject to independent channel coding are combined at the rate matching to form a unique flow. Yet related to rate matching, the discontinuous transmission (DTX) module indicates when the transmitter can be turned off (e.g., during alternating periods of silence of one party in a conversation), which allows reduction of interference.

In an equivalent base-band modeling, the modulation is as simple as a bit-to-symbol mapping. In this sense, the modulator simply converts an input bit sequence into an output symbol sequence, following a pre-specified mapping. In modern wireless systems, various modulation schemes and channel encoders are combined to create the so-called modulation and coding schemes (MCSs). Different MCSs are defined such that there is one MCS available for each typical propagation scenario. In other words, lower coding rates and modulation constellation sizes are used to achieve better reliability under unfavorable link conditions, and vice-versa. This technique is known as link adaptation (LA) or adaptive modulation and coding (AMC) and provides an alternative to power control (PC) techniques when dealing with changes on the link quality. A practical example of LA is described in Chapter 2 of this book for the enhanced data rate for GSM evolution (EDGE) system.

Besides the already mentioned blocks, intermediary segmentators, concatenators, and interleavers may be necessary to ensure the correct functioning of the whole chain. Likewise, some other functional blocks are present only in specific systems. For instance, spreading codes are used in code-division multiple–access (CDMA)-based technologies. In this regard, blocks performing the spreading and despreading spectrum functions are present in the chain.

7.3.2 Radio Channel Modeling

The radio channel model garners distinct characteristics of physical medium connecting transmitter and receiver antennas, such as atmosphere properties, position, and mobility of antennas and obstacles, path loss, shadowing, fading, noise and bandwidth. When dealing with mobile wireless communication systems, plain free space propagation and thermal noise are not enough to represent channel characteristics. There are also buildings, trees, foliages, and other obstacles in the medium that reflect or absorb signal energy. Hence, multiple reflective paths are created and the multipath propagation phenomenon is present.

At a receiver antenna, the resultant signal is the combination of all signals from the multiple paths. These delayed and damped signals interact with each other, either in a constructive or destructive manner, distorting the transmitted signal. This effect is known as multipath fading [46]. Thus, the time-spread resultant signal undergoes fluctuations in its amplitude, phase, and angle of arrival. Furthermore, relative motion between the transmitter and receiver antennas induces a time-variant behavior to the channel and determines the fading rate, causing a shift on Doppler frequency. Fades about 40 dB on signal envelope with nulls successively happening approximately a half carrier wavelength are common [24, 46].

On the mobile channel modeling, large-scale and small-scale fading should be distinguished accordingly. The large-scale term refers to the effects observed onto a long-time or long-distance scale, such as pathloss and shadowing. While pathloss refers to the mean signal attenuation, the shadowing refers to its fluctuation due to large obstacles. Small-scale fading instead is determined by the aforementioned time-spreading of the signal and time-variability of the channel [46], whose statistics obey, typically, a Rician or Rayleigh distribution, depending on whether there is line-of-sight component or not.

LL simulations, due to their fine time resolution, are usually focused on modeling small-scale fading only. The accurate simulation of small-scale fading in particular requires more elaborate models. The most popular option is the Jakes' model [24], based on the principle of sum of sinusoids, assuming that many paths arrive with different angles [24]. An alternative model is the Smith's one [50], which is based on four Gaussian processes that are passed throughout low-pass filters, transformed to time domain, and suitably combined.

7.3.3 Reception Chain

At the receiver side the radio channel effects must be undone and the transmission chain process reversed. The reception chain gathers the associated functional modules, some of them need additional knowledge about the transmission procedure such as synchronization.

The fading effects are the first to be mitigated as one wishes to approach performance under AWGN channel. At this point, some well-known methods can play an important role, such as Viterbi equalizer for global system for mobile communication (GSM) and rake receiver for CDMA-based systems [47].

Functional blocks should be available to implement demodulation, decoding, equalization, and so on. The DTX placeholders must be identified and then deleted. The processes of segmentation, concatenation, and interleaving must be inverted. The functional block reverse to rate matching should also be designed, especially when the puncturing mechanism is enabled.

For convolutional codes, the optimal decoding is given by a maximum-likelihood sequence estimation of signal with memory [40], by computing a certain metric for every possible sequence. Either Hamming or euclidean distance could be used as that metric, allowing respectively a hard or soft detection. For reducing the amount of calculation, the Viterbi detection algorithm is commonly used [49].

The turbo decoding is performed through a peculiar structure that combines two soft-input/soft-output (SISO) decoders with interleaver and de-interleaver devices. For details about turbo decoding the reader is referred to [45]. The extrinsic information is passed from the output of one decoder to the input of the other [27]. This task is repeated in several iterations, improving the data estimate. Basically, a SISO decoder can be accomplished through either the maximum a posteriori (MAP) algorithm or soft-output viterbi algorithm (SOVA) [42].

At the end of the receiver chain, the CRC module can verify whether the message is corrupted, using the same CRC polynomial used on the transmission chain. Moreover, if an error occurs, no action will be taken to fix it. However, automatic repeat request (ARQ) – a mechanism that triggers retransmissions of erroneous data blocks, aided by feedback from the CRC module – can be inserted.

H-ARQ is an enhanced version of ARQ based on soft combining schemes, that is, on the coherent superposition of multiple received redundancy versions of the same data block. H-ARQ is typically implemented as chase combining (CC) or incremental redundancy (IR). In chase combining (CC), the erroneous blocks are stored and combined to perform a more reliable detection. Incremental redundancy (IR) uses distinct retransmissions to send distinct versions of redundancy. In this case, only incremental redundancy is retransmitted improving incrementally the probability of correct detection of the data block.

Note that if any special block is used in the transmission chain, such as spreading in CDMA systems, the corresponding "decoding functionality" must be in the reception chain.

7.3.4 Single- and Multiuser Approaches

The performance assessment at the link level can be made following either *single-user* or *multiuser* approaches. In the single-user approach, the LL simulation is performed regarding the desired user only. This is typically understood as the link being part of a perfectly orthogonal multiple access scheme where co-channel interference is disregarded. The single-user approach is very popular when one wants to characterize the radio link performance as defined by the transceiver itself.

Conversely, in the multiuser approach, the LL performance is assessed under a scenario containing as many users in the SL. This approach can be interesting, for example, in studies on the mitigation of multiuser interference [56]. However, it may not be very useful to L2S interfaces as one would have to match the interfering scenario in the SL and LL. Since there exists a plethora of possible interfering scenarios in the SL, the applicability of LL multiple access simulations is limited. One notable exception, however, occurs in CDMA systems. In this case, multiuser interference may be approximated as a Gaussian noise [34, 41], which leads to a simple model that can be more easily dealt by L2S interface.

7.3.5 Case Study: HSDPA Simulator

When resorting to procedural programming paradigm, a link-level (LL) high-speed downlink packet access (HSDPA) simulator was designed in conformity with the 3rd. Generation Partnership Project (3GPP) technical specifications, based on the high-speed downlink shared channel (HS-DSCH) description. The references [3–6] are the central bibliographic sources which specify the transport channel processing (CRC, segmentations, turbo coding, puncturing, repetition, interleaving, and physical channel mapping), modulation, spreading, and scrambling.

The transmission time interval (TTI) the HS-DSCH is of 2 ms, achieving a short round-trip delay for the operation between the terminal and Node B for retransmissions [21]. The transmission chain of the HS-DSCH is shown in Fig. 7.4, with its transport channel processing and spreading/modulation chain segments. The functionalities of the reception chain and the mobile radio channel module are not shown, but are also implemented. Note that the spreading and modulation processes take place only after the *physical channel mapping*. Quadrature phase shift keying (QPSK) modulation and 16-quadrature amplitude modulation (16-QAM) are available. Each physical channel is direct-sequence spread in the spectrum with a code spreading factor (SF) of 16; afterward, it is multiplied by a complex scrambling code. One single scrambling code is used for all the physical channels.

Despite a fixed SF, a multicode transmission is allowed. Depending on the UE category, a user may use a maximum of 5, 10, or 15 codes. There are 12 terminal categories defined to allow different performances as well as levels of complexity [5, 36]. A more detailed description of the HSDPA radio technology is found in Chapter 3 of this book.

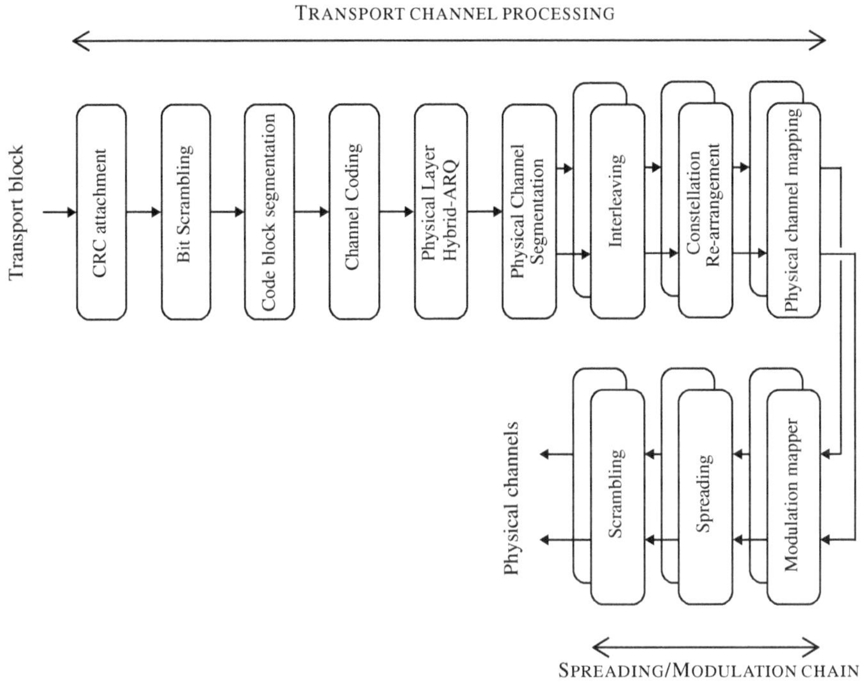

Fig. 7.4 Transmission chain of the HS-DSCH.

As the first processing on the transmission chain, a CRC attachment is generated in accordance with [3], which has a CRC size of 24 bits. If an error in the information block is detected while checking CRC code parity bits, the terminal requires a retransmission of the same packet from the Node B, and this process is repeated at SL until the packet is correctly received or until the number of retransmissions reaches the maximum value.

The *bit scrambling* is used in order to avoid problems with 16-QAM amplitude estimation in the receiver. By scrambling the data, the coded bit stream becomes sufficiently random to cause both inner and outer signal points in the 16-QAM constellation to be used uniformly.

The code block segmentation module is responsible for the adjustment of the bit sequence delivered by the CRC attachment module to the required input of the channel coding module. If the size of the bit sequence from this transport block is greater than 5,114 bits the segmentation must be performed before the turbo coding processing, where all segments are generated with the same size. If necessary, the addition of filler bits to the beginning of the first segment is performed before the segmentation; these filler bits are transmitted being set to logical zero. At the receiving side, the reverse processing can be divided into two stages: first, the segments are concatenated; and, later on, the filler bits are pruned, if there are any.

The channel coding of HS-DSCH is based on a powerful turbo coding [45], whose input length could range from 40 to 5,114 bits. Universal Mobile Telecommunication System (UMTS) turbo encoder consists of two parallel 8-state constituent encoders with coding rate of 1/3 and one internal interleaver. In turn, the turbo coder's internal interleaver consists of bits-input of a rectangular matrix with padding (insertion of dummy bits), intra-row and inter-row permutations of the rectangular matrix, and bits-output from the rectangular matrix with pruning (deletion of dummy bits). Trellis termination is accomplished by taking tail bits, which are padded latter on. Afterwards the channel coding outputs for each code block are serially concatenated. The turbo decoder used in the simulator is SOVA-based [42], which is attractive for its low complexity.

The H-ARQ functional block consists of two rate-matching stages and a virtual buffer between them [3]. The first rate-matching stage aligns the number of input bits to the virtual IR buffer. Its output is matched to the number of physical channel bits required by the high-speed physical downlink shared channel (HS-PDSCH) set in the TTI at the second stage, so that either puncturing or repetition strategies can be achieved. After the second rate-matching stage, the H-ARQ bit collection is performed, which is accomplished by using a rectangular interleaver.

Multicode transmission is supported and depends on the capability of the UE simulated, which is up to 15 HS-PDSCHs [3]. When more than one HS-PDSCH is used, the physical channel segmentation block divides the bit sequence into some physical channels (PhCHs). Each physical channel has a separated interleaver [3]. The UMTS interleavers consist of matrix, input bits with padding, the inter-column permutation for the matrix, and output bits from the matrix with pruning. The interleaver has a fixed size of 32×30, being single in case of QPSK modulation and double in case of 16-QAM.

In this simulator, both QPSK and 16-QAM modulations are available. Due to the usage of a turbo decoder algorithm, the demodulator must be able to work with soft bits. In case of 16-QAM modulation, the constellation rearrangement block can improve the performance by rearranging the symbol constellations between multiple transmission attempts as this provides an averaging effect among the reliability of the bits. However, note that this gain is available only for retransmissions and not for the initial transmission. Constellation rearrangement is obtained through bit manipulations at the output of the HS-DSCH interleaving block and it is controlled by a four-state bit mapping parameter with two independent operations: the bits can be swapped and/or have their logical values inverted [3].

The interleaved bits must be distributed into the physical channels structure and that is described in [4]. The bits are mapped to the PhCHs so that the bits for each PhCH are transmitted over the air in ascending order.

The channel quality indicator (CQI) value is an information carried through uplink direction using high-speed dedicated physical control channel (HS-DPCCH); it specifies the transport block size, number of HS-PDSCHs, modulation scheme, and reference power adjustment. For each UE category, a maximum number of bits available in the virtual IR buffer N_{IR} is determined following the table map comprising 30 CQI values, in [5].

After the modulation mapper, the spreading module takes part and, for each HS-PDSCH, there is one channelization code of SF of 16 obtained from the set of channelization codes reserved for HS-DSCH transmission [4]. This code is one of those generated by the orthogonal variable spreading factor (OVSF) code tree, as suggested in 3GPP standards [6].

The scrambling module is responsible for scrambling spread data with a complex-valued code. Each cell has one primary scrambling code, which is a segment of a gold sequence. At the reception chain, a conjugate-complex code is applied from the ones used at the transmission chain.

There are two radio channel models in the simulator: AWGN and multipath fading. For the multipath channel modeling, the modifications in Jakes' fading model proposed by Li & Huang [29] that generates multiple independent Rayleigh fading waveforms were realized. An independent fading process is generated for each multipath component according to the specified multipath power intensities and delays profile. Note that one needs a dedicated receiver for each HS-PDSCH. Furthermore, in case of multipath, a rake receiver is used, with maximal ratio combining (MRC), which has a finger perfectly synchronized to each resolvable path.

7.3.5.1 Simulation Results

This section presents the simulation results, a UE of category 5 is considered for some CQI values (see [5,Table 7A] and also Chapter 3). The turbo decoder uses the SOVA algorithm with eight iterations. Initially, the AWGN channel is considered. A range of 10,000 up to 20,000 blocks was considered to compose each BLER versus E_c/N_0 curve, as shown in Fig. 7.5. The E_c/N_0 is given by the ratio between received spread signal and noise power. In Fig. 7.6 the throughput for each CQI is illustrated. A detailed view of the throughputs for the first three tested CQI values is provided.

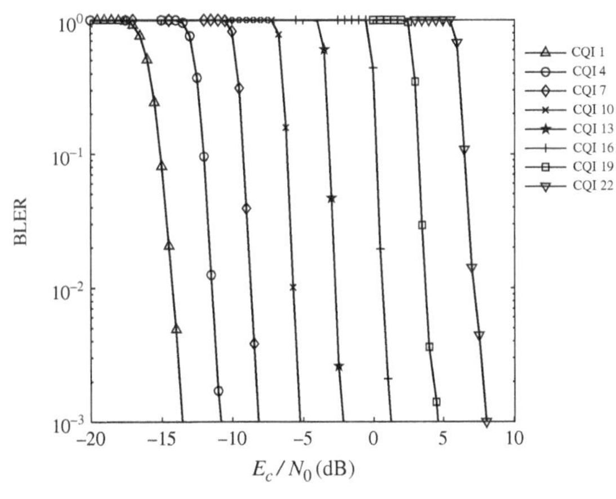

Fig. 7.5 BLER evaluated for HSDPA under AWGN channel, with UE category 5 and SOVA decoder.

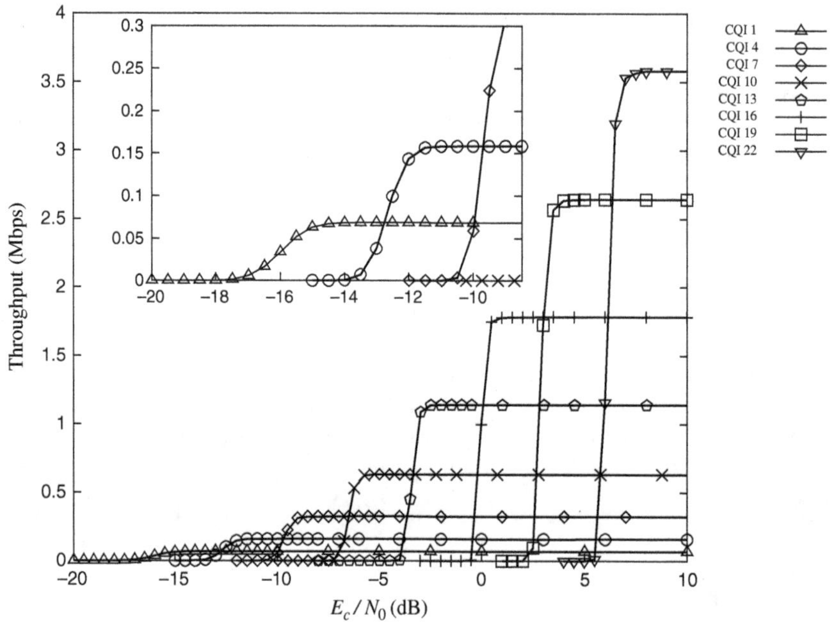

Fig. 7.6 Throughput evaluated for HSDPA under AWGN channel, with UE category 5 and SOVA decoder.

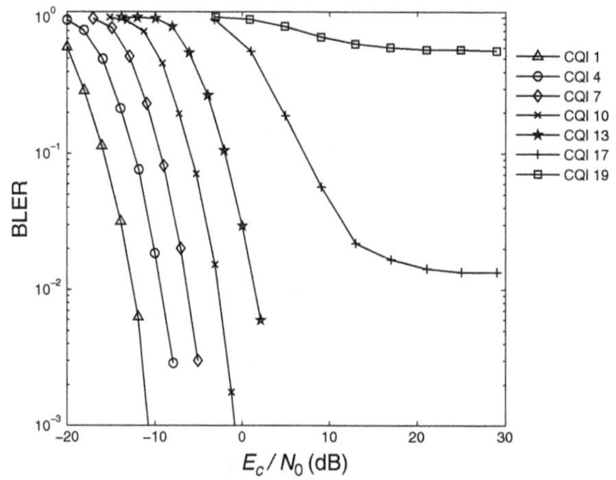

Fig. 7.7 BLER evaluated for HSDPA under Pedestrian B profile, with UE category 10 and SOVA decoder.

In a similar way, simulation results were obtained for the multipath channel, whose multipath power-delay profile follows the *Pedestrian B channel*, defined by UMTS [55, Appendix B].

In Fig. 7.7 the BLER for some CQI values are plotted assuming the UE category 10. The turbo decoder uses the SOVA algorithm, but now with 10 iterations. A rake receiver with MRC is used, which is able to deal with the six paths. Since a worse performance than under AWGN is expected, the number of blocks to be transmitted can be smaller for similar level of uncertainty of BLER estimates. A range of 8,000 up to 12,000 blocks was adopted to compose each BLER versus E_c/N_0 curve. Note the very bad performance at CQI 19. Indeed, this undesirable behavior is repeated for all CQIs whose employed modulation is the 16-QAM, being ascribed to the limited ability of rake in coping with interpath interference, which stands out for multilevel modulations.

7.4 Link-Level Software Development Framework

This section introduces a software development framework that aids in the design and implementation of the LL simulation tools. The proposed framework relies on the object-oriented programming (OOP) paradigm and well-established design pattern so as to define its creational, structural, and behavioral characteristics [18, 52]. The link-level software development framework (LSDF) has been put into effect using C++ language due to its inherent support to data abstraction, object-oriented concepts, and generic programming. However, the underlying programming techniques applied in the LSDF design and implementation are general and allow for the usage of other programming languages. Succinctly, the LSDF establishes a systematic procedure to build scalable and robust code, while the extension of functionalities is made viable by customizing the basic set of components straightforwardly.

The LSDF constitutes a skeleton for developing LL simulation tools, where modularity and reusability are of primarily concern. Operational entities are derived from basic components relying on the generic interface, while providing specific functionalities. Additionally, the unified higher-level interface allows incorporating external libraries seamlessly to the framework. For instance, many of the communication functionalities used to assess the downlink (DL) of wideband code division multiple-access (WCDMA) systems (see Section 7.4.3) are actually provided by ITPP library [16] and easily incorporated into LL simulators based on the proposed framework.

As aforementioned, the framework components are derived exploiting modularity and reusability as indispensable approaches. On one hand, modularity quickens the development activity, since distinct modules may be independently constructed relying solely on the common interface provided by the development framework. Equally important, the modular approach strengthens code reliability and facilitates the adoption of test-driven development techniques, such as xUnit testing framework [33]. On the other hand, reusability permits the utilization of previously implemented components with slight or no modification at all when developing either new functionalities or new software tools.

7.4.1 Generic Simulator Architecture

The LSDF encompasses both implementation procedures and basic structural components to lead the systematic development of LL simulation tools. Additionally, the framework structure constitutes a skeleton to derive more elaborated simulation tools extending the functionalities provided in the set of basic components. The LSDF consists of three fundamental components tailored to construct LL simulators: building modules, data blocks, and the bond container. Building modules are the fundamental "bricks" irremediably used to implement any operational component. Functional components are appropriately connected in a meaningful sequence so as to constitute the "transmission chain" of communication systems. The transmission and reception chains are established by a data structure referred to as bond container. The bond container is indeed a data abstraction implementing a logical container to store an arbitrary number of building modules in an object-oriented linked list. Additionally, the bond container provides functionalities to manipulate the stored data. Data blocks encapsulate the valid information that is indeed conveyed throughout the stack of functional modules by means of the exchange blocks.

7.4.1.1 Logical Structure

Following sections provide in-depth information about the operational structure of the LSDF and how the basic set of components may be suitably arranged to model communication systems.

Building Module

In Fig. 7.8 the building module logical structure is illustrated. Building modules are organized in two main parts: the generic interface and the self-contained functionality performing a specific task. In addition, there is a data compartment where data blocks may dwell amid undergone procedures, for instance, when data are either retrieved or encapsulated in exchange blocks for conveyance between adjacent modules, and during normal operation of functional components.

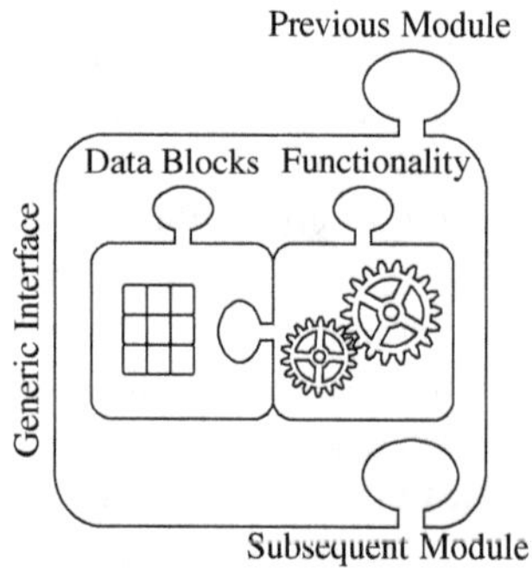

Fig. 7.8 Building module illustration presenting the generic interface, the functional part, and data compartment.

Regarding the software design, building modules may be implemented through multiple inheritances deriving from a virtual generic interface and an implementation wrapper enclosing the real functionality [52]. The building module may also be structured following the bridge pattern where abstraction and implementation are decoupled thus they can evolve independently [18].

The generic interface establishes an abstraction layer separating external interoperability from internal functionality. Actually, adjacent modules are insulated by the common interface and, therefore, it can only interact throughout this communication layer. The logical boundaries established by the generic interface allow for independent development of components and consequently quickens the implementation activity – building modules can be developed and validated independently on a demand basis. The modular design technique has been adopted not only to improve reusability, but also to reinforce maintainability of components.

Equally important, the specific functionality performed by a given module characterizes its purpose in the range of operations carried out by the transmission system and its interchangeability among similar components as well. For instance, a modulator stage that is in charge of modulating an input signal so as to convey information through the radio channel may apply several schemes interchangeably (either analog or digital) to perform such task, each one characterizing a specific modulator component. Similarly, the radio channel component, while performing the specific task of physical medium, may be implemented regarding distinct characteristics and impairments, such as noise, interference, and so on. Additionally, building modules are bound together using logical hooks that connect adjacent entities in order to compose a specific stack of modules. The logical hooks may be implemented as references – when considering C++ one may use either simple reference data type or pointers – to the adjacent levels of the stack [52].

Bond Container

When appropriately organized in a stack, the building modules actually constitute the several stages of a transmission chain. In Fig. 7.9 the overall structure of the bond container when modeling the transmission and reception chains of communication system is presented. The bond container is indeed a building module with distinguishable functionalities that make it work as a specialized container to store and manipulate building module data structures. Then, building modules are piled up in a meaningful sequence so as to constitute the transmission and reception chains of a specific communication system.

The bond container entity indeed implements the transmission and reception technologies of communication systems. Data blocks are fed into the bond container using the push functionality and retrieved afterward employing a pop routine. The bond container is also in charge of controlling the general updating procedure. During the updating procedure two main tasks are performed: (i) data are exchanged between adjacent components throughout the common interface and (ii) data blocks are modified in accordance with the functionality of each functional module. Notice that building modules carry out both the transmission and reception of data information when data blocks are transmitted or received.

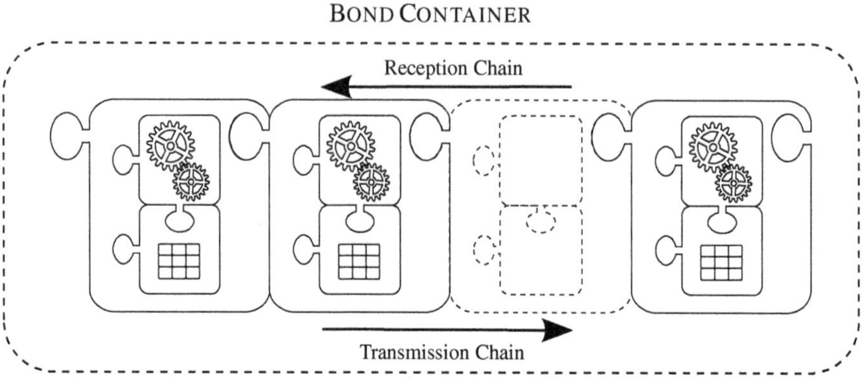

Fig. 7.9 Bond container implementing transmission and reception chains of communication systems.

As aforementioned, the bond container is a building module itself and therefore shares the generic interface. Then, once data are fed into the stack, the bond container triggers the transmission processes by updating itself. Consequently, data are propagated to the top most component in the stack and so forth deflagrating a reaction in chain. All in all, the updating procedure is initiated by the bond container, though each building module updates itself independently performing its own functionality and modifying the data block accordingly. Therefore, each building module is activated at the updates of the transmission and the reception chains. The transmission and reception chains are connected throughout the radio channel.

Data Blocks

Data are "packed" inside the data blocks so as to make consistent the exchanging interface among modules in the bond container. While propagating throughout the several stages of the stack of components, data should be further encapsulated inside exchange blocks, which literally guarantee a generic exchanging interface for data manipulation inside functional modules.

Data blocks provide further data abstraction to harmonize and facilitate the conveyance of information throughout the transmission and reception chains. In fact, data blocks establish a meaningful association between the data structure and the actual underlying concepts. For instance, in UMTS systems, control and data streams addressed to a given user may be conveniently modeled as arrays of symbols that are further wrapped by the corresponding data block so as to ensure a common exchanging interface and feed the stack of functional components initiating the transmission process. When updating the chain, the components can independently, though sequentially, access the content of data blocks and appropriately manipulate the enclosed data in accordance with their specific purposes.

While propagating through the stack, the structure of the enclosed data may be conveniently reorganized to reflect the undergone procedure. For example, a multidimensional rearrangement of the array of symbols may be used to model distinct data streams addressed to different users. In this way, not only the independence of data streams is preserved but the arrays are also kept sequentially ordered, while propagating through their corresponding transmission chains. Afterward, the streams are conveniently collapsed into a single sequence to emulate the expected interference among distinct users sharing the common air interface. Notice that even though there is a single bond container the independence among distinct transmission chains is emulated by reason of the data abstraction provided by data blocks. Moreover, spatial multiplexing techniques requiring multiple antennas transmission may be conceptually implemented as a multidimensional extension of the ordinary data block structure, where streams assigned to distinct antennas are organized in parallel arrays expanding the original time-dimension of the data flow.

An illustration of the applicability of multidimensional expansion of data blocks is shown in Fig. 7.10 for a MIMO-OFDM system. The data block in this case has three dimensions. For instance, in the data block delivered to the CRC building module the 3D array is composed of the information bits. Since the CRC building module only adds parity bits, the data block that is passed to the next building module is of the same type (only the number of bits packed in it will change). On the other hand, the symbol mapper building module changes the data block information type from bits to symbols.

In addition, the correspondence of each dimension in the data block may change, such as what happens after the multiple-input multiple-output (MIMO) building module where the "streams" dimension is mapped to the "transmit antennas" dimension.

Note that the use of data blocks to pack the actual information exchanged among the building modules contributes to the modularity of the system. One can, for instance, remove the orthogonal frequency division multiplexing (OFDM) module and the system can still function correctly, provided that this makes sense.[7.1]

However, even though the data block provides abstraction to the actual data exchanged by the building modules, there are still different types of data blocks, such as data blocks of bits and complex symbols. Therefore, further abstraction of exchange blocks was created and the data block of any type inherits from an exchange block. As a result, the data are in fact passed from one building module to another as a reference or a pointer of an exchange block, which must then be converted to the correct type of data block for subsequent processing. This further abstraction ensures that a single type of data is passed from one building module to another and it is an important aspect of the generic interface.

[7.1] The MIMO schemes are usually designed for flat fading channels and OFDM is able to "make the channel flat" for MIMO. Therefore, if the radio channel building module corresponds to a flat fading channel, then the system can work with or without the OFDM building module.

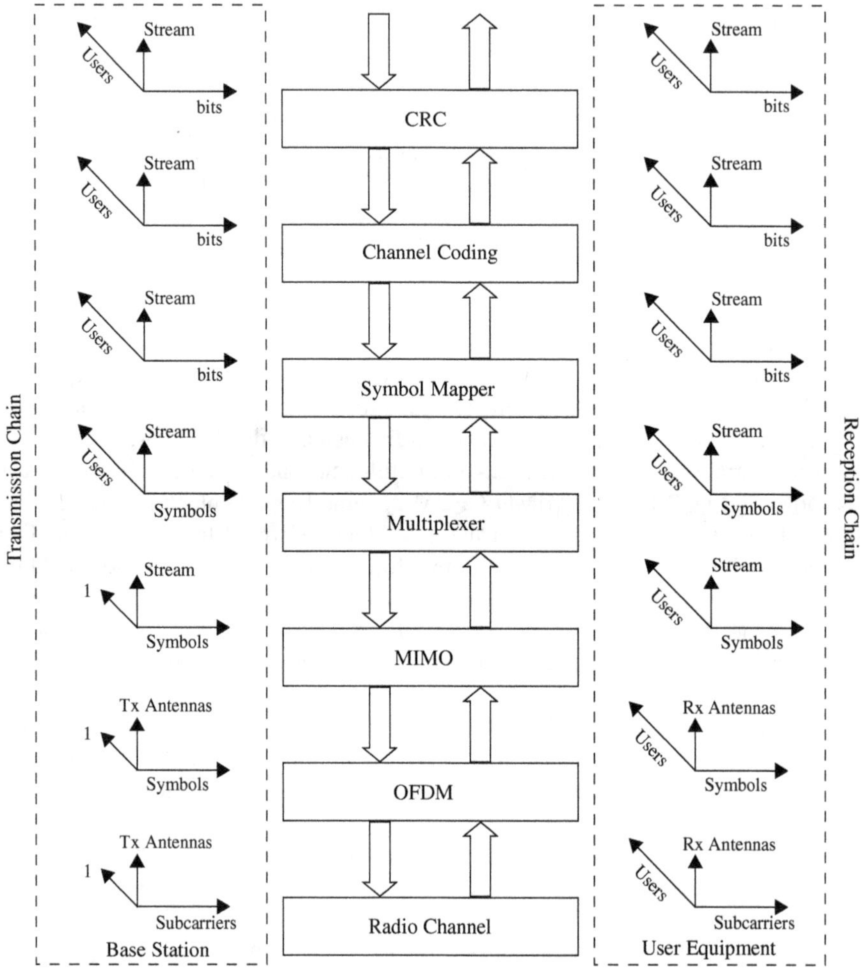

Fig. 7.10 Multidimensional expansion of data blocks for a MIMO-OFDM system.

7.4.2 Generic Information Flow

Once the transmission and reception chains are structured, the simulation can be properly launched. Simulations are executed following the Monte Carlo approach, where iterations correspond to updates of the bond container and underlying components. When the bond container is updated, the enclosed building modules are sequentially updated in a reaction in chain. The functional modules are updated according to the order they occupy in the transmission and reception chains so as to reflect the actual operation of the modeled communication system. In Fig. 7.11 the generic flow of information carried out when using the LSDF is illustrated.

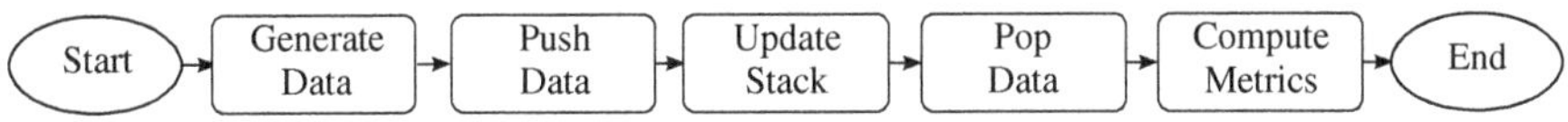

Fig. 7.11 Information flow carried out during iterations of a generic simulation tool implemented using the LSDF.

Typically, iteration is initiated by randomly generating the data payload and encapsulating the data inside data blocks afterward. Then, the data block is pushed into the transmission and reception chains by means of the interface provided by the bond container. Next, the bond container brings about the overall updating procedure deflagrating a chain reaction. Subsequently, the functional components update themselves independently. The overall update occurs in two stages reproducing an actual communication system: first, transmission process is undergone and thereafter the reception operations occur. The radio channel effects are also introduced in an intermediary stage between transmission and reception operations. During the transmission update, the radio frame is assembled and propagated throughout the radio channel. Conversely, the originally transmitted message is estimated by the completion of the reception chain. Additionally, the process dynamics may be assessed at execution time by attaching probe modules to the functional modules. The result metrics can be either regularly accompanied by assessing partial results or evaluated a posteriori when the simulation run terminates altogether.

7.4.3 Case Study: WCDMA-DL Simulator

This section illustrates the utilization of the LSDF to implement a WCDMA-DL simulation tool. Both the downlink physical layer and the radio link layer of the WCDMA dedicated channel (DCH) transport channel were implemented in accordance with the 3GPP specifications [3, 6]. The complete multiplexing and coding chains of the WCDMA DCH were also implemented following the link-level framework conventions. The WCDMA-DL simulator is implemented in C++ OOP language and has both modularity and reusability as foremost concerns. Modularity inherently provides development independence among programmers and tractability through development stages. Reusability mainly allows for time-saving during the implementation, since components previously developed can be readily re-utilized.

7.4.3.1 DL DCH Channel Structure

Transport channels specify how exchangeable information is actually conveyed through the medium. Transport channels are divided into two groups, namely, dedicated channels and common channels. In fact, common channels are network resources that are shared among group of users (maybe all of them) currently dwelling in a specific cell site. Conversely, dedicated channels are reserved resources, which are assigned to one user only, for instance, in a frequency or code basis [21]. Dedicated transport channels providing conversational services using adaptive multirate (AMR) codec with 12.2 kbps are addressed herein. The DCH modeling strictly follows the 3GPP Release 99 specifications.

The DCH conveys all the information addressed to a given user coming from higher layers, including service data (i.e., speech frames) and control information (e.g., handover commands or measurement reports). Moreover, the dedicated transport channel is characterized by specific functionalities such as fast power control, soft(er) handover, fast rate adaptation, and the possibility to support adaptive antenna techniques. The dedicated transport channel is mapped onto two physical channels. The dedicated physical data channel (DPDCH) transports higher layer information including user data, while the dedicated physical control channel (DPCCH) transports control information necessary for the physical channel [4].

In the presented model, each pair of bits represents a QPSK symbol. The frame structure consists of a sequence of radio frames. Succinctly, each radio frame encloses 15 slots (10 ms or 38,400 chips), whereas one slot corresponds to 2,560 chips (0.667 ms), which in turn equals one power control period.

The DL dedicated physical channel (DPCH) is composed of a downlink DPDCH and a downlink DPCCH, which are time-multiplexed with complex scrambling code. Thus, the dedicated data generated at higher layers and conveyed by the DPDCH are actually time-multiplexed with pilot bits, transmit power control (TPC) commands, and transport format combination indicator (TFCI) bits generated by the physical layer. The DPCH may or may not include the TFCI. When the TFCI bits are not transmitted, the DTX is used in the corresponding field. The I/Q branches of the modulator have equal power and the SFs range from 512 7,500 symbol/s down to 4 960,000 symbol/s [28]. The SF for the highest transmission rate determines the channelization code that should be reserved from the given code tree.

The case study assesses the provision of speech service when using the AMR codec with 12.2 kbps. The DPCH is implemented in the WCDMA-DL DCH simulator according to the following features:

- The DPCH TTI has fixed duration of 20 ms (two radio frames).
- The DPCH conveys only conversational service class users with data rate of 12.2 kbps.
- A fixed spreading factor (SF) of 128 was considered.
- Only the first transport channel (there are three in total) using AMR codec with 12.2 kbps is simulated. This is due to the fact that only the first transport channel performs CRC checksum, and therefore is essential for block error detection. The other two transport channels do not degrade the final voice quality (intelligibility) severely in case of block reception error.

7.4.3.2 WCDMA Transmission Chain

The transmission chain of the WCDMA-DL DCH comprises channel coding and data modulation functionalities. Additionally, interleaving, segmentation, and transport channels (TrCHs) multiplexing functionalities are performed as well. In Fig. 7.12

the overall transmission chain arrangement is illustrated. For more details about the DCH transmission chain see 3GPP technical specifications [3, 6].

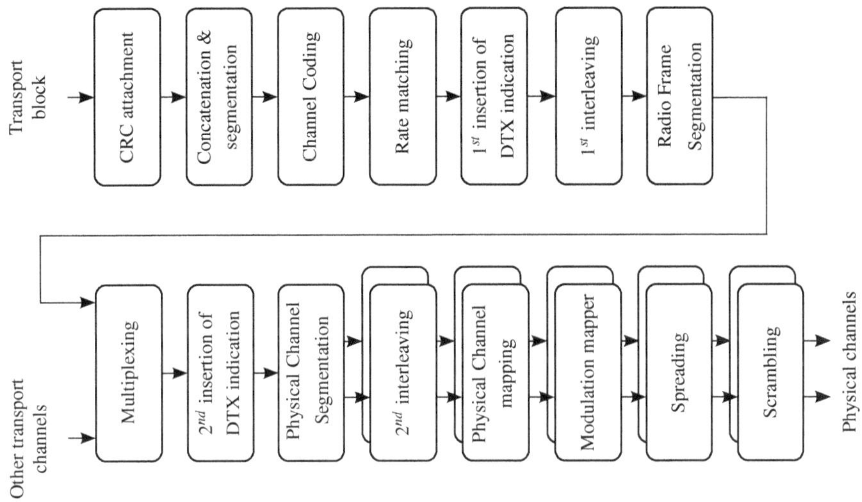

Fig. 7.12 Transport channel processing and spreading/modulation chains of the WCDMA-DL DCH.

At the beginning of the DCH transmission chain the CRC is attached to the transport block. Thus, the CRC is checked so as to verify the occurrence of errors for each detected block at the end of reception chain. The possible sizes for the CRC are 0, 8, 12, 16, and 24, which are signaled from higher layers. The error correction should be accomplished thereafter. Either a convolutional encoder with 9-constraint length or a turbo encoder [10] can be employed. The former employs $1/3$ or $1/2$ coding rates, while the latter makes use of $1/3$ rate only.

The rate-matching stage is performed using either puncturing or bits repetition. The WCDMA rate matching can simultaneously deal with many transport channels establishing a common operation point in order to differentiate quality-of-service (QoS) among them. Furthermore, both E_b/N_0 matching and unequal error protection control are carried out [51]. The rate matching can be controlled by means of semi-static parameter provided by higher layers. Additionally, DTX is implemented to bring about lower transmission rates for the downlink. The DTX indication is inserted in distinct points of the transmission chain so as to implement fixed or flexible positions (see Fig. 7.12).

Throughout the transmission chain, bits are interleaved, blocks are segmented, and channels are multiplexed. These functionalities are performed for all the processed transport channels accordingly. Afterward, the data modulation functionalities, i.e., modulation mapping, spreading, and scrambling, are performed to enable the communication through the radio channel. Notice that the WCDMA is modeled in equivalent base-band signal representation in order to prevent computational issues.

The WCDMA downlink scrambling code uses 38,400 chips of complex-valued long code, which is built from a gold sequence generated from 18-degree polynomials – the scrambling code has a period of 10 ms radio frame. The spreading code is based on the OVSF channelization codes. For downlink frequency division duplex (FDD), the SF ranges from 4 to 512 and does not vary with time. Typically, there exist one scrambling code and one spreading code tree per sector. Regarding multicode transmission for one user, the parallel code channels have different channelization codes, though the SFs are kept identical. DTX indications are tackled during demodulation appropriately.

7.4.3.3 Numerical Results

Table 7.1 summarizes the set of parameters utilized to configure the simulation campaign. In Fig. 7.13 the performance results in terms of BLER for the evaluated simulation scenario are illustrated: WCDMA-DL DCH considering the conversational service at 12.2 kbps.

Table 7.1 link-level WCDMA-DL simulator configuration parameters.

Parameter	**Value**
Transport block size (bits)	81
Transport block set sizes (bits)	81
Modulation scheme	QPSK
CRC size (bits)	12
Transport channel	DCH
Channel coding	Convolutional
Coding rate	1/3
TTI period (ms)	20
Spreading factor	128
Channel model	AWGN
Number of iterations per E_{chip}/N_0	15,000

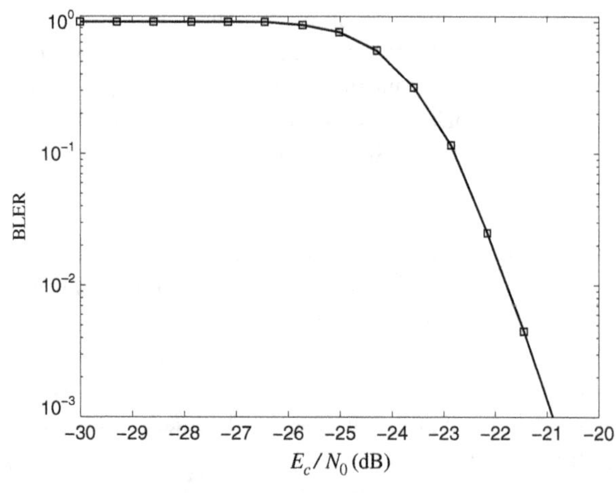

Fig. 7.13 BLER for the WCDMA-DL DCH transport channel.

7.5 Design of Link-to-System Interfaces

The main challenge for L2S interfaces is to approximate the performance assessment results obtained using a two-level simulator approach to the outcome of hypothetical single full simulator. Typically, L2S interfaces are conceived to provide estimates of BER, BLER, frame erasure rate (FER), or throughput for SL simulations.

It is well known that the BLER varies with SNR: the higher the SNR within one block of bits, the lower the BLER. Nevertheless, such relationship is usually non-linear and it changes with the radio link and the channel coding. Yet, the BLER to SNR relationship can be estimated through LL simulations, where a curve associating BLER values with distinct SNR conditions is stored onto a LuT. LuTs are composed of averaged BLER values obtained through extensive LL simulation campaigns, in which the effects of a radio link can be observed. On the LL simulator the values of BLER (or BER or FER) are calculated for distinct values of mean channel quality.

When employing AMC, the L2S interface must generate outputs for each MCS, which in turn identifies its corresponding set of LuTs. Thus, the L2S interface can be seen from the standpoint of SL part as a black-box, mapping current values of SNR to BLER obtained from LL simulations. Nevertheless, at the SL, the BLER is dealt in a probability sense, i.e., it is seen as a block error probability, so that a random test is performed in order to determine whether the transmitted frame or packet has been correctly received.

Similar procedures can be carried out to relate BLER, FER, or throughput with a vector of signal-to-noise ratio (SNR) or signal-to-interference ratio (SIR), whose elements may refer to a symbol instant or subcarrier in an OFDM link. That vector will be an input to the interface, which will generate a single output value (e.g., the BLER of those current radio conditions). The average value of the elements of the input vector may be representative for calculating the output, but depending on the characteristics of the system, a more accurate solution should be adopted. Indeed, the design of fine-tuned L2S interfaces aiming at specific systems is preferable, since a generic interface suffer from potential inaccuracies.

Note that this mapping is also dependent both on the environment (e.g., specific multipath channel) and the service (e.g., different bit rates), leading to large number of LuTs to be generated on the LL simulator and used on the SL. A two-step approach can alleviate it, making the first step service independent and the second environment independent, described in the following. Before that it is important to discern two kinds of bit error rate metrics that are used: bit error rate and raw bit error rate. The BER metric is taken after the detection and correction of error using FEC; whereas the raw bit error rate (RBER) is taken before it. In this chapter the RBER have always been taken right after the demodulation mapper.

In the following, the most influential L2S strategies are scrutinized. In Section 7.5.1, the basic average value interface (AVI) is presented, which is especially useful for well-contained scenarios, such as those characterized by AWGN or slow fading channel. Similarly, the actual value interface (AcVI) concept is described in Section 7.5.2, whose applicability extends to scenarios with significant variation in the channel quality, e.g., due to fading or frequency hopping. Furthermore, in Section 7.5.3 the central idea of variable orthogonality factor interface (VOFI) is presented, which takes into account the influence of each of the various multipath. Finally, the Effective SINR mapping (ESM) L2S interface – widely used for MIMO and OFDM technologies – is addressed in Section 7.5.4.

7.5.1 Average Value Interface

Assume that the input of an L2S interface corresponds to P $SINR$ values, whose ith element is γ_i, and the output is given by a unique $BLER$ value. The simplest approach to obtain a $BLER$ value for a given set of $SINR$ values consists in taking their arithmetic mean value, $\overline{\gamma}$, and mapping it into a suitable LuT. That LuT must be composed of pairs of $BLER$ and $\overline{\gamma}$ values. The aforementioned procedure corresponds to the average value interface (AVI), which is especially appropriate when the channel quality tends to remain uniform within the block length. Indeed, block length is given by P and it refers to the number of $SINR$ measures available within the transport block duration. Alternatively, the block length could be defined over temporal, spectral, spatial, or any other dimension of interest. In Fig. 7.14 a scheme of the AVI L2S is represented.

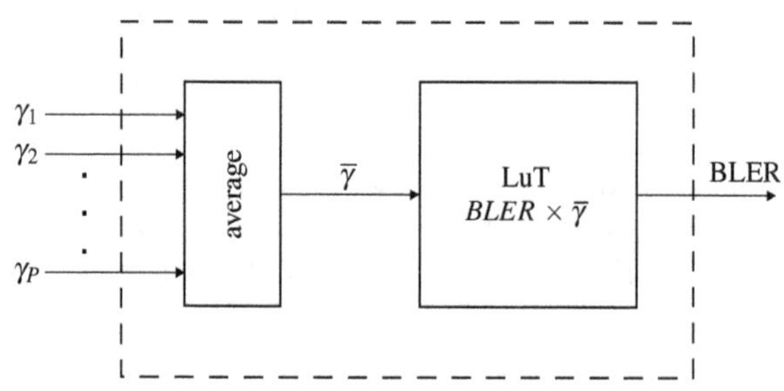

Fig. 7.14 AVI scheme.

Still, a two-step version of AVI L2S can be achieved: (i) to perform the mapping $\gamma \rightarrow RBER$; (ii) to perform the mapping $RBER \rightarrow BLER$. In doing so, the first mapping is performed independently for each element of γ, creating a new vector for RBER values. By using the two-step alternative approach, the average is computed over the RBER vector, instead of γ.

It is worth of notice that AVI is unable to capture variance of channel quality. Any variation on the channel quality within the block length is insufficiently detected by examining alterations on the mean value average. Nevertheless, AVI approach is appropriate to AWGN or even slow fading channels. In realistic scenarios, the neglect of the actual pattern of channel quality results in accuracy loss. Then, distinct disturbances would mislead the performance evaluation due to the averaging operation.

Example 7.1. Designing a single-step AVI for HSDPA system simulator could be a very clarifying exercise, especially if the H-ARQ operation with chase combining (CC) scheme is enabled. The block length is three, referring to the number of slots comprised within the TTI of 2 ms. Thus, three SNR samples will be taken for obtaining the BLER estimate. These operations must be performed for every transmission or retransmission, provided that adequate LuTs are employed. In CC scheme, a very simple modeling is allowable: there are not models for retransmission on the LL simulator and the H-ARQ model is inserted only in SL. For each look-up operation, a random test made in SL will say whether the block has been erroneously received. In affirmative case, a retransmission will be triggered. Now the look-up operation will use not only the current signal energy to compose the SNR inputs, but the energy accumulated since the original transmission. As further benefit that approach allows the SL to schedule independently each transmission replica.

7.5.2 Actual Value Interface

The principle of AcVI is similar to the AVI, but it captures abrupt changes in channel quality as well. There are infinite distributions of SNR values which possess the same SNR mean value. Hence, the same LL performance cannot be expected for a situation in which the SNR fluctuates considerably within the block length and another in which the SNR remains constant over the same mean value. In fact, the distribution of errors within a block length affects the channel decoder performance, even if the same mean channel quality is kept. For instance, many decoding algorithms do not perform well when raw bit errors occur in bursts.

The AcVI approach attempts to overcome such AVI drawbacks. AcVI is sensible to fast changes on the channel quality, for example, due to RRM decisions, fast fading, and sudden interference. There are a number of ways to implement the AcVI, from alternatives of average calculations to inclusion of auxiliary statistics. For instance, the geometric average might be employed to extract the quality representing $\bar{\gamma}$ [20], since any element diverting from the arithmetic mean value will penalize the BLER estimate.

Another alternative is to use direct estimates of standard deviation besides arithmetic mean value (see Fig. 7.15). Usually AcVI [35] is achieved in two steps: first, each SINR value is mapped to an RBER; afterward, estimates of both the mean RBER ($\hat{\mu}_{RBER}$) and corresponding standard deviation, $\hat{\sigma}_{RBER}$, of the RBER vector are evaluated. The mean and standard deviation estimates are then used for obtaining the corresponding FER or BLER value.

In this way, the effects of fast multipath fading and frequency hopping can be captured at SL with better accuracy when compared to AVI.

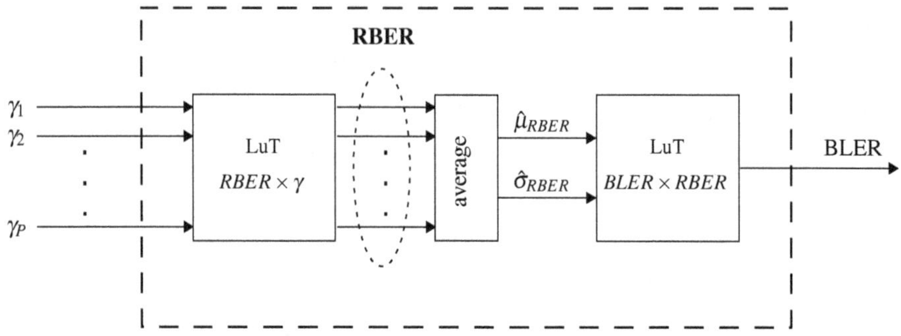

Fig. 7.15 Two-step AcVI scheme.

7.5.2.1 Case Study: AVI and AcVI for GSM

The GSM standard specifies the multiple access radio system based on a combination of time and frequency division multiple access technology, leading to a gross bit rate of around 270 kbps for a Gaussian minimum shift keying modulation with time–bandwidth product BT = 0.3. The carrier spacing is 200 kHz and the time division multiple access (TDMA) frame has a period of 4.615 ms, which is subdivided into eight slots. Herein, data within one slot correspond to the so-called "burst." Time-slots in a carrier constitutes physical channels; whereas a logical channel specifies the type of information carried by the physical channel. A more detailed description of the GSM physical layer is given in Chapter 2 of this book.

In this case study, the design of the AcVI in interference-limited scenarios is illustrated by using an LL simulator of GSM system, therefore, the SIR is used as the channel quality. The LL simulator is fully based on 3GPP standards [1, 2]. The AMR speech codec at 5.9 kbps (MR59FR) is employed and random frequency hopping is enabled.

The AVI is implemented in a single step so as to perform simple comparison. In Fig. 7.16 a LuT that can be used with AVI for scenarios of vehicular mobility on two distinct configurations is represented: with estimates of channel or perfect knowledge. The averaged SIR is mapped directly to FER without any other calculation or mapping. Note that the LuT of single-step L2S interface is dependent on both the environment and service.

When considering the two-step AcVI, in Fig. 7.17(a) a valid LuT to perform the first step of the mapping procedure is shown. The LuTs of the first step depend on the environment and physical layer improvements, but it is service independent. For the second step, it is necessary to calculate the mean and standard deviation estimates of the RBER. As derived in [35] for the GSM speech frame, those estimates can be given by

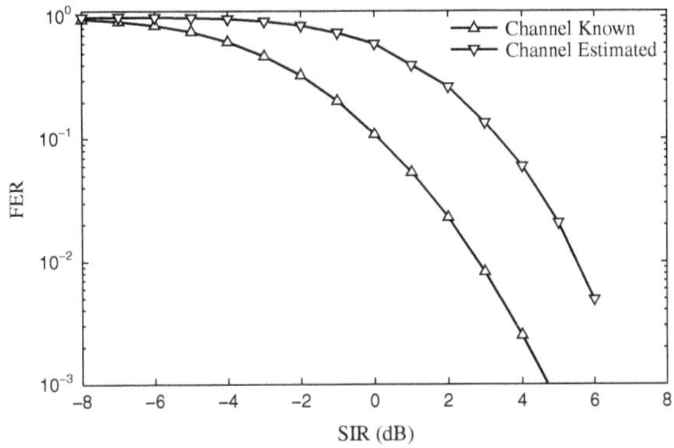

Fig. 7.16 Representations of LuT of single-step AVI in GSM.

$$\hat{\mu}_{RBER} = \frac{1}{8}\sum_{i=1}^{8} p_i, \tag{7.1a}$$

$$\hat{\sigma}_{RBER} = \sqrt{\frac{1}{912}\sum_{i=1}^{8} p_i(1-p_i) + \frac{1}{7}\sum_{i=1}^{8}(p_i - \hat{\mu}_{RBER})^2}, \tag{7.1b}$$

where p_i is the RBER value of the ith burst obtained in the first step.

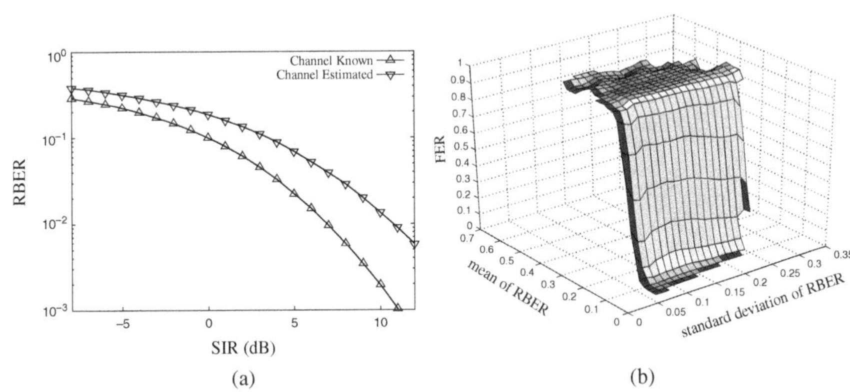

Fig. 7.17 LuTs for (**a**) the first and (**b**) the second steps of AcVI in GSM.

The sporadic behavior verified in the RBER values associated with each burst is exploited by channel coding schemes, whose performance depends on the stochastic characteristics of the errors introduced by the channel. When using AcVI, such effects can be taken into account on the SL evaluation as well. In Fig. 7.17(b) a smoothed map between the pairs of the mean and standard deviation of the RBER and FER when performing the second step is illustrated.

AcVI presents better accuracy at the expense of higher computational cost to obtain the LuT as well as in their usage by the SL part. The LL simulation campaigns must generate a large pattern of mean and standard deviation of RBER. The SL part must calculate the estimates of mean and standard deviation for each speech frame (or 8 bursts), which can become critical as the number of look-up operations grow.

The SL simulations can provide insightful elements when comparing AVI and AcVI. In this regard, a separate dynamic simulation tool was employed, which models a regular macro-cellular environment with several base-station subsystems that can be organized according to different frequency reuse. Both interfaces make use of the same set of LL results. More details of GSM system-level modeling can be obtained in Chapter 2.

The voice traffic model follows an on–off pattern given by a two-state Markov chain. A voice activity of 60% is assumed. In this work 12 hopping frequencies and an aggressive unitary frequency reuse pattern were used. In addition, the power control (PC) is implemented considering SIR target of 14 dB and updating period of 20 ms for the up-down algorithm.

In Fig. 7.18 the capacity results are illustrated. The AcVI and AVI curves are shown for both power-controlled and non-power-controlled voice bearers. It can be seen that AcVI presented a more conservative performance in both scenarios, achieving capacity results lower than those of the AVI.

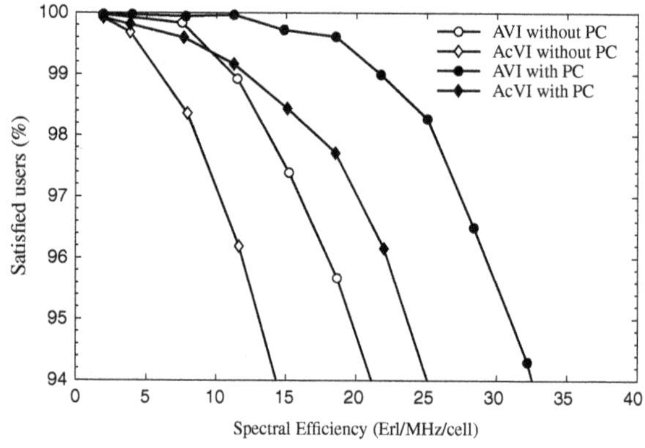

Fig. 7.18 System-level MR59FR performance results for the AcVI and AVI.

7.5.3 Orthogonality Factor-Based Interface

AcVI captures variability in the channel quality, however, the influence of each of the various multipath components cannot be taken into account. If the signal power distribution in each path is relevant to the whole performance of the receiver, a more elaborated L2S interface should be designed. A straightforward solution would be simulating the fading mechanism in both LL and SL with the same parameters and interchanging the complete information about channel state – all current channel complex coefficients together with noise and interference parameters – through the L2S interface. However, the LL simulator would depend on a multidimensional interpolation of potentially high complexity. The L2S interface based on orthogonality factor (OF) represents a simple and efficient manner to deal with multipath fading channels, especially for CDMA-based technologies.

Due to the synchronism of the downlink transmission among users in the same cell, it is common to use orthogonal codes for the user separation in WCDMA or high-speed packet access (HSPA). Nevertheless, the perfect orthogonality is lost in the presence of radio link distortions. By using rake receiver, the loss of orthogonality implies intracell and interpath interference. The effect of using orthogonal codes in WCDMA systems over fading channels was addressed in [14], assuming a conventional rake receiver with MRC.

The term orthogonality factor has been extensively used in the literature just as a measure of the degree of orthogonality between received signals [15, 22, 32, 38, 39]. Some authors use *orthogonality loss factor* and *orthogonality factor* interchangeably. The OF is a time-varying parameter that depends on the instantaneous multipath gain, thus, distinct users have independent OF for a given instant. Moreover, the time-averaged OF is useful when performing simple capacity assessments at SL [44].

The OF is defined as the ratio between the faded signal power and the faded interference power. Variants of such definition as well as several analytical expressions for the OF have been proposed [8, 15, 22, 31, 32, 38, 39], which may differ regarding the extent of simplifications. The mathematical expression adopted in this work for OF is based on the papers of Pedersen and Mogensen [39], Seeger et al. [43] and Passerini and Falciasecca [37], since they propose an advantageous trade-off between complexity and usefulness. Below the OF is represented as α_o:

$$\alpha_o = 1 - \left[\sum_{\ell} \frac{|h_\ell|^2}{\Psi - |h_\ell|^2} \right]^{-1}, \tag{7.2}$$

where h_ℓ is the complex coefficient of the ℓth path and Ψ is the instantaneous multipath gain given by $\Psi = \sum_m |h_m|^2$.

The α_o values are real and non-negative numbers and it can be analytically proved that (i) in the limit, in which only one path contributes to gain Ψ, the maximum value of α_o is one and (ii) the minimum value of α_o is $\alpha_o^{(min)} = 1/L$ and it occurs when all path gains are equal, i.e., $|h_\ell|^2 = \Psi/L, \forall \ell$, assuming a propagation channel with $L > 1$ paths. Notice that it is the same situation of maximal diversity.

A simple example that illustrates the meaning of the OF is depicted in Fig. 7.19: the α_o function for a channel with two paths is plotted for a generic Ψ. As explained earlier, the $\alpha_o^{(min)}$ value occurs when the two paths' gains are equal and does not depend on the Ψ value; therefore, a lower limit is determined by a straight line $[|h_1|^2 = |h_2|^2 = \Psi/2]$ with constant α_o value $(\alpha_o^{(min)} = 1/2)$.

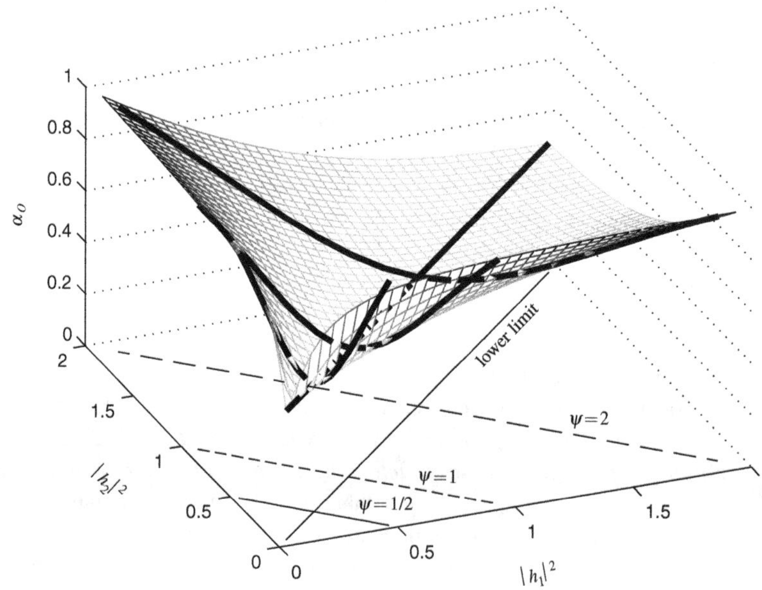

Fig. 7.19 Illustrative example of orthogonality factor function for radio profile embodying two paths.

Using this OF definition, it is possible to combine the envelope information of all paths in one single metric. The simultaneous occurrence of deep fadings in $L-1$ paths leads the OF value to be close to maximum, while OF value close to $1/L$ reflects a situation where all L paths have high gain. Intermediate cases are possible as well, the OF's behavior being mainly imposed by the strongest path.

VOFI is illustrated in Fig. 7.20. The LL simulator combines OF values and mean SNR into mapping pairs so as to derive the corresponding LuT. Such pairs characterize the channel state and identify error occurrences. An extensive LL simulation campaign will be necessary to place the channel state pairs (SNR,α_o) in a broad and dense range. From there on, the LuT of step one is obtained from the computation of RBER over narrow intervals of α_o values, whose center is given by $\overline{\alpha_o}$. The LuT of step two is obtained simply from the observed relation between the average RBER and BLER values.

7.5.3.1 Case Study: VOFI for HSDPA

The HSDPA LL simulator of Section 7.3.5 is used hereafter to derive the LuTs. Table 7.2 summarizes the configuration parameters of the HSDPA LL simulator. A rake receiver with MRC, which processes up to six paths, is used. Three HS-PDSCHs presenting CQI 10 of UE category 5 is employed as well. It is worthy to say that retransmissions are not modeled and perfect knowledge about the channel and receiver synchronism is assumed.

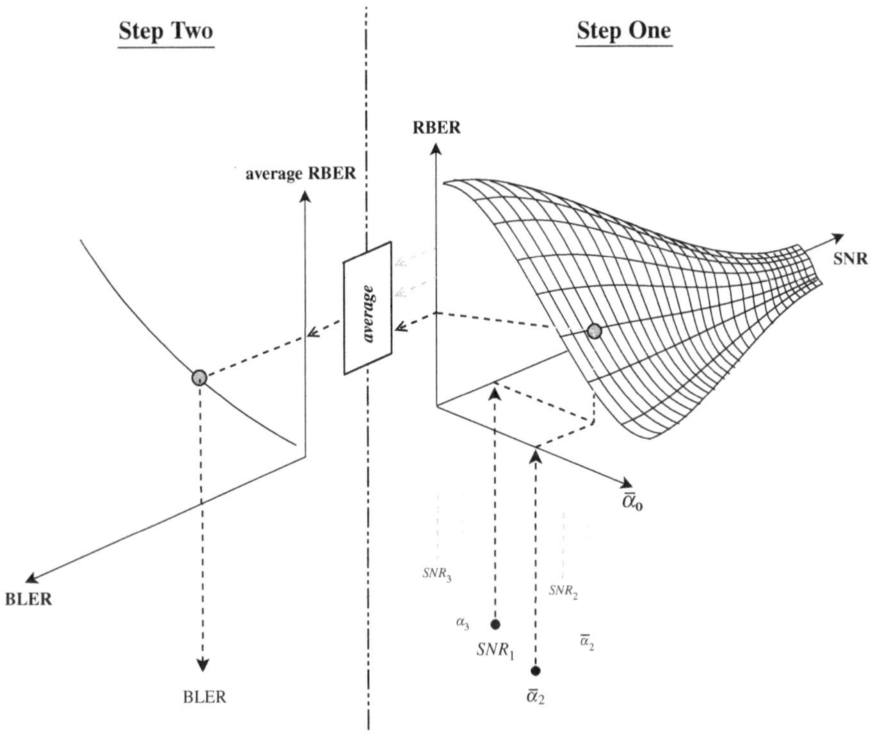

Fig. 7.20 Simplified mechanism of variable orthogonality factor interface.

Table 7.2 link-level HSDPA simulator configuration parameters

Parameter	Value
UE category	5
CQI value	10
Receiver	Rake MRC with 6 fingers
Turbo decoding algorithm	SOVA with 6 iterations
Channel model	Pedestrian B
Mobile speed (km/h)	3.0
Carrier frequency (GHz)	1.95
Number of iterations per E_c/N_0	10,000

In Fig. 7.21, the behavior of the RBER, for an average $E_c/N_0 = -10$ dB, and the average orthogonality factor taken at transmission time intervals (TTIs), where a close inverse relationship between them is notorious are shown.

A large range of OF values must be obtained repetitively so as to create LuTs. While high OF values are less common, the lower values are obtained with more precision. The histogram of OF samples obtained from the full simulation is shown in Fig. 7.22.

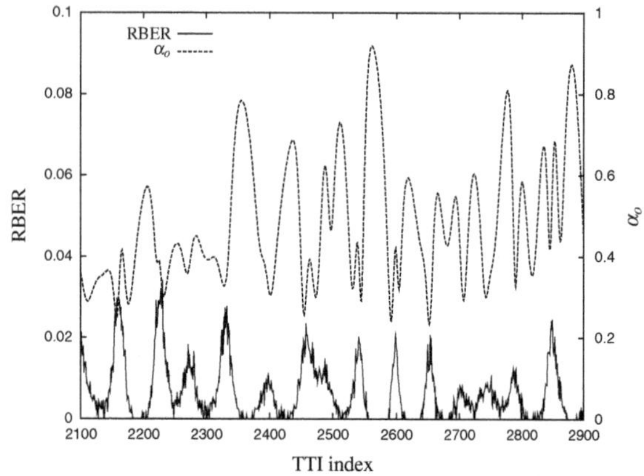

Fig. 7.21 Behavior of the RBER and average OF in time, for an average $E_c/N_0 = -10$ dB.

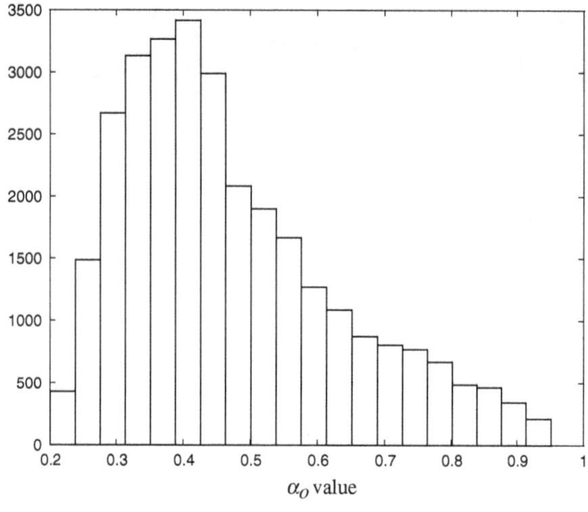

Fig. 7.22 Histogram of obtained OF samples.

In order to obtain the LuTs, the RBER is organized as a function of sorted pairs of SNR and OF that are measured in a time-slot basis. The resulting LuT for the performed simulations is shown in Fig. 7.23(a). The *BLER* $\times$ average RBER LuT is obtained following similar procedure. Figure 7.23(b) illustrates this LuT for the performed simulations. The monotonicity of RBER with respect to OF and SNR axes, together with the monotonicity of BLER with the average RBER axis, allows straightforward application of this kind of LuT to carry out L2S interfacing.

7.5.4 Effective SINR mapping (ESM)

ESM attempts to efficiently deal with multiple input variables in L2S interfaces. MIMO and OFDM systems are the main scenarios commonly seized by ESM, where an effective channel quality is used instead of multiple sub-channel or sub-

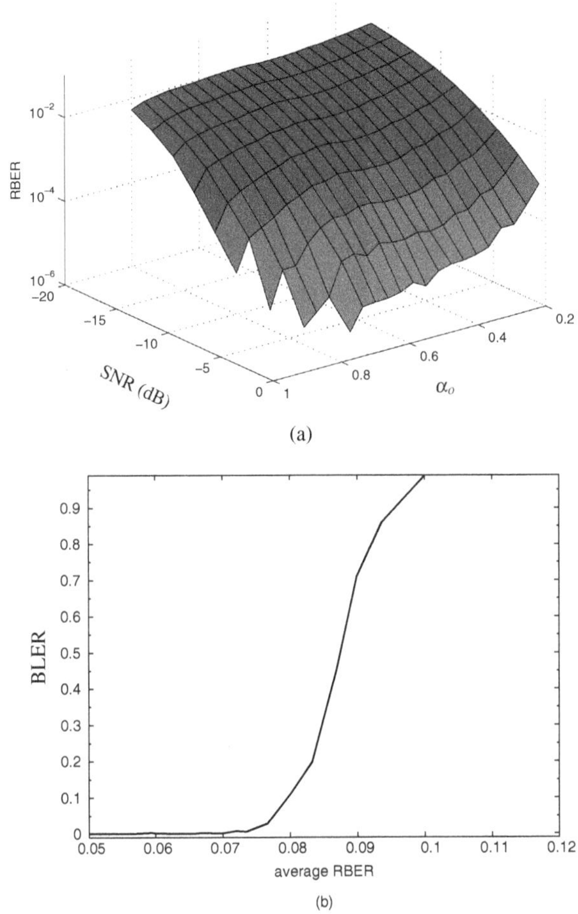

Fig. 7.23 LuTs for (**a**) the first and (**b**) the second steps of VOFI in HSDPA.

carrier qualities. Then, an artificial mapping can provide good BLER estimates [11]. Commonly, the curve of LL performance under an AWGN channel is sufficient for that purpose, as explained in the following paragraphs.

The operational principle of ESM can be organized in two stages, as it can be seen from Fig. 7.24: SINR compression and quality mapping. Initially, an effective SINR is found comprising all multiple states of the channel. Subsequently, the effective SINR value is mapped to a performance metric such as BLER or FER in a single-state channel, using, e.g., an AWGN channel performance curve.

The main goal of the SINR compression stage is to compact information about the multiple states of the channel into a single metric – the effective SINR, γ_{eff} is given by [11]

$$\gamma_{eff} = \alpha \cdot I^{-1}\left[\sum_{k=1}^{K} I\left(\frac{\gamma_k}{\beta}\right)\right], \tag{7.3}$$

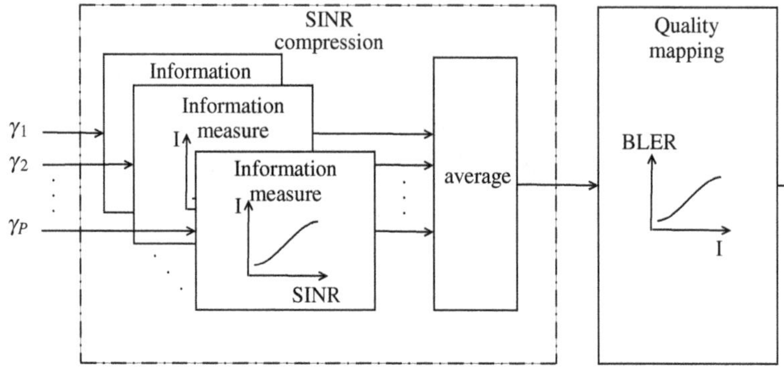

Fig. 7.24 Principle of effective SINR mapping.

where $I(\cdot)$ is the "information measure" $(\cdot)^{-1}$ is the inverse function, K is the number of subcarriers (or sub-channels), γ_k is the SNR (or SIR or SINR) of the kth sub-carrier, α and β are parameters that allow to adapt the model.

After the compression stage, a simple AWGN curve can be employed for the quality mapping stage [11, 54]. Few versions of ESM were published [11, 53], such as capacity ESM (CESM), cutoff rate ESM (CRESM), linear ESM (LiESM), exponential ESM (EESM), logarithmic ESM (LESM), and mutual-information ESM (MIESM).

For example, in the EESM the information measure is given by

$$I_{EESM}(\gamma_k) = 1 - \exp(-\gamma_k). \tag{7.4}$$

Then, replacing (7.4) in (7.3), γ_{eff} can be obtained by

$$\gamma_{eff} = -\alpha \ln \left[\frac{1}{K} \sum_{k=1}^{K} \exp\left(-\frac{\gamma_k}{\beta} \right) \right]. \tag{7.5}$$

In the process of the L2S interface specification, the scaling factor parameters α and β should be adjusted to calibrate the interface, increasing the accuracy of the whole composition of information measure and quality mapping. These parameters should be related to the modulation cardinality and the channel coding scheme. Thus, each pair of parameters is stipulated for only one MCS via suitable calibration of ESM.

Another approach that has been attracting much attention is the MIESM, which considers the mutual information of the channel. Presenting sigmoidal shape, the mutual information is suitable for models comprising efficient channel codes, since the amount of information that a channel can pass should be bounded [53]. For the sake of simplicity, only EESM is dealt in the following subsections. For details about MIESM, the reader is referred to [13, 19, 53, 57].

7.5.4.1 Calibration of ESM

The appropriate realization of L2S interface demands the compression and subsequent compilation of the artificial curve. Further calibration of the ESM L2S interface may be exploited in order to ensure the proper functioning of the interface. The calibration is achieved through the proper adjustment of its parameters.

Let a set of pairs $(BLER_{act}, \mathbf{SNR}_{act})$ denote the *actual performance surface*, where $\mathbf{SNR}_{act}$ is the vector of actual channel quality values, and another set given by $(BLER_{ref}, SNR_{ref})$ denote a *reference curve* obtained, for instance, under AWGN channel and using the same modulation and coding scheme (MCS). On the assessment of actual performance, the SNR_{eff} values are calculated, thus obtaining the *actual curve* given by the set of pairs $(BLER_{act}, SNR_{eff})$. Furthermore, the estimate $BLER_{pred}$ could be obtained by mapping SNR_{eff} into the reference curve. The set of pairs $(BLER_{pred}, SNR_{eff})$ comprises the *predicted curve*. The main goal is to obtain $BLER_{pred}$ values as close to $BLER_{act}$ ones as possible. For that matter, the L2S interface must be calibrated.

When using the EESM interface, the scaling factor parameters, α and β, are adjusted so as to reduce the mismatch between the actual and the predicted curves. In turn, it is important to highlight the dependence of the predicted curve with respect to scaling factor parameters, i.e., the predicted curve is drawn by the pairs

$$(BLER_{pred}(\alpha,\beta), SNR_{eff}(\alpha,\beta)).$$

The fitting of predicted curve to the actual one can be given in terms of a least squares metric:

$$\min_{\alpha,\beta} \sum_{c=1}^{C} |\Delta e_c(\alpha,\beta)|^2, \tag{7.6}$$

where C denotes the number of different realizations taken into account, which must be a large value, and Δe_c is the adjustment metric, which will determine the kind of fitting. The fitting can be driven either vertically by observing BLER [11]

$$\Delta e_c(\alpha,\beta) = BLER_{act} - BLER_{pred}(\alpha,\beta) \tag{7.7}$$

or horizontally by directly observing SNR values [58]:

$$\Delta e_c(\alpha,\beta) = SNR_{ref} - SNR_{eff}(\alpha,\beta). \tag{7.8}$$

Generally, not all collected data are relevant or sufficiently reliable and thus it is advisable to limit the set of data to a certain BLER interval [11].

Consider the BLER fitting to illustrate the process. Note that using (7.7), a poor fit at lower BLER values is obtained. Alternatively, two modified versions of such expression can be used: the normalized and the logarithmic ones, given by

$$\Delta e_c(\alpha,\beta)^{(n)} = \frac{BLER_{act} - BLER_{pred}(\alpha,\beta)}{BLER_{act}}, \tag{7.9a}$$

$$\Delta e_c(\alpha,\beta)^{(log)} = log(BLER_{act}) - log(BLER_{pred}(\alpha,\beta)). \tag{7.9b}$$

The modified versions can yield a better fit over the entire BLER region of interest [11]. Similar derivation can be also performed to address the SNR-based fitting (7.8) [58].

In Fig. 7.25(a) and (b) different effects on ESM calibration procedure considering log-BLER fitting (7.9b) and performing separate adjustment of α and β parameters are shown. The predicted curve obtained by adjusting each parameter separately is plotted besides actual curve and reference hypothetical curve. While the α parameter change is reflected in a horizontal shift of ESM curve, the β value can also slightly modify its shape.

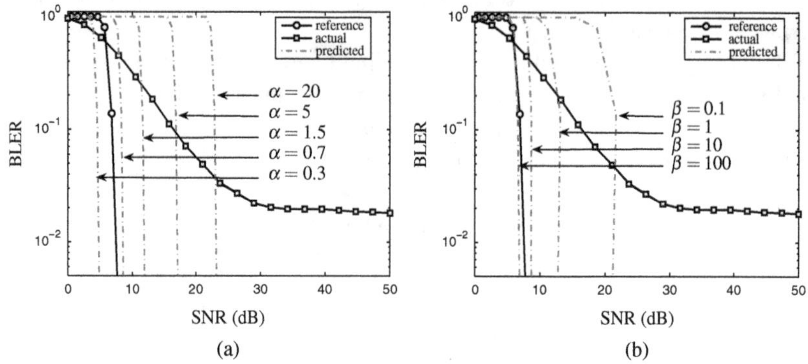

Fig. 7.25 Effect of variation of (**a**) α parameter while β is constant or (**b**) β parameter while α is constant on EESM interface.

7.5.4.2 Case Study: EESM for 3GPP's Long-Term Evolution

Here a simplified LL simulator of 3GPP's long-term evolution (LTE) is used, whose transmission chain is simply composed of CRC attachment, turbo encoder, and modulation mapper. The multicarrier channel is simulated by means of several uncorrelated fading channels, one for each subcarrier.

The EESM L2S interface is evaluated by means of extensive simulation campaign of a noise-limited scenario. The main parameters are shown in Table 7.3.

Messages are randomly generated rather than applying the standardized rate matching. Equation (7.10) is used to calculate the message length:

$$N_{message} = (N_s \cdot \log_2 M \cdot K \cdot R_c) - N_{CRC}, \tag{7.10}$$

where N_s is the number of OFDM symbols, M is the modulation order, K is the number of subcarriers, R_c is the code rate, and N_{CRC} is the CRC size.

Table 7.3 link-level 3GPP's LTE simulator configuration parameters

Parameter	Value
Size of signal constellation	4,16 or 64
Number of subcarriers	12
Symbols per block	7
CRC size (bits)	16 or 24
Channel coding	Turbo
Coding rate	1/3
Decoding algorithm	SOVA with 8 iterations
TTI (ms)	0.5
Channel models	AWGN and 3GPP ETU
Subcarrier bandwidth (kHz)	15
Carrier frequency (GHz)	2.0
Mobile speed (km/h)	3

First, CRC bits are attached to the original message. Afterward, the message passes through the turbo encoder. These bits are demultiplexed into K parallel streams, and each one mapped to a (most probably complex) symbol stream using some available modulation constellation, viz., QPSK, 16-QAM, and 64-quadrature amplitude modulation (QAM).

The simulated channel models are the 3GPP extended typical urban (ETU) [7] and AWGN models. The fading effect is generated for each subcarrier by performing 40 independent realizations each with 10 TTIs of 0.5 ms. In AWGN channel the instantaneous SNR of the K subcarriers are preserved, whereas in the ETU the instantaneous SNR changes each symbol, in both time and frequency domains. In each TTI, N_{simb} OFDM symbols are transmitted.

Figure 7.26 shows the performance in terms of BLER for QPSK modulation. The predicted curves are obtained after the EESM calibration, where $\alpha = \beta$ is 0.32. The effect of variation of parameter α or β at a time is illustrated in Fig. 7.27 for 64-QAM, starting on with $\alpha = \beta = 1.3$.

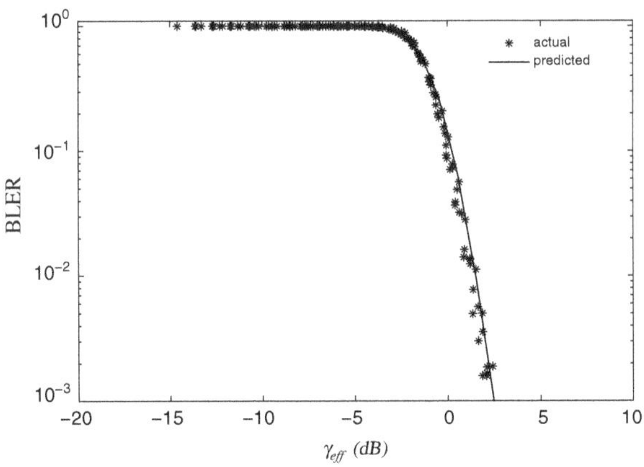

Fig. 7.26 Application of EESM on 3GPP LTE with QPSK modulation.

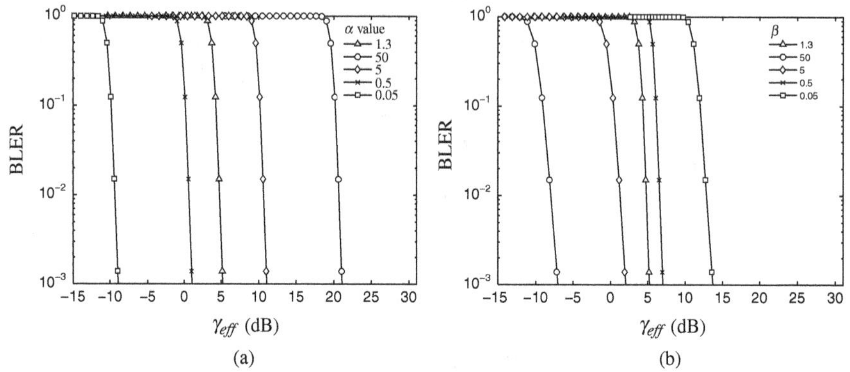

Fig. 7.27 Effect of variation of (**a**) α parameter with fixed $\beta = 1.3$ or (**b**) β parameter with fixed $\alpha = 1.3$ on EESM in LTE system using 64-QAM modulation.

7.6 Conclusions and Research Directions

This chapter not only addresses the development of link-level (LL) simulation tools but also investigates distinguished approaches for interfacing them with system-level (SL) simulators. Succinctly, the general aspects concerning the link-level (LL) modeling are discussed and the most common functional blocks of a typical simulator are detailed. In this regard, special attention is dedicated to models related to 3rd. Generation Partnership Project (3GPP) standardized systems. Indeed, a multitude of suitable approaches are available in order to pursue such simulation tools and link-to-system-level (L2S) interfaces, this work attempted to summarize the most prominent strategies to develop both processes.

The link-level software development framework (LSDF) is introduced as an effective strategy for developing LL simulator. The LSDF relies on object-oriented programming (OOP) concepts, intending to systematize the simulator implementation, and has both modularity and reusability as the foremost objectives. Modularity inherently provides development independence among programmers and tractability through development stages. Reusability mainly allows for time-saving during the implementation, since components previously developed can be readily re-utilized.

Additionally, the major types of L2S interfaces are described, viz., average value interface (AVI), actual value interface (AcVI), variable orthogonality factor interface (VOFI), and Effective SINR mapping (ESM), thereafter pertinent case studies are investigated, where performance results for the most influential interfaces are presented. Furthermore, it is observed that obtaining an efficient L2S interface demands more than simply choosing the L2S mapping accordingly. Decisions on the model details and in which part of the simulator the interface is inserted account for most of the incurred complexity and achievable accuracy of the complete communication chain simulation.

There are still few straightforward topics for further developments of wireless communication simulators. One of them is to generalize the LSDF to handle simultaneously multiple transmission and reception chains as interrelated processes. For instance, this will allow for simulating multiple asynchronous users (referring to the simulation of a typical uplink) in a more natural way. Yet there are advisable techniques that can be conveniently applied in order to streamline the LSDF overall: (i) the utilization of generic programming can be extended using parameterized data types and algorithms to improve performance and favor generality; (ii) regarding the creational patterns, abstract and prototype factories provide protection and further control when the instantiation of specific data types is assumed a critical task; (iii) considering the structural patterns, the bridge pattern may be effectively used to decouple abstraction and implementation, while allowing them to vary independently.

L2S interfaces were originally conceived to generate performance information to be used in system-level simulations. The applicability of such interfaces has been augmented recently and they also supply intermediate steps for radio resource management (RRM) or link adaptation (LA) algorithms. Consequently, intrinsic inaccuracies of these interfaces are more relevant, since they are propagated to subsequent steps of the simulation. All in all, the search for simple, flexible, and accurate methods to perform the whole communication chain remains a fertile research field.

References

1. 3GPP: Channel coding. Technical report, TS 45.003 v6.6.0 – Release 6 (2004)
2. 3GPP: Multiplexing and multiple access on the radio path. Technical report, TS 45.002 v6.7.0 – Release 6 (2004)
3. 3GPP: Multiplexing and channel coding (FDD). Technical Report, TS 25.212 v5.10.0 – Release 5, 3rd Generation Partnership Project, Sophia Antipolis, France (2005). URL `http://www.3gpp.org`
4. 3GPP: Physical channels and mapping of transport channels onto physical channels (FDD). Technical Report, TS 25.211 v5.8.0 – Release 5, 3rd Generation Partnership Project, Sophia Antipolis, France (2005). URL `http://www.3gpp.org`
5. 3GPP: Physical layer procedures (FDD). Technical Report, TS 25.214 v5.11.0 – Release 5, 3rd Generation Partnership Project, Sophia Antipolis, France (2005). URL `http://www.3gpp.org`
6. 3GPP: Spreading and modulation (FDD). Technical Report, TS 25.213 v5.6.0 – Release 5, 3rd Generation Partnership Project, Sophia Antipolis, France (2005). URL `http://www.3gpp.org`
7. 3GPP: User equipment (UE) radio transmission and reception. Technical Report, TS 36.101 v8.3.0 – Release 8, 3rd Generation Partnership Project, Sophia Antipolis, France (2008). URL `http://www.3gpp.org`
8. Awoniyi, O., Mehta, N., Greenstein, L.: Characterizing the orthogonality factor in WCDMA downlinks. IEEE Transactions on Wireless Communications **2**(4), 621–625 (2003)
9. Berrou, C., Glavieux, A.: Near optimum error correcting coding and decoding: turbo-codes. IEEE Transactions on Communications **44**(10), 1261–1271 (1996)
10. Berrou, C., Glavieux, A., Thitimajshima, P.: Near Shannon limit error-correcting coding and decoding: turbo-codes. IEEE International Conference on Communications **2**, 1064–1070 (1993)
11. Brueninghaus, K., Astély, D., Sälzer, T., Visuri, S., Alexiou, A., Karger, S., Seraji, G.A.: Link performance models for system level simulations of broadband radio access systems. IEEE International Symposium on Personal, Indoor and Mobile Radio Communications **4**, 2306–2311 (2005)

12. Cain, J.B., Clark, G.C., Geist, J.M.: Punctured convolutional codes of rate (n-1)/n and simplified maximum likelihood decoding. IEEE Transactions on Information Theory **IT-25**(1), 97–100 (1979)
13. Chen, X., Wan, L., Gao, Z., Fei, Z., Kuang, J.: The application of EESM and MI-based link quality models for rate compatible LDPC codes. In: IEEE Vehicular Technology Conference, pp. 1288–1292 (2007)
14. DaSilva, V., Sousa, E., Jovanović, V.: Performance of the forward link of a CDMA cellular network. In: IEEE International Symposium on Spread Spectrum Techniques and Applications pp. 213–217 (1994)
15. Droste, H., Beyer, H.: Distributions of orthogonality factor and multipath gain of the UMTS downlink obtained by measurement based simulations. IEEE Vehicular Technology Conference **1**, 411–415 (2005)
16. Free Software Foundation: IT++ – Scientific Library (2001). URL `http://itpp.sourceforge.net`. Accessed on August 17, 2008
17. Furuskär, A.: Radio resource sharing and bearer service allocation for multi-bearer service, multi-access wireless networks – methods to improve capacity. Ph.D. thesis, Royal Institute of Technology, Stockholm, Sweden (2003)
18. Gamma, E., Helm, R., Johnson, R., Vlissides, J.: Design patterns: elements of reusable object-oriented software, 1st edn. Addison-Wesley Professional (1995)
19. He, X., Niu, K., He, Z., Lin, J.: Link layer abstraction in MIMO-OFDM system. In: International Workshop on Cross Layer Design, pp. 41–44 (2007)
20. Holma, H.: A study of UMTS terrestrial radio access performance. Ph.D. thesis, Helsinki University of Technology, Espoo, Finland (2003)
21. Holma, H., Toskala, A. (eds.): WCDMA for UMTS: radio access for third generation mobile communications, 3rd edn. Wiley (2004)
22. Hunukumbure, M., Beach, M., Allen, B.: Downlink orthogonality factor in UTRA FDD systems. Electronics Letters **38**(4), 196–197 (2002)
23. Huy, D., Legouable, R., Kténas, D., Brunel, L., Assaad, M.: Downlink B3G MIMO OFDMA link and system level performance. In: IEEE Vehicular Technology Conference pp. 1975–1979 (2008)
24. Jakes, W.C.: Mobile microwave communication. Wiley, New York (1974)
25. Jeruchim, M.: Techniques for estimating the bit error rate in the simulation of digital communication systems. IEEE Journal on Selected Areas in Communications **2**(1), 153–170 (1984)
26. Jeruchim, M.C., Balaban, P., Shanmugan, K.S.: Simulation of communication systems: modeling, methodology and techniques, 1st edn. Kluwer Academic (2000)
27. Korhonen, J.: Introduction to 3G mobile communications, 2nd edn. Artech House, Inc., Norwood, MA, USA (2003)
28. Laiho, J., Wacker, A., Novosad, T.: Radio network planning and optimisation for UMTS, 1st edn. Wiley (2002)
29. Li, Y., Huang, X.: The simulation of independent Rayleigh faders. IEEE Transactions on Communications **50**(9), 1503–1514 (2002)
30. Malkamäki, E., de Ryck, F., Mourot, C., Urie, A.: A method for combining radio link simulations and system simulations for a slow frequency hopped cellular system. IEEE Vehicular Technology Conference **2**, 1145–1149 (1994)
31. Mehta, N.B., Greenstein, L.J., Willis, T.M., Kostic, Z.: Analysis and results for the orthogonality factor in WCDMA downlinks. IEEE Transactions on Wireless Communications **2**(6), 1138–1149 (2003)
32. Mehta, N., Molisch, A., Greenstein, L.: Orthogonality factor in WCDMA downlinks in urban macrocellular environments. In: IEEE Global Communications Conference **6** (2005)
33. Meszaros, G.: xUnit test patterns: refactoring test code, 1st edn. Addison-Wesley Signature Series. Addison-Wesley (2007)
34. Morrow, R.K., Lehnert, J.S.: Bit-to-bit error dependence in slotted DS/SSMA packet systems with random signature sequences. IEEE Transactions on Communications **37**(10), 1052 – 1061 (1989)

35. Olofsson, H., Almgren, M., Johansson, C., Höök, M., Kronestedt, F.: Improved interface between link level and system level simulations applied to GSM. IEEE 6th International Conference on Universal Personal Communications **1**, 79–83 (1997)
36. Parkvall, S., Englund, E., Malm, P., Hedberg, T., Persson, M., Peisa, J.: WCDMA evolved-high speed packet data services. Ericsson Review **2**, 56–65 (2003)
37. Passerini, C., Falciasecca, G.: Modeling of orthogonality factor using ray-tracing predictions. IEEE Transactions on Wireless Communications **3**(6), 2051–2059 (2004)
38. Passerini, C., Falciasecca, G., Bordoni, F.: Correlation between delay-spread and orthogonality factor in urban environments. Electronics Letters **37**(6), 384–386 (2001)
39. Pedersen, K., Mogensen, P.: The downlink orthogonality factors influence on WCDMA system performance. IEEE Vehicular Technology Conference **4**, 2061–2065 (2002)
40. Proakis, J.G.: Digital communications, 3rd edn. McGraw-Hill, USA (1995)
41. Pursley, M.B.: Performance evaluation for phase-coded spread-spectrum multiple-access communication – Part I: System analysis. IEEE Transactions on Communications **COM-25**(8), 795–799 (1977)
42. Robertson, P., Villebrun, E., Hoeher, P.: A comparison of optimal and sub-optimal MAP decoding algorithms operating in the log domain. IEEE International Conference on Communications **02**, 1009–1013 (1995)
43. Seeger, A., Sikora, M., Klein, A.: Variable orthogonality factor: a simple interface between link and system level simulation for high speed downlink packet access. IEEE Vehicular Technology Conference **4**, 2531–2534 (2003)
44. Sipila, K., Honkasalo, K., Laiho-Steffens, J., Wacker, A.: Estimation of capacity and required transmission power of WCDMA downlink based on a downlink pole equation. IEEE Vehicular Technology Conference **2**, 1002–1005 (2000)
45. Sklar, B.: A primer on turbo code concepts. IEEE Communications Magazine **35**(12), 94–102 (1997)
46. Sklar, B.: Rayleigh fading channels in mobile digital communication systems. I. Characterization. IEEE Communications Magazine **35**(7), 90–100 (1997)
47. Sklar, B.: Rayleigh fading channels in mobile digital communication systems. II. Mitigation. IEEE Communications Magazine **35**(7), 102–109 (1997)
48. Sklar, B.: Digital communications, 2nd edn. Prentice-Hall, USA (2001)
49. Sklar, B.: How I learned to love the trellis. IEEE Signal Processing Magazine **20**(3), 87– 102 (2003)
50. Smith, J.I.: A computer generating multipath fading simulation for mobile radio. IEEE Transactions on Vehicular Technology **24**(3), 39–40 (1975)
51. Sohn, I., Bang, S.C.: Performance studies of rate matching for WCDMA mobile receiver. In: IEEE VTC-Fall Vehicular Technology Conference, vol. 6, pp. 2661–2665 (2000)
52. Stroustrup, B.: The C++ programming language – special edition, 3rd edn. Addison-Wesley Professional (2000)
53. Tsai, S.S., Soong, A.C.K.: Effective-SNR mapping for modeling frame error rates in multiple-state channels. Tech. Rep. 3GPP2-C30-20030429-010, 3rd Generation Partnership Project 2 (2003)
54. Tuomaala, E., Wang, H.: Effective SINR approach of link to system mapping in OFDM/multi-carrier mobile network. 2nd International Conference on Mobile Technology, Applications and Systems **2** (2005)
55. UMTS: Selection procedures for the choice of radio transmission technologies of the UMTS. Technical Report, UMTS TR 101.112 v.3.2.0, ETSI (1998)
56. Verdú, S.: Multiuser detection. Cambridge University Press, USA (1998)
57. Wan, L., Tsai, S., Almgren, M.: A fading-insensitive performance metric for a unified link to system quality model. In: WCNC2006, Las Vegas, USA (2006)
58. Westman, E.: Calibration and evaluation of the exponential effective SINR mapping (EESM) in 802.16. Master's thesis, Royal Institute of Technology, Stockholm, Sweden (2006)

Chapter 8
Channel Equalization Techniques for Wireless Communications Systems

Cristiano M. Panazio, Aline O. Neves, Renato R. Lopes, and Joao M. T. Romano

8.1 Introduction and Motivation

In bandlimited, high data rate digital communication systems, equalizers are important devices. Their function is to restore the transmitted information, i.e., the information at the channel input, decreasing or eliminating channel interference. A large variety of techniques have been developed in the last 70 years, following the evolution of communication systems.

Initially, researchers were interested in guaranteeing the correct transmission of information between two points, leading to the so-called single-input/single-output (SISO) systems. The foundation of equalization and adaptive filtering was developed in this context. Considering that a communication channel can be modeled as a linear time-invariant (LTI) filter, whose output is added to a noise, the received signal is given by

$$x[n] = \sum_{k=-\infty}^{\infty} h[k]s[n-k] + v[n], \tag{8.1}$$

where $h[n]$ is the channel impulse response, $s[n]$ is the transmitted symbol, and $v[n]$ is the additive white Gaussian noise (AWGN). Rearranging terms to emphasize the presence of the symbol $s[n]$

$$x[n] = h[0]s[n] + \sum_{k=-\infty, k\neq 0}^{\infty} h[k]s[n-k] + v[n] \tag{8.2}$$

enables the observation that the received message is in fact given by the original signal added to noise and to a third term that is a function of delayed versions of the transmitted symbol. This term is the so-called intersymbol interference (ISI). One of the main tasks of an equalizer is to eliminate or at least to reduce its effect, and also that of the noise, so that the desired message can be recovered correctly. In fact, if the equalizer may be implemented as an LTI filter, then a perfect equalization is

F. Cavalcanti, S. Andersson (eds.), *Optimizing Wireless Communication Systems*,
DOI 10.1007/978-1-4419-0155-2_8,

achieved when the following equation is satisfied:

$$y[n] = As[n - \Delta], \tag{8.3}$$

where $y[n]$ is the equalizer output, A is a gain, and Δ is a delay. Note that this solution would only be possible if the convolution between the channel and the equalizer impulse responses resulted in a vector of the form $[0 \ldots 0\ 1\ 0 \ldots 0]$, that is, a null vector except for the position where $n = \Delta$. For this reason, this solution is known as the zero-forcing (ZF) solution. Unfortunately, this solution is often impossible to be attained, specially due to the structures used to model the channel and the equalizer filters. This linear equalization process is exemplified in Fig. 8.1. For channels with deep spectral nulls, only the use of non-linear structures may lead to satisfactory equalization results.

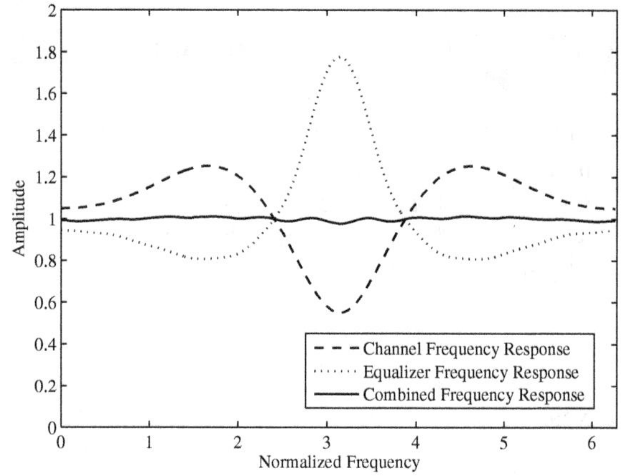

Fig. 8.1 Exemplifying the linear equalization of a channel.

When a wireless transmission is considered, the channel will not only introduce ISI but also something called fading, which results from the destructive interference between multiple paths. In such a context, it is important to take into account the user mobility, which causes a frequency offset due to the Doppler effect and that will cause phase and power fluctuations along the time. Equalizers must adapt to these channel variations. The exploitation of time diversity and/or frequency diversity becomes crucial for attaining good-quality higher data rate transmissions in lower signal-to-noise ratio (SNR). Soon enough, researchers found still another way of increasing quality: the exploitation of space diversity. Instead of transmitting through one antenna, why not using more than one? Or, similarly, if one antenna is used for transmission, why not use more than one to receive the information? This resulted in the so-called multiple-input single-output (MISO) and single-input multiple-output (SIMO) systems. New equalization techniques were proposed leading to important decreases in bit-error rate at the receiver output. Finally, generalizing the mentioned cases, we may consider several antenna for transmission and for reception, leading to the multiple-input multiple-output (MIMO) systems.

Still following the idea of increasing data rates and system capacity, depending on the problem at hand, equalization may not be sufficient to guarantee a good quality in reception. In fact, in practical systems, the use of error-correcting codes (ECC) is essential. In this case, equalization will be concerned with the recovery of the channel input signal, which is given by the coded transmitted symbols, and a decoder device must follow to ensure the data recovery. Forcing a certain interaction between these two devices, it is possible to achieve considerably better solutions than treating each one completely independently. This approach resulted in the so-called turbo-equalizers, which are very much related to turbo-codes.

This chapter is organized as follows. First, a wireless channel model that gives a good approximation of the impairments found in practice is described in Section 8.2. Then the next section gives an overview of equalization techniques, starting with a simple SISO system, where channel and equalizer are modeled by LTI filters. Next, the most commonly employed criteria and algorithms are described for situations in which a training sequence is available, named supervised techniques, and situations in which it is not, named unsupervised techniques. This study will be extended to other equalizer structures, such as the decision-feedback equalizer and the maximum-likelihood sequence estimator in Section 8.4. Section 8.5 will discuss equalization techniques in SIMO systems. Finally, Section 8.6 will extend the study to the joint use of equalization and error-correcting codes, discussing turbo-equalizers and its application.

8.2 Channel Modeling

Since equalizers are developed to deal with the interference inserted by a channel, it would be interesting to first understand how a wireless communication channel can be modeled, before starting the discussion on equalization techniques.

The most important interference in terms of data rate limitation is the ISI, which results from the fact that channels are band limited. Basically, the time response of the channel will be such that previously transmitted symbols will interfere on the current one. The first measure to reduce its effects is to consider a transmission and a receiver shaping filters that form a raised cosine pulse:

$$p(t) = \frac{\text{sinc}\,(t/T)\cos\,(\pi\alpha t/T)}{(1-4\alpha^2 t^2/T^2)}, \tag{8.4}$$

where α is the *roll*-off factor and T is the symbol period.

When considering a wireless communication system, the channel can be modeled using a multipath propagation model in which multipaths may be classified in two groups: those generated by local scatterers and those created by remote scatterers. The local scatterers generate paths that present small propagation delays when compared to the symbol period. For this reason they do not result in inter symbol interference (ISI), but since each path will have a different phase, a destructive interference may occur giving rise to the so-called fading.

In addition, this formulation also needs to account for the user mobility, which causes a frequency offset due to the Doppler effect and that will cause phase and power fluctuations along the time. In this case, some assumptions must be made. First, the local scatterers are disposed as a ring around the mobile user. Therefore, each scattered path will be perceived with a different Doppler frequency. The maximum Doppler frequency experienced is defined by

$$f_d = v f_c / c, \tag{8.5}$$

where v is the mobile speed, f_c is the carrier frequency, and c is the speed of light.

It is also assumed that the scatterers are uniformly distributed in this ring. The angle between the mobile direction of movement and the scatterer is defined as ϕ while the phase of each scattered path is defined as Φ. These two random variables are uniformly distributed over $[0, 2\pi)$. The perceived sum of N scattered paths at the receiver is a random process that is represented by

$$g(t) = N^{-1/2} \sum_{n=1}^{N} \mathrm{e}^{j\{2\pi f_d \cos(\phi[n])t + \Phi[n]\}}, \tag{8.6}$$

where $N^{-1/2}$ is a normalization value so that $\mathbb{E}\{|g(t)|^2\} = 1$.

The remote scatterers, which have their own local scatterers, reflect or diffract the transmitted signal. Due to the longer propagation paths, they generate signal sources with non-negligible delays τ, engendering ISI.

By assuming $L-1$ remote scatterers, the channel impulse response can be written as follows:

$$h(t) = \sum_{l=0}^{L-1} g_l(t) p(t) \delta(t - \tau[l]), \tag{8.7}$$

where $\tau[l]$ is the delay generated by the lth path.

The received signal is then given by

$$x(t) = \sum_{k=-\infty}^{\infty} s[k] h(t - kT) + v(t), \tag{8.8}$$

where $v(t)$ is a zero-mean Gaussian noise of variance σ_v^2.

Now that the channel model is known, the equalization problem and the study of techniques that will enable the reduction or elimination of ISI will be described in the following sections.

8.3 Equalization Criteria and Adaptive Algorithms

Equalization techniques can be classified as supervised or unsupervised. Supervised techniques use a known training sequence to firstly adapt the filter coefficients, searching for the minimum of the criterion given by the mean-squared error (MSE) between the filter output and the known training sequence. After a initial training period, usually the system is switched to a decision-directed mode so that possible channel variations can still be tracked. The main drawback in these techniques is the need of a training sequence, which consumes channel bandwidth and decreases the transmission data rate.

Unsupervised techniques were firstly proposed with the objective of overcoming these drawbacks, avoiding the need of transmitting a known sequence. In this case, criteria are based only on the received signal and on the knowledge of the statistical characteristics of the transmitted signal. Since higher order statistics are necessary, cost functions become multimodal and usually algorithms do not perform as well as in supervised cases.

The following sections describe a review of the most studied and used supervised and unsupervised equalization criteria and their corresponding adaptive algorithms. In all methods, a SISO scenario is considered, modeling the channel and the equalizer by LTI filters.

8.3.1 Supervised Techniques

The foundation of adaptive filtering is represented by two adaptive supervised algorithms that are derived from different but related criteria: the least mean square and the recursive least-squares algorithms. Before describing these two algorithms and others that are derived from them, it is important to describe the optimum linear filtering criteria.

8.3.1.1 The Least Mean Square Method

Consider a discrete time filter with coefficients w_i, $i = 0, \ldots, N_e - 1$. The input signal consists of a discrete wide-sense stationary process, $x[n]$. The filter output can be written as follows:

$$y[n] = \sum_{i=0}^{N_e-1} w_i^*[n]x[n-i] = \mathbf{w}^H[n]\mathbf{x}[n], \tag{8.9}$$

where $\mathbf{w}[n] = [w_0[n]\ w_1[n]\ \ldots\ w_{N_e-1}[n]]^T$ and $\mathbf{x}[n] = [x[n]\ x[n-1]\ \ldots\ x[n-N_e+1]]^T$.

The aim here is to find the filter taps $\mathbf{w}[n]$ so that the filter output signal will be as close as possible, in some sense that will be defined shortly, to a desired signal, $d[n-\Delta]$, where Δ is a constant delay. With this in mind, a natural idea would be to define an error between these two signals

$$e[n] = d[n-\Delta] - y[n], \tag{8.10}$$

and to obtain $\mathbf{w}$ that minimizes a function of this error. A simple and efficient choice is to use, as cost function, the MSE:

$$J_{MSE} = \mathbb{E}\left[|e[n]|^2\right], \tag{8.11}$$

which defines the minimum-mean-square-error (MMSE) criterion also known as the Wiener *criterion*.

Minimizing (8.11) with respect to the filter taps w_i results in the well-known Wiener–Hopf equations:

$$\mathbf{w} = \mathbf{R}_x^{-1}\mathbf{p}_{xd}, \tag{8.12}$$

where $\mathbf{R}_x$ is the autocorrelation matrix of $x[n]$ and $\mathbf{p}_{xd}$ is the cross-correlation vector between $x[n]$ and the desired signal $d[n-\Delta]$. Equation (8.12) gives the optimum coefficient values in the MMSE sense.

In practical situations, solving (8.12) directly may be difficult, since the exact statistics of $x[n]$ are not known, and may also be computationally costly since it involves a matrix inversion. In the search for a simple and efficient iterative way to solve (8.12), Widrow and Hoff, in 1960, proposed that which would become one of the most used and studied algorithms, the least mean square (LMS). The algorithm uses instantaneous estimates of $\mathbf{R}_x$ and $\mathbf{p}_{xd}$ through a stochastic approximation. It can be stated as

$$\mathbf{w}[n+1] = \mathbf{w}[n] + \mu\mathbf{x}[n]e^*[n], \tag{8.13}$$

where $e[n]$ is given by (8.10) and μ is the adaptation step size. Initialization is done considering the equalizer taps equal to zero.

Part of its success can be explained by its simplicity and low computational complexity. In addition, it has very good convergence properties, is robust to noise and to finite precision effects, and can be applied in a large variety of different problems. As expected, the algorithm also presents some limitations. Its convergence is not very fast and depends on the correlation of the input signal.

Observing the error surface generated by (8.11), it can be shown that the contour curves are elliptical and depend on the autocorrelation function of the input signal [23]. For uncorrelated signals, the contour curves will be circular which result in a faster convergence. This is illustrated in Figs. 8.2 and 8.3, where a simple system identification was simulated.

It is also important to mention a well-known modified version of the LMS algorithm, called the normalized least-mean-square algorithm (NLMS). This algorithm corrects a problem of gradient noise enhancement suffered by the original algorithm when the input signal is large. The solution divides the adaptation step size by the Euclidean square norm of $x[n]$ leading to

$$\mathbf{w}[n+1] = \mathbf{w}[n] + \frac{\mu}{\|\mathbf{x}[n]\|^2 + a}\mathbf{x}[n]e^*[n]. \tag{8.14}$$

This algorithm can be viewed as a variable step size least mean square algorithm. A small constant, a, is also usually added to the denominator in order to avoid a large

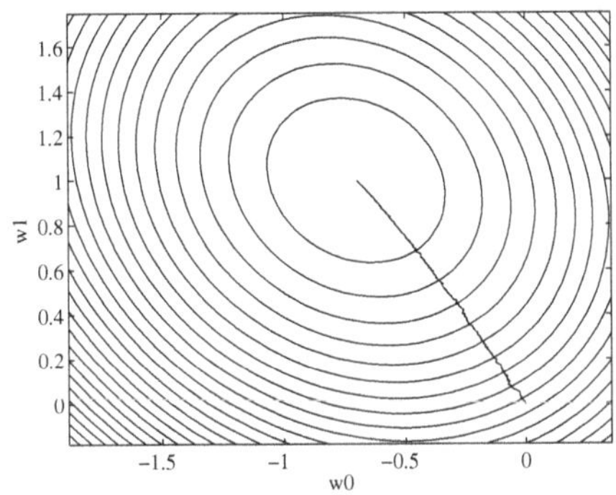

Fig. 8.2 LMS convergence when $x[n]$ is uncorrelated.

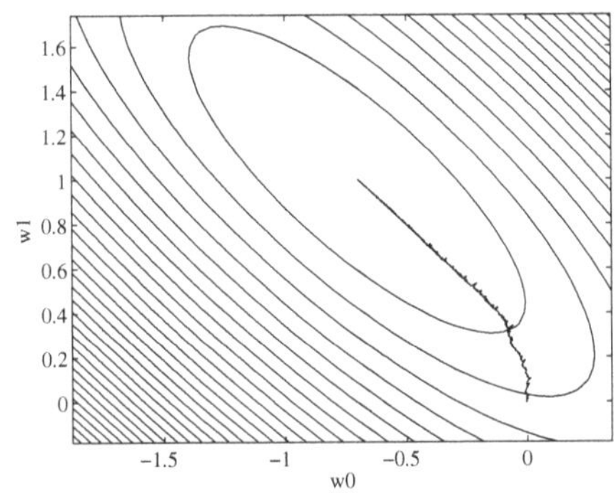

Fig. 8.3 LMS convergence when $x[n]$ is correlated.

step size when $x[n]$ is small. It is important to keep the resulting value within the bounds of stability. Usually, this algorithm presents better convergence properties than the original LMS.

8.3.1.2 The Least-Squares Method

The least-squares method can be viewed as an alternative to Wiener theory discussed above. The method is based on a window of observed data: $x[i]$ and $d[i-\Delta]$ for $i = 0, ..., n$. The goal is to find the filter taps $\mathbf{w}$ that minimize

$$\mathrm{J}_{LS}[n] = \sum_{i=0}^{n} |e[i]|^2, \tag{8.15}$$

where $e[i] = d[i-\Delta] - y[i] = d[i-\Delta] - \mathbf{w}^H[n]\mathbf{x}[n]$.

It is then possible to note that the least-squares method follows a deterministic approach. The cost function $\mathrm{J}_{LS}[n]$ depends on the data window being considered, changing with time. Thus, the optimum filter coefficients, $\mathbf{w}$, have to be recalculated at each time instant.

Usually, (8.15) is expressed with a weighting factor

$$J_{LS}[n] = \sum_{i=0}^{n} \lambda_f^{n-i} |e[i]|^2, \tag{8.16}$$

where λ_f is a positive constant smaller than 1. This criterion can also be called the exponentially weighted least squares and it opens the possibility of controlling the memory of the estimation, i.e., the size of the data window that will be considered. The constant λ_f is called the forgetting factor.

Searching for the minimum of $J_{LS}[n]$ with respect to the filter taps $\mathbf{w}$ results in

$$\mathbf{w}[n] = \mathbf{R_D}^{-1}[n]\mathbf{p_D}[n], \tag{8.17}$$

where

$$\mathbf{R_D}[n] = \sum_{i=0}^{n} \lambda_f^{n-i} \mathbf{x}[i]\mathbf{x}^H[i], \tag{8.18}$$

$$\mathbf{p_D}[n] = \sum_{i=0}^{n} \lambda_f^{n-i} d[i]\mathbf{x}[i] \tag{8.19}$$

and $\mathbf{x}[i] = [x[i]\; x[i-1]\; \ldots\; x[i-N_e+1]]^T$.

Solving (8.17) iteratively, $\mathbf{w}[n+1]$ is written as a function of $\mathbf{w}[n]$, the desired signal $d[n+1-\Delta]$ and the received signal $x[n+1]$ as

$$\mathbf{w}[n+1] = \mathbf{w}[n] + \mathbf{R_D}^{-1}[n+1]\mathbf{x}[n+1]e_a^*[n+1], \tag{8.20}$$

where $e_a[n]$ is the a priori error defined as $e_a[n] = d[n-\Delta] - \mathbf{w}^H[n-1]\mathbf{x}[n]$. Note that this is not the error that has to be minimized. As given by (8.16), (8.20) minimizes the a posteriori error defined by (8.10).

The difficulty presented by solving (8.20) at each time instant n is the need of inverting matrix $\mathbf{R_D}$, which has a high computational cost. To avoid this operation, it is possible to use the matrix inversion lemma [15, 23]. The resulting algorithm is the well-known recursive least squares (RLS) algorithm:

$$\begin{aligned}
\gamma[n+1] &= \frac{\lambda_f}{\lambda_f + \mathbf{x}^H[n+1]Q[n]\mathbf{x}[n+1]}, \\
\mathbf{g}[n+1] &= \lambda_f^{-1}\gamma[n+1]Q[n]\mathbf{x}[n+1], \\
\mathbf{Q}[n+1] &= \frac{1}{\lambda_f}\left(\mathbf{Q}[n] - \mathbf{g}[n+1]\mathbf{x}^H[n+1]\mathbf{Q}[n]\right), \\
e_a[n+1] &= d[n+1-\Delta] - \mathbf{w}^H[n]\mathbf{x}[n+1], \\
\mathbf{w}[n+1] &= \mathbf{w}[n] + \mathbf{g}[n+1]e_a^*[n+1],
\end{aligned} \tag{8.21}$$

where $\mathbf{Q}[n]$ is the inverse correlation matrix, $\mathbf{g}[n]$ is referred to as the *gain vector*, due to the fact that the filter taps are updated by this factor multiplied by the a priori error, and $\gamma[n]$ is the conversion factor which relates the a priori and the a posteriori errors: $e[n] = \gamma[n]e_a[n]$.

An analysis of this algorithm convergence behavior and numerical problems can be found in [15, 23]. The impact on the tracking of time-varying channels and the error misadjustment can be found in [29]. Further efficient and stable algorithms can be implemented using the QR decomposition method and lattice filtering [4].

8.3.1.3 Examples and Discussion

Supervised techniques have always been considered as being defined by convex cost functions presenting only one global minimum, that is, being given by unimodal criteria. A modern approach, however, takes into account the delay, Δ, and its importance in arriving at a good solution. Basically, this parameter is important in the context of equalization since the problem is solved when the filter output is a delayed version of the desired signal. If the problem involves transmission/reception of information, the delay depends on the unknown channel. Consequently, it is an unknown parameter that must also be optimized in the MMSE sense.

A simple example shows how an incorrect choice for Δ may lead to poor solutions. Consider the transmission of a binary phase-shift keying (BPSK)[8.1] modulated signal $s[n]$ through a channel given by $h(z) = 1 - 2.5z^{-1} + z^{-2}$, without the addition of noise. An equalizer with 15 coefficients is used in the receiver, to correct the distortions introduced by this channel. In Fig. 8.4 is shown the minimum MSE value obtained through the optimum Wiener solution for several choices of the delay Δ. The choice of the delay is related to the channel's phase: minimum phase channels require none or small delays, maximum phase channels need large delays, and mixed phase channels are somewhere between the two previous kinds. As the SNR decreases, the optimal delay will tend to an intermediate value, since the Wiener solution will tend to the matched filter.

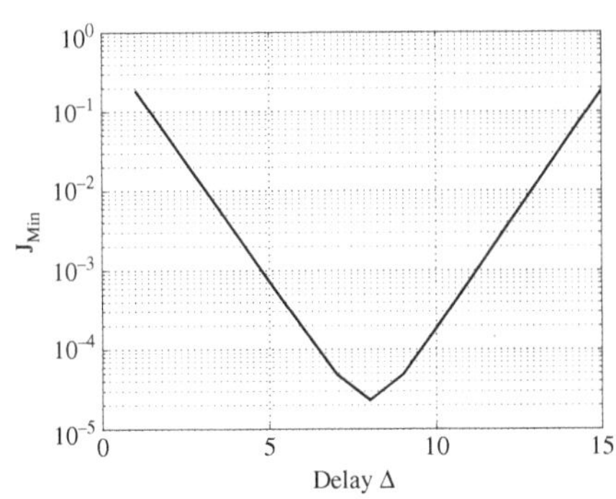

Fig. 8.4 J_{min} for several delay values.

[8.1] Symbols belong to the alphabet $\{-1, +1\}$.

The MSE during convergence for the LMS and RLS algorithms, considering two different values of Δ, are illustrated in Fig. 8.5. The results show that it is possible to obtain a much smaller MSE after convergence when the correct value of delay is used.

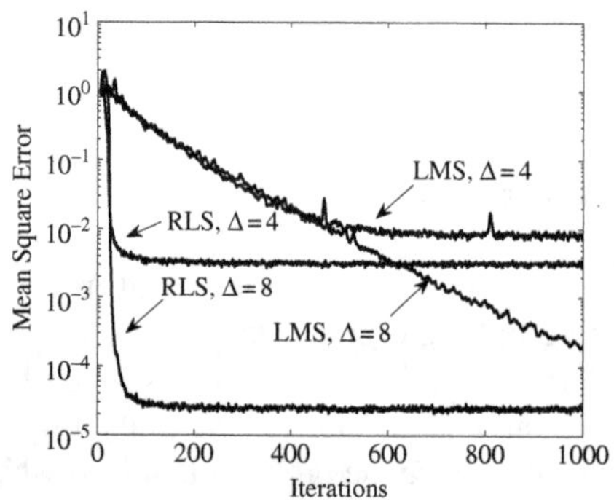

Fig. 8.5 Mean square error for LMS and RLS algorithms for $\Delta = 4$ and $\Delta = 8$.

In addition, Fig. 8.5 shows the difference in performance between both algorithms. The LMS step size μ was set at 0.008, the highest value for which the algorithm is still stable. The RLS forgetting factor λ_f was set at 0.99 and the matrix $\mathbf{Q}[n]$ was initialized with $\delta = 0.1$. The obtained result illustrates how the LMS algorithm converges slowly when the input signal is correlated, while the RLS is not affected.

An analysis of the influence of the step size in the tracking of time-varying channel can be found in [29].

8.3.2 Unsupervised Techniques

Differently from supervised techniques, that are based on the second-order statistics of the signals involved and on the use of a known training sequence, unsupervised or blind techniques need to recur to higher order statistics in order to cope with the absence of further information about the desired signal. This leads to non-convex cost functions and convergence to local minima becomes an issue to be dealt with.

Our study of unsupervised methods will start with the statement of the two most important theorems which explain the context in which blind filtering is possible.

8.3.2.1 Unsupervised Equalization Theorems

Benveniste–Goursat–Ruget (BGR) theorem was first stated in 1980 [12], searching for a criterion where only the statistical characteristics of the desired signal were known. The authors already knew that second-order statistics were not sufficient since they do not carry phase information. The idea was then to consider the probability density function of the involved signals. Consider that the following conditions are met: the transmitted signal has independent and identically distributed (i.i.d.) symbols, the channel and the equalizer are linear filters and no noise is added, perfect channel inversion is possible, that is, zero-forcing solutions are attainable. Thus, the theorem is stated as follows:

Theorem 8.1. *If the probability density function of $y[n]$ equals that of $s[n]$, posed that $s[n]$ is non-Gaussian, a zero-forcing solution is guaranteed.*

The restriction of having non-Gaussian transmitted signals comes from the fact that a filtered Gaussian signal is still Gaussian. Thus, the problem would resume to a power adjustment.

Ten years after BGR theorem was stated, Shalvi and Weinstein (SW) were able to refine it, using the cumulant[8.2] of $y[n]$ and $s[n]$. Defining $C^y_{p,q}$ as being the (p,q)-order cumulant of $y[n]$, Shalvi and Weinstein stated the following [41].

Theorem 8.2. *Under the conditions specified above, if $\mathbb{E}\left[|y[n]|^2\right] = \mathbb{E}\left[|s[n]|^2\right]$ then $|C^y_{p,q}| \leq |C^s_{p,q}|$, for $p+q \geq 2$, with equality if and only if perfect (zero-forcing) equalization is attained.*

While BGR theorem considered the probability density function, which indirectly involves all the moments of the signals $s[n]$ and $y[n]$, SW theorem reduces the dependence to the variance and one higher order moment of these signals.

All blind equalization criteria depend, implicitly or explicitly, on these two theorems. The SW theorem is of particular interest since it is the basis for two of the most studied criteria in this domain: the constant modulus criterion and the Shalvi–Weinstein criterion.

8.3.2.2 Criteria and Algorithms

The first family of blind deconvolution algorithms proposed in the literature is known as Bussgang algorithms, since the statistics of the deconvolved signal are approximately Bussgang. In general, these algorithms are developed to minimize a cost function defined by

$$J_B(n) = \mathbb{E}\left[|y[n] - \hat{s}[n]|^2\right], \tag{8.22}$$

where $y[n]$ is the filter output given by (8.9) and $\hat{s}[n]$ is the estimated transmitted symbol, obtained through a nonlinear, zero memory function $\hat{s}[n] = g(y[n])$.

[8.2] The *cumulant* is a statistic measure derived from the natural logarithm of the characteristic function of a random variable [33]. It is equal to the value of moments until third order. As an example, the cumulant of a random variable x, with zero mean, and its conjugate x^* is equal to its variance: $cum(x,x^*) = \mathbb{E}\left[|x|^2\right]$. Here, the following notation for the (p,q)-order of x will be used: $cum(\underbrace{x,x,\ldots,x}_{p};\underbrace{x^*,x^*,\ldots,x^*}_{q}) = C^x_{p,q}$.

The decision-directed algorithm, proposed by Lucky [32], was one of the first *Bussgang algorithms* and is one of the most used blind algorithms, specially since it is used together with supervised techniques. Usually, systems present an initial training phase to reduce ISI and switch to decision-directed mode to keep tracking channel variations. In this case, the nonlinear function $g(y[n])$ is given by the decision device, depending on the modulation being used.

The constant modulus criterion is also a Bussgang method. Proposed by Godard [21], it is one of the most studied algorithms in the context of unsupervised techniques. The cost function penalizes deviations of the filter output from a constant modulus:

$$J_{CM} = \mathbb{E}\left[\left||y[n]|^2 - R_2\right|^2\right], \tag{8.23}$$

where $R_2 = \frac{\mathbb{E}\left[|s[n]|^4\right]}{\mathbb{E}\left[|s[n]|^2\right]}$. The resulting algorithm, known as the constant modulus algorithm (CMA), is given by

$$\mathbf{w}[n+1] = \mathbf{w}[n] - \mu \mathbf{x}^*[n]e[n], \tag{8.24}$$

$$e[n] = y[n]\left(|y[n]|^2 - R_2\right).$$

Another important family of criteria is obtained directly from the Shalvi–Weinstein theorem. The criterion is stated as follows [41, 42]:

$$max|C^y_{p,q}| \text{ subject to } C^y_{1,1} = C^s_{1,1}, \tag{8.25}$$

which is known as the Shalvi–Weinstein (SW) criterion.

The algorithm that searches for the maximum of (8.25) results from a non-linear mapping which converges to the stationary points of the criterion. Consider the use of a (2,2)-order cumulant, which reduces to the kurtosis that can be defined as a function of moments as

$$\mathrm{K}(y) = \mathbb{E}\left[|y|^4\right] - 2\mathbb{E}^2\left[|y|^2\right] - \left|\mathbb{E}\left[y^2\right]\right|^2. \tag{8.26}$$

The algorithm can be stated as follows:

$$\mathbf{w}[n+1] = \mathbf{w}[n] + \frac{\beta}{\delta}\mathbf{Q}[n]\mathbf{x}[n]^* y[n]\left[|y[n]|^2 - \frac{\mathbb{E}\left(|s[n]|^4\right)}{\mathbb{E}\left(|s[n]|^2\right)}\right], \tag{8.27}$$

where β is a constant, $\delta = C^s_{2,2}/C^s_{1,1}$, and $\mathbf{Q}$ is proportional to the inverse autocorrelation matrix of $x[n]$:

$$\mathbf{Q}[n+1] = \frac{1}{1-\beta}\left(\mathbf{Q}[n] - \frac{\beta \mathbf{Q}[n]\mathbf{x}^*[n]\mathbf{x}[n]^T \mathbf{Q}[n]}{1-\beta+\beta \mathbf{x}[n]^T \mathbf{Q}_n \mathbf{x}^*[n]}\right). \tag{8.28}$$

The algorithm stated above is known as the super exponential algorithm (SEA) due to the fact that it converges at an exponential rate [42].

8.3.3 Case Study: Channel Identification and Tracking

Channel identification and tracking is important in several applications. Often, receivers use this information to recover the transmitted message. Specially in wireless systems, where receivers are usually moving, tracking channel variations is crucial for a good performance. In this case study, the supervised techniques discussed in Section 8.3.1 will be applied to the problem of channel identification and tracking.

First a time division multiple access (TDMA) cellular system defined by the IS-136 standard is discussed. Transmitted symbols are modulated using a $\pi/4$-differential quadrature phase-shift keying (DQPSK) modulation, i.e., symbols are given by $\sqrt{2}e^{j\theta}$, where θ is obtained adding the previous symbol phase with an angle chosen randomly from $\{\pi/4,\ 3\pi/4,\ -3\pi/4,\ -\pi/4\}$. Data are transmitted in frames of 162 symbols, from which the first 14 are available for training. As stated in Section 8.2, the transmission/receiver filters form a raised cosine pulse with roll-off equal to 0.35. The symbol rate of this system is equal to 24.3 kbauds, which usually renders the delay spread less than one symbol period. The channel is considered to have a length $L = 2$. A propagation model with two Rayleigh paths with equal power (-3 dB), and a relative delay equal to one symbol period T were assumed. It is also assumed that the mobile is moving at 100 km/h and the carrier frequency is 900 MHz, resulting in a normalized Doppler frequency of $f_d T = 3.4 \times 10^{-3}$. An SNR of 19 dB was considered.

The symbol recovery was done using a maximum-likelihood sequence estimation (MLSE) receiver. More details about it will be given in Section 8.4, where this example will be resumed. For the moment, it is only important to know that this receiver needs the channel information and a good estimation is important to result in a good overall performance.

The LMS, NLMS, and RLS algorithms were tested in this context. After the first 14 available training symbols, the algorithms were switched to a decision-directed mode. Initial conditions are stated in Fig. 8.6(a).

Algorithm Parameters		
$\pi/$ 4-DQPSK modulation 2-tap filters initialized with zero		
	Training Mode	Decision Directed Mode
LMS	$\mu = 0.15$	$\mu = 0.1$
NLMS	$a = 0.01$	$a = 0.01$
RLS	$\lambda_f = 0.65$ $\delta = 4e-6$	$\lambda_f = 0.9$

(a)

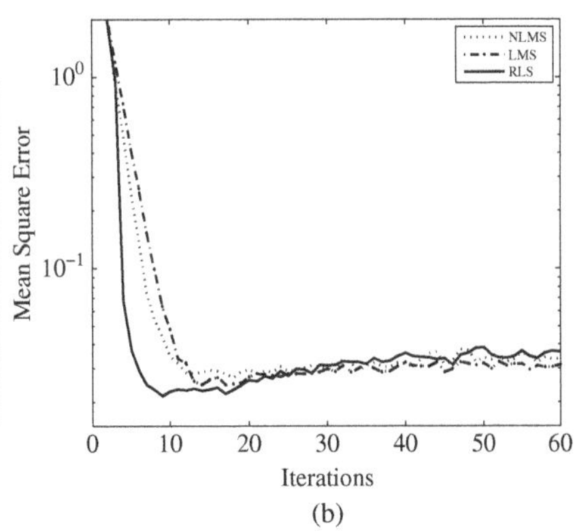

(b)

Fig. 8.6 Channel tracking case study: (a) algorithm parameters and (b) MSE performance for LMS, NLMS, and RLS.

In Fig. 8.6(b) the MSE during the algorithms adaptation, considering 1000 independent trials, is shown. It is interesting to note that, in this case, the convergence speed of the LMS and RLS algorithms is similar, different from the result shown in Fig. 8.5. This was expected since here the filter input is an uncorrelated signal.

8.4 Improving Equalization Performance Over Time Dispersive Channels

In the previous section, iterative adaptation algorithms that are used to optimize the equalizer parameters based on a chosen criterion were presented. For the sake of simplicity, only linear time-domain filtering structures were treated. In this section, non-linear filtering techniques that can provide superior performance when compared to linear filtering are presented.

Wireless communication channels are described by a multipath propagation model that is normally simulated using a time-varying finite impulse response (FIR) filter. This filter introduces ISI that distorts the transmitted signal. The ISI can be removed by another filter that equalizes the received signal. A simple and robust approach is to use a linear filter as the equalizer. It can assume a FIR or an infinite impulse response (IIR) form. The IIR filter can lead to a more efficient implementation but its adaptation is non-linear and it presents local minima and stability problems [38, 43].

A clever modification of the IIR structure can provide a more efficient technique in terms of bit-error rate also with the advantage of avoiding the adaptation problems of the IIR filter in supervised adaptation mode. It is the so-called decision-feedback equalizer (DFE) [8], depicted in Fig. 8.7.

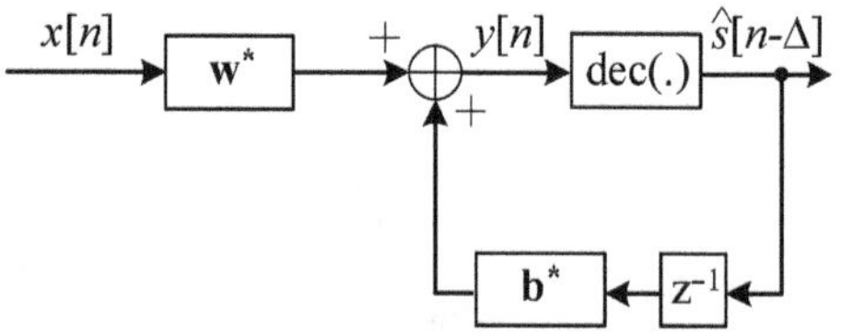

Fig. 8.7 The decision-feedback equalizer (DFE).

The feedforward filter **w** of the DFE is responsible for eliminating the pre-cursor response of the channel, where the cursor is the element of the channel impulse response with the largest energy. The feedback filter **b** uses the past decisions to eliminate the post-cursor response of the equivalent channel created by the convolution of the real channel with the feedforward filter. It is important to observe the insertion of a delay z^{-1} in the feedback loop to make it strictly causal.

The main advantage of the DFE in comparison to a linear filter resides in the fact that, by using a decision device in the feedback loop, it can eliminate the noise enhancement that occurs in linear filtering. Such characteristic is specially important in channels that present spectral nulls, where the noise enhancement is more pronounced. Furthermore, it does not pose the stability problems that may arise in an IIR equalizer, since the decision device limits the amplitude of the signal in the feedback loop. Although the addition of the decision device in the feedback loop has these two beneficial effects, it may cause an error burst, also known as error propagation, when incorrect decisions are fed back. The length of the bursts depends on the noise realizations, channel, modulation, and transmitted sequence. A detailed study of this phenomenon and its impact on the performance can be seen in [3, 11, 24, 25]. In [6, 28, 31] ECC is jointly used with the equalizer in order to mitigate the error propagation phenomenon.

The filter coefficients can be obtained by using the MMSE criterion, using the assumption that only correct symbols are fed back, which is true during the equalizer training phase. In this context, the output of the DFE can be written as

$$y[n] = \begin{bmatrix} \mathbf{w}^H & \mathbf{b}^H \end{bmatrix} \begin{bmatrix} \mathbf{x}[n] \\ \mathbf{s}[n-1-\Delta] \end{bmatrix}, \tag{8.29}$$

where $\mathbf{x}[n] = [x[n]\ x[n-1]\ \ldots\ x[n-N_w+1]]$, N_w is the length of the feedforward filter, $\mathbf{s}[n-1-\Delta] = [s[n-1-\Delta]\ s[n-2-\Delta]\ \ldots\ s[n-N_b-\Delta]]$, N_b is the length of the feedback filter, and Δ is the training delay. Then, by defining the error as in (8.10) and the MMSE criterion as in (8.11) the Wiener–Hopf solution is described by

$$\begin{bmatrix} \mathbf{w} \\ \mathbf{b} \end{bmatrix} = \begin{bmatrix} \mathbf{R}_x & \mathbf{M} \\ \mathbf{M}^H & \sigma_s^2 \mathbf{I} \end{bmatrix}^{-1} \begin{bmatrix} \mathbf{p} \\ \mathbf{0} \end{bmatrix}, \tag{8.30}$$

where $\mathbf{R}_x = \mathbb{E}\{\mathbf{x}[n]\mathbf{x}^H[n]\}$, $\mathbf{M} = \mathbb{E}\{\mathbf{x}[n]\mathbf{s}^H[n-1-\Delta]\}$, and $\mathbf{p} = \mathbb{E}\{\mathbf{x}[n]s^*[n-\Delta]\}$.

Like the linear equalizer, the adaptation of the DFE can be carried out by both least mean square or least-squares algorithms.

Even if the DFE filtering structure presents a considerable advantage over the linear filtering solution, there is still another receiver that achieves higher performance. By assuming that the transmitted symbols are equiprobable and independent, the optimal solution is to maximize the likelihood function of the received sequence:

$$\hat{\mathbf{s}} = \arg\max_{\mathbf{s}} \mathrm{p}(\mathbf{x}|\mathbf{s}) = \arg\max_{\mathbf{s}} \frac{1}{(2\pi\sigma_n^2)^{D/2}} \exp\left\{ \frac{-\|\mathbf{x} - \mathbf{H}_c\mathbf{s}\|^2}{2\sigma_n^2} \right\}, \tag{8.31}$$

where $\mathbf{H}_c$ is the channel matrix convolution and D is the length of the observed received sequence. This kind of receiver is known as the MLSE.[8.3]

To maximize (8.31), the argument of the exponential must be minimized, i.e., the squared Euclidean distance between $\mathbf{x}$ and $\mathbf{H}_c\mathbf{s}$ represented by $\|\mathbf{x} - \mathbf{H}_c\mathbf{s}\|^2$. Rewriting (8.31) gives

---

[8.3] The MLSE is also referred in the literature as the maximum-likelihood sequence detector (MLSD).

$$\hat{\mathbf{s}} = \arg\min_{\mathbf{s}} \sum_{n=0}^{D-1} \left| x[n] - \sum_{j=0}^{L-1} h[j]s[n-j] \right|^2, \tag{8.32}$$

where L is the channel impulse response length.

A direct way to find the most likely transmitted sequence $\hat{\mathbf{s}}$ is to make an exhaustive search among all possible M^D sequences, where M is the cardinality of the modulation. It is clear that the complexity becomes too high even for a small D.

However, there is a more efficient way to perform this search. The ISI generated by the channel can be seen as the output of a finite state machine with M^{L-1} states. Therefore, the channel output may be represented by a trellis diagram and the maximum-likelihood sequence for the received sequence $\mathbf{x}$ is the sequence of state transitions, i.e., a path that minimizes the squared Euclidean distance. In such context, the Viterbi algorithm is able to efficiently execute this path search [17, 44, 48]. Using this algorithm, each decoded symbol needs M^L metrics to be calculated. In comparison to the brute-force search, the complexity of this method does not grow with the sequence length.

The Viterbi algorithm does not need to keep track of all the received sequence, since the survivor path,[8.4] associated with each state, tends to converge as we go back in time in the trellis. This reduces both the memory cost and the latency needed to obtain the symbol estimation. A rule of thumb is that a decision delay Δ of five times the channel memory is enough to obtain reliable decisions.

Note that the channel must be estimated in order to calculate the metrics. A first estimation may be obtained using a training sequence that is later switched to tentative decisions with a tentative delay $\Delta' < \Delta$. This tentative delay should be small enough to keep track of time-varying channels with a good accuracy and provide decisions with sufficient reliability. The maximum-likelihood sequence estimator technique is illustrated in Fig. 8.8.

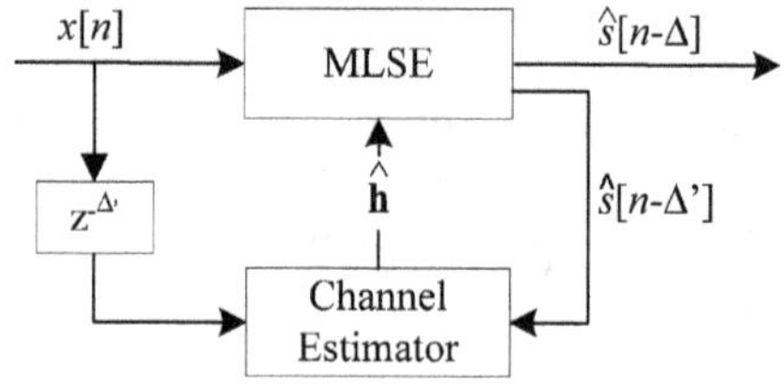

Fig. 8.8 The maximum-likelihood sequence estimator (MLSE).

An example of the performance differences among the different equalization techniques is shown in Example 8.1.

Example 8.1 (Performance comparison). Consider the Proakis B channel $h(z) = 0.407 + 0.815z^{-1} + 0.407z^{-2}$ [37]. This channel presents two close zeros that are next to the unitary circle, producing a very frequency-selective channel. Figure 8.9

[8.4] There are M^{L-1} paths that arrive at one state. The path with the lowest squared Euclidean distance is called the survivor path.

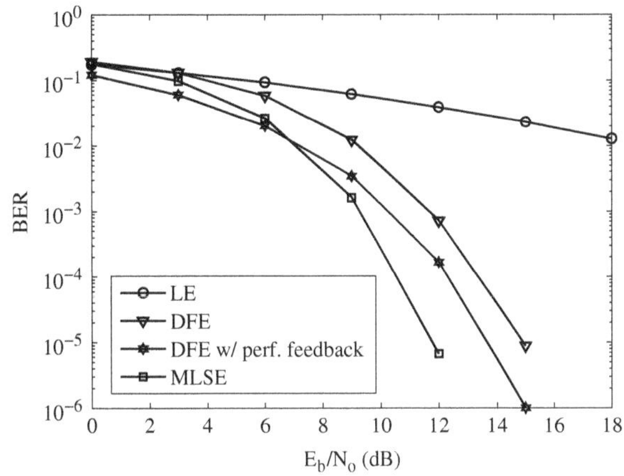

Fig. 8.9 BER comparison for different equalization techniques for the Proakis (b) channel $h(z) = 0.407 + 0.815z^{-1} + 0.407z^{-2}$.

shows the bit-error rate (BER) for QPSK modulation as a function of the E_b/N_o. The linear equalizer (LE) is a FIR filter with 17 coefficients. The DFE has eight coefficients for the feedforward filter and two coefficients for the feedback filter. All the coefficients were obtained using the MMSE criterion and with perfect channel knowledge. The training delay Δ for the LE was 9 and for the DFE was 7. Both delays minimize the MSE for the E_b/N_o region around 10–16 dB. The DFE with perfect feedback was also simulated to observe the performance degradation caused by error propagation. As expected, the DFE provides a far superior performance in comparison to the LE. This equalizer suffers from the noise enhancement phenomenon that is intensified due to the high-frequency selectivity of the selected channel. The error propagation in the DFE imposes a performance penalty around 1 dB for this channel. It is worth noting that lengthier and more powerful post-cursor responses will cause much higher degradation. Finally the MLSE with a decision delay of 10 provides more than 3 dB gain over the DFE.

8.4.1 Case Study: Maximum-Likelihood Sequence Estimation for the IS-136 Cellular System

Resuming the case study presented in Section 8.3.3, in this section, the system performance will be analyzed in terms of BER.

An IS-136 TDMA system will be considered, with differential modulation $\pi/4$-DQPSK. The symbol rate $1/T$ of this system is equal to 24.3 kbauds, the roll-off $\alpha = 0.35$ and the considered channel length is equal to $L = 2$.

A two-path propagation model with equal power (-3 dB) was adopted, with a relative delay different from zero. An LMS algorithm was used to identify and track the channel. For IS-136, a 14-symbol training sequence is available. The tracking was done using a tentative delay of two symbols and the decision delay is equal to five symbols. In this analysis, it is assumed that the mobile is moving at 30 km/h and the carrier frequency is equal to 900 MHz, resulting in a normalized Doppler frequency of $f_d T = 10^{-3}$. The performance of the MLSE receiver is shown in Fig. 8.10. In this figure, the performance of the differential receiver alone is also presented. The relative delay of T provides the best MLSE performance since the channel coefficients are uncorrelated in this scenario. The relative delay of $0.25T$ generates less ISI and beneficiates the differential decoder. Nevertheless, it must be noted that even in an AWGN channel the MLSE can provide additional performance improvements, since it can take into account the memory present in the differential modulation $\pi/4$-DQPSK.

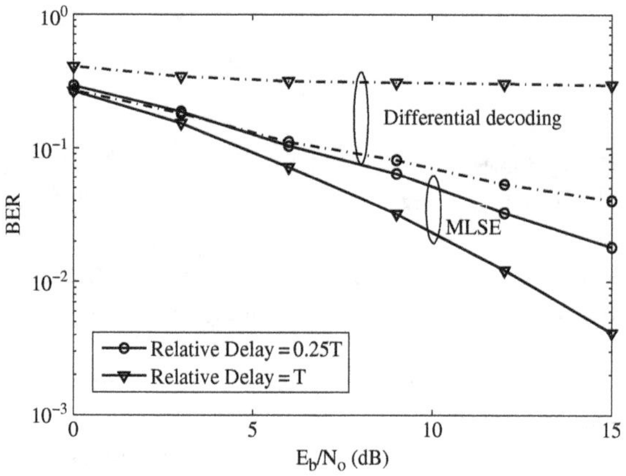

Fig. 8.10 BER comparison for different relative delays between the two paths and a normalized Doppler frequency of $f_d T = 10^{-3}$.

It is also important to emphasize that the MLSE is used in practice in the GSM/EDGE system (e.g., [19]).

8.5 Equalization with Multiple Antennas

The ever-growing demand for improved performance in terms of higher network capacity and per user bit rates has made the use of multiple antenna techniques increasingly interesting. It allow us to combat the two most important problems that plagues wireless communications: co-channel interference and fading.

Multiple antennas can be used in both transmitter and receiver. When the system has multiple antennas only in the transmitter, the system is considered a MISO system. A well-known technique that uses this approach is the Alamouti space–time block-coding scheme [2], but it must be noted that it can also use multiple antennas in the receiver to provide additional robustness. In the case of multiple antennas used only in the receiver, a SIMO system is obtained. Finally, a MIMO system is defined when multiple antennas are used in both transmitter and receiver [20]. This chapter will focus on the study of SIMO systems.

8.5.1 Beamforming

One array configuration that is widely studied in wireless communication is the uniform linear array (ULA), where the antennas are aligned in one direction and equally spaced.

Due to propagation characteristics, two different approaches are used: beamforming and diversity. In order to better understand the principles involved in this technique, this section presents the propagation model for the ULA.

Let us consider a ULA with isotropic antennas that has no coupling between them and that is mounted on the y-axis of a cartesian plane. An incident plane wave impinges the array with an angle of arrival θ_a that is measured with respect to the x-axis. Consider also that this plane wave is modulated by the complex baseband signal $s(t)$. Therefore, taking the first antenna of the array as the time reference and being Δd the spacing between the antennas, the input of the mth element of the array can be written as follows:

$$x_m(t) = s\left(t - \frac{m\Delta d}{c} sin\theta_a\right) \mathrm{e}^{-j\frac{2\pi}{\lambda} m\Delta d \sin\theta_a}, 0 \le m \le M_r - 1, \tag{8.33}$$

where λ is the wavelength, given by c/f_c, where c is the speed of light, f_c is the carrier frequency, and M_r is the number of elements in the ULA.

In telecommunications, it is commonly assumed that the bandwidth B of $s(t)$ is small enough so that $\frac{M_r\Delta d}{c} B \ll 1$. This allows us to ignore the time delay in (8.33), i.e., $s(t - \frac{m\Delta d}{c} sin\theta_a) \approx s(t)$ for every value of m and θ_a.

The input signals $x_m(t)$ are weighted by a coefficient w_m^* and then summed to generate the array output $y(t)$. The ULA is illustrated in Fig. 8.11.

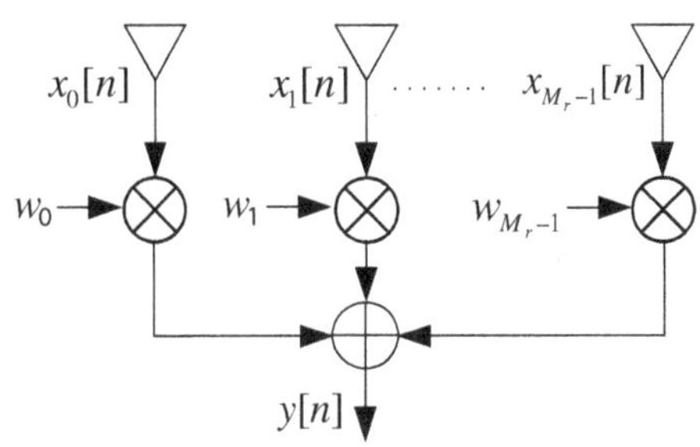

Fig. 8.11 An antenna array with M_r elements.

It is convenient to represent it in vectorial form:

$$\begin{aligned} y(t) &= \mathbf{w}^H \mathbf{x}(t) \\ &= s(t)\mathbf{w}^H \mathbf{f}(\theta_a) \end{aligned}, \tag{8.34}$$

where

$$\mathbf{w} = \begin{bmatrix} w_0 & w_1 & \cdots & w_{M_r-1} \end{bmatrix}^T \tag{8.35}$$

is the weight vector and

$$\mathbf{f}(\theta_a) = \left[1 \ \mathrm{e}^{-j\frac{2\pi}{\lambda}\Delta d \sin(\theta_a)} \ \cdots \ \mathrm{e}^{-j\frac{2\pi}{\lambda}(M_r-1)\Delta d \sin(\theta_a)}\right]^T \tag{8.36}$$

is the so-called steering vector of the array.

Assuming a beamforming processing, the usual choice for the antenna spacing is $\Delta d = \lambda/2$. Such choice is justified by the fact that if $\Delta d < \lambda/2$, spatial resolution is lost. The opposite happens for $\Delta d > \lambda/2$ but, in this case, an ambiguity occurs for $|\theta_a| < \pi/2$, which can be seen as the equivalent of the spectral aliasing phenomenon.

The multipath channel model is similar to the one presented in Section 8.2. In this context, the local scatterers may introduce a perturbation in the angle of arrival which must be taken into account. Then, the perceived normalized sum of N scattered paths at the ULA can be written as follows:

$$\mathbf{g}(t) = N^{-1/2} \sum_{n=1}^{N} \mathrm{e}^{j\{2\pi f_d \cos\phi[n]t + \Phi[n]\}} \mathbf{f}(\theta_a + \vartheta[n]), \tag{8.37}$$

where $\vartheta[n]$ is a random variable uniformly distributed over $[-\theta_{spread}/2, \theta_{spread}/2]$, where θ_{spread} is known as the angle spread.

Then, considering $L-1$ remote scatterers with their own local scatterers, the space–time impulse response can be written as follows:

$$\mathbf{h}(t) = \sum_{l=0}^{L-1} \mathbf{g}_l(t) p(t) \delta(t - \tau[l]), \tag{8.38}$$

where $\tau[l]$ is the delay generated by the lth path and $p(t)$ is the modulation pulse.

Finally, the received signal is given by

$$\mathbf{x}(t) = \sum_{k=-\infty}^{\infty} s[k]\mathbf{h}(t - kT) + \mathbf{v}(t), \tag{8.39}$$

where $\mathbf{v}(t)$ is the noise vector of dimension M_r and each element has variance σ_v^2.

It is worth noting that a more advanced channel model can be found in [1].

There are many criteria that can be used to calculate the weights $\mathbf{w}$. An important criteria that should be taken into account is the MMSE criterion:

$$\mathrm{J_{MSE}} = \mathbb{E}\left\{|s[n-\Delta] - \mathbf{w}^H \mathbf{x}[n]|^2\right\}, \tag{8.40}$$

where Δ is the training delay. The optimum coefficients are obtained by the Wiener–Hopf equation described in (8.12).

The greatest limitation of the beamforming technique is that the degree of freedom to cancel interferers is limited to $M_r - 1$. This is easily explained by inspecting the array's steering vector, described in (8.36). If $\mathrm{e}^{-j\frac{2\pi}{\lambda}m\Delta d \sin\theta_a}$ is replaced by $z^{-m}, z = \mathrm{e}^{j\frac{2\pi}{\lambda}\Delta d \sin(\theta_a)}$, it is easy to notice that the ULA provides $M_r - 1$ zeros that can be used to cancel interferers. This can be illustrated with two examples for

Table 8.1 Desired user and interferers configuration.

	Desired user, scenario I Path #1	Desired user, scenario I Path #2	Desired user, scenario II Path #1	Desired user, scenario II Path #2	Interferer #1 Path #1	Interferer #2 Path #1
AOA	30°	−15°	30°	−15°	60°	0°
Delay	0	0	0	T	0	0
Power (dB)	−3	−3	−3	−3	0	0

which the user and interferers configurations are described in Table 8.1. Let us consider $M_r = 3$, 10 dB of SNR per antenna and both user and interferers transmit using QPSK modulation. The array coefficients are obtained using the MMSE criterion with $\Delta = 0$. The radiation diagram, obtained by evaluating $y[n] = \mathbf{w}^H \mathbf{f}(\theta)$ for $0 \leq \theta < 2\pi$, and the ULA output $y[n] = \mathbf{w}^H \mathbf{x}[n]$ are depicted in Figs. 8.12 and 8.13.

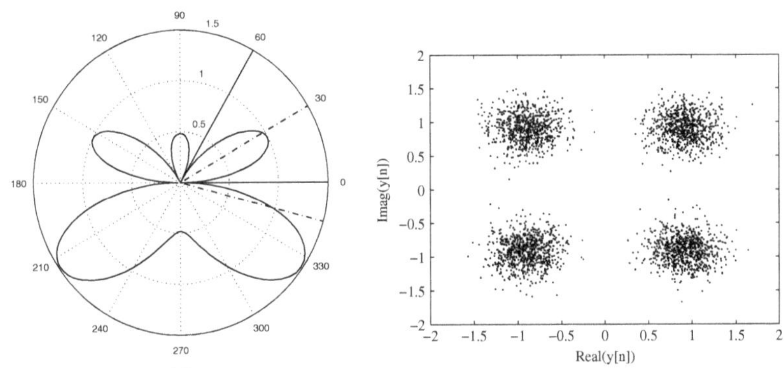

(a) (b)

Fig. 8.12 (a) Radiation diagram for the user in scenario I and interferers configuration described in Table 8.1: (−·) desired user paths and (–) interferers. (b) ULA output.

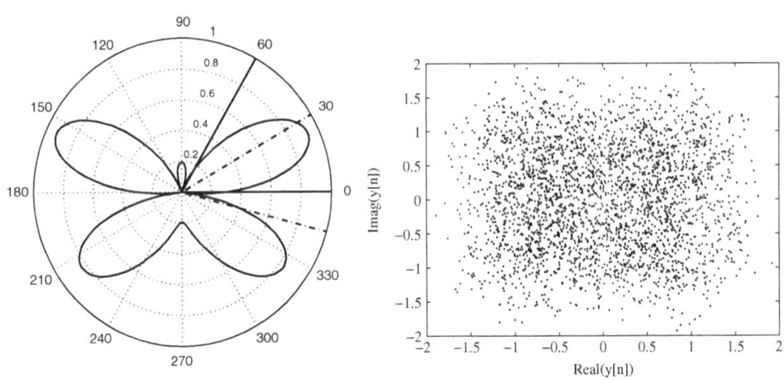

(a) (b)

Fig. 8.13 (a) Radiation diagram for the desired user in scenario II and interferers configuration described in Table 8.1: (−·) desired user paths and (–) interferers. (b) ULA output.

For the desired user in scenario I, described in Table 8.1, the array is able to combine both desired user paths and can perfectly cancel both interferers, as shown in Fig. 8.12. However, for the scenario II, the delayed path of the desired user is ISI. In this scenario, the array must cancel three interferers and not only two as compared to the former. Nevertheless, the array does not have enough degrees of freedom to do so and the performance is largely affected as shown in Fig. 8.13. Furthermore, it must be noted that even if it had enough degrees of freedom to cancel the delayed path, it is not the best approach, specially when the paths are considered to be affected by fading, where every desired signal component should be used to improve signal-to-noise ratio. In the next section, techniques that can better cope with this type of environment are presented.

8.5.2 Space-Time Equalizer Structures

The presence of delayed multipaths from the desired user and interferers may outnumber the available degrees of freedom of an antenna array. Another problem is due to the fact that canceling the desired user-delayed multipaths is not a good strategy, since this would not take advantage of the available signal diversity, which is essential to combat fading channels. However, with some modifications, an antenna array can provide better performance in this context.

One possible solution consists in adding adaptive filters for each antenna branch of the array. This solution, depicted in Fig. 8.14, is the so-called broadband array or simply space–time linear equalizer (ST-LE), since it can now deal with the frequency selectivity generated by the delayed paths. These filters allow to capture and coherently combine desired user-delayed paths as well as cancel delayed paths from the same interferer by doing exactly the opposite.

Fig. 8.14 Space–time linear equalizer.

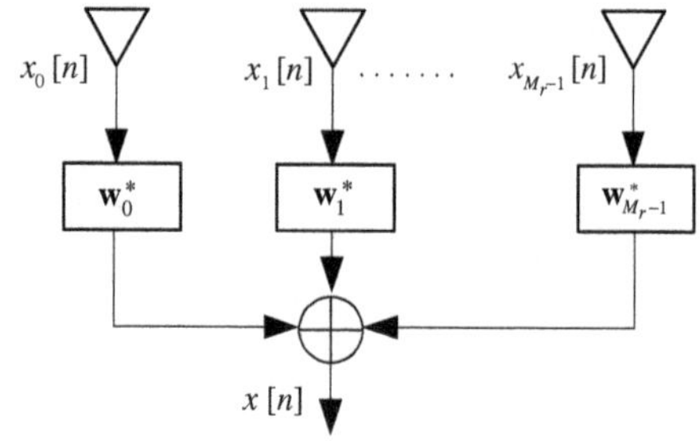

The output of the ST-LE at the nth time instant can be described as the linear combination of the filter weights and the correspondent inputs that can be written as follows:

$$y[n] = \mathbf{w}^H \mathbf{x}[n], \tag{8.41}$$

where

$$\mathbf{w} = \left[\mathbf{w}_0^T \ \mathbf{w}_1^T \ \cdots \ \mathbf{w}_{M_r-1}^T\right]^T, \tag{8.42}$$

$\mathbf{w}_k$ are the N_e weights of the FIR filter attached to the kth antenna and

$$\mathbf{x}[n] = \left[\mathbf{x}_0^T[n]\ \mathbf{x}_1^T[n]\ \cdots\ \mathbf{x}_{M_r-1}^T[n]\right]^T \tag{8.43}$$

is the correspondent filter inputs. The MSE is defined as in (8.20).

Now, the operation of the space–time equalization structure will be illustrated. Consider the desired user in scenario II, presented in Table 8.1, and no interferers at all. ST-LE with $M_r = 3$ and $N_e = 2$ is used, the SNR per antenna is 10 dB and the training delay is $\Delta = 1$. In Fig. 8.15 thc radiation diagram for each weight bank[8.5] of the ST-LE is shown. Note that for the first bank, the delayed path is captured and the other one, at 30°, is suppressed. In the second bank, occurs exactly the contrary. In this example, the ST-LE acts like a RAKE receiver [37].

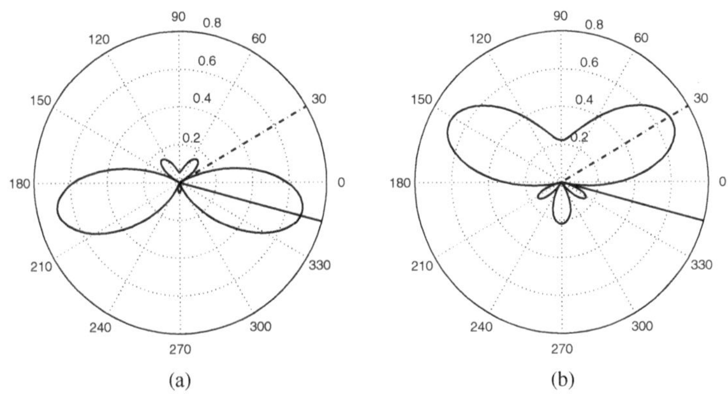

Fig. 8.15 Desired user configuration in scenario II, presented in Table 8.1, path #1 shown by (−·) and path #2 shown by (−), an SNR per antenna equal to 10 dB. (a) Radiation diagram for the first weight bank and (b) radiation diagram for the second weight bank.

However, the additional degrees of freedom may not suffice for other situations. For instance, consider again the previous configuration with the desired user in scenario II but now including the interferers. With $M_r = 3$, each weight bank does not have enough degrees of freedom to cancel both interferers and one of the user paths as shown in Fig. 8.16(a). In comparison to the ULA with $M_r = 3$ (see Fig. 8.13), the time dimension gives an additional degree of freedom that allows the ST-LE to perform slightly better. Nevertheless, since the equalization in time dimension is more important in such a case, a more efficient time-domain equalization structure can be used, such as the ST-DFE :

$$y[n] = \mathbf{w}^H \mathbf{u}[n] + \mathbf{b}^H \hat{\mathbf{s}}[n-1-\Delta] \tag{8.44}$$

[8.5] The weight bank is formed by the ith coefficient of every equalizer $\mathbf{w}_k$.

or an ST-MLSE filtering structure. The coefficient solution for the ST-DFE has the same form as that in (8.30). For the ST-MLSE, the optimal performance is obtained by adding a whitening filter after the space–time front end. For high SNR, the coefficient solution can be approximated by the ST-DFE solution [7]. A detailed derivation of the solutions can also be found in [7], together with the analyses of the minimum time-domain filter size. Figure 8.16(b) illustrates the ST-DFE output for the desired user in scenario II, in Table 8.1, SNR per antenna equal to 10 dB, $M_r = 3$, $N_e = 2$, $N_b = 1$ and $\Delta = 1$. Its performance is far better than that achieved by the ST-LE (see Fig. 8.16(a)) with the same parameters.

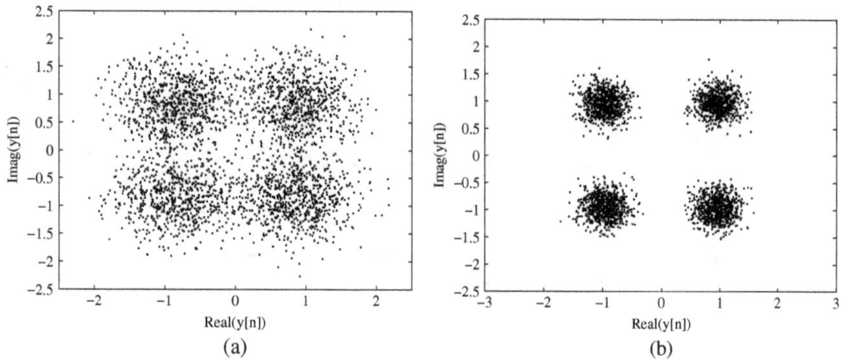

Fig. 8.16 Equalizer output for desired user and interferers configuration described in Table 8.1: (a) ST-LE output and (b) ST-DFE output.

Besides putting a filter in each antenna receiver branch, there is another possible way to obtain an array with more degrees of freedom. By assuming that the ISI can be treated by an equalizer, a pure spatial antenna array can spend its degrees of freedom on canceling the co-channel interference. Since the spatial and temporal signal equalizations are performed separately but not disjointly, this approach is called decoupled space–time (DST) equalization. Many variations of this approach have been proposed (e.g., [18, 22, 26, 35, 45]).

In comparison to the ST approach, the DST presents lower performance but, on the other hand, it can offer lower computational complexity.

Figure 8.17 shows a comparison of the radiation pattern between the conventional antenna array (AA) and the decoupled space–time technique for the desired user in scenario II and the interference presented in Table 8.1, with $M_r = 3$ and 10 dB of SNR per antenna. It is clear that the DST can mitigate the interferers and the AA cannot. Also, for comparison, Fig. 8.18 shows the output of the AA-DFE and DST-DFE, both using a DFE with parameters $N_e = 3$ and $N_b = 1$. Comparing Figs. 8.13(b) and 8.18(a), the DFE can enhance the output of the conventional AA, but it is not nearly as good as the DST-DFE output, shown in Fig. 8.18(b).

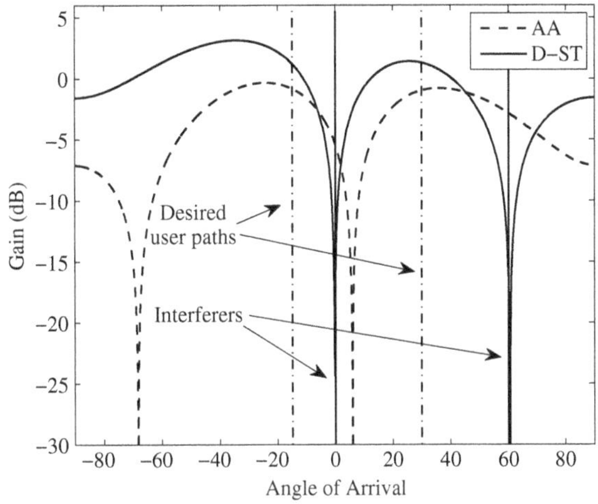

Fig. 8.17 Diagram pattern for the antenna array (AA) and the decoupled space–time (DST) technique with $M_r = 3$ and SNR=10 dB for the desired user in scenario II and interferers configuration shown in Table 8.1.

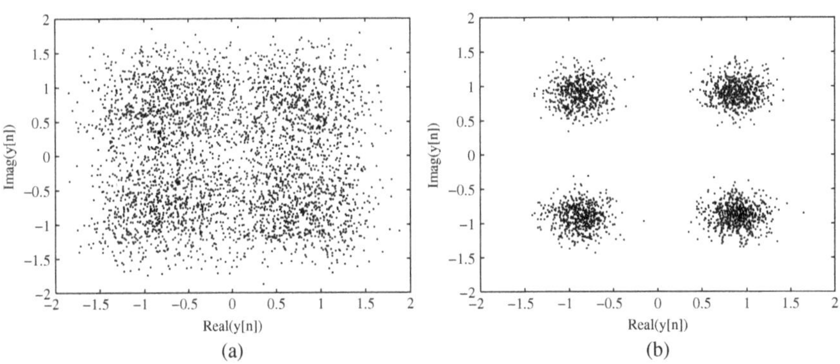

Fig. 8.18 Time-domain equalizer output for the desired user in scenario II and interferers configuration described in Table 8.1: (a) AA-DFE output and (b) DST-DFE output.

8.5.2.1 Case Study: Space–Time Equalization in the Uplink of an EDGE Cellular System

To illustrate the performance difference among the space–time equalizer structures, an EDGE-based system is considered. The modulation is an 8-PSK with a signaling rate of 270.833 kbauds and a roll-off factor equal to 0.35, assuming a typical urban (TU) power and delay profile, presented in Table. 8.2, and 30 km/h for both user and interferer. The signal-to-interference ratio (SIR) is 6dB. All receivers have $M_r = 3$ antennas and assuming a full diversity scenario, i.e., an angle spread equal to 360°. The DFE in both AA-DFE and DST-DFE receivers have $N_e = 3$ and $N_b = 5$. The ST-DFE has three taps per antenna and $N_b = 5$. The channel estimator has 10 coefficients, from which 2 are used to estimate the pre-cursor response and the others are used to calculate the post-cursor response. These coefficients are used to calculate

Table 8.2 Typical urban (TU) relative delay and power profile.

	Path #1	Path #2	Path #3	Path #4	Path #5	Path #6
Relative delay (μs)	0.2	0	0.3	1.4	2.1	4.8
Relative mean power (dB)	−3	0	−2	−6	−8	−10

the DFE solution. All structures are adapted by an RLS algorithm. Each time-slot has a training sequence of 26 symbols and 116 data symbols. It is also assumed that both user and interferer time-slots are time aligned. The BER at the equalizer output is shown in Fig. 8.19. The AA-DFE cannot deal with the abundance of delayed multipaths from both user and interferer and has the worst overall performance. The other two structures can better handle the interference and are able to extract more of the channel diversity. However, the ST-DFE presents superior performance for higher E_b/N_o values.

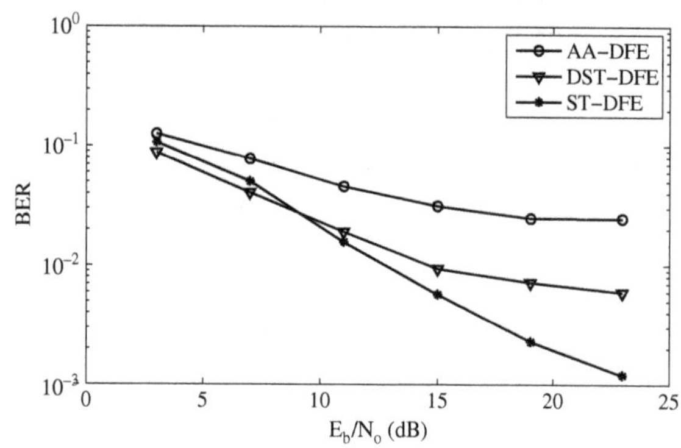

Fig. 8.19 Space–time equalizers performance.

8.6 Turbo-equalization: Near Optimal Performance in Coded Systems

The equalizers described in the previous sections of this chapter are essentially techniques that try to recover the signal at the channel input, based on the observation of the channel output. However, in most communication systems, the channel input is not the bit sequence of interest. In fact, practical systems employ error-correcting codes (ECC) [27]. These codes introduce redundancy into the information bits, thus increasing the system resilience to transmission errors. However, because of the redundancy, the channel input is not equal to the information bits.

In systems employing ECC, the detection strategy that minimizes the probability of error is similar to the maximum-likelihood equalizer. However, in this case, the receiver should seek the information sequence, i.e., the ECC input, that maximizes the likelihood of the channel output. On the other hand, the ML equalizer seeks the channel input, i.e., the ECC output, that maximizes the likelihood of the observation. Unfortunately, the search for the most likely information sequence requires a brute-force strategy, wherein every possible sequence is tested. If the message is transmitted in blocks of 1000 bits, this results in a search over 2^{1000} possible sequences, which is well above the number of atoms in the observable universe. (Current estimates place this number at 2^{266}.) Clearly, the resulting complexity is infeasible.

In practical systems, the receivers employ a low-complexity, suboptimal strategy for equalization and ECC decoding. First, the received sequence is equalized with any of the equalizers described in the previous sections of this chapter. Note that to mitigate the intersymbol interference the equalizers ignore the fact that the channel input is actually a coded sequence. In the second stage, the equalizer output is fed to a decoder for the ECC. This decoder exploits the structure of the ECC to recover some transmission errors, providing a generally good estimate of the information symbols. However, the decoder assumes that the equalizer completely eliminated ISI. In other words, equalizer and decoder operate independently.

To see why the independent approach is suboptimal, consider the example of a system employing a DFE, where the estimates of past symbols are used to cancel their interference and, hopefully, to improve the performance of the equalizer. Consider that a given symbol estimate is in error. If this wrong symbol is used in a DFE, its interference will not be canceled. Instead, it will be made worse, causing error propagation. The ECC may be able to recover this symbol correctly, and error propagation could be mitigated if the ECC could help the equalizer. However, since the structure of the ECC is not exploited by the DFE in the independent approach, the wrong symbol will be fed back, and error propagation will occur.

Turbo-equalizers provide a middle-ground solution between the infeasible exhaustive search approach and the independent approach. While keeping a complexity that is a constant multiple of the independent approach, it allows the equalizer to exploit the ECC to improve its performance. This is achieved through iterations between the equalizer and the decoder. In the first pass, the equalizer and the decoder work as in the independent approach, unaware of each other. In the ensuing iterations, the equalizer uses the decoder output to, hopefully, improve its estimates of the transmitted symbols. Given these better estimates, the decoder may then improve its own estimates of these symbols. The iterations then repeat, leading to an overall improved performance. In fact, the ISI introduced by the channel may be completely removed by the turbo-equalizer.

Turbo-equalizers rely on two key concepts, also found in turbo-codes: soft information and extrinsic information. Soft information means that the equalizer and the decoder exchange real numbers that may be used to estimate the transmitted symbol, and also measure how reliable a given estimate is. Usually, the a posteriori probability of the bits given the channel output is a great choice for soft information. In particular, the a posteriori probability may be computed by an algorithm similar to the Viterbi equalizer that was proposed by Bahl, Cocke, Jelinek and Raviv (BCJR) [9]. More importantly, the BCJR algorithm can easily incorporate a priori probabilities on the transmitted bits. This fact is exploited by turbo-equalizers: the equalizer output is used as a priori probabilities by the decoder, whereas the decoder output is used as a priori probabilities by the equalizer. This is how the equalizer benefits from the decoder output, and vice versa. Extrinsic information is harder to define, and a precise definition is left for later parts of this section.

Given their significant performance gains over traditional, non-iterative receivers, turbo-equalizers seem like attractive candidates for the receivers of future generation systems. Unfortunately, these gains come at a price: computational complexity. The BCJR algorithm is the equalizer of choice for turbo-equalization, but its computational cost grows exponentially with the channel memory. This has sparked a research interest on low-complexity alternatives to the BCJR equalizer. Fortunately, some unique characteristics of the ISI channel can be exploited to derive lower-complexity alternatives to the traditional BCJR algorithm.

In this section, turbo-equalizers will be explained in detail. In Section 8.6.1, the general concepts of turbo-equalization are described. In Section 8.6.2, the BCJR algorithm is described. In Section 8.6.3 some low-complexity alternatives to the BCJR algorithm are described. Finally, in Section 8.6.4, some simulation results that verify the performance improvements brought about by turbo-equalization are presented.

8.6.1 Principles

In this section, some of the principles behind turbo-equalization will be reviewed. First, the general setup of a turbo-equalizer is described. Then, the a posteriori probability is defined, and its merits for being the information to be exchanged between the equalizer and the decoder are discussed. Finally, the concept of extrinsic information is defined. A description of an algorithm for computing the a posteriori probability and the extrinsic information is deferred to the next section.

Turbo-equalizers are employed in coded systems. In general, it is assumed that the encoder is a block code or a terminated convolutional code [27], and a whole codeword will be recovered. This is in contrast to traditional equalizers, where symbol-by-symbol decisions are made. Also, it is assumed that an interleaver is inserted between the encoder and the channel. It is important to emphasize that its presence is crucial for turbo-equalizers. The resulting transmitter, for which a turbo-equalizer will be employed, is shown in Fig. 8.20. Note that the variables involved in this figure correspond to a whole codeword. Thus, **m** represents a block of

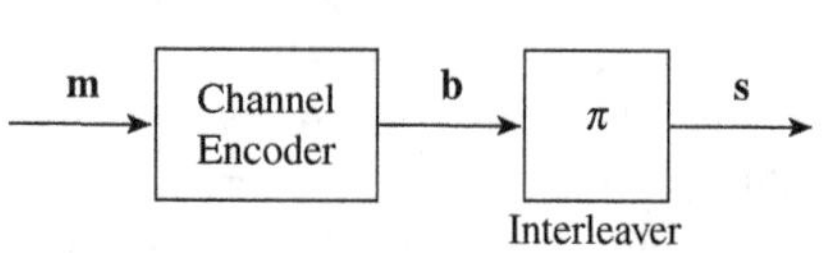

Fig. 8.20 The transmitter for a system with a turbo-equalizer. The channel encoder can be any code for which a soft-output decoder exists.

information bits, **b** represents a codeword and **s** represents the transmitted symbols after interleaving.

The general setup of a turbo-equalizer is shown in Fig. 8.21. The first block in this figure is the soft-input soft-output equalizer. Its inputs are the received sequence **x** corresponding to the transmission of a whole codeword, and the extrinsic information from the decoder, $\boldsymbol{\lambda}^{\mathrm{e}}$. Its output after deinterleaving, $\boldsymbol{\lambda}^{\mathrm{d}}$, is the extrinsic information. The decoder then uses $\boldsymbol{\lambda}^{\mathrm{d}}$ to compute improved values of $\boldsymbol{\lambda}^{\mathrm{e}}$, and the iterations repeat. Both the equalizer and the decoder may be based on the BCJR algorithm, which is described in the next section. In the remainder of this section, some variables in Fig. 8.21 are explained in more detail.

Fig. 8.21 Diagram of a turbo-equalizer.

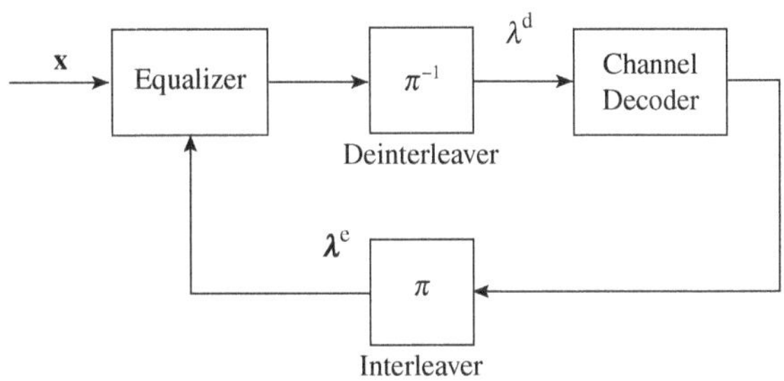

The information exchanged between the blocks of a turbo-equalizer must be soft, carrying at the same time an estimate of the transmitted bits and a measure of how reliable this estimate is. Turbo-equalizers exploit the reliability of the symbol estimates to decide how they will be used. Symbols with low reliability are practically ignored, whereas symbols with high reliability are treated as if they were the actual transmitted symbols.

Traditionally, the a posteriori probability is the soft information of choice for turbo-systems. For a BPSK modulation, the a posteriori probability is fully captured by the logarithm of the ratio of a posteriori probabilities (APP), which is loosely referred to as the log-likelihood ratio (LLR), defined as

$$L_n = \log\left(\frac{\Pr(s[n] = +1|\mathbf{x})}{\Pr(s[n] = -1|\mathbf{x})}\right), \tag{8.45}$$

where $s[n]$ refers to the nth transmitted symbol and **x** refers to the received sequence, corresponding to the transmission of one codeword. Note that L_n is actually the logarithm of the ratio of a posteriori probabilities (APP), not of likelihoods; however, the term LLR is now standard. In this chapter, for ease of notation, it is assumed that a BPSK modulation is used. Extension of turbo-equalization to higher order modulations can be found in [14, 47].

The LLR has several properties that make it useful for turbo-equalization. First, its sign gives the bit estimate that minimizes the probability of error [10]. Indeed, if $L_n > 0$, then the APP that the transmitted bit was 1 is larger, so this decision minimizes the probability of error. A similar reasoning holds when $L_n < 0$. More importantly, the magnitude of L_n measures the reliability of the estimate.

Now, applying Bayes' rule, L_n can be written as follows:

$$L_n = \log\left(\frac{\Pr(\mathbf{x}|s[n] = +1)}{\Pr(\mathbf{x}|s[n] = -1)}\right) + \log\left(\frac{\Pr(s[n] = +1)}{\Pr(s[n] = -1)}\right). \tag{8.46}$$

The second term in this equation, called a priori information (API), represents the log of the ratio of the a priori probabilities on the transmitted symbol. In general, $\Pr(s[n] = +1) = \Pr(s[n] = -1)$, so that the API should be zero. In turbo-equalization, however, the extrinsic information is treated as API, which forces this term to be non-null. In other words, the equalizer makes

$$\lambda_n^{\mathrm{e}} = \log\left(\frac{\Pr(s[n] = +1)}{\Pr(s[n] = -1)}\right). \tag{8.47}$$

Note that this is an approximation imposed by the iterative algorithm of a turbo-equalizer: the transmitted symbols *are* equally likely.

Equation (8.46) also highlights another important point. The LLR is the sum of the extrinsic information plus another term. If the LLR is fed directly to the decoder, then the extrinsic information provided by the decoder would return to it, causing positive feedback. However, a simple subtraction can eliminate the direct dependence of the LLR on the extrinsic information. This is how the equalizer output is computed: first the BCJR algorithm computes L_n, then the equalizer outputs $L_n - \lambda_n^{\mathrm{e}}$. The interleaver further improves the independence between the extrinsic information and the a priori information, hence its importance.

Figure 8.21 explains most of the turbo-equalization algorithm. The equalizer runs the BCJR algorithm, computing the LLR assuming that the a priori probabilities of the symbols are given by λ_n^{e}. The extrinsic information at the equalizer input is subtracted from the LLR, generating the extrinsic information that is fed to the decoder. The decoder then computes its LLR and extrinsic information, which is fed back to the equalizers. The iterations then repeat, until a stopping criterion is met. Note that the computational cost of each iteration is the same as of a traditional, noniterative, system. Thus, turbo-equalizers increase the complexity by a factor equal to the number of iterations, which is normally below 10. Also, at the first iteration the extrinsic information at the equalizer input is set to zero, and the equalizer operates as in a traditional system.

To finish the description of the turbo-equalizer, the BCJR algorithm is described in the following section.

8.6.2 The BCJR Algorithm

In this section, the BCJR algorithm, which is used to compute the LLR at the equalizer output, is described. The BCJR algorithm is based on a trellis description of the

ISI channel, similar to the Viterbi algorithm. Before describing a general form of the BCJR algorithm, a specific example is given. Suppose that the channel is given by $h(z) = 1 + z^{-1}$, so that its output at time n is $x[n] = s[n] + s[n-1] + \nu[n]$, where $\nu[n]$ is additive white Gaussian noise. Then, applying the definition of conditional probability followed by a marginalization on $s[n-1]$:

$$\Pr(s[n] = q|\mathbf{x}) = \sum_{p\in\pm 1} \Pr(s[n-1] = p, s[n] = q, \mathbf{x})/p(\mathbf{x}), \tag{8.48}$$

where q and p can assume the values $+1$ or -1. The advantage of the term on the right is that it can be decomposed in three independent terms, which can be easily calculated. It is also important to highlight that in computing ratios of probabilities, the term $p(\mathbf{x})$ can be ignored.

Now, let $\mathbf{x}_{k<n}$ and $\mathbf{x}_{k>n}$ denote vectors containing the past and future channel outputs, respectively. Then, using conditional probabilities:

$$\begin{aligned}\Pr(s[n-1] = p, s[n] = q, \mathbf{x}) =& \Pr(s[n] = q, x[n], \mathbf{x}_{k>n}|s[n-1] = p, \mathbf{x}_{k<n})\alpha_n(p)\\ =& \Pr(s[n] = q, x[n]|\mathbf{x}_{k>n}, s[n-1] = p, \mathbf{x}_{k<n})\\ &\times \Pr(\mathbf{x}_{k>n}|s[n-1] = p, \mathbf{x}_{k<n})\alpha_n(p),\end{aligned} \tag{8.49}$$

where

$$\alpha_n(p) = \Pr(s[n-1] = p, \mathbf{x}_{k<n}). \tag{8.50}$$

Furthermore, given $s[n-1] = p$, the joint probability of $s[n]$ and $x[n]$ depends only on the noise at time n. As such, it is independent of the past and future observations. Furthermore, given $s[n] = q$, the future observations are independent of the past. Then, (8.49) can be rewritten by

$$\Pr(s[n-1] = p, s[n] = q, \mathbf{x}) = \gamma_n(p,q)\beta_{n+1}(q)\alpha_n(p), \tag{8.51}$$

where

$$\begin{aligned}\gamma_n(p,q) =& \Pr(s[n] = q, x[n]|s[n-1] = p),\\ \beta_{n+1}(q) =& \Pr(\mathbf{x}_{k>n}|s[n] = q).\end{aligned} \tag{8.52}$$

Finally, the three terms in (8.51) can be computed as follows. First, use the definition of conditional probability to write

$$\gamma_n(p,q) = \Pr(x[n]|s[n] = q, s[n-1] = p)\Pr(s[n] = q|s[n-1] = p). \tag{8.53}$$

The first term on the right can be easily computed by noting that, given $s[n] = q, s[n-1] = p$, then $x[n]$ is a Gaussian random variable with mean $q + p$ and variance equal to the noise variance. Also, assuming that the bits are independent, the second term on the right is simply the probability that $s[n] = q$, i.e., the a priori probability of $s[n]$. These are computed from the extrinsic information defined in (8.47). Indeed, noting that $\Pr(s[n] = +1) + \Pr(s[n] = -1) = 1$, the a priori probabilities of $s[n]$ can be written as

$$\begin{aligned}\Pr(s[n]=+1) &= \frac{\exp(\lambda_n^{\mathrm{e}})}{1+\exp(\lambda_n^{\mathrm{e}})},\\ \Pr(s[n]=-1) &= \frac{1}{1+\exp(\lambda_n^{\mathrm{e}})}.\end{aligned} \tag{8.54}$$

The values of $\alpha_n(p)$ and $\beta_{n+1}(q)$ are computed by a forward and a backward recursion, respectively. Indeed, exploiting again the Markov structure of the channel:

$$\begin{aligned}\alpha_n(p) &= \sum_{q\in\pm 1} \alpha_{n-1}(q)\gamma_n(q,p),\\ \beta_n(q) &= \sum_{p\in\pm 1} \beta_{n+1}(p)\gamma_n(q,p).\end{aligned} \tag{8.55}$$

The initialization of these recursions will be discussed later.

To describe the BCJR algorithm for a general channel, firstly note that the channel is associated with a finite state machine (FSM), whose state is the symbols in the channel memory. For instance, the channel $h(z) = 1+z^{-1}$ has one memory element, and the state of the FSM is thus given by the symbol $s[n-1]$. A transition in the FSM is caused by the transmission of a symbol $s[n]$. The output of the FSM depends on the state and the transition, and is equal to the noiseless channel output. Again, in the example, the output corresponding to a state $s[n-1]$ and transition $s[n]$ is given by $s[n]+s[n-1]$. The actual channel output is the output of the FSM plus the noise term. These definitions are the same as those leading to the Viterbi equalizer.

Now, let $\psi[n]$ denote a possible state in the trellis at time n. The APP $\Pr(s[n] = a|\mathbf{x})$ can be computed from the APPs of the transitions, by summing over all transitions caused by the transmission of $s[n] = a$:

$$\Pr(s[n]=a|\mathbf{x}) = \sum_{p,q|a^{(p,q)}=a} \Pr(\psi[n]=p,\psi[n+1]=q|\mathbf{x})/p(\mathbf{x}), \tag{8.56}$$

where $a^{(p,q)}$ is the symbol that causes a transition from state p to state q. As in the example, using the fact that an FSM generates a Markov chain, the numerator in the summand of (8.57) can be written as

$$\Pr(\psi[n]=p,\psi[n+1]=q|\mathbf{x}) = \alpha_n(p)\gamma_n(p,q)\beta_{n+1}(q), \tag{8.57}$$

where

$$\begin{aligned}\alpha_n(p) =& \Pr(\psi[n]=p,\mathbf{x}_{k<n}),\\ \gamma_n(p,q) =& \Pr(\psi[n+1]=q,x[n]|\psi[n]=p),\\ \beta_{n+1}(q) =& \Pr(\mathbf{x}_{k>n}|\psi[n+1]=q).\end{aligned} \tag{8.58}$$

As before, rewritting $\gamma_n(p,q)$ results in

$$\gamma_n(p,q) = \Pr(x[n]|\psi[n+1]=q,\psi[n]=p)\Pr(\psi[n+1]=q|\psi[n]=p), \tag{8.59}$$

where $\Pr(x[n]|\psi[n+1]=q,\psi[n]=p)$ is a Gaussian density function with variance equal to that of the noise term. Its mean is the FSM output corresponding to a transition from state p to q. The second term is simply the a priori probability that the channel input at time n is $a^{(p,q)}$, i.e., the input that causes a transition between states $\psi[n] = p$ and $\psi[n+1] = q$. Again, this value is computed from the extrinsic information coming from the decoder.

In the general setting, the values of $\alpha_n(p)$ and $\beta_{n+1}(q)$ are also computed by forward and backward recursions given by

$$\alpha_n(p) = \sum_q \alpha_{n-1}(q)\gamma_n(q,p) \quad \text{and} \quad \beta_n(q) = \sum_p \beta_{n+1}(p)\gamma_n(q,p). \tag{8.60}$$

Note that these sums are over all possible states. However, it is important to emphasize that not all state transitions are possible; for these transitions, it is necessary to set $\gamma = 0$. Thus, the invalid transitions may be ignored in the recursions. The recursions are initialized according to some assumptions. If the channel is flushed before and/or after transmission of a codeword by the transmission of L known symbols, the corresponding value of $\alpha_{-1}(p)$ and/or $\beta_{M+1}(q)$ is set to 1, while the remaining values are set to zero. Otherwise, the initial values of these variables are set to be equal.

It is important to point out that the recursions for α and β may lead to underflow in finite precision computers. However, ratios of probabilities must be calculated, so that multiplicative factors are irrelevant in our computations. Thus, after computing the recursions at time instant n, $\alpha_n(p)$ and $\beta_n(q)$ may be normalized so that $\sum_p \alpha_n(p) = 1$ and $\sum_q \beta_n(q) = 1$. This normalization avoids the underflow problem.

The BCJR algorithm can also be used to compute the APP for convolutional codes, since they can also be represented by an FSM. However, in the case of turbo-equalization, the decoder does not have access to a channel observation, only to the equalizer output. Thus, the probability of a transition is determined solely from the API. In other words, for the decoder, $\gamma_n(p,q)$ can be computed as

$$\gamma_n(p,q) = \Pr(s[n] = a^{(p,q)}). \tag{8.61}$$

The other steps of the BCJR algorithm for the convolutional decoder are the same as the equalizer.

As is well known, the complexity of the BCJR algorithm grows exponentially with the channel memory and constellation size. As a result, the BCJR equalizer may be infeasible for channels with long memory or for high-order modulations. In the next section, some alternatives to the BCJR equalizer, with reduced complexity are described.

8.6.3 Structures and Solutions for Low-Complexity Implementations

Low-complexity alternatives to the BCJR algorithm are highly desirable, and may even be a necessity for the practical employment of turbo-equalizers. In this section, two strategies to reduce the complexity of the BCJR algorithm are described. These can be grouped into two categories.

The first strategy is based on reduced search algorithms [13, 16, 40]. These are similar to the BCJR algorithm and use similar recursions. However, they reduce complexity by ignoring some state transitions or by ignoring some states altogether. For instance, the algorithm in [16] retains only the states with the largest values of α and/or β, and considers only transitions stemming for these states. Although these strategies provide a good compromise between performance and complexity, they normally fail to completely eliminate the ISI, as will be shown in the simulation results. Therefore, they will not be described in further detail in this section.

The second strategy is based on linear filters and interference cancelation. Seminal works in this area include [39, 46, 47, 49]. Essentially, these algorithms compute a linear minimum mean square error estimate of the transmitted symbols, using the a priori information to compute the means and variances of the interfering symbols. The resulting estimator depends on the specific value of the API of the interfering symbols, and thus is different for every transmitted symbol. Thus, the equalizer is time varying.

Before describing the linear filter techniques in detail, it is interesting to consider an ideal situation wherein all the transmitted symbols but one are known to the detector. Let the unknown symbol be $s[n]$. In this case, the influence of the remaining symbols on the received sequence $\mathbf{x}$ can be computed and canceled. Then, the resulting sequence, containing only the influence of the desired symbol, goes through a matched filter, whose output is used to estimate $s[n]$. The resulting detector achieves the matched-filter bound [10].

In turbo-equalization, the interfering symbols are not known with certainty. However, the decoder provides their a priori probabilities, so that tentative estimates of these symbols can be made. These can be used to make tentative estimates of their interference, which is then canceled. The resulting sequence, with hopefully less interference than the received sequence $\mathbf{x}$, is then filtered. The resulting scheme is depicted in Fig. 8.22. If the quality of the tentative estimates is good, i.e., if the soft output provided by the decoder has large reliability, most of the interference was successfully canceled, and this filtering operation should be performed by a matched filter. If, on the other hand, the tentative estimates are poor, very little interference should be canceled, so that the filter input is similar to the received sequence $\mathbf{x}$. In this case, the filter should be a traditional equalizer used to mitigate ISI, such as the MMSE or ZF equalizers.

Two points must be emphasized about the structure shown in Fig. 8.22. First, the extrinsic information is used to estimate the *interference* term on $\mathbf{x}$. In other words, the contribution of $s[n]$ to $\mathbf{x}$ is not eliminated. As a consequence, the extrinsic information related to $s[n]$ is not used when $s[n]$ is being estimated. That is to say that the equalizer output at time n is independent of λ_n^{e}. In other words the equalizer output corresponds to extrinsic information.

After this intuitive motivation for the use of linear equalizers for turbo-equalization, a rigorous description of a strategy based on MMSE equalization is

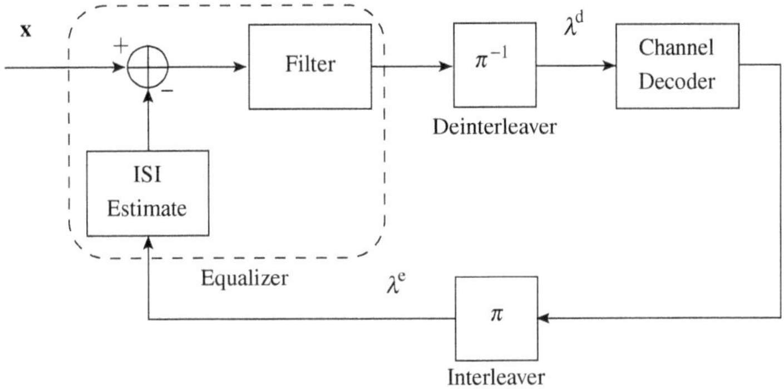

Fig. 8.22 Diagram of a turbo-equalizer based on linear filters, showing some details of the equalization block.

presented. From this point on, the derivation is restricted to BPSK modulations; extension to other modulations can be found in [14, 47]. To incorporate the API into the derivation of the MMSE equalizer, it is important to observe that this equalizer depends on the first and second moments of the variables involved. In a turbo-equalizer, we can use the extrinsic information to estimate a posteriori values of these statistics, conditioned on the received signal **x**. For a BPSK modulation, (8.54) is used to compute the mean:

$$\begin{aligned}\mathbb{E}[s[n]|\mathbf{x}] &= +1 \times \Pr(s[n] = +1|\mathbf{x}) - 1 \times \Pr(s[n] = +1|\mathbf{x}) \\ &= \frac{\exp(\lambda_n^{e})}{1+\exp(\lambda_n^{e})} - \frac{1}{1+\exp(\lambda_n^{e})} \\ &= \tanh(\lambda_n^{e}/2).\end{aligned} \tag{8.62}$$

Likewise, the variance can be computed as follows:

$$\begin{aligned}\mathrm{var}[s[n]|\mathbf{x}] &= \mathbb{E}[s[n]^2|\mathbf{x}] - \mathbb{E}[s[n]|\mathbf{x}]^2 \\ &= 1 - \tanh^2(\lambda_n^{e}/2).\end{aligned} \tag{8.63}$$

As with traditional equalizers, a delay is introduced to ensure causality so that at time n, $s[n-\Delta]$ is estimated. Now, let $\mathbf{w}[n]$ and $\mathbf{x}[n]$ be length N_e vectors of equalizer coefficients and inputs at time n, respectively. It is well known [49] that the MMSE linear estimate of $s[n-\Delta]$ based on $\mathbf{x}[n]$ is given by

$$\hat{s}[n-\Delta] = \mathbf{w}[n]^H \left(\mathbf{x}[n] - \mathbb{E}[\mathbf{x}[n]]\right), \tag{8.64}$$

where $\mathbf{w}[n] = \left(\mathbb{E}[\mathbf{x}[n]\mathbf{x}[n]^H]\right)^{-1} \mathbb{E}[\mathbf{x}[n]b^*[n-\Delta]]$.

To compute the expected values required by $\hat{s}[n-\Delta]$, let $\mathbf{H}_c$ be the $N_e \times (N_e + L+1)$ channel convolution matrix and $\mathbf{s}[n]$ be the vector of channel inputs of length

$N_e + L + 1$, where L is the channel memory. These are assumed to be independent random variables with mean and variance given by (8.62) and (8.63), except for the entry corresponding to the desired symbol. For this entry, the API is not used, so $s[n]$ is still assumed to have zero mean and unit variance, resulting in

$$\mathbb{E}[\mathbf{x}[n]] = \mathbf{H}_c \bar{\mathbf{s}}[n], \tag{8.65}$$

where $\bar{\mathbf{s}}[n]$ is a length $(N_e + L + 1)$ vector containing the expected values of the channel inputs, whose ith entry is given by

$$[\bar{\mathbf{s}}[n]]_i = \begin{cases} 0, & i = \Delta, \\ \tanh(\lambda^{\mathrm{e}}_{n-i}/2), & \text{otherwise.} \end{cases} \tag{8.66}$$

The covariance matrix of $\mathbf{x}[n]$, $\mathbf{R}_{\mathbf{x}}[n]$ is given by

$$\begin{aligned} \mathbf{R}_x[n] &= \mathbb{E}[\mathbf{x}[n]\mathbf{x}[n]^H] \\ &= \mathbf{H}_c \mathbb{E}[\mathbf{s}[n]\mathbf{s}[n]^H]\mathbf{H}_c^H + \sigma^2 \mathbf{I}. \end{aligned} \tag{8.67}$$

Let $\mathbf{R}_s[n] = \mathbb{E}[\mathbf{s}[n]\mathbf{s}[n]^H]$. Note that the transmitted symbols are still assumed to be independent, so that $\mathbb{E}[s[n]b^*[m]] = 0$ when $n \neq m$. Thus, $\mathbf{R}_s[n]$ is a diagonal matrix. Its diagonal element corresponding to $\mathbb{E}[|s[n-\Delta]|^2]$ is equal to 1, since the statistics of the symbol of interest based on the API are not changed. The remaining values are computed according to (8.63):

$$[\mathbf{R}_s[n]]_{i,i} = \begin{cases} 1, & i = \Delta, \\ 1 - \tanh^2(\lambda^{\mathrm{e}}_{n-i}/2), & \text{otherwise.} \end{cases} \tag{8.68}$$

Finally, as in traditional MMSE equalization, $\mathbb{E}[\mathbf{x}[n]b^*[n-\Delta]] = \mathbf{p}$, where $\mathbf{p}$ is the Δth column of $\mathbf{H}_c$, with counting beginning at 0.

In summary, the MMSE estimate of $s[n-\Delta]$ given $\mathbf{x}[n]$ is given by

$$\hat{s}[n-\Delta] = \mathbf{p}^H \left(\mathbf{H}_c \mathbf{R}_s[n]\mathbf{H}_c^H + \sigma^2 \mathbf{I}\right)^{-1} \left(\mathbf{x}[n] - \mathbf{H}_c \bar{\mathbf{s}}[n]\right), \tag{8.69}$$

and the equalizer coefficients are given by $\mathbf{w}[n] = \left(\mathbf{H}_c \mathbf{R}_s[n]\mathbf{H}_c^H + \sigma^2 \mathbf{I}\right)^{-1} \mathbf{p}$. Also note that the equalizer coefficients depend on the variance of the interfering symbols, which change with time. This results in a, a time-varying equalizer (TVE) whose coefficients must be computed anew for every time instant.

Now that the equalizer output was calculated, it is necessary to write it in the form of an LLR. To that end, the equalizer output in (8.69) is rewritten as

$$\hat{s}[n-\Delta] = A[n]s[n-\Delta] + v'[n], \tag{8.70}$$

where $A[n]$ represents the bias of the MMSE equalizer and $v'[n]$ represents the mean-squared error. Now, using standard MMSE techniques, it can be shown [46] that

$$A[n] = \mathbf{w}^H[n]\mathbf{p}, \tag{8.71}$$

and the error $v'[n]$ is a zero-mean random variable with variance

$$\sigma_{v'}^2 = \mathbf{w}^H[n]\mathbf{p}(1 - \mathbf{p}^H\mathbf{w}[n]). \tag{8.72}$$

Now, a crucial approximation is made [36]: it is assumed that $v'[n]$ is Gaussian. In this case, the equalizer output can be seen as the output of an AWGN channel. Computation of the LLR in this case is straightforward:

$$\log\left(\frac{\Pr(s[n] = +1|\hat{s}[n])}{\Pr(s[n] = -1|\hat{s}[n])}\right) = \log\left(\frac{\frac{1}{\sqrt{2\pi\sigma_{v'}^2}}\exp\left(-\frac{1}{2\sigma_{v'}^2}(\hat{s}[n] - A[n])\right)}{\frac{1}{\sqrt{2\pi\sigma_{v'}^2}}\exp\left(-\frac{1}{2\sigma_{v'}^2}(\hat{s}[n] + A[n])\right)}\right), \tag{8.73}$$

$$= \frac{2A[n]}{\sigma_{v'}^2}\hat{s}[n].$$

Equation (8.69) has some interesting interpretations. At the first iteration, the API on all symbols is zero. Thus, all symbols are assumed to have zero mean and unit variance, so the equalizer coefficients correspond to the traditional MMSE equalizer. On the other hand, if the API is of high quality, then the interfering bits are estimated with almost certainty. In other words, the variance of the interfering bits is zero, and their expected value is equal to their actual value. In this case, the matrix inversion lemma may be used to show that the equalizer reverts to an interference canceler with matched filter, as expected [46].

As mentioned before, the equalizer in (8.69) is time varying, so that its complexity is in the order of N_e^2. Even though this may be smaller than the complexity of the BCJR, it can still be prohibitive for long channels. Thus, some alternatives to further reduce the complexity of the TVE were proposed in the literature. The first alternative was proposed by the same authors of the TVE. Based on the limiting behavior of the equalizer analyzed in the previous paragraph, the authors in [46] proposed a hybrid equalizer (HE) that switches between the MMSE and the interference canceler. The choice is based on a measure of the quality of the API, proposed in [46]: if the API is good according to this measure, the interference canceler is used. If the API is bad, the MMSE equalizer is used.

The hybrid equalizer abruptly changes between two extreme scenarios: one that considers no API and another that considers perfect API. An interesting alternative with similar complexity is the soft-feedback equalizer (SFE) [31]. The SFE is based on two ideas. The first is to consider that the a priori information provided by the decoder, λ_n^{e}, is not a sequence of deterministic values known beforehand by the equalizer. Instead, the SFE considers λ_n^{e} to be a random variable with a given mean and variance, and it minimizes the mean-squared error based on this assumption. The result is a time-invariant equalizer, with linear complexity.

The second idea behind the SFE is similar in principle to the DFE. The TVE uses the API to compute tentative estimates of the interfering symbols. Now, at time n, $s[n-\Delta]$ is estimated; however, at this time instant the equalizer has already computed the extrinsic information on the symbols that precede $s[n-\Delta]$. This can be combined with the API from the decoder to produce a posteriori probabilities on these symbols, $\mathbf{L}^{\mathrm{e}}$, as in (8.46). These APPs should provide more reliable symbol estimates than the API alone.

The structure of the SFE is depicted in Fig. 8.23. In this figure, the received sequence $\mathbf{x}$ first goes through a linear filter with impulse response $\mathbf{w}$. The output of this filter contains a contribution from the desired symbol, $s[n-\Delta]$, plus residual interference from both past and future symbols. The a priori information from the decoder, $\boldsymbol{\lambda}^{\mathrm{e}}$, is used to produce tentative estimates of the future interfering symbols, based on (8.62). These symbol estimates then go through a filter with impulse response $\mathbf{s}_1$, whose output is an estimate of the residual interference at the output of $\mathbf{w}$ caused by future symbols. The interference from past symbols is canceled similarly. The difference is that the tentative estimates are based on the full LLR $\mathbf{L}^{\mathrm{e}}$.

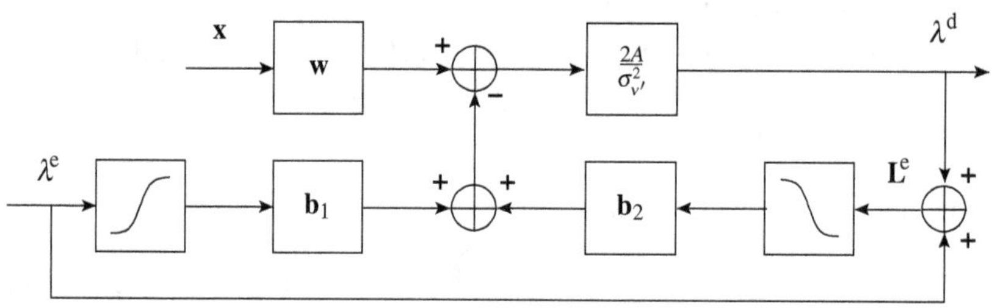

Fig. 8.23 Diagram of the soft-feedback equalizer.

It should be pointed out that the structure depicted in Fig. 8.23 can also be used to represent the TVE and the HE. The main difference is in the choice of the filters. The other difference is that the feedback loop, connecting the equalizer output to the filter $\mathbf{s}_2$, does not exist in the TVE and the HE.

The SFE coefficients can be computed using standard MSE minimization techniques, similar to the derivation of the DFE. Indeed, these coefficients are given by

$$\begin{aligned}\mathbf{w} &= \left(\mathbf{H}_c\mathbf{H}_c^H - \alpha_1\mathbf{H}_1\mathbf{H}_1^H - \alpha_2\mathbf{H}_c\mathbf{H}_c^H + \sigma^2\mathbf{I}\right)^{-1}\mathbf{p},\\ \mathbf{s}_1 &= -\mathbf{H}_1^H\mathbf{w},\\ \mathbf{s}_2 &= -\mathbf{H}_2^H\mathbf{w}.\end{aligned} \tag{8.74}$$

As before, $\mathbf{p}$ is the Δth column of $\mathbf{H}_c$, with counting beginning at 0. The matrices $\mathbf{H}_1$ and $\mathbf{H}_2$ are submatrices of $\mathbf{H}_c$, which are defined by writing

$$\mathbf{H}_c = \left[\mathbf{H}_1 \quad \mathbf{p} \quad \mathbf{H}_2\right]. \tag{8.75}$$

Finally,

$$\begin{aligned}\alpha_1 &= \mathbb{E}\left[\tanh\left(\frac{\lambda_n^{\mathrm{e}}}{2}\right)s[n]\right],\\ \alpha_2 &= \mathbb{E}\left[\tanh\left(\frac{L_n^{\mathrm{e}}}{2}\right)s[n]\right].\end{aligned} \tag{8.76}$$

These expected values are estimated before each iteration of the SFE. More details on how to estimate α_1 and α_2 can be found in [31].

8.6.4 Simulation Results

This section presents some simulation results attesting the good performance of turbo-equalizers, and also compares several different equalization strategies.

In the first simulation, the performance of a BCJR-based turbo-equalizer is compared with turbo-equalizers based on linear filters: the TVE and the HE of [46], the SFE of [31], and the reduced-state (RS) equalizer of [16]. To that end, the transmission of 2^{15} bits through a channel with impulse response $\mathbf{h} = [0.227, 0.46, 0.688, 0.46, 0.227]$ is simulated. The bits are first encoded with a rate of $1/2$ recursive systematic convolutional encoder with generator polynomials $[7\ 5]$ in octal representation. The results, shown in Fig. 8.24, are averaged over 100 trials and after 14 iterations of the turbo-equalizer. The TVE, SFE, and HE use forward equalizers with 15 coefficients and a delay of $\Delta = 6$. As seen in the figure, the more complex the equalizer, the better the performance. However, for a BER of 10^{-3}, the SFE is only 0.33 dB away from the TVE, while its complexity is similar to the HE. Note that the results for the TVE, the SFE, the HE, and the BCJR were already presented in [30].

The RS equalizer uses only eight states, half of those of a full-complexity BCJR algorithm. The output saturation parameter specified in [16] was set to $\gamma = \exp(-5)$. As shown in Fig. 8.24, both the RS and the TVE turbo-equalizers have waterfall regions[8.6] around 4.75 dB. However, as seen in this figure, the RS equalizer fails to eliminate ISI for the range of SNR considered. In fact, RS is eventually outperformed by all other turbo-equalizers.

In Fig. 8.24 the performance of the code in an AWGN channel, which does not introduce any intersymbol interference, is also plotted. This curve shows one of the most striking features of turbo-equalizers: after a few iterations, and for a high-enough SNR, the equalizers perform as if there were no channel. In other words, turbo-equalization is capable of completely removing the ISI. Also, Fig. 8.24 shows the smallest value of E_b/N_o required for error-free transmission of a BPSK signal with a rate 1/2 code on the channel $\mathbf{h}$, as predicted by Shannon's results. This rate was computed using the results in [5]. As seen in the figure, for a BER of 10^{-3},

[8.6] In turbo-systems, the waterfall region is the range of SNR where the BER decreases quickly.

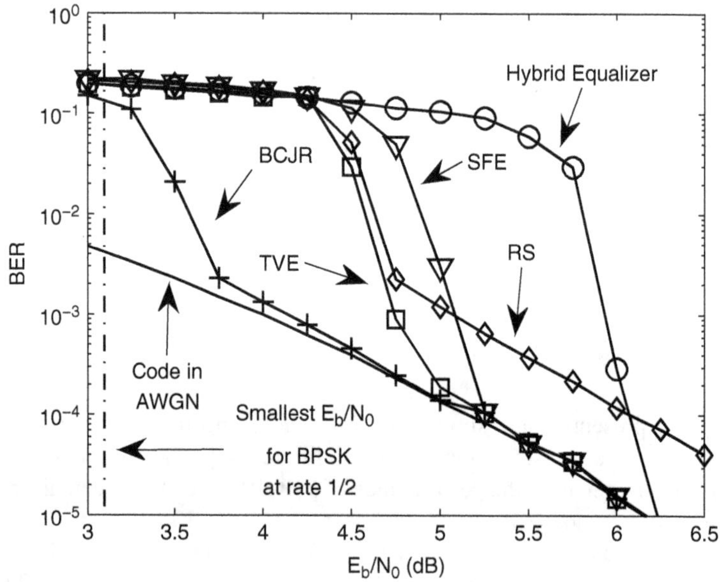

Fig. 8.24 BER performance of several turbo-equalizers.

the BCJR-based turbo-equalizer operates at only 1 dB from the Shannon limit. Note that this performance is achieved with a fairly simple code.

8.7 Conclusions

There is a great variety of equalization techniques reported in the literature. In this chapter a few expressive techniques exploring different structures and algorithms were selected, analyzed and illustrated. First, simple SISO systems were described to review classical adaptive equalization techniques, discussing different supervised and unsupervised optimization criteria and possible algorithms, taking into account computational cost, speed of convergence, and misadjustment. Nonlinear equalization techniques that can provide an additional performance gain were also introduced.

Next, the SIMO equalization structures were analyzed by incorporating the space dimension through the use of multiple receive antennas. This kind of structure presents important advantages when combating ISI, fading, and multiuser interference.

Finally, the turbo-equalization techniques, which represent the state of art in equalization, are presented. Through the joint use of filtering and error correction codes, it is able to achieve a near optimal performance with a much smaller computational complexity when compared to the optimal solution.

References

1. 3GPP TR 25.996: Spatial channel model for multiple input multiple output (MIMO) simulations, available online at http://www.3gpp.org. 3GPP (2003)
2. Alamouti, S.: A simple transmit diversity technique for wireless communications. IEEE Journal on, Selected Areas in Communications, **16**(8), 1451–1458 (1998)
3. Altekar, S., Beaulieu, N.: Upper bounds to the error probability of decision feedback equalization. IEEE Transactions on Information Theory, **39**(1), 145–156 (1993)
4. Apolinário, Jr, J.A.: QRD-RLS Adaptive Filtering. 1st edn. Springer (2009)
5. Arnold, D.M., Loeliger, H.-A., Vontobel, P.O., Kavcic, A., Zeng, W.: Simultlaion-based computation of Information rates for channels with memory. IEE Transactions of Information Theory, **52**(8), 3498–3508 (2006). DOI 10.1109/TIT.2006.878110
6. Ariyavisitakul, S., Li, Y.: Joint coding and decision feedback equalization for broadband wireless channels. IEEE Journal on Selected Areas in Communications, **16**(9), 1670–1678 (1998)
7. Ariyavisitakul, S., Winters, J., Lee, I.: Optimum space–time processors with dispersive interference: unified analysis and required filter span. IEEE Transactions on Communications, **47**(7), 1073–1083 (1999)
8. Austin, M.: Decision feedback equalization for digital communications over dispersive channels. MIT Research Laboratory of Electronics Technical Report (461) (1967)
9. Bahl, L., Cocke, J., Jelinek, F., Raviv, J.: Optimal decoding of linear codes for minimizing symbol error rate (corresp.). IEEE Transactions on Information Theory, **20**(2), 284–287 (1974)
10. Barry, J.R., Messerschmitt, D.G., Lee, E.A.: Digital Communications, 3rd edn. Springer: New York (2003)
11. Beaulieu, N.: Bounds on variances of recovery times of decision feedback equalizers. IEEE Transactions on Information Theory, **46**(6), 2249–2256 (2000)
12. Benvenist, A., Goursat, M., Ruget, G.: Robust identification of a nonminimum phase system: blind adjustment of a linear equalizer in data communications. IEEE Transactions on Automatic Control, **AC-25**(3), 385–399 (1980)
13. Colavolpe, G., Ferrari, G., Raheli, R.: Reduced-state BCJR-type algorithms. IEEE Journal on Selected Areas in Communications, **19**(5), 848–859 (2001)
14. Dejonghe, A., Vandendorpe, L.: Turbo-equalization for multilevel modulation: an efficient low-complexity scheme. IEEE International Conference on Communications, ICC 2002 **3**, 1863–1867 (2002)
15. Diniz, P.: Adaptive Filtering: Algorithms and Practical Implementation. Kluwer Academic Publishers: Dordrecht (1997)
16. Fertonani, D., Barbieri, A., Colavolpe, G.: Reduced-complexity BCJR algorithm for turbo equalization. IEEE Transactions on Communications, **55**(12), 2279–2287 (2007)
17. Forney G., J.: Maximum-likelihood sequence estimation of digital sequences in the presence of intersymbol interference. IEEE Transactions on Information Theory, **18**(3), 363–378 (1972)
18. Fujii, M.: Path diversity reception employing steering vector arrays and sequence estimation techniques for ISI channels. IEEE Journal on Selected Areas in Communications, **17**(10), 1735–1746 (1999)
19. Gerstacker, W., Schober, R.: Equalization concepts for EDGE. IEEE Transactions on Wireless Communications, **1**(1), 190–199 (2002)
20. Gesbert, D., Shafi, M., Shan Shiu, D., Smith, P., Naguib, A.: From theory to practice: an overview of MIMO space–time coded wireless systems. IEEE Journal on Selected Areas in Communications, **21**(3), 281–302 (2003)
21. Godard, D.: Self-recovering equalization and carrier tracking in two-dimensional data communication systems. IEEE Transactions on Communications, **28**(11), 1867–1875 (1980)
22. Hanaki, A., Ohgane, T., Ogawa, Y.: A novel cost function for cascaded connection of adaptive array and MLSE. IEEE VTS 50th Vehicular Technology Conference, 1999. VTC 1999 - Fall, vol. 1, 6–10 (1999)
23. Haykin, S.: Adaptive Filter Theory, 3rd edn. Prentice Hall: Englewood Cliffs, NJ (1996)

24. Kennedy, R.A., Anderson, B.D.O.: Recovery times of decision feedback equalizers on noiseless channels. IEEE Transactions on Communications, **35**, 1012–1021 (1987)
25. Kennedy, R.A., Anderson, B.D.O., Bitmead, R.R.: Tight bounds on the error probabilities of decision feedback equalizers. IEEE Transactions on Communications, **35**, 1022–1029 (1987)
26. Leou, M.L., Yeh, C.C., Li, H.J.: A novel hybrid of adaptive array and equalizer for mobile communications. IEEE Transactions on Vehicular Technology, **49**(1), 1–10 (2000)
27. Lin, S., Costello, D.J.: Error Control Coding, 2nd edn. Prentice Hall: Englewood Cliffs, NJ (2004)
28. Liu, J.T., Gelfand, S.: Optimized decision-feedback equalization for convolutional coding with reduced delay. IEEE Transactions on Communications, **53**(11), 1859–1866 (2005)
29. Ljung, L., Gunnarsson, S.: Adaptation and tracking in system identification—a survey. Automatica **26**(1), 7–21 (1990)
30. Lopes, R., Barry, J.R.: Soft-output decision-feedback equalization with a priori information. IEEE Global Telecommunications Conference, 2003. GLOBECOM '03, vol. 3, 1705–1709 (2003)
31. Lopes, R., Barry, J.R.: The soft-feedback equalizer for turbo equalization of highly dispersive channels. IEEE Transactions on Communications **54**(5), 783–788 (2006)
32. Lucky, R., Salz, J., Weldon, E.: Principles of Data Communication. McGraw-Hill: Nova York (1968)
33. Nikias, C., Petropulu, A.: Higher-order Spectra Analysis: A Nonlinear Signal Processing Framework. Prentice Hall: Englewood Cliffs, NJ (1993)
34. Paulraj, A., Papadias, C.: Space–time processing for wireless communications. IEEE Signal Processing Magazine, **14**(6), 49–83 (1997)
35. Pipon, F., Chevalier, P., Vila, P., Monot, J.J.: Joint spatial and temporal equalization for channels with ISI and CCI-theoretical and experimental results for a base station reception. 1997 First IEEE Signal Processing Workshop on Signal Processing Advances in Wireless Communications, 309–312 (1997)
36. Poor, H.V., Verdu, S.: Probability of error in MMSE multiuser detection. IEEE Transactions on Information Theory, **43**(3), 858–871 (1997)
37. Proakis, J.: Digital Communications, 4th edn. McGraw-Hill: New York (2001)
38. Regalia, P.A.: Adaptive IIR filtering in signal processing and control. Marcel Dekker: New York (1995)
39. Reynolds, D., Wang, X.: Low-complexity turbo-equalization for diversity channels. IEEE Transactions on Communications, **81**(5), 989–995 (2001)
40. Rusek, F., Loncar, M., Prlja, A.: A comparison of Ungerboeck and Forney models for reduced-complexity ISI equalization. IEEE Global Telecommunications Conference, GLOBECOM '07, 1431–1436 (2007)
41. Shalvi, O., Weinstein, E.: New criteria for blind deconvolution of nonminimum phase systems (channels). IEEE Transactions on Information Theory, **36**(2), 312–321 (1990)
42. Shalvi, O., Weinstein, E.: Blind Deconvolution, chap. Universal Methods for Blind Deconvolution. Prentice Hall: Englewood Cliffs, NJ (1994)
43. Shynk, J.: Adaptive IIR filtering. IEEE ASSP Magazine, **6**(2), 4–21 (1989)
44. Sklar, B.: How I learned to love the trellis. IEEE Signal Processing Magazine, **20**(3), 87–102 (2003)
45. Tomisato, S., Matsumoto, T.: A joint spatial and temporal equalizer using separated spatial and temporal signal processing for broadband mobile radio communications. IEEE Third Workshop on Signal Processing Advances in Wireless Communications, 2001 (SPAWC '01), 298–301 (2001)
46. Tuchler, M., Koetter, R., Singer, A.: Turbo equalization: principles and new results. IEEE Transactions on Communications, **50**(5), 754–767 (2002)
47. Tuchler, M., Singer, A., Koetter, R.: Minimum mean squared error equalization using a priori information. IEEE Transactions on Signal Processing, **50**(3), 673–683 (2002)
48. Ungerboeck, G.: Adaptive maximum-likelihood receiver for carrier-modulated data-transmission systems. IEEE Transactions on Communications, **22**(5), 624–636 (1974)
49. Wang, X., Poor, H.: Iterative (turbo) soft interference cancellation and decoding for coded CDMA. IEEE Transactions on Communications **47**(7), 1046–1061 (1999)

Chapter 9
Channel Estimation for OFDM Systems: Techniques, Algorithms, and Performance

Rui F. Vigelis, Darlan C. Moreira, and Charles C. Cavalcante

9.1 Introduction

The use of OFDM is gaining more and more interest from communications and signal processing communities due to its great potential to cope with frequency-selective channels. Further, OFDM also allows to reduce the complexity for tasks such as equalization and channel estimation in systems where the bandwidth of the channel is wide, when compared to classical (non-OFDM) approaches.

This technology, in conjunction with when associated with other diversity methods such as multiple-input multiple-output (MIMO) and space–time block code (STBC), has been shown to be the logical choice for consideration in Beyond-3G wireless systems.

The rest of the chapter is organized as follows. Section 9.2 describes the model of an orthogonal frequency division multiplexing (OFDM) system, including a model for the time-varying radio propagation channel. Section 9.3 is dedicated to non-adaptive methods for time-varying channels. Adaptive methods are then investigated in Section 9.4. Different channel estimation strategies for MIMO-OFDM system are presented in Section 9.5. In Section 9.6, the ideas and proposals are summarized and future research directions within this area are identified.

9.2 OFDM Fundamentals

"Divide and conquer" is usually considered a good strategy to solve difficult problems. The idea is to split the problem into smaller and simpler problems that taken separately can be more easily solved. These solutions are then combined to solve the original problem.

In wideband transmission "the hard problem" consists of transmitting a signal through a frequency-selective channel, which requires the use of equalization and the channel transfer function to be estimated. Performing this equalization in practice with data rates of several megabits per second is not an easy task.

F. Cavalcanti, S. Andersson (eds.), *Optimizing Wireless Communication Systems*,
DOI 10.1007/978-1-4419-0155-2_9, © Springer Science+Business Media, LLC 2009

Following the divide-and-conquer paradigm, OFDM modulation divides the channel into narrowband subchannels. By making all subchannels narrowband, each subchannel can be considered to be frequency flat channel, as shown in Fig. 9.1. Each subchannel has a constant amplitude gain and a constant phase shift over the subchannel's bandwidth when passing a radio wave through it.[9.1] The information is then transmitted through these subchannels and the receiver only has to multiply the detected information in each subchannel by a complex gain to compensate the attenuation and phase shift of the whole channel. This approach makes estimation of the channel and decoding the received information simpler.

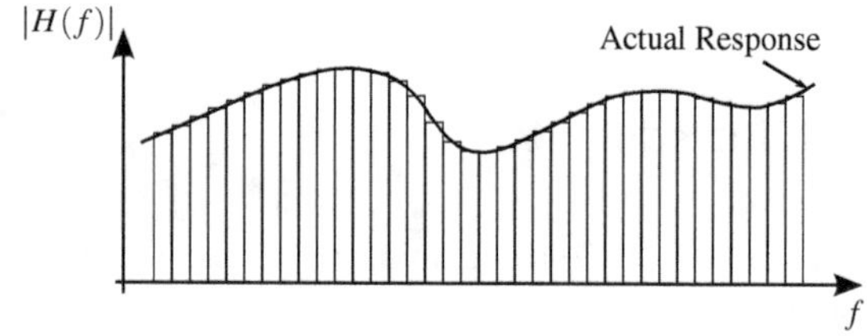

Fig. 9.1 A staircase model approximation of the gain from an arbitrary channel.

9.2.1 OFDM Transceiver Architecture and Multicarrier Channel Models

The OFDM technique started as an evolution of frequency division multiplexing (FDM). Instead of using guard bands to separate the subcarriers in the signal reception as in FDM, there is a special superposition of the subcarriers in OFDM that makes them orthogonal. This allows the receiver to decode the information in each subcarrier.

Figure 9.2 shows an OFDM spectrum for a five subcarrier system, where T is the symbol duration period and the frequency is normalized with respect to $1/T$ with the zero in the OFDM central carrier frequency.[9.2] Note that each subcarrier is carefully centered in the "zero crossing points" of the other subcarriers, which corresponds to the earlier mentioned "special superposition".

In the original idea of the OFDM [2], a set of coherent oscillators in parallel was used, one for each subcarrier. This resulted in a difficult and expensive implementation, especially when the number of subcarriers is large. In Fig. 9.3 the baseband model of an OFDM system is illustrated. The s_k is the symbol to be transmitted by the kth subcarrier.

[9.1] The greater the number of subchannels, the closer this approximation will be to the actual channel.

[9.2] Each subcarrier has a pulse shape typical of a quadrature amplitude modulation (QAM) symbol. That is, a $\mathrm{sinc}(f_c T)$ function, where f_c is the carrier frequency.

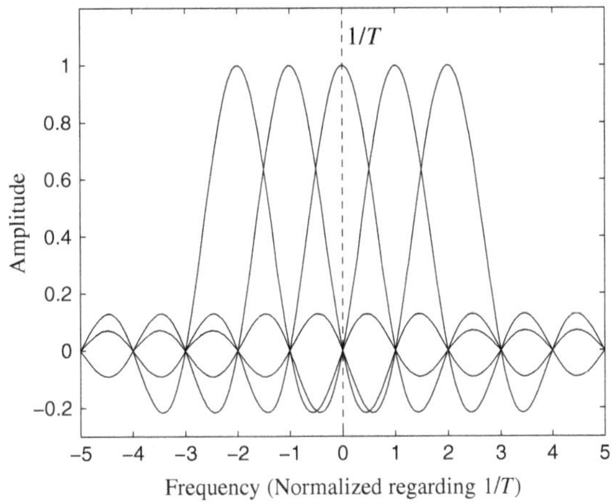

Fig. 9.2 OFDM spectrum.

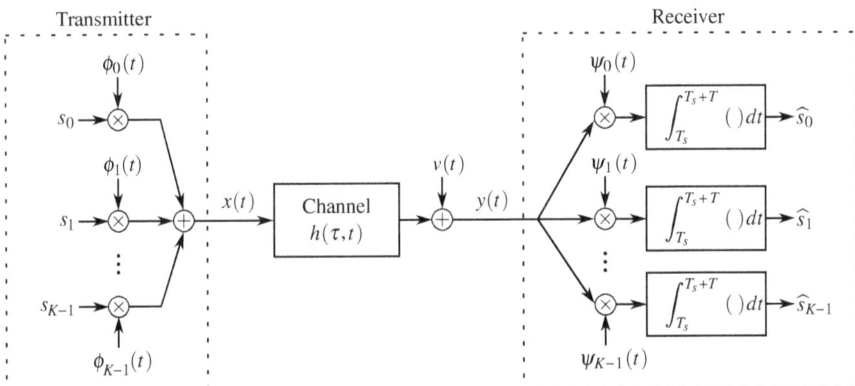

Fig. 9.3 Baseband OFDM continuous model.

The transmit filter ϕ_k for one OFDM symbol in Fig. 9.3 is given by

$$\phi_k(t) = e^{j2\pi \frac{k}{T}(t-T_{\mathrm{cp}})} w(t - T_{\mathrm{cp}}), \tag{9.1}$$

where $k = 0, 1, \ldots,$ K is the number of subcarriers, T_{cp} is the guard time duration, and $w(t)$ is a time window given by[9.3]

$$w(t) = \begin{cases} 1 & \text{if } t \in [0, T], \\ 0 & \text{otherwise.} \end{cases} \tag{9.2}$$

Note that the time duration of the OFDM symbol is in fact $T_{\mathrm{Total}} = T + T_{\mathrm{cp}}$.

[9.3] Note that the rectangular form of the time window $w(t)$ is responsible for the sinc spectrum for each subcarrier in Fig. 9.2.

After passing through the frequency-selective channel $h(\tau,t)$, the OFDM signal $x(t)$ is deformed and corrupted by white Gaussian noise $v(t)$, such that

$$y(t) = x(t) * h(\tau,t) + v(t), \tag{9.3}$$

where "$*$" denotes linear convolution.

Since the added guard time interval T_{cp} is designed to be greater than the channel delay spread, there is no inter symbol interference (ISI) from one OFDM symbol to another. One way to eliminate ISI and avoid introducing Inter-carrier interference (ICI) is to use a cyclic extension of the OFDM symbol as the guard time interval [3, 5, 10], commonly referred to as cyclic prefix (CP) and is shown in Fig. 9.4. This makes the linear convolution in (9.3) equivalent to a cyclic convolution, which eliminates the ICI.

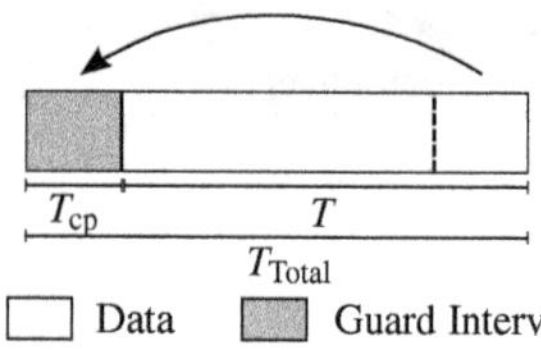

Fig. 9.4 Guard interval in OFDM as a cyclic extension.

Transmitting the cyclic prefix as a guard interval results in a loss of power, which degrades the signal-to-noise ratio (SNR) by (in dB)

$$\text{SNR}_{\text{loss}} = -10\log_{10}\left(1 - \frac{T_{\text{cp}}}{T_{\text{Total}}}\right). \tag{9.4}$$

However, the ISI – ICI-free transmission motivates the SNR loss, which is less then 1 dB for $\frac{T_{\text{cp}}}{T_{\text{Total}}} < 0.2$.

Since a circular convolution in the time domain is equal to a multiplication in the frequency domain, the transmitted symbol in each subcarrier in the received OFDM signal will be equivalent to the transmitted signal multiplied by a complex number representing the impact from the passage through the subcarrier channel[9.4] and corrupted by additive white Gaussian noise (AWGN).

From (9.1) and Fig. 9.3, the OFDM signal including the cyclic prefix is given by

$$x(t) = \begin{cases} \sum_{k=0}^{K-1} s_k \phi_k(t) & \text{if } t \in [T_{\text{cp}}, T_{\text{Total}}], \\ x(t+T) & \text{if } t \in [0, T_{\text{cp}}]. \end{cases} \tag{9.5}$$

At the receiver, the information of each subcarrier can be separated using a matched filter or a correlator, as shown in Fig. 9.3. The receiver filter ψ_k, $k =$

[9.4] Provided that the channel in each subcarrier can be assumed to be a flat channel.

$0,1,\ldots,K-1$, is given by

$$\psi_k(t) = \phi(-t) = e^{-j2\pi\frac{k}{T}(t-T_{\text{cp}})} w(t - T_{\text{cp}}). \tag{9.6}$$

9.2.1.1 OFDM Discrete Time Model

From (9.1) and (9.6) one can see that the transmitter and receiver in an OFDM system are similar to the inverse and direct Fourier transform, respectively. With the advent of the fast fourier transform (FFT) and inverse fast fourier transform (IFFT) algorithms, the set of oscillators in OFDM modulation and the correlator in OFDM demodulation was replaced by the IFFT and FFT algorithms, respectively [17]. This resulted in the simpler digital implementation for OFDM that is used today, shown in Fig. 9.5, where S/P stands for serial-to-parallel and P/S for the parallel-to-serial conversions.

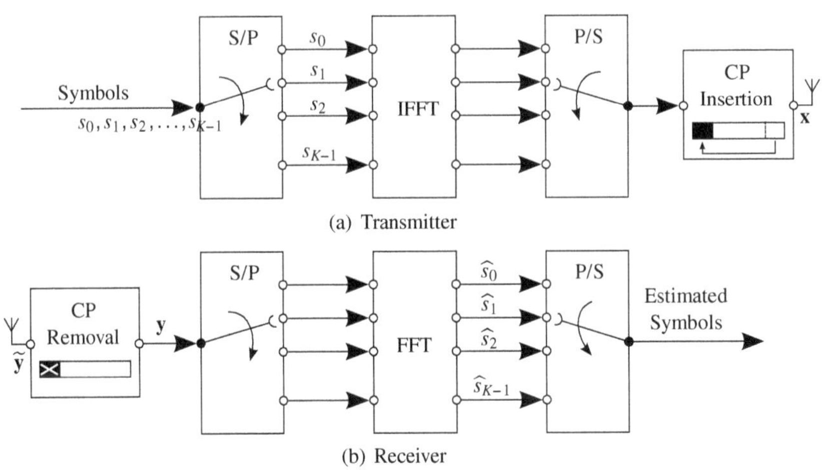

Fig. 9.5 OFDM system.

First, the symbols $s_0, s_1, \ldots, s_{K-1}$ are grouped as a vector $\mathbf{s}$. Then the inverse Fourier transform can be calculated using IFFT($\mathbf{s}$), which results in a vector with dimension $K \times 1$. Adding the cyclic prefix is equivalent to the simple task of concatenating the last N_{cp} samples of the vector IFFT($\mathbf{s}$) with itself, where N_{cp} is the size of the cyclic prefix in the samples. Therefore, the vector $\mathbf{x}$ with dimension $(K+N_{\text{cp}}) \times 1$ is obtained and corresponds to the samples of the OFDM transmit symbol which will have length $N_L = K + N_{cp}$.

The received OFDM signal $\tilde{\mathbf{y}}$, also with dimension $(K+N_{\text{cp}}) \times 1$, is given by

$$\tilde{\mathbf{y}} = \text{FFT}(\text{IFFT}(\mathbf{x}) * \mathbf{h} + \mathbf{v}), \tag{9.7}$$

where $\mathbf{h}$ is the vector representation of the channel parameters in (9.3). The effect of the cyclic prefix inclusion and removal is that the linear convolution in (9.7) is changed to a circular convolution. Consequently, the received signal after CP removal is given by

$$\mathbf{y} = \mathrm{FFT}(\mathrm{IFFT}(\mathbf{s}) \circledast \mathbf{h} + \mathbf{v}), \tag{9.8}$$

where $\circledast$ denotes circular convolution. Since the Fourier transform of a circular convolution of two vectors is equivalent to the multiplication of their individual Fourier transforms, (9.8) becomes

$$\mathbf{y} = \mathbf{s} \odot \mathrm{FFT}(\mathbf{h}) + \eta, \tag{9.9}$$

where "$\odot$" denotes element-by-element multiplication[9.5] and $\mathbf{v}$ and η are the noise in time and frequency domain, respectively. In (9.9) it is clearly shown that each received symbol is equal to the transmitted symbol multiplied by the corresponding subcarrier channel frequency response and corrupted by AWGN, as it was previously stated. Therefore, the receiver only needs to divide the received symbol by the corresponding channel frequency response,[9.6] instead of performing a full equalization in the received OFDM symbol.

9.2.1.2 Trade-Off Between OFDM System Parameters

The design of an OFDM system requires a trade-off between several conflicting requirements. For instance, the CP must be greater than the channel memory but the transmission of the CP results in an SNR loss as shown in (9.4). For that reason, the CP length is usually designed as one-fifth or less of the total OFDM symbol time so that the SNR loss is less then 1 dB [12].

For a given bandwidth B, the symbol time increases as the number of subcarriers K increases. Also, K must be sufficiently large so that the channel for each subcarrier may be considered as a flat channel (see Fig. 9.1). However, when the number of subcarriers is increased for a fixed bandwidth the subcarrier spacing Δf is decreased and the OFDM system becomes more sensitive to carrier frequency offset and phase noise, which requires a more complex and expensive implementation. The effect of a frequency offset is twofold. It rotates and attenuates the useful signal component, and introduces ICI. In Fig. 9.6 an OFDM system with ICI due to a frequency offset δf is illustrated, which causes a loss of the orthogonality among the subcarriers.

As in single carrier systems, time synchronism is also important in OFDM systems. Nevertheless, timing requirements are somewhat relaxed in OFDM systems because of the cyclic prefix and the fact that the symbol time is greater than in single carrier systems. Therefore, there is a relationship between the requirements on time and frequency synchronization. The former can be relaxed by increasing the

[9.5] The element-by-element operation makes it clear that each element of $\mathbf{y}$ depends on only one element of $\mathbf{s}$. That is, there is no ICI.

[9.6] This division is not shown in Fig. 9.5.

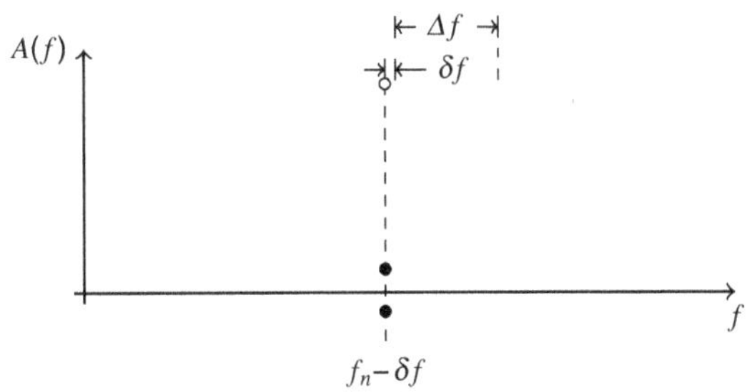

Fig. 9.6 Inter-carrier interference due to frequency offset δf from the subcarrier frequency f_n.

number of subcarriers (and thus the symbol time), while the latter can be relaxed by reducing the number of subcarriers.

9.2.2 Time-Varying Channel Models

When the channel varies in time, the orthogonality between subcarriers is lost. The subcarrier attenuation and the ICI depend on how the OFDM translates the time-varying channel from time to frequency domain, and on how the channel varies in time, i.e., on the channel statistics. In the following the subcarrier attenuation and the ICI are formulated in terms of the time domain channel response.

Let $\mathbf{F}$ denote the normalized Fourier matrix, whose kl-entry is $\frac{1}{\sqrt{K}}\omega_K^{kl}$, where $\omega_K = \exp(-j2\pi/K)$. With this notation, the received signal at time instant n, given by (9.9), is rewritten as

$$\mathbf{y}[n] = \mathbf{F}\mathbf{H}_v[n]\mathbf{F}^H\mathbf{s}[n] + \boldsymbol{\eta}[n], \tag{9.10}$$

for a time-varying channel matrix $\mathbf{H}_v[n]$ with kl-entry $(\mathbf{H}_v[n])_{kl} = h^n_{k,\langle k-l\rangle_K}$, where

$$h^n_{k,l} = (\mathbf{h}^0[(n-1)N_L + N_{\text{cp}} + k])_l, \tag{9.11}$$

and $\mathbf{h}^0[m] = (h[m,0],\ldots,h[m,L-1],0,\ldots,0)^T$ is a K-dimensional vector.

Now denote $\mathcal{H}_v[n] = \mathbf{F}\mathbf{H}_v[n]\mathbf{F}^H$. In this matrix, the diagonal $\text{diag}(\mathcal{H}_v[n])$ attenuates the symbols $\mathbf{s}[n] = (s[n,0],\ldots,s[n,K-1])^T$, and the other elements in $\mathcal{H}_{\text{ICI}}[n] = \mathcal{H}_v[n] - \text{diag}(\mathcal{H}_v[n])$ provide interference between subcarriers. An expansion of $\mathcal{H}_v[n]$ shows that its diagonal corresponds to the vector

$$\mathbf{H}[n] = \text{FFT}(\bar{\mathbf{h}}^0[n]), \tag{9.12}$$

where

$$\bar{\mathbf{h}}^0[n] = \frac{1}{K}\sum_{k=0}^{K-1}\mathbf{h}^0[nN_L + N_{\text{cp}} + k]. \tag{9.13}$$

Using $\mathbf{u}[n] = \mathcal{H}_{\mathrm{ICI}}[n]\mathbf{s}[n]$ to represent the ICI, (9.9) is rewritten as

$$\mathbf{y}[n] = \mathrm{diag}(\mathbf{H}[n])\mathbf{s}[n] + \mathbf{u}[n] + \boldsymbol{\eta}[n]. \tag{9.14}$$

Notice that if the channel does not vary in time, i.e., $\mathbf{h}[n] = \mathbf{h}$ for all n, then $\bar{\mathbf{h}}^0[n] = \mathbf{h}^0$ and $\mathbf{H}[n] = \mathbf{H} = \mathrm{FFT}(\mathbf{h}^0)$, for all n. Moreover, $\mathbf{u}[n] = \mathbf{0}$.

In the following section, the subcarrier correlations for a wide-sense stationary (WSS)–uncorrelated scattering (US) channel model are provided.

9.2.2.1 Subcarrier Correlation

The considered channel model is the WSS–US one with a constant number of paths. In this case, the base-band impulse response is given by

$$h[m,l] = \sum_{i=0}^{L_p-1} \gamma_i[m] g_i[l], \tag{9.15}$$

where $\gamma_i[m]$ is the complex amplitude of the ith path, and $g_i[l] = g(lT_s - \tau_i)$, where T_s is the sample period, τ_i is the delay of the ith multipath, and $g(\tau)$ is the shaping filter impulse response that satisfies the Nyquist criterion. The WSS–US assumption implies that

$$\mathbb{E}\{\gamma_{i'}^*[m']\gamma_i[m'+m]\} = \begin{cases} \rho_i r_i[m] & \text{if } i' = i, \\ 0 & \text{if } i' \neq i, \end{cases} \tag{9.16}$$

where ρ_i and $r_i[m]$ denote the mean power and normalized correlation of the ith path, respectively.

From (9.15), the channel response in the frequency domain (subcarrier attenuation) can be expanded as

$$H[n,k] = \sum_{l=0}^{L-1} \bar{h}[n,l]\omega_K^{kl} = \sum_{i=0}^{L_p-1} \bar{\gamma}_i[n] \left(\sum_{l=0}^{L-1} g_i[l]\omega_K^{kl} \right), \tag{9.17}$$

where

$$\bar{\gamma}_i[n] = \frac{1}{K} \sum_{k=0}^{K-1} \gamma_i[(n-1)N_L + N_{\mathrm{cp}} + k]. \tag{9.18}$$

With the assumption $g_i(lT_s - \tau_i) \approx 0$, for $l \neq 0, \ldots, L-1$, and due to the Nyquist criterion, the term in parentheses in (9.17) can be approximated as

$$\approx \sum_{l=-\infty}^{\infty} g(lT_s - \tau_i) \exp(-j2\pi kl/K) = \exp(-j2\pi k\Delta f \tau_i) \tag{9.19}$$

where $\Delta f = 1/T_s K$. Then (9.17) is rewritten as

$$H[n,k] = \sum_{i=0}^{L_p-1} \bar{\gamma}_i[n] \exp(-j2\pi k\Delta f \tau_i). \tag{9.20}$$

Now the subcarrier correlation can be rewritten as follows:

$$r_H[n,k] = \mathbb{E}\{H^*[n',k']H[n'+n,k'+k]\}$$
$$= \sum_{i=0}^{L_p-1} \mathbb{E}\{\overline{\gamma}_i^*[n']\overline{\gamma}_i[n'+n]\}\exp(-j2\pi k\Delta f\tau_i). \tag{9.21}$$

From the expansion of $\overline{\gamma}_i[n]$ in the expectation above, a straightforward computation results in

$$\mathbb{E}\{\overline{\gamma}_i^*[n']\overline{\gamma}_i[n'+n]\} = \frac{\rho_i}{K^2}\sum_{i_1=0}^{K-1}\sum_{i_2=0}^{K-1} r_i[nN_L+i_2-i_1] = \kappa_i\rho_i\overline{r}_i[n], \tag{9.22}$$

where $\overline{r}_i[n]$ is the normalized correlation of $\overline{\gamma}_i[n]$ and κ_i is the normalization factor, defined as

$$\kappa_i = \frac{1}{K^2}\sum_{i_1=0}^{K-1}\sum_{i_2=0}^{K-1} r_i[i_2-i_1]. \tag{9.23}$$

Observe that this factor satisfies $0 \leq \kappa_i \leq 1$ and can be interpreted as the power loss ratio of the ith path, since $\kappa_i = \mathbb{E}|\overline{\gamma}_i[n]|^2/\rho_i$. Hence, with this notation, (9.21) can be rewritten as follows:

$$r_H[n,k] = \sum_{i=0}^{L_p-1} \kappa_i\rho_i\overline{r}_i[n]\exp(-j2\pi k\Delta f\tau_i). \tag{9.24}$$

Just when all paths have the same correlation function $r_t[n]$, the separability property [7] is valid:

$$r_H[n,k] = \kappa\sigma_h^2\overline{r}_t[n]r_f[k], \tag{9.25}$$

where $\sigma_h^2 = \sum_{i=0}^{L_p-1}\rho_i$ and $r_f[k]$ is the normalized frequency correlation, defined as

$$r_f[k] = \sum_{i=0}^{L_p-1}\frac{\rho_i}{\sigma_h^2}\exp(-j2\pi k\Delta f\tau_i). \tag{9.26}$$

Observe that all paths have the same factor κ. The channel power σ_h^2 is attenuated by the factor κ, such that the subcarrier power is given as $\sigma_H^2 = \kappa\sigma_h^2$.

The approximation in (9.19) implies that the subcarrier correlation in (9.21) does not depend on k'. As a consequence, all subcarriers have the same power $\sigma_H^2 = \mathbb{E}|H[n,k]|^2$. Such an approximation is perfectly reasonable if the interference between OFDM symbols is negligible.

In what follows, a relationship between the Fourier transforms of $r_t[n]$ and $\overline{r}_t[n]$ is reached, denoted by $p_t[n]$ and $\overline{p}_t[n]$, respectively. Applying the Fourier transform to $\overline{r}_t[n]$:

$$\kappa\overline{p}_t(\nu) = \frac{1}{K^2}\sum_{i_1,i_2=0}^{K-1}\sum_{n=-\infty}^{\infty} r_t[nN_L+i_2-i_1]\exp(-j2\pi n\nu), \tag{9.27}$$

where ν represents the frequency variable. The last summation in the above equation is identified as [13]

$$\frac{1}{N_L} p_t\left(\frac{\nu}{N_L}\right) \exp\left(j2\pi \frac{i_2 - i_1}{N_L} \nu\right) \tag{9.28}$$

for $-1/2 \leq \nu \leq 1/2$, whose substitution implies

$$\kappa \overline{p}_t(\nu) = \frac{1}{N_L} p_t\left(\frac{\nu}{N_L}\right) \left\{ \frac{1}{K^2} \sum_{i_1, i_2 = 0}^{K-1} \exp\left(j2\pi \frac{i_2 - i_1}{N_L} \nu\right) \right\}. \tag{9.29}$$

If the term in the brackets in (9.29) is denoted by $m_t(\nu/N_L)$:

$$\kappa \overline{p}_t(\nu) = \frac{1}{N_L} p_t\left(\frac{\nu}{N_L}\right) m_t\left(\frac{\nu}{N_L}\right), \tag{9.30}$$

where further simplifications result in

$$m_t(\nu) = \frac{\operatorname{sinc}^2(K\nu)}{\operatorname{sinc}^2(\nu)}. \tag{9.31}$$

The last two equations provide the desired relationship. The function $m_t(\nu)$ is even and strictly decreasing in $[0, 1/2]$ with maximum $m_t(0) = 1$. Such a property justifies the appearance of the attenuation factor $0 \leq \kappa \leq 1$, and shows how the power spectral density of $H[n,k]$ is attenuated. In addition, the factor κ can be alternatively expressed as follows:

$$\kappa = \int_{-1/2}^{1/2} p_t(\nu) m_t(\nu) d\nu. \tag{9.32}$$

The following section will show that the remaining power $\sigma_h^2 - \sigma_H^2$ appears as the ICI power.

9.2.2.2 ICI Power

Since the transmitted symbols $s[n,k]$ are i.i.d. (independent and identically distributed), and so $H[n,k]s[n,k]$ is uncorrelated from its inter-carrier interference, power of $y[n]$, given by (9.14), for the kth subcarrier $\sigma_y^2[k]$ is given by

$$\sigma_y^2[k] = \sigma_H^2 \sigma_s^2 + \sigma_{\text{ICI}}^2[k] \sigma_s^2 + \sigma_\eta^2 = \sigma_H^2 + \sigma_{\text{ICI}}^2[k] + \sigma_\eta^2, \tag{9.33}$$

where $\sigma_{\text{ICI}}^2[k]$ is the ICI power over the kth subcarrier.

For all subcarriers, $\sigma_{\text{ICI}}^2[k]$ is the same. In order to find the ICI power, to prove this affirmation, $\sigma_y^2[k]$ is calculated and can be shown that the ICI power does not depend on k. $\sigma_{\text{ICI}}^2[k]$ was first calculated because its calculation is less tedious and more suitable for the calculus of ICI power. As demonstrated in Appendix 1 for $\sigma_s^2 = 1$, the ICI power is given by

$$\sigma_{\text{ICI}}^2 = \sigma_h^2 - \sigma_H^2. \tag{9.34}$$

This result allows to conclude that the received signal $y[n,k]$ has equal power for all k. Then, for $\sigma_s^2 = 1$, (9.33) can be rewritten as

$$\sigma_y^2 = \sigma_h^2 + \sigma_\eta^2. \tag{9.35}$$

In (9.34) and (9.35), the index k was omitted.

The expression in (9.34) has broad applicability, since few restrictions were assumed. In the derivation of (9.34), only the following assumptions were used: (a) the symbols $s[n,k]$ are i.i.d., (b) the interference between OFDM symbols is negligible, and (c) the WSS–US channel model holds. Additionally, the tapped delay line (TDL) assumption can easily be removed.

For the case when the channel paths have the same Doppler spectrum, the factor κ can be introduced:

$$\sigma_{\text{ICI}}^2 = (1-\kappa)\sigma_h^2. \tag{9.36}$$

The above equation together with (9.32) is similar to that found in [9], except for the use of $\text{sinc}^2(K\nu)$ in the place of $m_t(\nu)$. In [9], an infinite number of subcarriers was regarded, resulting in an upper bound for the ICI power. Here, instead, the expression in (9.36) is exact, since a finite number of interfering subcarriers was considerated.

9.2.2.3 Upper Bound for the ICI Power

In this section, an upper bound for σ_{ICI}^2 for certain maximum Doppler frequency ν_d is derived. According to (9.36), σ_{ICI}^2 is maximized when κ is minimized. Then, from κ defined in (9.32)

$$\kappa = \int_{-\nu_d}^{\nu_d} p_t(\nu)m_t(\nu)d\nu \geq \left(\min_{\nu_d \leq \nu \leq \nu_d} m_t(\nu)\right) \cdot \int_{-\nu_d}^{\nu_d} p_t(\nu)d\nu = m_t(\nu_d), \tag{9.37}$$

where $m_t(\nu) \in [-\nu_d, \nu_d]$ is minimized at $\nu = \nu_d$. Indeed, $m_t(\nu)$ is concave in $[-1/2, 1/2]$ with maximum at $\nu = 0$. Therefore, the maximum σ_{ICI}^2 is given by

$$\sigma_{\text{ICI,max}}^2 = [1 - m_t(\nu_d)]\sigma_h^2. \tag{9.38}$$

In Fig. 9.7 the ICI power σ_{ICI}^2 as a function of the maximum Doppler frequency ν_d for the two-path, Jakes, and uniform spectra is illustrated. The probability density function $p_t(\nu)$ and the correlation function $r_t[n]$ of each spectrum are given in Table 9.1. In all simulations, the values $\sigma_h^2 = \sigma_s^2 = 1$ and $K = 128$ were employed. The simulation shows that for the studied spectra, the two-path spectrum presents the largest ICI power.

9.2.3 *Generic MIMO-OFDM Model*

The use of multiple antennas in the transmitter and/or in the receiver as in MIMO systems is a key technique to facilitate the high spectral efficiency and data rates

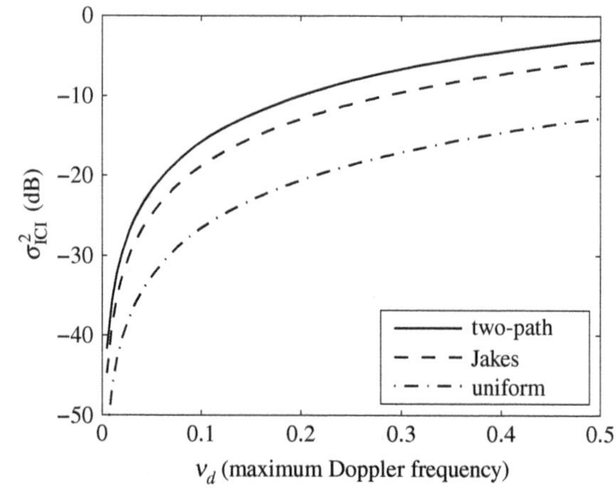

Fig. 9.7 Comparison of exact ICI power for two-path, Jakes, and uniform spectra.

Table 9.1 Characteristics of studied spectra.

Type	PDF	Correlation function
Two-path	$p_t(\nu) = \frac{1}{2}[\delta(\nu+\nu_d)+\delta(\nu-\nu_d)]$	$r_t[n] = \cos(2\pi\nu_d n)$
Jakes	$p_t(\nu) = \begin{cases} \dfrac{1}{\pi\nu_d}\dfrac{1}{\sqrt{1-(\nu/\nu_d)^2}} & \text{for } \lvert\nu\rvert < \nu_d \\ 0 & \text{otherwise} \end{cases}$	$r_t[n] = J_0(2\pi\nu_d n)$ [9.7]
Uniform	$p_t(\nu) = \begin{cases} \dfrac{1}{2\nu_d} & \text{for } \lvert\nu\rvert < \nu_d \\ 0 & \text{otherwise} \end{cases}$	$r_t[n] = \mathrm{sinc}(2\nu_d n)$

required in next generation wireless systems. While it is possible to take advantage of MIMO in a frequency-selective channel, most of the MIMO architectures proposed in the literature are designed for flat fading channels. In this context, the combination of MIMO with OFDM seems as a logical step to overcome the ISI caused by a frequency-selective channel and takes advantage of the spatial dimension provided by MIMO.

A generic MIMO-OFDM architecture is shown in Fig. 9.8. The channel matrix $\mathbf{H}(\tau,t)$ has dimension $M_r \times M_t$, where M_r is the number of receive antennas and M_t is the number of transmit antennas. Note that this is a generic frequency-selective and time-varying channel, since the channel depends on the time delay τ and time instant t.

As can be seen in (9.9), the resulting effect of the OFDM system is "changing a wideband frequency-selective channel into a set of narrowband flat channels". This means that the OFDM transmitter, the channel, and the OFDM receiver seem as an equivalent set of narrowband flat channel matrices $\mathbf{H}[k]_{M_r\times M_t}$ by the MIMO

[9.7] $J_0(\cdot)$ is the zeroth-order Bessel function of the first kind

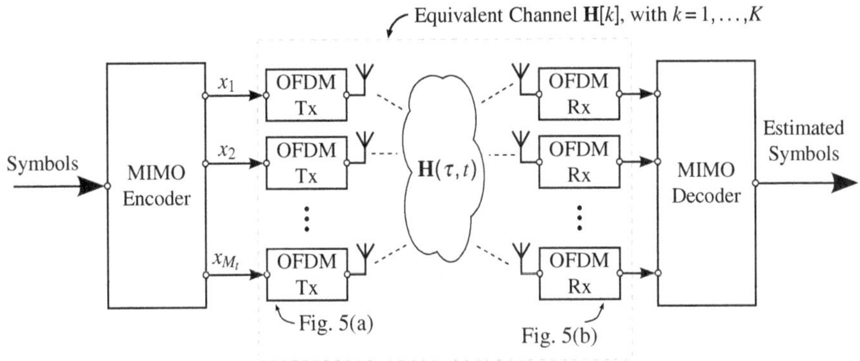

Fig. 9.8 Generic MIMO-OFDM model.

system.[9.8] Hence, Fig. 9.8 becomes similar to Fig. 10.1 and the MIMO schemes presented in Chapter 10 can be employed.

Expanding (9.9) to the MIMO-OFDM case, the signal received at the jth antenna branch is given by[9.9]

$$y_j[k] = \sum_{i=1}^{M_t} H_{ij}[k]x_i[k] + \eta_j[k], \tag{9.39}$$

where H_{ij} is the channel coefficient between the ith transmit antenna and jth receive antenna. The mapping from data symbols to x_i, $i = 1, 2, \ldots, M_t$, depends on the used MIMO encoder. However, any known information (pilot tones) used for channel estimation is inserted after this mapping. Also, the channel estimation and removal of any pilot tone is performed before MIMO decoder. Therefore, the particularly used MIMO scheme can be abstracted and what really matters regarding channel estimation is the number of transmit and receive antennas.

9.3 Channel Estimation for Time-Varying Channels

The OFDM transforms a channel selective in frequency into a set of flat fading sub-channels. Although flat channels allow us a simpler approach, they should be estimated in order to recover coherently the transmitted symbols $s[n,k]$ from the received signal $y[n,k]$. The solution for the estimation problem is found by defining pilot subcarriers where the transmitted symbols are known at the receiver side. Clearly the number of pilot subcarriers must be kept as low as possible, in order to obtain a larger number of data subcarriers, while still guaranteeing good-enough

[9.8] Note that the channel seen by the MIMO system is the channel frequency response in a subcarrier.

[9.9] On account of the compounding with OFDM, the channel H in the equations from now on is in fact in the frequency domain and depends on the subcarrier index k instead of a delay τ.

channel estimation quality for obtaining the required data detection performance. During this analysis, it was assumed that the pilot subcarriers are disposed in *grid*, as illustrated in Fig. 9.9. The initial step is the least squares (LS) estimation of the pilot subcarriers, which provides a relatively poor estimate. Based on the channel correlations in time and frequency, the LS estimate can be improved by means of a filter that exploits this correlation. The remaining subcarriers are estimated via interpolation, which can be designed under various criteria, such as minimum-mean-square-error (MMSE) or low-pass interpolator. In this section the MMSE estimator is derived, considering the infinite impulse response (IIR) and finite impulse response (FIR) cases. In this section a robust version of the MMSE estimator is also derived.

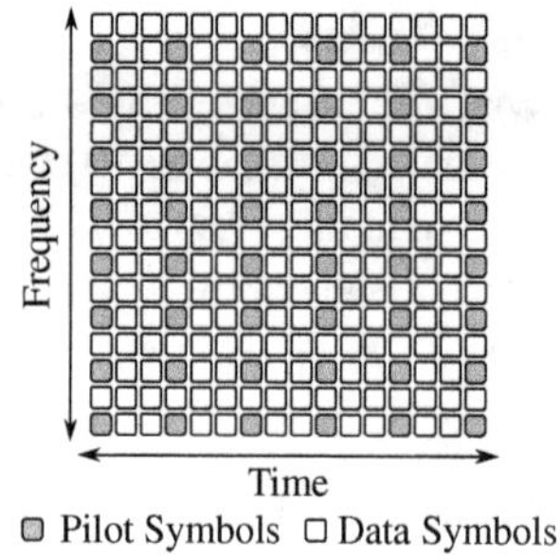

Fig. 9.9 Grid arrangement of the pilot subcarriers.

9.3.1 MMSE Estimator

For simplifying the LS estimation, the pilot symbols are selected from the phase-shift keying (PSK) constellation. For the remaining subcarriers, the symbols can be chosen from any constellation. In most cases, the data symbols are selected from a QAM constellation. Since the transmitted symbols at the pilot subcarriers are known at the receiver, the LS estimation can be realized by

$$\begin{aligned}\widehat{H}_{LS}[n,k] &= y[n,k]s^*[n,k] = H[n,k] + (u[n,k] + \eta[n,k])s^*[n,k] \\ &= H[n,k] + z[n,k], \end{aligned} \tag{9.40}$$

where $z[n,k] = (u[n,k] + \eta[n,k])s^*[n,k]$. Although the ICI, which corresponds to the term $u[n,k]$, is correlated for different n's and k's, the term $z[n,k]$ satisfies

$$\mathbb{E}\{z^*[n_1,k_1]z[n_2,k_2]\} = 0 \qquad \text{for } n_1 \neq n_2 \text{ and/or } k_1 \neq k_2, \tag{9.41a}$$

$$\mathbb{E}\{H^*[n_1,k_1]z[n_2,k_2]\} = 0 \qquad \text{for any } n_1, n_2, k_1, k_2. \tag{9.41b}$$

This is due to the independence between the symbols $s[n,k]$. These correlation properties simplify the derivation of the *Wiener filter*. Furthermore, for the power ρ of the ICI-plus-noise term:

$$\rho = \mathbb{E}|z[n,k]|^2 = \mathbb{E}|u[n,k]|^2 + \mathbb{E}|\eta[n,k]|^2 = \sigma_{\mathrm{ICI}}^2 + \sigma_\eta^2 = (1-\kappa)\sigma_h^2 + \sigma_\eta^2. \quad (9.42)$$

In this formulation, it is assumed that the pilot subcarriers are placed at positions $n = mM_{pt}$, for $m \in \mathbb{Z}$, and $k = lM_{pf}$, for $l = 0, 1, \ldots, N_p$, where it is assumed that K is multiple of M_{pf}.

9.3.1.1 IIR Case

The Wiener filter coefficients $c[m,l;n',k]$ are selected such that the estimator given by

$$\widehat{H}[n,k] = \sum_{m=-\infty}^{\infty} \sum_{l=(k-k'-K)/M_{pf}+1}^{(k-k')/M_{pf}} c[m,l;n',k]\widehat{H}_{LS}[n-n'-mM_{pt}, k-k'-lM_{pf}] \quad (9.43)$$

minimizes the mean-squared error (MSE):

$$\mathbb{E}|\widehat{H}[n,k] - H[n,k]|^2. \quad (9.44)$$

The indices $n' = \langle n \rangle_{M_{pt}}$ and $k' = \langle k \rangle_{M_{pf}}$ are here inserted in order to simplify notation. Note that the Wiener filter coefficients are indexed by n' in the place of n, which does not occur with respect to the index k. In Fig. 9.10 the relationship between these indices is illustrated.

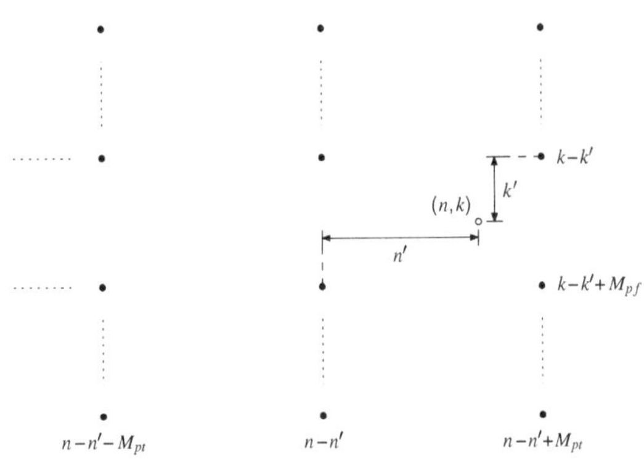

Fig. 9.10 Indexes used in the derivation of the MMSE estimator.

The minimization of the error in (9.44) leads to the orthogonality principle [14]:

$$\mathbb{E}\{(\widehat{H}[n,k] - H[n,k])\widehat{H}_{LS}^*[n-n'-mM_{pt}, k-k'-lM_{pf}]\} = 0, \quad (9.45)$$

for $m \in \mathbb{Z}$ and $l = (k-k'-K)/M_{pf}+1,\dots,(k-k')/M_{pf}$. Replacing (9.43) in (9.45)

$$\sum_{m_1,l_1} c[m_1,l_1;n',k] r_H[(m-m_1)M_{pt},(l-l_1)M_{pf}] - r_H[n'+mM_{pt},k'+lM_{pf}] + \rho c[m,l;n',k] = 0, \quad (9.46)$$

The expression in (9.46) is a consequence of the correlation properties stated in (9.41). Using the properties given in (9.25) that characterize the channel correlations, with $\sigma_h^2 = 1$, (9.46) can be rewritten as

$$\sum_{m_1,l_1} c[m_1,l_1;n',k] \kappa \bar{r}_t[(m-m_1)M_{pt}] r_f[(l-l_1)M_{pf}] - \kappa \bar{r}_t[n'+mM_{pt}] r_f[k'+lM_{pf}] + \rho c[m,l;n',k] = 0, \quad (9.47)$$

where $\mathbf{r}_f[k] = (r_f[k], r_f[k-M_{pf}],\dots,r_f[k-K+M_{pf}])$, $\mathbf{R}_{pf} = (\mathbf{r}_f[0],\mathbf{r}_f[M_{pf}],\dots,\mathbf{r}_f[K-M_{pf}])$, and $\mathbf{c}_2[m;n',k]$ is given by

$$\mathbf{c}_2[m;n',k] = \begin{pmatrix} c[m,(k-k')/M_{pf};n',k] \\ c[m,(k-k')/M_{pf}-1;n',k] \\ \vdots \\ c[m,(k-k'-K)/M_{pf}+1;n',k] \end{pmatrix}. \quad (9.48)$$

The sub-index "2" in $\mathbf{c}_2[m;n',k]$ indicates that its elements $c[m,l;n',k]$ are organized along l. Then, (9.47) can be rewritten as

$$(\rho \mathbf{I} + \kappa \mathbf{R}_{pf} \bar{r}_t[mM_{pt}]) * \mathbf{c}_2[m;n',k] = \kappa \bar{r}_t[n'+mM_{pt}] \mathbf{r}_f[k]. \quad (9.49)$$

Applying a Fourier transform to the equation above results in

$$(\rho \mathbf{I} + \kappa \mathbf{R}_{pf} \bar{\bar{p}}_t(\nu)) \mathbf{c}_2(\nu;n',k) = \kappa \bar{\bar{p}}_t(\nu) \phi(\nu;n') \mathbf{r}_f[k], \quad (9.50)$$

where $\phi(\nu;n') = \exp\left(j2\pi \frac{n'}{M_{pt}} \nu\right)$, $\bar{\bar{p}}_t(\nu)$ is the Fourier transform of $\bar{r}_t[mM_{pt}]$, related to $\bar{p}_t(\nu)$ as follows:

$$\bar{\bar{p}}_t(\nu) = \frac{1}{M_{pt}} \bar{p}_t\left(\frac{\nu}{M_{pt}}\right). \quad (9.51)$$

Assuming $\widetilde{\mathbf{R}}_f = (\mathbf{r}_f[0],\dots,\mathbf{r}_f[K-1])$ and $\mathbf{C}(\nu;n') = (\mathbf{c}_2(\nu;n',0),\dots,\mathbf{c}_2(\nu;n',K-1))$, (9.51) can be rewritten as

$$(\rho \mathbf{I} + \kappa \mathbf{R}_{pf} \bar{p}_t(\nu)) \mathbf{C}(\nu;n') = \kappa \bar{\bar{p}}_t(\nu) \phi(\nu;n') \widetilde{\mathbf{R}}_f. \quad (9.52)$$

Let the eigendecomposition of $\mathbf{R}_{pf}$ be given as

$$\mathbf{R}_{pf} = \mathbf{U}\mathbf{D}\mathbf{U}^H, \quad (9.53)$$

where $\mathbf{U}$ is a unitary matrix and $\mathbf{D}$ is the diagonal matrix containing the eigenvalues d_l. The elements of the pseudo-inverse $\mathbf{D}^\dagger$ of $\mathbf{D}$ are calculated according to

$$d_l^\dagger = \begin{cases} 1/d_l & \text{if } d_l \neq 0, \\ 0 & \text{if } d_l = 0. \end{cases}$$

From (9.53), the inverse of the matrix multiplying the term $\mathbf{C}(\nu;n')$ in (9.52) can be easily found. Let $\boldsymbol{\Phi}(\nu)$ be a diagonal matrix whose entries are

$$\Phi_l(\nu) = \frac{\kappa d_l \overline{\overline{p}}_t(\nu)}{\rho + \kappa d_l \overline{\overline{p}}_t(\nu)}. \tag{9.54}$$

Then, (9.52) can be finally rewritten as

$$\begin{aligned} \mathbf{C}(\nu;n') &= \mathbf{U}\boldsymbol{\Phi}(\nu)\phi(\nu;n')\mathbf{D}^\dagger\mathbf{U}^H\widetilde{\mathbf{R}}_f \\ &= \mathbf{U}\boldsymbol{\Phi}(\nu)\phi(\nu;n')\mathbf{U}^H\mathbf{R}_{pf}^\dagger\widetilde{\mathbf{R}}_f. \end{aligned} \tag{9.55}$$

In (9.55) three filtering blocks are provided, each one with a specific function. The first one $\mathbf{U}\boldsymbol{\Phi}(\nu)\mathbf{U}^H$ minimizes the noise over $\widehat{H}_{LS}[mM_{pt}, lM_{pf}]$. This filter is illustrated in Fig. 9.11.

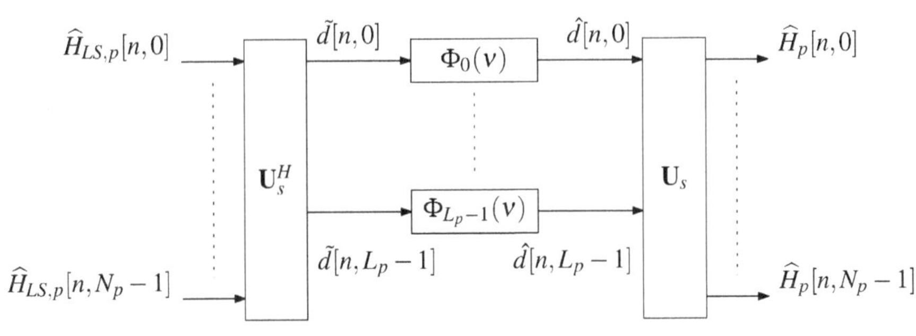

Fig. 9.11 Estimation of the pilot subcarriers based on the MMSE criterion.

The matrices $\mathbf{U}_s$ are the L_p first columns of $\mathbf{U}$ and $\widehat{H}_{LS,p}[n,k] = \widehat{H}_{LS}[nM_{pt}, kM_{pf}]$. The second block is constituted by the filters $\phi^*(\nu;0), \ldots, \phi^*(\nu;M_{pt}-1)$. Note that the inverse Fourier transform of $\phi^*(\nu;n') = \exp(-j2\pi \frac{n'}{M_{pt}}\nu)$ is $\mathrm{sinc}(n+n'/M_{pt})$ and these filters constitute an interpolation filter by sinc's. The interpolation filter is shown in Fig. 9.12.

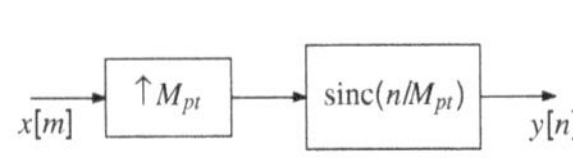

Fig. 9.12 Low-pass interpolator.

This block can be implemented by a polyphase filter bank $\mathbf{P}$. Finally, the block that implements the interpolation in frequency can be written as follows:

$$\mathbf{R}^{\dagger}_{pf}\widetilde{\mathbf{R}}_f, \tag{9.56}$$

whose input has dimension $N_p \times 1$ and an output of dimension $L_p \times 1$. These three blocks are shown in Fig. 9.13.

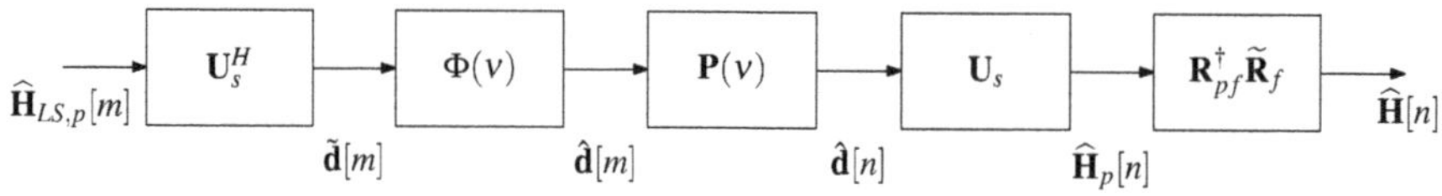

Fig. 9.13 MMSE estimator.

9.3.1.2 FIR Case

In (9.43), the index m assumes values over the integers. This leads to impractical implementation of the Wiener filter found in the last section. As an alternative, the range of the index m is limited. Considering $m = -M, \ldots, M$ and following the steps above, an expression analogous to (9.55) is obtained

$$\mathbf{C}(\nu;n') = \mathbf{U}\boldsymbol{\Phi}^*(\nu;n')\mathbf{U}^H\mathbf{R}^{\dagger}_{pf}\widetilde{\mathbf{R}}_f, \tag{9.57}$$

where $\boldsymbol{\Phi}(\nu;n')$ is a diagonal matrix whose lth element is the frequency domain response, a filter of length $2M+1$ whose coefficients are

$$\mathbf{c}[l;n'] = \begin{cases} \left(\frac{\rho}{\kappa d_l}\mathbf{I} + \mathbf{R}_t\right)^{-1} \mathbf{r}_t[n'] & \text{if } d_l \neq 0, \\ 0 & \text{otherwise,} \end{cases} \tag{9.58}$$

where $\mathbf{r}_t[n'] = (\overline{r}_t[n' + M \cdot M_{pt}], \ldots, \overline{r}_t[n' - M \cdot M_{pt}])^T$ and $\mathbf{R}_t = (\mathbf{r}_t[-M \cdot M_{pt}], \ldots, \mathbf{r}_t[M \cdot M_{pt}])$. Here a substantial difference is that $\boldsymbol{\Phi}(\nu;n')$ cannot be separated, while in the IIR case, the pilot subcarriers are first estimated, and a sinc interpolation in time is realized after. If the correlations $\overline{r}_t[mM_{pt}]$ are nearly null for $|m| > M$, an approximation to the IIR case can be applied. First, the pilot subcarriers are estimated using the FIR filter $\mathbf{c}[l;0]$ of length $2M+1$ as above, and a sinc interpolation in time is applied. This is the approach employed in the robust estimator.

9.3.2 *Robust Estimator*

In order to obtain the optimum channel estimation, as seen previously, it is required that the channel correlations in time and frequency are known at the receiver side. As a drawback, the estimation of the channel correlations demands large computational load. Additionally, the channel statistics may change in time. A sub-optimal solution is to make a choice for the correlations $\overline{\overline{r}}_t[n]$ and $r_f[n]$ that, even if they differ from the real correlations, lead to a performance close to the optimum case. Such an estimator is said to be robust in the sense of not depending on the channel statistics.

In what follows, the estimation of the pilot subcarriers is described, i.e., the filter:

$$\mathbf{C}(\nu) = \mathbf{C}(\nu;0) = \mathbf{U}\boldsymbol{\Phi}(\nu)\mathbf{U}^H. \tag{9.59}$$

To simplify notation, $c[m,l;k]$ will be used in the place of $c[m,l;0,kM_{pt}]$, for $k = 0,\dots,N_p-1$. Let $\overline{\text{MSE}}$ be the MSE averaged over pilot subcarriers:

$$\overline{\text{MSE}} = \frac{1}{N_p}\sum_{k=0}^{N_p-1} \mathbb{E}|\widehat{H}[n,kM_{pf}] - H[n,kM_{pf}]|^2. \tag{9.60}$$

The following analysis will be relative to the IIR Wiener filter. Let $\bar{\bar{r}}_{tr}[n]$ and $r_{fr}[n]$ be the correlations used in the filters whose coefficients $\tilde{c}[m,l;k]$ are expressed as follows:

$$\widetilde{\mathbf{C}}(\nu) = \widetilde{\mathbf{U}}\widetilde{\boldsymbol{\Phi}}(\nu)\widetilde{\mathbf{U}}^H, \tag{9.61}$$

where $\widetilde{\mathbf{U}}$, $\widetilde{\boldsymbol{\Phi}}(\nu)$, and $\widetilde{\mathbf{C}}(\nu)$ are given in (9.53), (9.54), and (9.59), respectively, with the difference that $\bar{\bar{r}}_t[n]$ and $r_f[n]$ are replaced by $\bar{\bar{r}}_{tr}[n]$ and $r_{fr}[n]$. The MSE attained by the filter whose coefficients are $\tilde{c}[m,l;k]$ is given by [7]

$$\overline{\text{MSE}} = \frac{1}{N_p}\int_{-1/2}^{1/2} \kappa\bar{\bar{p}}_t(\nu)\,\text{tr}\{\widetilde{\mathbf{U}}^H\mathbf{R}_{pf}\widetilde{\mathbf{U}}(\widetilde{\boldsymbol{\Phi}}(\nu)-\mathbf{I})(\widetilde{\boldsymbol{\Phi}}(\nu)-\mathbf{I})^H\}d\nu$$
$$+\frac{1}{N_p}\int_{-1/2}^{1/2}\rho\,\text{tr}\{\widetilde{\boldsymbol{\Phi}}^H(\nu)\widetilde{\boldsymbol{\Phi}}(\nu)\}d\nu, \tag{9.62}$$

where $\text{tr}\{\cdot\}$ denotes the trace of the matrix. Initially, it is assumed that the estimator is exactly matched in frequency, i.e., the coefficients $\tilde{c}[m,l;k]$ are selected as

$$\mathbf{R}_{pf} = \mathbf{U}\mathbf{D}\mathbf{U}^H, \tag{9.63}$$

where the entries of the diagonal matrix $\mathbf{D}$ are given by

$$(\mathbf{D})_{ll} = \begin{cases} d_l, & \text{if } 0 \le l \le L_p - 1, \\ 0, & \text{if } L_p \le l \le N_p - 1. \end{cases} \tag{9.64}$$

For each l, a $\bar{p}_{t,3}(\nu;l)$ is chosen such that

$$\widetilde{\Phi}_l(\nu) = \frac{\kappa d_l \bar{\bar{p}}_{tr}(\nu;l)}{\rho + \kappa d_l \bar{\bar{p}}_{tr}(\nu;l)}, \tag{9.65}$$

where the arbitrary terms $\bar{\bar{p}}_{tr}(\nu;l)$ satisfy the constraints

$$\int_{-1/2}^{1/2} p_{tr}(\nu;l)m_t(\nu)d\nu = \kappa, \qquad \int_{-1/2}^{1/2} p_{tr}(\nu;l)d\nu = 1. \tag{9.66}$$

Hence, (9.62) simplifies to

$$\overline{\text{MSE}} = \overline{\text{MSE}}'' + \overline{\text{MSE}}', \tag{9.67}$$

where

$$\overline{\text{MSE}}'' = \frac{1}{N_p}\sum_{l=0}^{N_p} \kappa d_l \int_{-1/2}^{1/2} \{\overline{\overline{p}}_t(\nu) - \overline{\overline{p}}_{tr}(\nu;l)\}|\widetilde{\Phi}_l(\nu) - 1|^2 d\nu, \tag{9.68}$$

$$\overline{\text{MSE}}' = \frac{1}{N_p}\sum_{l=0}^{N_p} \rho \int_{-1/2}^{1/2} \widetilde{\Phi}_l(\nu) d\nu. \tag{9.69}$$

The functions $\overline{\overline{p}}_{tr}(\nu;l)$ will be found under the constraints in (9.66) that maximize $\overline{\text{MSE}}'$. Then, it will be shown that the "residual" $\overline{\text{MSE}}''$ is nulled for these $\overline{\text{MSE}}'$ found. This problem is formulated as follows:

$$\text{maximize:} \quad \int_{-1/2}^{1/2} \frac{d_l y(\nu) m_t(\nu)}{\overline{\overline{\rho}} + d_l y(\nu) m_t(\nu)} d\nu, \tag{9.70}$$

$$\text{constrained to:} \quad \int_{-1/2}^{1/2} y(\nu) m_t(\nu) d\nu = \kappa, \tag{9.71}$$

$$\int_{-1/2}^{1/2} y(\nu) d\nu = 1, \tag{9.72}$$

where $\overline{\overline{\rho}} = N_s M_{pt} \rho$. Applying the Lagrange multipliers technique, this optimization problem leads to

$$-\frac{d_l m_t(\nu)}{[\overline{\overline{\rho}} + d_l y(\nu) m_t(\nu)]^2} + \lambda_1 m_t(\nu) + \lambda_2 = 0, \tag{9.73}$$

where λ_1 and λ_2 are selected so that $y(\nu)$ satisfies the above constraints.

Inserting $\widetilde{\Phi}(\nu)$, given in (9.65), (9.68) can be rewritten as

$$\overline{\text{MSE}}'' = \frac{1}{N_p}\sum_{l=0}^{N_p} \overline{\overline{\rho}}^2 \int_{-1/2}^{1/2} [p_t(\nu) - p_{tr}(\nu;l)] \cdot \frac{d_l m_t(\nu)}{[\overline{\overline{\rho}} + d_l p_{tr}(\nu;l) m_t(\nu)]^2} d\nu. \tag{9.74}$$

Hence, the result found in (9.73) implies

$$\overline{\text{MSE}}'' = \frac{1}{N_p}\sum_{l=0}^{N_p} \overline{\overline{\rho}}^2 \int_{-1/2}^{1/2} [p_t(\nu) - p_{tr}(\nu;l)] \cdot (\lambda_1 m_t(\nu) + \lambda_2) d\nu = 0. \tag{9.75}$$

Using this result, (9.67) can be reduced to

$$\overline{\text{MSE}} = \overline{\text{MSE}}' \tag{9.76}$$

and then the filter performance does not depend on the channel statistics.

Some constraints can be discarded or inserted in the problem. If one of the constraints given in (9.71) and (9.72) is discarded, the term $\overline{\text{MSE}}''$ continues to be equal to zero; however, the performance worsens, since the resulting $\overline{\text{MSE}}'$ increases. If the maximum Doppler frequency ν_d is supposedly known, depending on the discarded constraint the following cases will occur:

$\underline{\lambda_1 = 0:}$ In this case, (9.73) results in

$$\kappa_l \overline{\overline{p}}_{tr}(\nu;l) = \left[\frac{1}{I_1}\left(1+\frac{\overline{\overline{\rho}}}{d_l}I_2\right)m_t^{1/2}\left(\frac{\nu}{N_s M_{pt}}\right)-\frac{\overline{\overline{\rho}}}{d_l}\right], \tag{9.77}$$

where $I_1 = \int_{-\nu_d}^{\nu_d} \frac{1}{\sqrt{m_t(\nu)}} d\nu$, $I_2 = \int_{-\nu_d}^{\nu_d} \frac{1}{m_t(\nu)} d\nu$ and κ_l is inserted for making equal to unit the power of $\overline{\overline{p}}_{tr}(\nu;l)$ in the interval $2\nu_d$. With (9.77), the expression for $\overline{\text{MSE}}'$ can be written as

$$\overline{\text{MSE}}' = 2\nu_d \overline{\overline{\rho}} \frac{K}{N_p} - \frac{1}{N_p}\sum_{l=0}^{K-1} \frac{I_1^2 \overline{\overline{\rho}}^2}{I_2 \overline{\overline{\rho}} + d_l}, \tag{9.78}$$

where K is the number of eigenvalues d_l different from zero.

$\underline{\lambda_2 = 0:}$ In this case, (9.73) results in

$$\overline{\overline{p}}_{tr}(\nu;l) = \frac{1}{2(N_s M_{pt} \nu_d)}, \tag{9.79}$$

which provides

$$\overline{\text{MSE}}' = 2\nu_d \overline{\overline{\rho}} \frac{K}{N_p} - \frac{1}{N_p}\sum_{l=0}^{K-1} \frac{(2\nu_d)^2 \overline{\overline{\rho}}^2}{(2\nu_d)\overline{\overline{\rho}} + \kappa d_l}. \tag{9.80}$$

The expression for the $\overline{\text{MSE}}'$ found in these cases only differ on the summations, which result in an $\overline{\text{MSE}}'$ lower than $2\nu_d \overline{\overline{\rho}} \frac{K}{N_p}$.

The estimators found above are robust in the sense that their performances, expressed as $\overline{\text{MSE}}'$, do not depend on the time channel correlations $\overline{\overline{r}}_t[n]$. It is required to know $\mathbf{R}_{pf}$ and ρ. For eliminating the dependence on $\mathbf{R}_{pf}$, (9.61) is rewritten as

$$\widetilde{\mathbf{C}}(\nu) = \mathbf{F}\widetilde{\boldsymbol{\Phi}}(\nu)\mathbf{F}^H, \tag{9.81}$$

where $\mathbf{F}$ is the normalized Fourier matrix and $\widetilde{\Phi}_l(\nu) = \widetilde{\Phi}(\nu)$ is given according to (9.65), with $\overline{\overline{p}}_{tr}(\nu;l) = \overline{\overline{p}}_{tr}(\nu)$ and

$$d_l = \begin{cases} N_p/L_p & \text{for } 0 \le l \le L_p - 1, \\ 0 & \text{for } L_p \le l \le N_p - 1, \end{cases} \tag{9.82}$$

where L_p is the channel length. Then, (9.62) is reduced to

$$\overline{\text{MSE}} = \frac{1}{N_p}\int_{-1/2}^{1/2} \kappa \overline{\overline{p}}_t(\nu)|\widetilde{\Phi}(\nu) - 1|^2 \cdot \text{tr}(\mathbf{F}^H \mathbf{R}_{pf} \mathbf{F} \cdot \mathbf{I}')d\nu + \frac{L_p}{N_p}\int_{-1/2}^{1/2} \rho |\widetilde{\Phi}(\nu)|^2 d\nu, \tag{9.83}$$

where $\mathbf{I}' = \text{diag}\{(\mathbf{1}_{L_p}, \mathbf{0}_{N_p - L_p})\}$. For $\sigma_h^2 = 1$, one can easily show that $\text{tr}(\mathbf{F}^H \mathbf{R}_{pf} \mathbf{F} \cdot \mathbf{I}') = N_p$. Therefore, this equation can be rewritten as

$$\overline{\text{MSE}} = \overline{\text{MSE}}'' + \overline{\text{MSE}}', \tag{9.84}$$

where

$$\overline{\text{MSE}}'' = \kappa \int_{-1/2}^{1/2} \{\overline{\overline{p}}_t(\nu) - \overline{\overline{p}}_{tr}(\nu)\} |\widetilde{\Phi}(\nu) - 1|^2 d\nu, \tag{9.85}$$

$$\overline{\text{MSE}}' = \frac{L_p}{N_p} \int_{-1/2}^{1/2} \rho |\widetilde{\Phi}(\nu)|^2 d\nu. \tag{9.86}$$

The analysis for the choice of $\overline{\overline{p}}_{tr}(\nu)$ is analogous.

The robust estimator derived above depends on ρ and L_p, since $\widetilde{\Phi}(\nu) = \widetilde{\Phi}_l(\nu)$ in (9.65) is given in terms of these parameters.

Since in practice the filters have finite impulse response length, (9.58) is used, which is rewritten below for $n' = 0$:

$$\mathbf{c}[l] = \left(\frac{\rho}{\kappa d_l} \mathbf{I} + \mathbf{R}_t \right)^{-1} \mathbf{r}_t, \tag{9.87}$$

where the indices of $\mathbf{c}[l;0]$ and $\mathbf{r}_t[0]$ were omitted. The correlations found in (9.77) and (9.79) can be inserted in (9.87). This turns out into a practical implementation of an FIR robust filter.

The dependence on ρ in $\overline{\overline{p}}_{tr}(\nu;l)$, given in (9.77), is eliminated by making $\overline{\overline{\rho}} = 0$, such that

$$\overline{\overline{p}}_{tr}(\nu) = \kappa_l^{-1} I_1^{-1} m_t^{1/2}(\nu / N_s M_{pt}). \tag{9.88}$$

Observe that $\overline{\overline{p}}_{tr}(\nu)$ given in (9.79) does not depend on ρ. The dependence on ρ given in (9.87) can be eliminated and $\rho / \kappa d_l$ is substituted for a small constant δ, such that the inverse existing in this equation results in

$$\mathbf{c}[l] = (\delta \mathbf{I} + \mathbf{R}_t)^{-1} \mathbf{r}_t. \tag{9.89}$$

9.3.3 Performance Evaluation

The parameters used in the simulations are given in Table 9.2.

Figure 9.14 illustrates the filtering strategies developed and analyzed in this section. The algorithms are described in what follows.

Initially, a "semi-robust" case is considerated, where the matrix **U**, the number of multipaths L_p, and the parameters κ, ρ, and d_l are known. The algorithms taken into account are the following:

- **U-P**: The LS estimate is projected onto the subspace spanned by the first L_p columns of **U**.

Table 9.2 Simulation parameters.

Parameter	Value
Bandwidth	800 kHz
Number of subcarriers	$K = 128$
Length of cyclic prefix	$N_{\text{cp}} = 15$
Number of pilot subcarriers	$N_p = 16$
Pilot symbols constellation	4-PSK
Data symbols constellation	16-QAM
Channel model	TDL with Jakes spectrum
Channel power	$\sigma_h^2 = 1$
Symbol power	$\sigma_s^2 = 1$
Number of multipaths	$L_p = L_{p,\max} = 4$

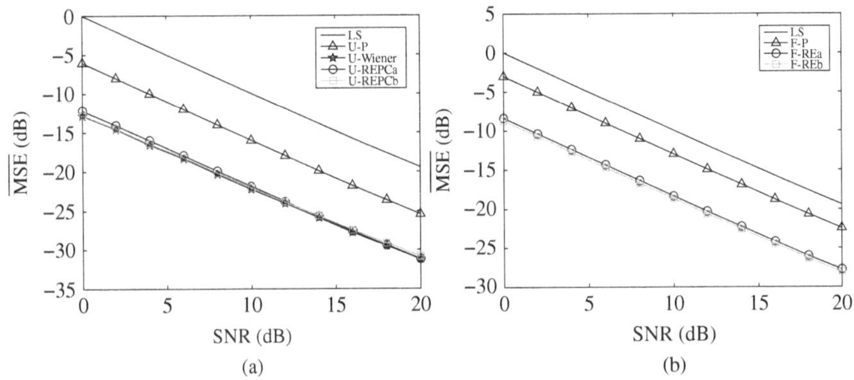

Fig. 9.14 $\overline{\text{MSE}} \times$ SNR curves with (a) **U** assumed known and (b) with **F** in the place of **U**, for the parameters in Table 9.2, and the values $f_d = 200$ Hz and $M_{pt} = 3$.

- **U-Wiener**: The coefficients in the temporal section are selected according to (9.58), where it is supposed that the channel correlations and parameters κ, ρ, and d_l are known.
- **U-REPCa**: The parameters κ and ρ are also known, and the coefficients are selected from (9.58), with the channel correlations given according to (9.77).
- **U-REPCb**: The same as U-REPCa, with the channel correlations given in (9.79).

For the totally robust case, the normalized Fourier matrix **F** is considered in the place of **U**, and it is assumed that the channel length L is known. The filtering cases taken into account are the following:

- **F-P**: The LS estimate is projected onto the subspace spanned by the first L_p columns of **F**.
- **F-REa**: The coefficients from the temporal section are selected according to (9.89), which does not require the knowledge of parameters κ, ρ, and d_l. The channel correlations are given according to (9.88).
- **F-REb**: The same as F-REa, with the channel correlations given in (9.79).

9.4 Recursive Methods

Section 9.3 illustrated that MMSE estimator is divided into three filtering blocks, as shown in Fig. 9.11. First, the input vector is projected onto the subspace defined by $\mathbf{U}_s$. In the sequel, the components of this projection are filtered. And finally the signal vector is recovered from the filtered components. The adaptive structure estimates the subspace given by $\mathbf{U}_s$, and filter the noise over the components of the projections. In this section, the low-rank adaptive filter (LORAF) [16] and the projection approximation subspace tracking (PAST) [1, 18] algorithms are applied for the subspace estimation. The following section describes two proposed algorithms. First, an adaptive algorithm, based on QR decomposition, that can filter the noise over the components and second an algorithm that estimates the ICI-plus-noise power, and the dimension of the subspace.

9.4.1 Subspace Estimation

This section describes the estimation of matrix U. As described in Section 9.3

$$\begin{aligned}\widetilde{\mathbf{R}}_f &= \mathbb{E}\{\widehat{\mathbf{H}}_{LS,p}[n]\widehat{\mathbf{H}}_{LS,p}^H[n]\} = \mathbb{E}\{\mathbf{H}_p[n]\mathbf{H}_p^H[n]\} + \mathbb{E}\{\mathbf{z}_p[n]\mathbf{z}_p^H[n]\} \\ &= \kappa\mathbf{R}_{pf} + \rho\mathbf{I} = \mathbf{U}\mathbf{\Lambda}\mathbf{U}^H,\end{aligned} \tag{9.90}$$

where

$$\mathbf{H}_p[n] = (H[nM_{pt},0], H[nM_{pt},M_{pf}],\ldots,H[nM_{pt},K-M_{pf}])^T, \tag{9.91}$$

$$\widehat{\mathbf{H}}_{LS,p}[n] = (\widehat{H}_{LS}[nM_{pt},0], \widehat{H}_{LS}[nM_{pt},M_{pf}],\ldots,\widehat{H}_{LS}[nM_{pt},K-M_{pf}])^T, \tag{9.92}$$

$$\mathbf{z}_p[n] = (z[nM_{pt},0], z[nM_{pt},M_{pf}],\ldots,z[nM_{pt},K-M_{pf}])^T, \tag{9.93}$$

and

$$\mathbf{\Lambda} = \kappa\mathbf{D} + \rho\mathbf{I}. \tag{9.94}$$

Therefore, we have a subspace estimation problem, where the L_p-dominating eigenvectors of $\mathbf{R}_{pf}$ are estimated. This problem can be written as

$$\widetilde{\mathbf{R}}_f = (\mathbf{U}_s\ \mathbf{U}_n)\begin{pmatrix}\mathbf{\Lambda}_s & \mathbf{0} \\ \mathbf{0} & \mathbf{\Lambda}_n\end{pmatrix}(\mathbf{U}_s\ \mathbf{U}_n)^H, \tag{9.95}$$

where $\mathbf{\Lambda}_s = \kappa \cdot \mathrm{diag}\{d_0,\ldots,d_{L_p-1}\} + \rho\mathbf{I}_{L_p}$, $\mathbf{\Lambda}_n = \rho\mathbf{I}_{N_p-L_p}$ and it is assumed that $d_0 \geq d_1 \geq \cdots \geq d_{L_p-1}$. The subspace spanned by the columns of $\mathbf{U}_s$ is the signal subspace we are interested in (the subspace where the vectors $\mathbf{H}_p[n]$ lie). The subspace spanned by the columns of $\mathbf{U}_n$ is the noise subspace. Equation (9.20) can be rewritten as follows:

$$\mathbf{H}_p[n] = \mathbf{W}_{\tau,p}\bar{\boldsymbol{\gamma}}[n], \tag{9.96}$$

where $\bar{\boldsymbol{\gamma}}[n] = (\bar{\gamma}_0[n], \ldots, \bar{\gamma}_{L_p-1}[n])^T$ and $(\mathbf{W}_{\boldsymbol{\tau},p})_{m,i} = \exp(-j2\pi m M_{pf} \Delta f \tau_i)$ show us that

$$\widetilde{\mathbf{R}}_f = \mathbf{W}_{\boldsymbol{\tau},p} \mathbb{E}\{\bar{\boldsymbol{\gamma}}[n] \bar{\boldsymbol{\gamma}}^H[n]\} \mathbf{W}_{\boldsymbol{\tau},p}^H + \rho \mathbf{I} = \mathbf{W}_{\boldsymbol{\tau},p} \boldsymbol{\Sigma} \mathbf{W}_{\boldsymbol{\tau},p}^H + \rho \mathbf{I}, \tag{9.97}$$

where $\boldsymbol{\Sigma} = \kappa \cdot \mathrm{diag}\{\rho_0, \ldots, \rho_{L_p-1}\}$. And, therefore

$$\mathbf{U}[\kappa \mathbf{D}]\mathbf{U}^H = \mathbf{W}_{\boldsymbol{\tau},p} \boldsymbol{\Sigma} \mathbf{W}_{\boldsymbol{\tau},p}^H. \tag{9.98}$$

Since the matrix in the right-hand side has rank L_p, the dimension of the signal subspace is the number of paths of the channel. From (9.96), that was derived from the TDL channel model, the signal subspace spanned by $\mathbf{H}_p[n]$ is constituted by column subspace of $\mathbf{W}_{\boldsymbol{\tau},p}$.

The matrix $\mathbf{R}$ can be estimated recursively as

$$\widehat{\mathbf{R}}[n] = \alpha \widehat{\mathbf{R}}[n-1] + (1-\alpha) \widehat{\mathbf{H}}_{LS,p}[n] \widehat{\mathbf{H}}_{LS,p}^H[n], \tag{9.99}$$

where α is a forgetting factor. The eigendecomposition of $\widehat{\mathbf{R}}[n]$ can be calculated by

$$\widehat{\mathbf{R}}[n] = \mathbf{U}[n] \boldsymbol{\Lambda}[n] \mathbf{U}^H[n] \tag{9.100}$$

and, hence, select $\mathbf{U}_s$ as the eigenvectors corresponding to the L_p largest eigenvalues found in this decomposition.

A straightforward computation of the eigendecomposition of $\widehat{\mathbf{R}}[n]$ requires a high computational load, whose implementation in practice is inviable. As alternative, $\mathbf{U}[n]$ is computed recursively from the previous matrix $\mathbf{U}[n-1]$, by means of some subspace tracking (ST) algorithm. Two examples of ST algorithms currently available in the literature are the LORAF [16] in its versions 1, 2, and 3; and the PAST [18] and its orthogonal version, the OPAST [1].

9.4.2 Temporal Filter Estimation

The coefficients of the filters that exploit the channel correlations in time, given in (9.58), can be rewritten, for $n' = 0$, as

$$\mathbf{c}[l] = (d_l \kappa \mathbf{R}_t + \rho \mathbf{I})^{-1} d_l \kappa \mathbf{r}_t, \tag{9.101}$$

where the indices of $\mathbf{c}[l;0]$ and $\mathbf{r}_t[0]$ were omitted.

For the correlations defined by (9.101), the recursive expressions will be established. Let

$$\mathbb{E}\{\widehat{\mathbf{H}}_{LS,p}[n+n_1] \widehat{\mathbf{H}}_{LS,p}^H[n_1]\} = \kappa \bar{r}_t[n] \mathbf{R}_{pf} + \rho \delta[n] \mathbf{I}. \tag{9.102}$$

If both sides are multiplied by $\mathbf{U}_s^H$ and $\mathbf{U}_s$, results in

$$\mathbb{E}\{\tilde{\mathbf{d}}[n+n_1] \tilde{\mathbf{d}}^H[n_1]\} = \kappa \bar{r}_t[n] \mathbf{D} + \rho \delta[n] \mathbf{I} \tag{9.103}$$

or

$$\mathbb{E}\{\tilde{d}[n+n_1,l]\tilde{d}^*[n_1,l]\} = d_l\kappa\tilde{r}_t[n] + \rho\delta[n]\mathbf{I}, \tag{9.104}$$

where $\tilde{\mathbf{d}}[n] = (\tilde{d}[n,0],\ldots,\tilde{d}[n,L_p-1])^T = \mathbf{U}_s^H\widehat{\mathbf{H}}_{LS,p}[n]$. If we denote $\tilde{\mathbf{d}}[n;l] = (\tilde{d}[n+M,l],\ldots,\tilde{d}[n-M,l])^T$, we can write

$$\widetilde{\mathbf{R}}_t[l] = \mathbb{E}\{\tilde{\mathbf{d}}[n;l]\tilde{\mathbf{d}}^H[n;l]\} = d_l\kappa\mathbf{R}_t + \rho\mathbf{I}, \qquad \tilde{\mathbf{r}}_t[l] = \mathbb{E}\{\tilde{\mathbf{d}}[n;l]\tilde{d}^*[n,l]\} = d_l\kappa\mathbf{r}_t + \rho\mathbf{e}_M,$$

where $\mathbf{e}_M = (0,\ldots,0,1,0,\ldots,0)^T$, with 1 in the $(M+1)$th entry. Then, (9.101) is rewritten as

$$\begin{aligned}\mathbf{c}[l] &= \widetilde{\mathbf{R}}_t^{-1}[l](\tilde{\mathbf{r}}_t[l] - \rho\mathbf{e}_M)\\ &= \mathbf{e}_M - \rho\widetilde{\mathbf{R}}_t^{-1}[l]\mathbf{e}_M,\end{aligned} \tag{9.105}$$

where $\widetilde{\mathbf{R}}_t^{-1}[l]\tilde{\mathbf{r}}_t[l] = \mathbf{e}_M$.

We can use the following recursive estimative for $\widetilde{\mathbf{R}}_t[l]$:

$$\widetilde{\mathbf{R}}_t[n;l] = \alpha\widetilde{\mathbf{R}}_t[n-1;l] + (1-\alpha)\tilde{\mathbf{d}}[n;l]\tilde{\mathbf{d}}^H[n;l]. \tag{9.106}$$

And, hence, the estimative for $\mathbf{c}[l]$ in time n is given by

$$\mathbf{c}[n;l] = \mathbf{e}_M - \rho\widetilde{\mathbf{R}}_t^{-1}[n;l]\mathbf{e}_M. \tag{9.107}$$

A recursive updating of $\mathbf{P}[n;l] = \widetilde{\mathbf{R}}_t^{-1}[n;l]$ using the Woodbury identity [6] leads to the recursive estimate of $\hat{d}[n;l]$ given in Algorithm 9.1.

Algorithm 9.1 Updating of the Temporal Filter.

Initialization: $\mathbf{P}[0;l] \leftarrow \mathbf{I}$ and $0 < \alpha < 1$;
for all n **do**
$\mathbf{P}[0;l] \leftarrow \mathbf{I}$
$\mathbf{k}[n;l] \leftarrow \dfrac{\boldsymbol{\pi}[n;l]}{\alpha\cdot(1-\alpha)^{-1} + \tilde{\mathbf{d}}^H[n;l]\boldsymbol{\pi}[n;l]}$
$\mathbf{P}[n;l] \leftarrow \alpha^{-1}\mathbf{P}[n-1;l] - \alpha^{-1}\mathbf{k}[n;l]\boldsymbol{\pi}^H[n;l]$
$\mathbf{p}[n;l] \leftarrow \mathbf{P}[n;l]\mathbf{e}_M$
$\hat{d}[n;l] \leftarrow \tilde{d}[n;l] - \rho(\mathbf{p}^H[n;l]\tilde{\mathbf{d}}[n;l])$
end for

An explicit computation of the matrix $\mathbf{P}[n;l]$ can be avoided. There exist fast algorithms that provide an updating for $\mathbf{k}[n;l]$ requiring a computational complexity of $O(2M+1)$ [6]. The L_p filters require a computational load of $O((2M+1)L_p)$.

Due to numerical instabilities and changes in the distribution of the eigenvalues of the correlation matrix of $\tilde{\mathbf{d}}[n;l]$, the updating of $\mathbf{P}[n;l]$, and consequently a fast computation for $\mathbf{k}[n;l]$, can present explosive divergence. To avoid this problem, an algorithm based on QR decomposition [6] is adopted. In Appendix 2, a derivation of this algorithm is found, which is outlined in Algorithm 9.2. This algorithm also supports a fast version, with a computational complexity or order $O(2M+1)$ [6].

Algorithm 9.2 Algorithm Based on QR Decomposition.

Initialization: $\mathbf{P}[0;l] = \mathbf{I}$; $\mathbf{p}[0;l] = \mathbf{e}_M$; $0 < \alpha < 1$

for all n **do**

$$\begin{pmatrix} \alpha^{1/2}\tilde{\mathbf{R}}_t^{1/2}[n-1;l] & (1-\alpha)^{1/2}\tilde{\mathbf{d}}[n;l] \\ \mathbf{0}^H & 1 \end{pmatrix} \boldsymbol{\Theta}[n;l] \leftarrow \begin{pmatrix} \mathbf{B}_{11}[n;l] & \mathbf{0} \\ \mathbf{b}_{21}^H[n;l] & b_{22}[n;l] \end{pmatrix}$$

$\tilde{\mathbf{R}}_t^{1/2}[n;l] \leftarrow \mathbf{B}_{11}[n;l]$

$\mathbf{k}[n;l] \leftarrow (1-\alpha)^{1/2}\{\tilde{\mathbf{R}}_t^{-H/2}[n;l]\mathbf{b}_{21}[n;l]\}$ (via back-substitution)

$\mathbf{p}[n;l] \leftarrow \alpha^{-1}\mathbf{p}[n-1;l] - \alpha^{-1}\mathbf{k}[n;l](\tilde{\mathbf{d}}^H[n;l]\mathbf{p}[n-1;l])$

$\hat{d}[n;l] \leftarrow \tilde{d}[n;l] - \rho(\mathbf{p}^H[n;l]\tilde{\mathbf{d}}[n;l])$

end for

9.4.3 Estimation of the ICI-Plus-Noise Power and the Number of Multipaths

In this section, the power of the interference plus noise present in the OFDM symbol is calculated. Further, the number of multipaths of the channel model will be estimated as well.

It is assumed that the system operates with a maximum number of multipaths $L_{p,\max} < N_p$. We define

$$\tilde{\mathbf{d}}[n] = \mathbf{U}_s^H \widehat{\mathbf{H}}_{LS,p}[n], \tag{9.108}$$

where $\tilde{\mathbf{d}}[n]$ has dimension $L_{p,\max} \times 1$ and $\mathbf{U}_s$ contains the $L_{p,\max}$-dominating eigenvalues of $\mathbf{U}$. We can write

$$\begin{aligned} P_{\widehat{H}_{LS}} &= \mathbb{E}\{\widehat{\mathbf{H}}_{LS,p}^H[n]\widehat{\mathbf{H}}_{LS,p}[n]\} & P_{\tilde{d}} &= \mathbb{E}\{\tilde{\mathbf{d}}^H[n]\tilde{\mathbf{d}}[n]\} \\ &= N_p\kappa\sigma_h^2 + N_p\rho, & &= N_p\kappa\sigma_h^2 + L_{p,\max}\rho. \end{aligned} \tag{9.109}$$

Then, the ICI-plus-noise power ρ can be calculated using

$$\rho = \frac{P_{\widehat{H}_{LS}} - P_{\tilde{d}}}{N_p - L_{p,\max}}. \tag{9.110}$$

By means of updating the recursive estimates of $P_{\widehat{H}_{LS}}$ and $P_{\tilde{d}}$, according to

$$P_{\widehat{H}_{LS}}[n] = \alpha P_{\widehat{H}_{LS}}[n-1] + (1-\alpha)\widehat{\mathbf{H}}_{LS,p}^{H}[n]\widehat{\mathbf{H}}_{LS,p}[n], \tag{9.111}$$

$$P_{\tilde{d}}[n] = \alpha P_{\tilde{d}}[n-1] + (1-\alpha)\tilde{\mathbf{d}}^{H}[n]\tilde{\mathbf{d}}[n], \tag{9.112}$$

the following recursive estimation method for ρ is obtained:

$$\begin{aligned}\widehat{\rho}[n] &= (P_{\widehat{H}_{LS}}[n] - P_{\tilde{d}}[n])/(N_p - L_{p,\max}) \\ &= \alpha \cdot \widehat{\rho}[n-1] + (1-\alpha)\cdot(\widehat{\mathbf{H}}_{LS,p}^{H}[n]\widehat{\mathbf{H}}_{LS,p}[n] - \tilde{\mathbf{d}}^{H}[n]\tilde{\mathbf{d}}[n])/(N_p - L_{p,\max}).\end{aligned} \tag{9.113}$$

Concerning the estimation of the dimension L_p of the signal space, we know that

$$\mathbb{E}|\tilde{d}[n;l]|^2 = \kappa d_l + \rho. \tag{9.114}$$

Therefore, L_p can be found by choosing the largest l such that $\mathbb{E}\{|\tilde{d}[n;l]|^2\} > \rho$. The estimation of the mean power of the lth coordinate is realized according to

$$p[n;l] = \alpha \cdot p[n-1;l] + (1-\alpha)\cdot|\tilde{d}[n;l]|^2. \tag{9.115}$$

In the sequel, L_p is estimated:

$$\widehat{L}_p[n] = \#\{p[n;l]; p[n;l] > \beta \cdot \widehat{\rho}[n]\}, \tag{9.116}$$

where "#" denotes the number of elements of a set. The parameter $\beta > 1$ was inserted in order to minimize the probability of a wrong selection of a coordinate l for which $\mathbb{E}|\tilde{d}[n;l]|^2 = \rho$, which can occur due to estimation errors.

The algorithm defined by (9.113), (9.115), and (9.116) is summarized in Algorithm 9.3.

Algorithm 9.3 Estimation of Parameters ρ and $\mathbf{L_p}$.

Initialization: $L_{p,\max}$; $\widehat{\rho}[n] = 0$; $p[n;l] = 0$; $0 < \alpha < 1$; $\beta > 1$;
for all n **do**
$\widehat{\rho}[n] \leftarrow \alpha \cdot \widehat{\rho}[n-1] + (1-\alpha)\cdot(\widehat{\mathbf{H}}_{LS,p}^{H}[n]\widehat{\mathbf{H}}_{LS,p}[n] - \tilde{\mathbf{d}}^{H}[n]\tilde{\mathbf{d}}[n])/(N_p - L_{p,\max})$
$p[n;l] \leftarrow \alpha \cdot p[n-1;l] + (1-\alpha)\cdot|\tilde{d}[n;l]|^2$
$\widehat{L}_p[n] \leftarrow \#\{p[n;l]; p[n;l] > \beta \cdot \widehat{\rho}[n]\}$
end for

In the expression for $\hat{d}[n;l]$ in Algorithm 9.1 or 9.2, the estimator $\widehat{\rho}[n]$ given in (9.113) can be used in the place of ρ.

9.4.4 Performance Evaluation

The parameters used in the simulations are found in Table 9.2. Figure 9.15 shows the learning curves of the filtering algorithms developed and analyzed in this section.

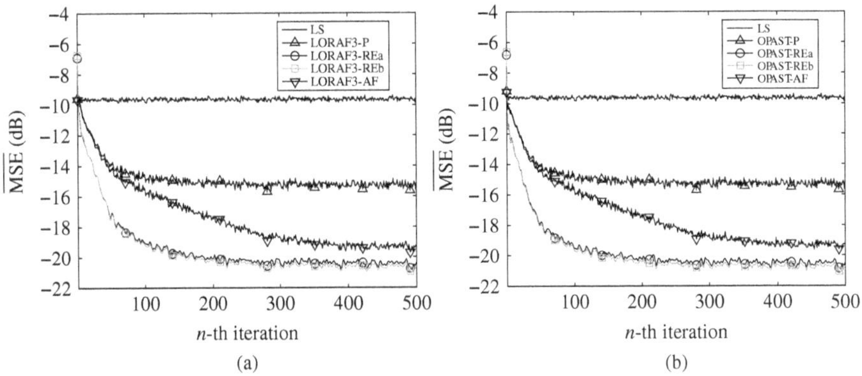

Fig. 9.15 Learning curves with **U** estimated by algorithms: (a) LORAF3 and (b) OPAST, for the parameters in Table 9.2, and SNR = 10 dB, $f_d = 500\,\text{Hz}$, and $M_{pt} = 1$.

The matrix **U** is estimated by the LORAF3 [16] or OPAST [1, 18] algorithms. For the temporal section, the following algorithms are considered:

- $*$-**P**: The LS estimate is projected onto the subspace spanned by the first L_p columns of **U**.
- $*$-**REa**: The coefficients from the temporal section are selected according to (9.89), which does not require the knowledge of the parameters κ, ρ, and d_l. The channel correlations are given by (9.88).
- $*$-**REb**: Similar to $*$-REa, but with the channel correlations given in (9.79).
- $*$-**AF**: The estimates are provided by the QR decomposition-based algorithm outlined in Algorithm 9.2, with the parameters ρ and L_p estimated by the algorithm given in Algorithm 9.3.

9.5 Channel Estimation for MIMO-OFDM Wireless Systems

When multiple antennas are used, an additional dimension is added to the channel estimation. Each pair of transmit and receive antennas corresponds to a channel link that must be estimated. It is typically required that only one transmit antenna is allowed to transmit a pilot symbol in a given subcarrier to avoid interference from other antennas in the channel estimation process. This will be discussed in Section 9.5.2. Two different strategies for channel estimation in MIMO-OFDM systems are presented in Sections 9.5.1 and 9.5.2.

9.5.1 Block-Type Channel Estimation

In the block-type channel estimation (BTCE) strategy, the pilot symbols are arranged in a training block where all subcarriers in the OFDM symbol are used for channel estimation. The channel is estimated by using this training block, followed by the transmission of a given number of OFDM symbols with only data where no channel estimation is performed. After that a new training block is transmitted and the process is repeated. Figure 9.16 illustrates an example of block-type channel estimation for a case where six data symbols are sent for each training block.

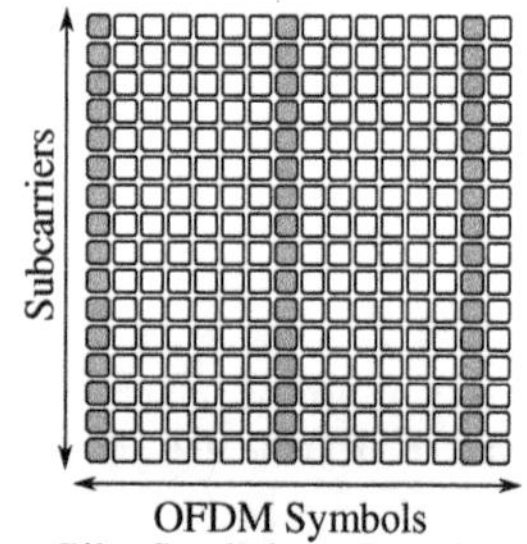

Fig. 9.16 Block-type channel estimation strategy.

In [8] a BTCE strategy is proposed for OFDM systems with transmit diversity where special training sequences are designed so that it is possible to transmit using all antennas at the same time and still estimate the channel in the receiver. This in contrast to other channel estimation strategies for system with multiple transmit antennas where whenever a subcarrier is used to send a pilot symbol in one transmit antenna, it may not be used by other transmit antennas to avoid interference in the estimation process.

The estimation approach used in [8] is to minimize the MSE. Let $y[k]$ be the kth subcarrier of a received OFDM symbol, then

$$y[k] = \sum_{i=1}^{M_t} H_i[k]x_i[k] + \eta[k], \tag{9.117}$$

where i is the transmit antenna index, M_t is the number of transmit antennas, $\eta[k]$ is the noise in the kth subcarrier, and $H[k]$ is the channel frequency response in the kth subcarrier[9.10], that is,

$$H[k] = \sum_{\ell=0}^{L-1} h[\ell]\omega_K^{k\ell}, \tag{9.118}$$

where $h[\ell]$ is the ℓth channel tap (channel with L taps) and $\omega_K = \exp(-j2\pi/K)$. The MMSE cost function to be minimized is given by

$$C\left(\left\{\widehat{h}_i[\ell];\ i = 1,2,\ldots,M_t\right\}\right) = \sum_{k=0}^{K-1} \left| y[k] - \sum_{i=1}^{M_t}\sum_{\ell=0}^{L-1} \widehat{h}_i[\ell]\omega_K^{k\ell}x_i[k] \right|^2. \tag{9.119}$$

[9.10] It is assumed that the number of subcarriers is large enough so that the bandwidth of each subcarrier is much less than the coherence bandwidth of the channel and $H[k]$ is an scalar.

Taking the derivative of (9.119) with respect to $\widetilde{h}_i[\ell]$ yields [8]

$$\sum_{k=0}^{K-1}\left(y[k]-\sum_{i=1}^{M_t}\sum_{\ell=0}^{L-1}\widetilde{h}_i[\ell]\omega_K^{k\ell}x_i[k]\right)\omega_K^{-k\ell_0}x_j^*[k]=0, \tag{9.120}$$

for $j=1,2,\ldots,M_t$ and $\ell_0=0,1,\ldots,L-1$.

Define

$$p_j[\ell]=\sum_{k=0}^{K-1}y[k]x_j^*[k]\omega_K^{-k\ell}\quad\text{and}\quad q_{ij}[\ell]=\sum_{k=0}^{K-1}x_i[k]x_j^*[k]\omega_K^{-k\ell}. \tag{9.121}$$

Equation (9.120) is then equivalent to

$$\sum_{i=1}^{M_t}\sum_{\ell=0}^{L-1}\widetilde{h}_i[\ell]q_{ij}[\ell_0-\ell]=p_j[\ell_0], \tag{9.122}$$

for $j=1,2,\ldots,M_t$ and $\ell_0=1,2,\ldots,L-1$. In matrix notation, the equation can be rewritten by

$$\mathbf{Q}\widetilde{\mathbf{h}}=\mathbf{p}\quad\Rightarrow\quad\widetilde{\mathbf{h}}=\mathbf{Q}^{-1}\mathbf{p}, \tag{9.123}$$

where

$$\widetilde{\mathbf{h}}=\begin{pmatrix}\widetilde{\mathbf{h}}_1\\ \widetilde{\mathbf{h}}_2\\ \vdots\\ \widetilde{\mathbf{h}}_{M_t}\end{pmatrix}_{L\cdot M_t\times 1},\ \mathbf{p}=\begin{pmatrix}\mathbf{p}_1\\ \mathbf{p}_2\\ \vdots\\ \mathbf{p}_{M_t}\end{pmatrix}_{L\cdot M_t\times 1},\ \text{and}\ \mathbf{Q}=\begin{pmatrix}\mathbf{Q}_{11} & \mathbf{Q}_{21} & \cdots & \mathbf{Q}_{M_t1}\\ \mathbf{Q}_{12} & \mathbf{Q}_{22} & \cdots & \mathbf{Q}_{M_t2}\\ \vdots & \vdots & \ddots & \vdots\\ \mathbf{Q}_{1M_t} & \mathbf{Q}_{2M_t} & \cdots & \mathbf{Q}_{M_tM_t}\end{pmatrix}_{L\cdot M_t\times L\cdot M_t}$$

with $\widetilde{\mathbf{h}}_i=\left(\widetilde{h}_i[0],\widetilde{h}_i[1],\ldots,\widetilde{h}_i[L-1]\right)^T$, $\mathbf{p}_i=(p_i[0],p_i[1],\ldots,p_i[L-1])^T$, and

$$\mathbf{Q}_{ij}=\begin{pmatrix}q_{ij}[0] & q_{ij}[-1] & \cdots & q_{ij}[-L+1]\\ q_{ij}[1] & q_{ij}[0] & \cdots & q_{ij}[-L+2]\\ \vdots & \vdots & \ddots & \vdots\\ q_{ij}[L-1] & q_{ij}[L-2] & \cdots & q_{ij}[0]\end{pmatrix}_{L\times L}.$$

To avoid the necessity of calculating the inverse of matrix $\mathbf{Q}$ in (9.123), optimum training sequences ($x_i[k]$, $i=1,2,\ldots,M_t$) are also proposed in [8] in such a way that the term $q_{ij}[\ell]$ takes the form

$$q_{ij}[\ell]=\begin{cases}K\delta[\ell] & \text{for } i=j,\\ 0 & \text{for } i\neq j,\end{cases}$$

and the matrices $\mathbf{Q}_{ij}$ become diagonal matrices multiplied by the constant K. Therefore, the estimated channel for the ith transmit antenna is then given by

$$h_i[\ell]=\frac{1}{K}p_i[\ell]. \tag{9.124}$$

The design of the optimum sequences is possible for different numbers of employed transmit antenna $M_t \leq \frac{K}{L}$ and is given by

$$x_i[k] = x_1[k]\omega_K^{-\overline{K}_0(i-1)k}, \qquad i = 2,\ldots,M_t \tag{9.125}$$

where $\overline{K}_0 = \lfloor \frac{K}{M_t} \rfloor$ and $\lfloor x \rfloor$ denotes the largest integer lower than x, and $x_1[k]$ is a known sequence with constant modulus such as, for instance, a sequence of PSK symbols.

9.5.2 Pilot-Assisted Channel Estimation

In the pilot-assisted channel estimation (PACE) strategy, N_p pilot tones are inserted in each OFDM symbol (usually uniformly distributed among the subcarriers) to allow channel estimation. The receiver knows the location of these pilot tones and their values so that the channel may be estimated for these subcarriers and, after that, for the remaining subcarriers by using an interpolation method. In Fig. 9.17 an example of PACE for a case with two transmit antennas is illustrated.

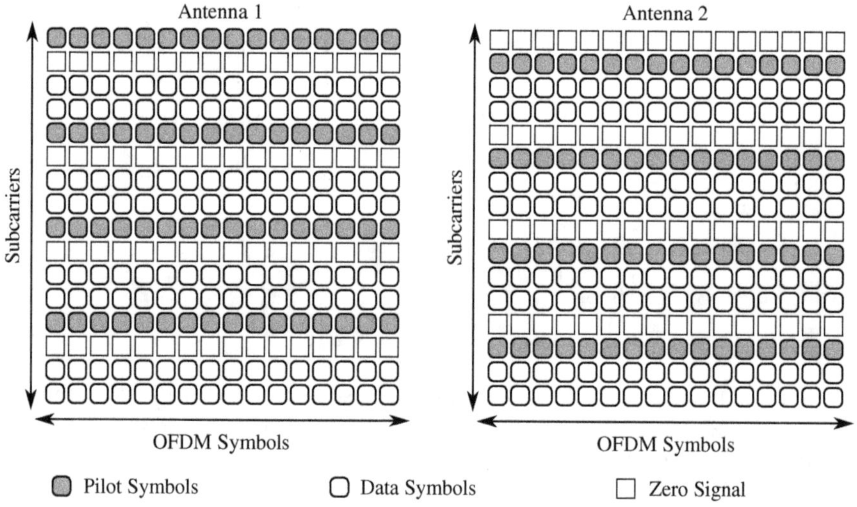

Fig. 9.17 pilot-assisted channel estimation strategy.

Note that the receiver uses the fact that the pilot tones are known to estimate the channel. That is, the modification of the known information must be caused only by the channel and, therefore, the transmitter is not allowed to transmit either pilot tones or data symbols through the other transmit antennas in the same subcarrier.

Hence the zero signal which is shown in Fig. 9.17.

In [15] a PACE strategy is proposed for OFDM systems with transmit diversity where an LS estimator is used to estimate the channel for the pilot tones (frequency domain) and then an iterative algorithm is employed to interpolate and reduce the noise of the channel estimate by means of time domain filtering.

Let k_p be the subcarrier indexes with pilot tones, where $p = 0, 1, 2, \ldots, N_p - 1$ and N_p is the number of pilot tones, the received signal is given by

$$y[k_p] = x_i[k_p] H_i[k_p] + \eta[k_p]. \tag{9.126}$$

Note that even though there are multiple transmit antennas there is no sum operator in (9.126), since only the ith antenna is transmitting information while all other "transmit the zero signal" as shown in Fig. 9.17. For the same reason, the actual value of k_p is different for the different transmit antennas and no antenna index is used for k_p for simplicity of notation.

An initial estimate of the channel can be obtained as

$$\widehat{H}[k_p] = \frac{y[k_p]}{c} = H[k_p] + \frac{\eta[k_p]}{c}, \tag{9.127}$$

where c is the pilot symbol value. An N_p-point IFFT is then applied to obtain the estimated channel in the time domain $\widehat{h}[\ell]$, with $\ell = 0, 1, N_p - 1$. Since the channel has a length $L < N_p$, the elements $\ell \geq L$ are the result of only noise and a filtering in time domain may be performed by simply eliminating these elements to obtain the new channel estimate in time domain $\widehat{h}^1[\ell]$, with $\ell = 0, 1, \ldots, L$. However, the channel length L is not always known. Provided that the cyclic prefix (CP) was correctly designed, a practical approach is to assume that the channel length is equal to the CP length, that is $L = N_{\text{cp}}$. Therefore, the noise component is reduced to the fraction N_{cp}/N_p of its original value.

Now, the below iterative procedure can be applied for the iterations over m ($m > 1$):

- Apply a K-point FFT to $h^m[\ell]$ to obtain the estimated estimate channel for all subcarriers[9.11]

$$\widehat{H}_i^m[k] = \sum_{\ell=0}^{N_{\text{cp}}} \widehat{h}_i^m[\ell] \exp\left(-j2\pi \frac{k\ell}{K}\right), \qquad k = 0, 1, 2, \ldots, K-1. \tag{9.128}$$

- Replace the frequency response estimates in the pilot tones with the ones obtained from (9.127) (not necessary in the last iteration).
- Compute the metric

$$\Delta = \max\{|\widehat{H}_i^m[k] - \widehat{H}_i^{m-1}[k]|\}, \quad k = 0, 1, \ldots, K-1.$$

- If Δ is below a specified threshold the iteration is terminated. Otherwise, the new estimation of $\widehat{h}^{m+1}[\ell]$ is obtained by converting $\widehat{H}^m[k]$ to time domain and performing the filtering in the domain again.

[9.11] This process corresponds to an interpolation of the estimated channel through the Fourier transform.

9.5.3 Comparison of BTCE and PACE

Since BTCE and PACE use completely different approaches for channel estimation, it is intuitive to expect that they have different performances depending on the considered scenario. Particularly, one can expect by comparing Figs. 9.16 and 9.17 that PACE is able to track channel variations better than BTCE. That is, the channel variation introduces an error floor when using BTCE.

In Figs. 9.18 and 9.19 the block error rate (BLER) of a MIMO-OFDM system with different values of Doppler frequency (mobility range) is compared for different MIMO schemes, where the system parameters are described in Table 9.3.

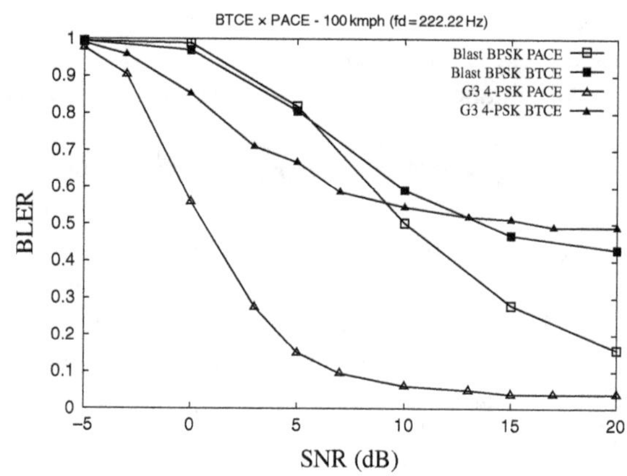

Fig. 9.18 BTCE × PACE with Doppler frequency $f_d =$ 100 Hz.

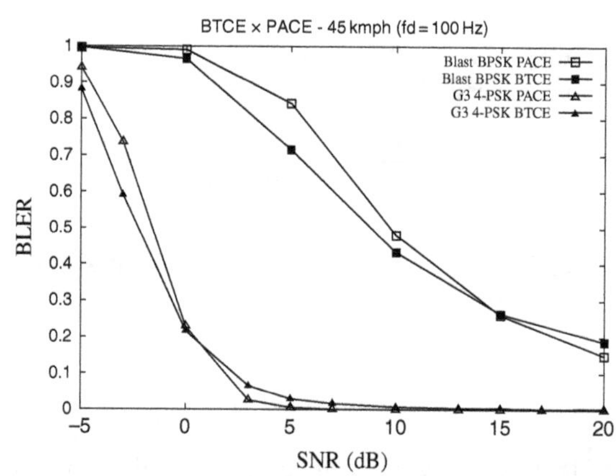

Fig. 9.19 BTCE × PACE with Doppler frequency $f_d =$ 222.22 Hz

Figure 9.18 illustrates that BTCE performs better than PACE, in terms of BLER, for low SNR values (lower than 0 dB for the G3 MIMO scheme and lower than 15 dB for the BLAST MIMO scheme). See Chapter 10 for definitions and references on MIMO structures. This is explained by the fact that the channel estimate pro-

Parameter	Value
Number of subcarriers	1024
CP size (in samples)	20
Channel	COST259 typical urban
Modulation	PSK
Number of Rx antennas	3

Table 9.3 Simulation parameters.

vided by BTCE is more robust against noise. However, when the SNR increases, the channel time variation becomes more important than noise and PACE performs better.

This is specially true when using the G3 MIMO scheme. Since it requires that the channel remains constant for eight (OFDM) symbol periods, as described in Chapter 10, it is more sensible to channel variations.

Furthermore, in a higher mobility scenario as shown in Fig. 9.19, BTCE does not perform better for any SNR value for the G3 MIMO scheme. The degradation due to channel variation in BTCE can be decreased if the channel is estimated more frequently, but this will also increase the channel estimation overhead. In [11] both techniques are compared taking this overhead into account. It is suggested that the channel estimation strategy itself can be a parameter to be adapted when performing link adaptation.

9.6 Conclusions and Research Directions

This chapter provided an overview of some methods and algorithms as well as some important results for channel estimation in orthogonal frequency division multiplexing (OFDM) systems, considering time-varying channels and multiple-input multiple-output (MIMO) technology of relevance for Beyond-3G wireless systems.

The problem of inter-carrier interference was analyzed and an upper bound for the power of the interference due to time-varying channels was derived.

Furthermore, a set of estimators were discussed and analyzed for the case when the channel presents time variation and also for the case when MIMO technology is employed.

The estimators were based on different strategies and presented different behaviors with respect to performance and – for the recursive versions – convergence time.

The topics covered in this chapter provide some issues which can be extended and/or further investigated.

The always present problem of performance versus complexity is a major issue to be addressed. Regarding the robust estimator approach, the optimization of λ_1 and

λ_2 is an important point and may lead to a higher performance with an increase on the computational burden. On the other hand, if the knowledge about the matrices required for the estimators is a quantized version of the real ones, the performance would be more similar to the one expected in real-world systems.

Regarding implementation issues, aspects of channel tracking is a main problem for high-mobility scenarios. In addition, for the use of MIMO systems, the overhead of signaling must be evaluated in order to allow a solution that does not provide a high increasing on the issue. Those are points that must be covered in order to fully demonstrate the capabilities of the methods in practical (commercial) systems.

Appendix 1

In this appendix, the expression given in (9.34) is derived.

If the symbols at the subcarriers are i.i.d., the Inter-carrier interference (ICI) power at the kth subcarrier is given by

$$\sigma_{\text{ICI}}^2[k] = \mathbb{E}\left|\sum_{i=0}^{K-1} (\mathcal{H}_{\text{ICI}}[n])_{ki} s[n,i]\right|^2 = \sum_{i=0}^{K-1} \mathbb{E}|(\mathcal{H}_{\text{ICI}}[n])_{ki}|^2. \qquad (9.129)$$

Let $e_{ml} = \left(h^n_{m,\langle m-l\rangle_K} - \frac{1}{K}\sum_{i=0}^{K-1} h^n_{i,\langle m-l\rangle_K}\right)$ be the (m,l)th entry of $\mathcal{H}_{\text{ICI}}[n]$. Since the (k,i)th entry of $\mathcal{H}_{\text{ICI}}[n]$ is given by

$$(\mathcal{H}_{\text{ICI}}[n])_{ki} = \frac{1}{K}\sum_{m=0}^{K-1}\sum_{l=0}^{K-1} e_{ml}\,\omega_K^{km-il}, \qquad (9.130)$$

after some simplifications, (9.129) can be rewritten as

$$\sigma_{\text{ICI}}^2[k] = \frac{1}{K}\sum_{l,m_1,m_2=0}^{K-1} \mathbb{E}\{e^*_{m_1 l} e_{m_2 l}\}\,\omega_K^{k(m_2-m_1)}. \qquad (9.131)$$

The expansion of the summation above in terms of $h^n_{m,l}$ and further simplifications result in

$$\sigma_{\text{ICI}}^2[k] = \frac{1}{K}\sum_{l,m_1,m_2=0}^{K-1} \mathbb{E}\{h^{n*}_{m_1,\langle m_1-l\rangle_K} h^n_{m_2,\langle m_2-l\rangle_K}\}\cdot\omega_K^{k(m_2-m_1)} - \mathbb{E}|H[n,k]|^2. \quad (9.132)$$

Writing $h^n_{m_2,\langle m_2-l\rangle_K}$ as a function of $\gamma_k[m]$ and $g_k[l]$, the expectation in (9.132) can be expressed as

$$\sum_{i=0}^{L_p-1} \rho_i r_i[m_2-m_1] g_i^*[\langle m_1-l\rangle_K] g_i[\langle m_2-l\rangle_K]. \qquad (9.133)$$

Then, for the summation in (9.132) results in

$$\frac{1}{K}\sum_{i=0}^{L_p-1}\sum_{l,q=0}^{K-1}\rho_i r_i[q]\cdot\left(\sum_{\langle m_2-m_1\rangle_K=q} g_i^*[\langle m_1-l\rangle_K]g_i[\langle m_2-l\rangle_K]\right)\cdot\omega_K^{kq}. \tag{9.134}$$

The last summation above is recognized as the qth element of the circular convolution of $g_i[m]$ with itself. Since the Fourier transform of $g_i[m]$ is approximated by $\exp(-j2\pi k\Delta f\tau_i)$, we have that this convolution is a Dirac pulse, i.e., equal to 1, for $q=0$, and 0, otherwise. Then, (9.134) results in

$$\frac{1}{K}\sum_{i=0}^{L_p-1}\sum_{l=0}^{K-1}\rho_i r_i[0]=\sigma_h^2. \tag{9.135}$$

Finally, the desired result is obtained

$$\sigma_{\mathrm{ICI}}^2=\sigma_h^2-\sigma_H^2, \tag{9.136}$$

where the index k was omitted in $\sigma_{\mathrm{ICI}}^2[k]$, since this term has the same value for all k.

Appendix 2

In this appendix, the algorithm based on QR decomposition given in Algorithm 9.2 is derived. Let $\widetilde{\mathbf{R}}_t^{1/2}[n;l]$ be the square root of $\widetilde{\mathbf{R}}_t[n;l]$, such that

$$\widetilde{\mathbf{R}}_t[n;l]=\widetilde{\mathbf{R}}_t^{1/2}[n;l]\widetilde{\mathbf{R}}_t^{H/2}[n;l], \tag{9.137}$$

where $\widetilde{\mathbf{R}}_t^{H/2}[n;l]$ is an upper triangular matrix. The terms in (9.106) are organized as follows:

$$\mathbf{G}[n]=\begin{pmatrix}\alpha\widetilde{\mathbf{R}}_t[n;l] & (1-\alpha)^{1/2}\tilde{\mathbf{d}}[n;l]\\ (1-\alpha)^{1/2}\tilde{\mathbf{d}}^H[n;l] & 1\end{pmatrix}. \tag{9.138}$$

Using the decomposition given in (9.137), this equation can be rewritten as

$$\mathbf{G}[n]=\begin{pmatrix}\alpha^{1/2}\widetilde{\mathbf{R}}_t^{1/2}[n-1;l] & (1-\alpha)^{1/2}\tilde{\mathbf{d}}[n;l]\\ \mathbf{0}^H & 1\end{pmatrix}\begin{pmatrix}\alpha^{1/2}\widetilde{\mathbf{R}}_t^{H/2}[n-1;l] & \mathbf{0}\\ (1-\alpha)^{1/2}\tilde{\mathbf{d}}^H[n;l] & 1\end{pmatrix}. \tag{9.139}$$

Applying a sequence of Givens rotations, a unitary matrix $\boldsymbol{\Theta}[n;l]$ is obtained and satisfies

$$\underbrace{\begin{pmatrix}\alpha^{1/2}\widetilde{\mathbf{R}}_t^{1/2}[n-1;l] & (1-\alpha)^{1/2}\tilde{\mathbf{d}}[n;l]\\ \mathbf{0}^H & 1\end{pmatrix}}_{\mathbf{A}}\boldsymbol{\Theta}[n;l]=\underbrace{\begin{pmatrix}\mathbf{B}_{11}[n;l] & \mathbf{0}\\ \mathbf{b}_{21}^H[n;l] & b_{22}[n;l]\end{pmatrix}}_{\mathbf{B}}, \tag{9.140}$$

where $\mathbf{B}_{11}[n;l]$ is a lower triangular matrix. Since $\boldsymbol{\Theta}[n;l]$ is unitary, i.e., $\boldsymbol{\Theta}[n;l]\boldsymbol{\Theta}^H[n;l] = \mathbf{I}$, we have

$$\underbrace{\begin{pmatrix} \alpha^{1/2}\widetilde{\mathbf{R}}_t^{1/2}[n-1;l] & (1-\alpha)^{1/2}\tilde{\mathbf{d}}[n;l] \\ \mathbf{0}^H & 1 \end{pmatrix}}_{\mathbf{A}} \underbrace{\begin{pmatrix} \alpha^{1/2}\widetilde{\mathbf{R}}_t^{H/2}[n-1;l] & \mathbf{0} \\ (1-\alpha)^{1/2}\tilde{\mathbf{d}}^H[n;l] & 1 \end{pmatrix}}_{\mathbf{A}^H}$$
$$= \underbrace{\begin{pmatrix} \mathbf{B}_{11}[n;l] & \mathbf{0} \\ \mathbf{b}_{21}^H[n;l] & b_{22}[n;l] \end{pmatrix}}_{\mathbf{B}} \underbrace{\begin{pmatrix} \mathbf{B}_{11}^H[n;l] & \mathbf{b}_{21}[n;l] \\ \mathbf{0}^H & b_{22}^*[n;l] \end{pmatrix}}_{\mathbf{B}^H}. \tag{9.141}$$

Expanding the matrix products and comparing the terms in both sides of (9.141), the following identities are obtained:

$$\begin{aligned} \widetilde{\mathbf{R}}_t[n;l] &= \mathbf{B}_{11}[n;l]\mathbf{B}_{11}^H[n;l], \\ (1-\alpha)^{1/2}\tilde{\mathbf{d}}[n;l] &= \mathbf{B}_{11}[n;l]\mathbf{b}_{21}[n;l], \\ 1 &= \mathbf{b}_{21}^H[n;l]\mathbf{b}_{21}[n;l] + b_{22}[n;l]b_{22}^*[n;l], \end{aligned}$$

which result in

$$\mathbf{B}_{11}[n;l] = \widetilde{\mathbf{R}}_t^{1/2}[n;l],$$
$$\mathbf{b}_{21}[n;l] = (1-\alpha)^{1/2}\widetilde{\mathbf{R}}_t^{-1/2}[n;l]\tilde{\mathbf{d}}[n;l], \tag{9.142}$$
$$|b_{22}[n;l]|^2 = 1-(1-\alpha)\mathbf{d}_M^H[n;l]\widetilde{\mathbf{R}}_t^{-1}[n;l]\tilde{\mathbf{d}}[n;l]. \tag{9.143}$$

Inserting in (9.143) the updating expression for $\mathbf{P}[n;l] = \widetilde{\mathbf{R}}_t^{-1}[n;l]$ as given in Algorithm 9.1, and further simplifications, results in

$$|b_{22}[n;l]|^2 = (1-\alpha)^{-1}\gamma[n;l], \tag{9.144}$$

where

$$\gamma[n;l] = \frac{1}{(1-\alpha)^{-1}+\alpha^{-1}\tilde{\mathbf{d}}^H[n;l]\mathbf{P}[n-1;l]\tilde{\mathbf{d}}[n;l]}. \tag{9.145}$$

And, hence,

$$b_{22}[n;l] = (1-\alpha)^{1/2}\gamma^{1/2}[n;l]. \tag{9.146}$$

Thus (9.140) can be rewritten as

$$\begin{pmatrix} \alpha^{1/2}\widetilde{\mathbf{R}}_t^{1/2}[n-1;l] & (1-\alpha)^{1/2}\tilde{\mathbf{d}}[n;l] \\ \mathbf{0}^H & 1 \end{pmatrix} \boldsymbol{\Theta}[n;l]$$
$$= \begin{pmatrix} \widetilde{\mathbf{R}}_t^{1/2}[n;l] & \mathbf{0} \\ (1-\alpha)^{1/2}\tilde{\mathbf{d}}^H[n;l]\widetilde{\mathbf{R}}_t^{-H/2}[n;l] & (1-\alpha)^{1/2}\gamma^{1/2}[n;l] \end{pmatrix}. \tag{9.147}$$

By now expanding the updating expression for $\mathbf{k}[n;l]$ given in Algorithm 9.1 results in

$$(1-\alpha)^{-1}\mathbf{k}[n;l] = \alpha^{-1}\mathbf{P}[n-1;l]\tilde{\mathbf{d}}[n;l] - \alpha^{-1}\mathbf{k}[n;l]\tilde{\mathbf{d}}^H[n;l]\mathbf{P}[n-1;l]\tilde{\mathbf{d}}[n;l], \quad (9.148)$$

and, hence

$$\mathbf{k}[n;l] = (1-\alpha)\mathbf{P}[n;l]\tilde{\mathbf{d}}[n;l] = (1-\alpha)\widetilde{\mathbf{R}}_t^{-1}[n;l]\tilde{\mathbf{d}}[n;l]. \quad (9.149)$$

From (9.142), we have

$$\begin{aligned}\widetilde{\mathbf{R}}_t^{H/2}[n;l]\mathbf{k}[n;l] &= (1-\alpha)^{1/2}\{(1-\alpha)^{1/2}\widetilde{\mathbf{R}}_t^{-1/2}[n;l]\tilde{\mathbf{d}}[n;l]\} \\ &= (1-\alpha)^{1/2}\mathbf{b}_{21}[n;l]. \end{aligned} \quad (9.150)$$

Since $\widetilde{\mathbf{R}}_t^{H/2}[n;l]$ is upper triangular, the solution of the system in (9.150) can be found using the *back-substitution* method [4]. Therefore, we obtain $\mathbf{k}[n;l]$. Since $\mathbf{p}[n;l] = \mathbf{P}[n;l]\mathbf{e}_M$, we have the following recursive expression for $\mathbf{p}[n;l]$:

$$\mathbf{p}[n;l] = \alpha^{-1}\mathbf{p}[n-1;l] - \alpha^{-1}\mathbf{k}[n;l](\tilde{\mathbf{d}}^H[n;l]\mathbf{p}[n-1;l]). \quad (9.151)$$

Therefore, $\mathbf{p}[n;l]$ can be updated since $\mathbf{k}[n;l]$ is known in time $n-1$.

The algorithm thus obtained, based on QR decomposition, is constituted by (9.147)–(9.151) and is summarized in Algorithm 9.2.

References

1. Abed-Meraim, K., Chkeif, A., Hua, Y.: Fast orthonormal PAST algorithm. IEEE Signal Processing Letters **7**(3), 60–62 (2000). DOI 10.1109/97.823526
2. Chang, R.W., Gibby, R.A.: A theoretical study of performance of an orthogonal multiplexing data transmission scheme. IEEE Transactions on Communication Technology **16**(4), 529–540 (1968). DOI 10.1109/TCOM.1968.1089889
3. Glisic, S.: Advanced Wireless Communications. John Wiley & Sons: New York (2004)
4. Golub, G.H., van Loan, C.F.: Matrix Computations, 3 edn. The Johns Hopkins University Press, Baltimore, Maryland (1996)
5. Harada, H., Prasad, R.: Simulation and Software Radio for Mobile Communications. The Artech House Universal Personal Communication Series. Artech House (2002)
6. Haykin, S.: Adaptive Filter Theory. Prentice Hall: Englewood Cliffs, NJ (2002)
7. Li, Y.: Pilot-symbol-aided channel estimation for OFDM in wireless systems. IEEE Transactions on Vehicular Technology **49**(4), 1207–1215 (2000). DOI 10.1109/25.875230
8. Li, Y.: Simplified channel estimation for OFDM systems with multiple transmit antennas. IEEE Transactions on Wireless Communications **1**(1), 67–75 (2002). DOI 10.1109/7693.975446
9. Li, Y., Cimini Jr., L.J.: Bounds on the interchannel interference of OFDM in time-varying impairments. IEEE Transactions on Communications **49**(3), 401–404 (2001). DOI 10.1109/26.911445
10. Manton, J.H.: Dissecting OFDM: the independent roles of the cyclic prefix and the IDFT operation. IEEE Communications Letters **5**(12), 474–476 (2001). DOI 10.1109/4234.974490
11. Moreira, D.C., Cavalcante, C.C.: Channel estimation in link adaptation strategies for MIMO-OFDM systems. In: Wireless World Research Forum, Meeting (WWRF), vol. 17, Heidelberg (2006)

12. Nee, R.V., Prasad, R.: OFDM for Wireless Multimedia Communications. Artech House (2000)
13. Oppenheim, A.V., Schafer, R.W., Buck, J.R.: Discrete-Time Signal Processing, 2nd edn. Prentice Hall: Englewood Cliffs, NJ (1999)
14. Papoulis, A., Pillai, S.U.: Probability, Random Variables and Stochastic Processes, 4 edn. McGraw-Hill: New York (2001)
15. Qiao, Y., Yu, S., Su, P., Zhang, L.: Research on an iterative algorithm of ls channel estimation in MIMO OFDM systems. IEEE Transactions on Broadcasting **51**(1), 149–153 (2005). DOI 10.1109/TBC.2004.842524
16. Strobach, P.: Low-rank adaptive filters. IEEE Transactions on Signal Processing **44**(12), 2932–2947 (1996). DOI 10.1109/78.553469
17. Weinstein, S., Ebert, P.: Data transmission by frequency-division multiplexing using the discrete Fourier transform. IEEE Transactions on Communication Technology **19**, 628–634 (1971). DOI 10.1109/TCOM.1971.1090705
18. Yang, B.: Projection approximation subspace tracking. IEEE Transactions on Signal Processing **43**(1), 95–107 (1995). DOI 10.1109/78.365290

Chapter 10
Link Adaptation for MIMO-OFDM Systems

Darlan C. Moreira, Walter C. Freitas Jr., Cibelly A. de Araújo, and Charles C. Cavalcante

10.1 Introduction

The paradigm of the design of a wireless system has changed. Since the use of the dimensioning for the "worst case", which means to design the system to work on the fading margin available when the channel has its poorest behavior, the driver of the optimization has evolved to a more suitable use of the available resources for performing a reliable communication.

This approach is then called link adaptation (LA), when the system chooses the parameters which are the most suitable for usage in a certain channel condition. The always increasing demand for higher data rates, lower energy consumption, etc., requires that the system resources are utilized as efficiently as possible and LA techniques are already a reality in any modern wireless communication systems to achieve that goal.

While many aspects of LA, such as usage of different modulations and code rates for providing better "protection" to data streams according to the channel condition, have already been understood, each system has a different set of "interesting parameters" to be adapted in multiple dimensions and the trade-off between LA gains and signaling overhead still provides challenges to be answered.

Typical dimensions used in LA procedure are *modulation* and *coding*. The choice of the modulation allows the system to improve/decrease the spectral efficiency and the code rate impacts the amount of redundancy inserted for error protection into data frames. However, it is possible to envisage the exploitation of other features of the wireless system, for instance the spatial and frequency domains.

This chapter describes the use of transmission modes considering parameters which are important to the performance of a wireless system, in particular the extension of LA to the MIMO-OFDM case in fourth generation (4G) systems.

The rest of the chapter is organized as follows. The fundamentals of multiple-input multiple-output (MIMO) systems are presented in Section 10.2, where classical MIMO schemes are described. Section 10.3 discusses the trade-off between the diversity and multiplexing gains that can be extracted from the MIMO channel

F. Cavalcanti, S. Andersson (eds.), *Optimizing Wireless Communication Systems*,
DOI 10.1007/978-1-4419-0155-2_10,

and describes some hybrid MIMO schemes that are able to achieve both gains simultaneously. The fundamentals of LA are described in Section 10.4 for a soft-input/soft-output (SISO) system and then extended to the MIMO-OFDM case. The summary of the chapter and envisaged research directions are discussed in Section 10.5.

10.2 Fundamentals of MIMO Transceiver Architectures

Higher transmission rates and better reliability are always desirable in communication systems. However, according to information theory, to increase one, the other has to be decreased [27], unless we are willing to use more system resources, such as power or bandwidth, which is not always possible.

In this context, the use of the spatial dimension through MIMO strategies is mandatory in the next generation systems, such as long-term evolution (LTE) and LTE-Advanced [3, 10]. By using the spatial dimension, more degrees of freedom can be used to increase the data rates and/or the reliability of the system without the need of more system resources.

10.2.1 Space Diversity/Spatial Dimension Gains

There are different types of gains that can be extracted from the spatial dimension, such as *array gain*, *coding gain*, *diversity gain*, *multiplexing gain*, etc. A MIMO system has the generic form shown in Fig. 10.1 and the gains actually extracted from the spatial dimension depend on how the MIMO encoder maps the input symbols into the coded symbols sent by each transmitter antenna and/or how the reception processing is done.

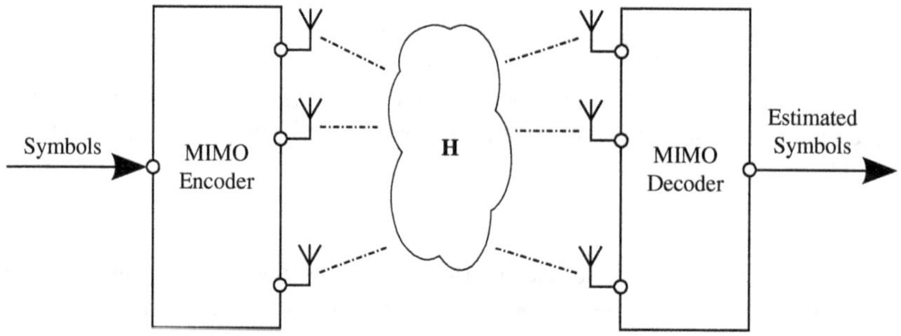

Fig. 10.1 Generic MIMO system.

The array gain refers to an average increase in signal-to-noise ratio (SNR) and is obtained when the receiver coherently combines the signal that arrives at the multiple receiver antennas. A similar gain, the coding gain, can be obtained when the transmitter encodes the symbols using, for instance, space–time codes, but not all space–time codes yield a coding gain. Both the array gain and the coding gain are seen in a graphic of bit error rate (BER), or similar measure, versus SNR as a shift to left in the curve when compared to a system without multiple transmitter antennas,[10.1] as depicted in Fig. 10.2.

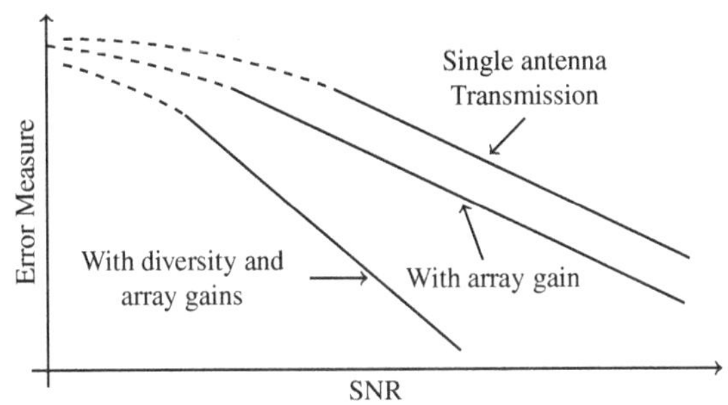

Fig. 10.2 Diversity gain versus array gain.

The diversity gain consists of increasing the reliability of the received information by combining the different versions of the faded signals. That is, since in MIMO systems there are multiple links corresponding to the pairs of transmit and receiver antennas, the probability that all links are in a fade is low and by combining all links the resultant signal exhibits a lower fading (amplitude variation) when compared to a signal from a single link.

Different from the array/coding gain, the effect of the diversity gain is an increase in the curve slope for high SNRs as shown in Fig. 10.2. The more degrees of freedom/orders-of-diversity the MIMO system can take advantage of, the more is the curves' relative slope increase. Several MIMO strategies can be employed to extract a diversity gain from the channel and a common strategy in the literature is the space–time codes, which will be presented in Section 10.2.2. It should be noted that the maximum diversity gain that the (spatially uncorrelated) channel can provide is given by $M_r \times M_t$, where M_r is the number of receiver antennas [23, 38] and M_t is the number of transmitter antennas [23, 38].

The multiplexing gain consists of increasing the transmission data rate by using the spatial dimension to separate multiple data streams. That is, two or more data streams are transmitted at the same time and frequency through the different transmitter antennas. Due to degree of freedom provided by the spatial domain the channels of the different data streams are different and can be seen as signatures used in the receiver to separate the streams as it will be described in Section 10.2.3.

However, a trade-off exists between diversity and multiplexing gains such that increasing one will decrease the other [38]. Also, it should be noted that the MIMO

[10.1] Since their effect is the same (in the perceived SNR), coding gain and array gain will be used interchangeably.

architecture presented in most of the literature assumes a flat fading channel. This assumption can be well-motivated when using MIMO jointly with orthogonal frequency division multiplexing (OFDM), as discussed in Section 8.7.

10.2.2 Space–Time Coding

Considering a single-input multiple-output (SIMO) system and assuming a flat fading channel, the channel vector **h** is given by

$$\mathbf{h} = [h_1, h_2, \ldots, h_{M_r}]^T. \tag{10.1}$$

For a transmitted symbol s, the received symbol vector is then given by

$$\mathbf{y} = \mathbf{h}s + \mathbf{v}, \tag{10.2}$$

where $\mathbf{v}$ is the considered noise, usually assumed additive white Gaussian noise (AWGN). In order to realize a *receiver diversity gain* and maximize the SNR the receiver can perform a maximal ratio combining (MRC) [23, Chapter 5], i.e.,[10.2]

$$\begin{aligned} \widetilde{\mathbf{y}} &= \mathbf{h}^H\mathbf{h}s + \mathbf{h}^H v \\ &= \|\mathbf{h}\|^2 s + \mathbf{h}^H v, \end{aligned} \tag{10.3}$$

where $\widetilde{\mathbf{y}}$ is the $M_r \times 1$ receiver output.

On the other hand, in a multiple-input single-output (MISO) system with multiple antennas at the transmitter, the simplistic approach of transmitting the same signal from all transmitter antennas does not provide any diversity gain at the transmitter [23]. In this case, the received signal is a linear superposition of the transmitted signals from all transmitter antennas plus noise. However, since the total transmit power has to be divided among all antennas, no diversity gain is obtained and, therefore, a more elaborate approach has to be used to extract a *transmitter diversity gain.*

The approach will depend on whether there is channel information available at the transmitter or not. When the channel is known to the transmitter the power in each transmitter antenna can be weighted appropriately by a weight vector **w** chosen subject to $\|\mathbf{w}\|^2 = M_t$, where this restriction is necessary to ensure that the total transmit energy is not changed. The weight vector that maximizes the SNR is then given by [21]

$$\mathbf{w} = \sqrt{M_t}\,\frac{\mathbf{h}^H}{\|\mathbf{h}\|}. \tag{10.4}$$

This solution is called maximal ratio transmission (MRT) and is similar to the MRC. The obtained gain corresponds in fact to a diversity gain plus a coding gain instead of only a diversity gain. However, the assumption that the channel is known at the transmitter can typically not be fully satisfied.

10.2 The channel **h** is assumed to be known at the receiver.

When the channel is not known to the transmitter, the most common approach in the literature is the space–time block codes (STBCs), which require channel information only at the receiver. The transmit signal is encoded in a way to extract the spatial diversity while at the same time it can be decoded by the receiver using simple linear processing. However, due to the lack of channel knowledge at the transmitter no coding gain is obtained and STBCs obtain only the diversity gain.[10.3]

A well-known STBC is the Alamouti scheme described in [4]. It encodes the input signal onto two transmitter antennas and has the advantage of being a full-diversity code that requires only linear processing in the receiver. The Alamouti code has the advantage that it does not decrease the spectral efficiency compared to the SISO case.

A full-diversity code extracts the maximum diversity gain that the MIMO channel can provide, that is, $M_t \times M_r$. On the other hand, a full-rate code achieves the maximum multiplexing gain corresponding to $\min(M_t, M_r)$ [23].

Given two transmit symbols s_1 and s_2, the space–time codeword matrix for the Alamouti scheme is given by

$$\mathbf{S} = \begin{bmatrix} s_1 & s_2 \\ -s_2^* & s_1^* \end{bmatrix},$$

where the rows correspond to the time slots, the columns correspond to the transmitter antennas, and* stands for complex conjugate. Hence, in the first time slot the first antenna sends the symbol s_1 while the second one sends the symbol s_2. After that, the first antenna sends $-s_2^*$ and the second one sends s_1^* in the subsequent time slot.

The signal in the single receiver antenna for the two time slots is then given by, respectively, y_1 and y_2 such that $y_1 = h_1 s_1 + h_2 s_2 + v_1$ and $y_2 = -h_1 s_2^* + h_2 s_1^* + v_2$, where h_1 is the channel between the first transmitter antenna and the receiver antenna, h_2 is the channel between the second transmitter antenna and the receiver antenna, and v_1 and v_2 are zero mean circularly symmetric complex gaussian (ZMCSCG) noise. In matrix notation, the received signal can be written as

$$\mathbf{y} = \begin{bmatrix} y_1 \\ y_2^* \end{bmatrix} = \begin{bmatrix} h_1 & h_2 \\ h_2^* & -h_1^* \end{bmatrix} \begin{bmatrix} s_1 \\ s_2 \end{bmatrix} + \begin{bmatrix} v_1 \\ v_2^* \end{bmatrix}. \tag{10.5}$$

To decode the transmitted information the receiver only needs to multiply $\mathbf{y}$ by the conjugate transpose of the channel matrix in (10.5) yielding

$$\begin{bmatrix} \widetilde{y}_1 \\ \widetilde{y}_2 \end{bmatrix} = \begin{bmatrix} \|h_1\|^2 + \|h_2\|^2 & 0 \\ 0 & \|h_1\|^2 + \|h_2\|^2 \end{bmatrix} \begin{bmatrix} s_1 \\ s_2 \end{bmatrix} + \begin{bmatrix} \widetilde{v}_1 \\ \widetilde{v}_2 \end{bmatrix}. \tag{10.6}$$

Note that this simple matrix multiplication is enough to decode the transmitted information, since $\widetilde{y}_1$ depends only on s_1 and $\widetilde{y}_2$ depends only on s_2. Also, the term $\|h_1\|^2 + \|h_2\|^2$ clearly shows that each transmitted symbol is amplified by both channel gains resulting in a diversity gain of two (two diversity branches).

[10.3] The array gain can still be obtained if multiple antennas are used in the receiver.

In [33] STBCs for more than two transmitter antennas are proposed, but they have a code rate lower than one for any complex symbol constellation. The space–time codeword matrices for the H3, G3, and G4 schemes, presented in [33], and for the Alamouti scheme, addressed as G2, are reproduced below.

$$\mathbf{S}_{G2} = \begin{bmatrix} s_1 & s_2 \\ -s_2^* & s_1^* \end{bmatrix}, \qquad \mathbf{S}_{H3} = \begin{bmatrix} s_1 & s_2 & \frac{s_3}{\sqrt{2}} \\ -s_2^* & s_1^* & \frac{s_3}{\sqrt{2}} \\ \frac{s_3^*}{\sqrt{2}} & \frac{s_3^*}{\sqrt{2}} & \frac{-s_1-s_1^*+s_2-s_2^*}{2} \\ \frac{s_3^*}{\sqrt{2}} & -\frac{s_3^*}{\sqrt{2}} & \frac{s_2+s_2^*+s_1-s_1^*}{2} \end{bmatrix},$$

$$\mathbf{S}_{G3} = \begin{bmatrix} s_1 & s_2 & s_3 \\ -s_2 & s_1 & -s_4 \\ -s_3 & s_4 & s_1 \\ -s_4 & -s_3 & s_2 \\ s_1^* & s_2^* & s_3^* \\ -s_2^* & s_1^* & -s_4^* \\ -s_3^* & s_4^* & s_1^* \\ -s_4^* & -s_3^* & s_2^* \end{bmatrix}, \qquad \mathbf{S}_{G4} = \begin{bmatrix} s_1 & s_2 & s_3 & s_4 \\ -s_2 & s_1 & -s_4 & s_3 \\ -s_3 & s_4 & s_1 & -s_2 \\ -s_4 & -s_3 & s_2 & s_1 \\ s_1^* & s_2^* & s_3^* & s_4^* \\ -s_2^* & s_1^* & -s_4^* & s_3^* \\ -s_3^* & s_4^* & s_1^* & -s_2^* \\ -s_4^* & -s_3^* & s_2^* & s_1^* \end{bmatrix}.$$

Another option to take advantage of MIMO channel properties and increase the reliability is to employ space–time trellis codes (STTCs) [32, 33], where the functions of symbol mapper and space–time encoder are combined into a single block. In fact, STBCs and STTCs are to space–time codes as block codes and trellis-coded modulation are to channel coding. While STBCs extract only a diversity gain from the MIMO channel, STTCs can extract both diversity and coding gains yielding a better BER performance.[10.4] The disadvantage of STTCs is that they are more complex to encode (and more difficult to construct good codes) and decode compared to the case for STBCs. While a linear decoder is used for STBCs, STTCs rely on a Viterbi decoder. These aspects explain the greater interest in STBCs compared to STTCs.

10.2.3 Spatial Multiplexing

In the previous section the main goal was to increase reliability by using the spatial dimension to obtain a diversity gain. Herein the objective is to maximize the spectral efficiency by using the spatial dimension to obtain a multiplexing gain. The idea is to split the information and send it into M_t streams, where M_t is the number of transmitter antennas and each stream is transmitted in a separate antenna.

[10.4] Note that in both cases an array gain is also obtained when multiple receiver antennas are employed.

This idea was initially proposed with different names. The structure of transmission/reception with multiple antennas is described in [24], and the vertical Bell Labs layered space–time (VBLAST), described in [13], is an architecture to realize a multiplexing gain, with the constraint to operate with the same number of antennas at the transmitter and the receiver. At the transmitter, the information is divided into streams and sent through the different antennas with no special processing required. That is, the space–time codeword for the VBLAST scheme is given by

$$\mathbf{s} = \begin{bmatrix} s_1 \\ s_2 \\ s_3 \end{bmatrix}. \tag{10.7}$$

It is assumed that, for all detection algorithms, the received signal vector $\mathbf{x}$, with dimension $M_r \times 1$, is expressed by

$$\mathbf{x} = \mathbf{H}\mathbf{s} + \mathbf{v}, \tag{10.8}$$

where $\mathbf{H}$ is the $M_r \times M_t$ MIMO channel matrix, $\mathbf{s}$ is the $M_t \times 1$ transmitted signal vector, and $\mathbf{v}$ is the $M_r \times 1$ noise vector. Since all streams are transmitted at the same time and frequency, each element of the received signal vector $\mathbf{x}$ has contributions from all transmitter antennas. Consequently, when decoding each stream the receiver has to eliminate the interference from the other streams by using the spatial dimension. Some linear or nonlinear detection algorithms can be employed for this task. Linear receivers are described in Section 10.2.3.1, which change only in the optimization criterion for the filter calculation. Nonlinear receivers are described in Section 10.2.3.2, where the main idea is to cancel the interference of already detected streams, addressed as layers.

Another MIMO scheme similar to the VBLAST scheme and worth mentioning here is diagonal Bell Labs layered space–time (DBLAST). Instead of M_t different streams, M_t copies of the same stream are transmitted where each copy is shifted one time slot from the previous one. The space–time codeword for the DBLAST scheme is given by

$$\mathbf{S} = \begin{bmatrix} s_1 & s_2 & s_3 & - & - & \cdots \\ - & s_1 & s_2 & s_3 & - & \cdots \\ - & - & s_1 & s_2 & s_3 & \cdots \end{bmatrix}. \tag{10.9}$$

Even though the DBLAST scheme is similar to Bell Labs layered space–time (BLAST), it yields in fact a diversity gain instead of a multiplexing gain. Nevertheless, it is described in this section instead of in Section 10.2.2 because of this similarity. As it can be seen in (10.9), there are some gap elements in the matrix $\mathbf{S}$, which represent an absence of transmission. Because of these gaps not all symbols see the same diversity.

10.2.3.1 Linear Detection

The linear receiver model is given by

$$\mathbf{y} = \mathbf{W}\mathbf{x}, \tag{10.10}$$

where $\mathbf{x}$ is the received signal vector before filtering, $\mathbf{y}$ is the filtered signal vector, and the weight matrix $\mathbf{W}$ may be obtained by several optimization criteria.

The most direct optimization criterion is to nullify the interference when the noise vector is a null vector, i.e.,

$$\begin{aligned} \mathbf{W} &= \arg\min \; \mathbb{E}\{\|\mathbf{W}\mathbf{x} - \mathbf{s}\|^2\} \\ \text{s.t.: } & \mathbf{W}\mathbf{x}|_{\mathbf{v}=\mathbf{0}} = \mathbf{0}, \end{aligned} \tag{10.11}$$

which yields

$$\mathbf{W} = (\mathbf{H}^H\mathbf{H})^{-1}\mathbf{H}^H. \tag{10.12}$$

This is the well-known zero-forcing (ZF) receiver, which works well in a relative noiseless channel (high SNR). However, because of the constraint $\mathbf{W}\mathbf{x}|_{\mathbf{v}=\mathbf{0}} = \mathbf{0}$ the ZF receiver has a problem of enhancing the noise, which degrades its performance in noisy channels (lower SNR).

To overcome this limitation, a good strategy is to change the optimization criterion to the minimum-mean-square-error (MMSE), that is,

$$\mathbf{W} = \arg\min \; \mathbb{E}\{\|\mathbf{W}\mathbf{x} - \mathbf{s}\|^2\}, \tag{10.13}$$

which results in the following weight vector $\mathbf{W}$ [23]:

$$\mathbf{W} = \left(\mathbf{H}^H\mathbf{H} + \frac{M_t}{\gamma}\mathbf{I}_{M_t}\right)^{-1}\mathbf{H}^H, \tag{10.14}$$

where γ is the SNR value and $\mathbf{I}_{M_t}$ is the $M_t \times M_t$ identity matrix. Named after its optimization criterion, this receiver is known as the MMSE receiver. While the MMSE receiver does not usually eliminate the interference completely, it does not suffer from the noise enhancement problem and it is more balanced than the ZF receiver.

10.2.3.2 Nonlinear Detection

It is possible to substantially increase the performance of the receiver if nonlinear detection is employed. The successive interference cancellation (SIC) [23] detection is an example of a nonlinear receiver where the first layer is detected with a linear receiver, such as the ZF or MMSE receivers, producing an estimate $\widehat{s}_1$. After that, the contribution from layer 1 on the receive signal is estimated and cancelled, resulting in a signal $\mathbf{x}_2$.

In general, at the ith layer, the signal $\mathbf{x}_i$ is expected not to have interference caused by the previous layers $j < i$. Therefore, based on the symbol estimate for ith layer, $\hat{\mathbf{s}}_i$, its contribution on the receive signal is estimated and subtracted from the receive signal $\mathbf{x}_i$. This procedure results in a modified signal called $\mathbf{x}_{i+1}$ expressed as

$$\mathbf{x}_{i+1} = \mathbf{x}_i - \hat{s}_i \mathbf{h}_i, \tag{10.15}$$

where $\mathbf{h}_i$ is the ith column from the channel matrix $\mathbf{H}$ corresponding to the channel gains associated with the ith layer and $\hat{s}_i \mathbf{h}_i$ represents the estimated interference from the ith layer. Finally, the receive signal $\mathbf{x}_{i+1}$ is interference free from the layers $1,\ldots,i$. This signal acts as a feedback of the spatial filter for the next layer $(i+1)$. Figure 10.3 depicts the decoding of each layer by the SIC receiver.

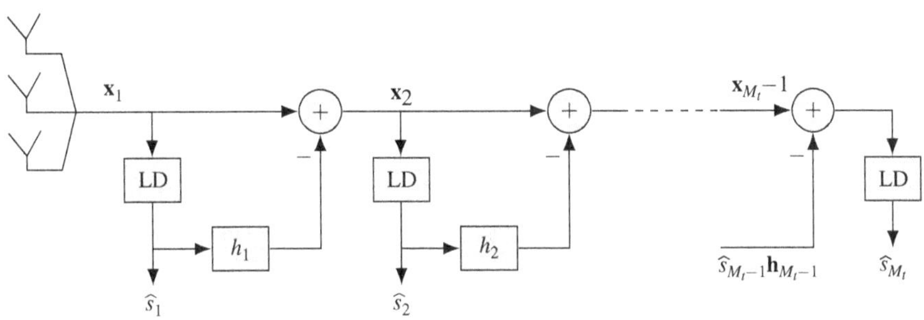

Fig. 10.3 SIC receiver.

If all decisions are correct, the interference is totally eliminated from the previous detected symbols, resulting in better predictions for the following symbols. In Fig. 10.4 a comparison of the MMSE linear detection (LD) and SIC receivers is shown for each layer. As it can be seen from the figure, SIC has a better performance and each successive detection iteration yields a different BER value. On the other hand, LD is performed in a single step and all layers have the same BER value.

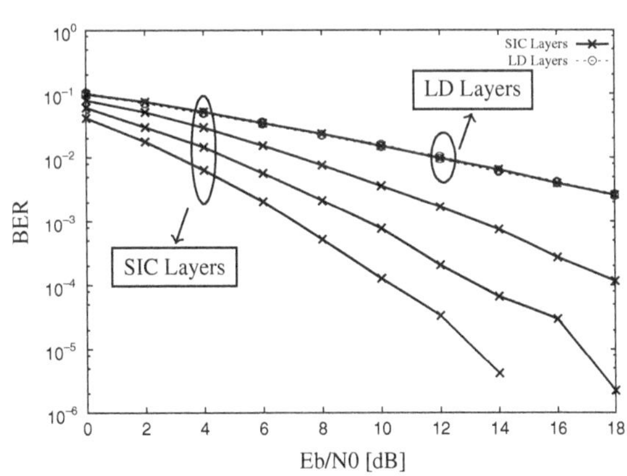

Fig. 10.4 Comparison of the LD and SIC layers for theVBLAST scheme with $M_t = M_r = 4$.

Furthermore, in the SIC detector, if a layer has a low SNR and an error occurs in its detection, this error will be propagated to the subsequent layers even if they have higher SNR. This problem can be mitigated by properly ordering the layer detection in SIC, which is then denoted as ordered successive interference cancellation (OSIC). The layers are ordered in decreasing order of SNR such that the first detected layer corresponds to the layer with highest SNR. Figure 10.5 compares the BER behavior for different SNR values using LD, SIC, and OSIC detectors.

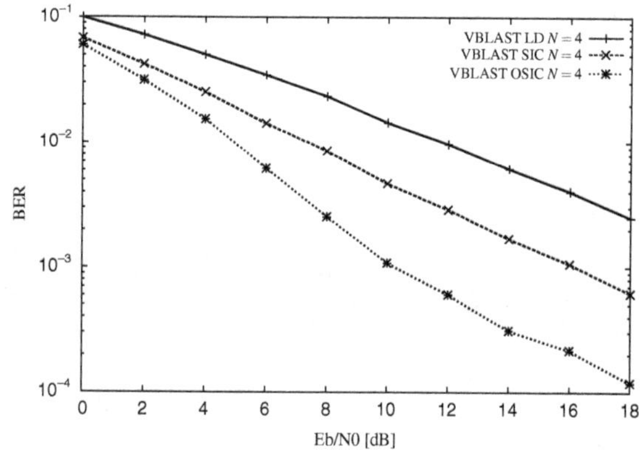

Fig. 10.5 Comparison of LD, SIC, and OSIC receivers for theVBLAST scheme with $M_t = M_r = 4$.

10.2.4 Channel State Information (CSI)

As already mentioned in Sections 10.2.2 and 10.2.3, the channel must be known at the receiver to decode the coded information or to separate the different data streams when STBC or multiplexing is used, respectively. On the other hand, when the channel is also known in the transmitter, the capacity can be further increased by a non-uniform power allocation among the transmitter antennas or some kind of precoder technique [23]. However, due to the channel variation in time, frequency, and space inherent to wireless systems, it is difficult to have this information available at the transmitter.

In time division duplex (TDD) systems it is usually assumed that the channel is approximately the same in both ways: downlink and uplink. Since in these systems the downlink and uplink channels usually correspond to the same frequency bands with only a time separation, this assumption is justified provided that the time separation is lower than the coherence time of the channel. Therefore, the transmitter can acquire information about the direct channel by using information from the reverse channel.[10.5]

On the other hand, this assumption is not valid in frequency division duplex (FDD) systems, since the downlink and uplink channels have a frequency sepa-

[10.5] It should be noted, however, that while the propagation channel is the same the RF circuitry will differ between the receive and transmit branches. Hence, RX/TX branch calibration is required for TDD systems to be able to exploit this channel reciprocity.

ration greater than the coherence bandwidth of the channel. In such systems the channel must be estimated in the receiver and some kind of feedback channel must exist to report the channel conditions to the transmitter. However, this requires that additional control information must be sent to the transmitter using available resources.[10.6]

In this context, the methods can be classified regarding the amount of CSI information: full CSI where the channel is required to be known at the receiver and at the transmitter [8] and partial CSI where the channel is known only at the transmitter or receiver [25]. The latter method is also referred to as side information. Full CSI methods require a large amount of feedback information in order to provide to the transmitter all channel characteristics estimated at the receiver. The partial CSI methods decrease the quantities by using limited versions of the feedback information. Some methods use a set of precoders from which, according to a suitable criterion, one is selected to be used at some time [35]. Other strategies are based on the use of statistical measurements which can be passed to the transmitter less often. These methods are also known as dynamic CSI [36].

Another way to reduce the required feedback is the method based on channel quality indicator (CQI), which is a measure that comprises the information about the channel state in order to transmit a reduced amount of data for a selection/estimation of the best parameters to use the channel [20]. Some commercial systems, such as high-speed packet access (HSPA), use methods for prediction of CQI in order to reduce the interference in the uplink due to those frequent transmissions [11].

Other systems under development, such as the 3GPP LTE, are going to use some kind of schemes which do not require calibration. These transceivers have a set of precoders, called codebook, and use a criterion to select, according to the channel estimate, the best precoder [1]. Hence, only the index of the precoder is transmitted. This reduces a lot of the feedback information and makes the process completely adaptive. The precoder is selected every time the channel is estimated. Strategies of this type are discussed in detail in Chapter 12.

With channel state information available in both the transmitter and the receiver the capacity can be increased with a non-uniform power allocation among the transmitter antennas [23] optimized to maximize the capacity, such as the *water filling* algorithm. Alternatively, the reliability can be increased if the non-uniform power allocation is optimized to maximize the SNR, such as the weight vector, given by (10.4), mentioned in Section 10.2.2.

10.3 Advanced MIMO Transceiver Architectures

The potential for multiple antennas to provide link robustness can be traced back to Marconi's experiments in Pohdu [7]. The same idea of Marconi is still considered in current wireless systems. The maximal diversity order of a MIMO channel can be

[10.6] Feedback reduction is an intense research topic to reduce the drawback of expensive channel state information (CSI) reporting. See [22] for an overview of limited feedback in wireless communication systems.

achieved by space–time code schemes. However, in such schemes a capacity loss is necessarily present. In counterpart, strategies designed to obtain capacity increase in MIMO channels are far away from the maximal diversity order. The inter-relation about these two main possible gains in the MIMO channel was shown in the seminal paper of Zheng and Tse [38].

10.3.1 The Trade-Off Between Multiplexing and Diversity Gains

In [38], Zheng and Tse provided a simple expression relating the two main possible gains in MIMO wireless channels given by

$$d(r) = (M_t - r)(M_r - r), \tag{10.16}$$

where $r \in \mathbb{Z}$ is the multiplexing gain represented in a high SNR scenario and defined as

$$r = \lim_{SNR \to \infty} \frac{R(SNR)}{\log SNR}, \tag{10.17}$$

and d is the diversity gain defined as the error probability of a given space–time code. A fixed rate R is related with the multiplexing gain r by

$$R = r \log SNR. \tag{10.18}$$

Thus, the diversity gain could be expressed as

$$d = - \lim_{SNR \to \infty} \frac{P_e(SNR)}{\log SNR}. \tag{10.19}$$

As a consequence of (10.16), the maximal diversity and multiplexing gains in a MIMO wireless channel are, $r_{\max} = \min(M_t, M_r)$ and $d_{\max} = M_t M_r$, respectively, as illustrated in Fig. 10.6. In the literature most of the space–time code transceivers were proposed aiming to obtain just one of these upper bounds.

By the analysis of the trade-off proposed by Zheng and Tse transceivers that achieve higher diversity gain necessarily achieve a lower multiplexing gain, and vice versa. Furthermore, due to the variation of the wireless channel caused by fading the system could benefit from a specific gain depending on the channel state, suggesting that schemes adapting between diversity and multiplexing gains should be considered.

10.3.2 MIMO Transceiver Structure Design

The use of multiple transmit and receiver antennas may result in great capacity gains. Indeed, in a rich scattering environment the deployment of antenna arrays at both link ends results in a capacity that increases almost linearly with the minimum

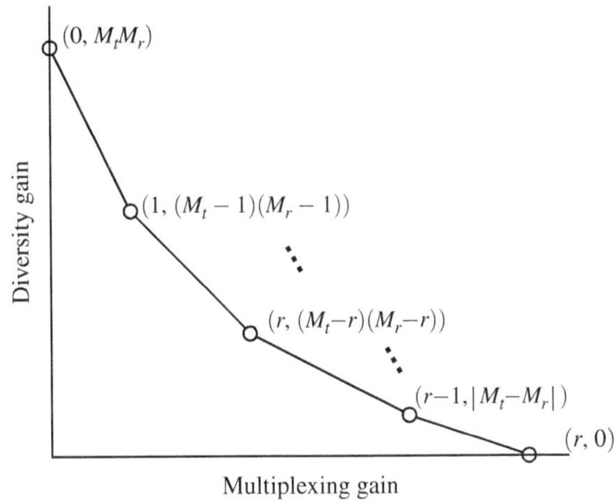

Fig. 10.6 Illustrative results of the MIMO channel trade-off of [38].

number of antennas [13, 34]. MIMO antenna systems may also provide diversity gain, which is a measure of robustness against fading [30]. There is, however, the trade-off discussed in Section 10.3.1, in which the diversity gain can only be increased if the multiplexing gain is decreased, see [38]. The conventional systems described in Sections 10.2.2 and 10.2.3 lie in extreme points in the trade-off curve, Fig. 10.6, as they only provide multiplexing or diversity gains. In this section, we describe some approaches which lie in intermediate points in the trade-off curve, providing both types of gains at the same time.

The idea of a transceiver structure aiming to achieve both spatial gains, diversity and multiplexing, was first proposed by Tarokh et al. in [31]. The authors combined STTC and array processing by partitioning antennas at the transmitter into small groups. The signal transmitted in each group of antennas goes through a given STTC. At the receiver, the signals from each STTC are separated by a nonlinear processing technique that suppresses signals transmitted from other groups of antennas, by treating them as interference. Then, the STTCs are individually decoded. Tarokh et al.'s idea involves a fixed transmission structure in [31], where the authors did not consider adapting the transmitter to the channel conditions.

Since wireless channel is random, the use of a fixed structure designed to the worst-case propagation scenarios would represent a waste of the resources in more favorable situations. In the following, a structure that combines traditional space–time codes and multiplexing schemes to capture both diversity and multiplexing gains as presented in [31] is described, where a family of structures makes the adaptation to more (or less) diversity and multiplexing in accordance with the channel state. These structures are called hybrid MIMO transmit scheme (HMTS).

In general, the transmission process of HMTSs can be divided in layers, similar to VBLAST. However, in contrast to VBLAST, in the HMTS case a layer may consist of the stream of symbols at the output of a STBC, which is sent to a group of antennas, or of an uncoded stream, which is transmitted from a single antenna. Based on this concept of layers, HMTS transceiver schemes combine pure diversity schemes (e.g., STBC) with pure spatial multiplexing schemes (e.g., VBLAST). In HMTS, some layers are space–time coded across two, three, or four antennas. For the remaining layers, a VBLAST approach is used. With this idea, hybrid MIMO schemes achieve a compromise between spatial multiplexing and transmit diversity gains. The basic idea behind these structures is to combine array processing and space–time coding, as first presented in [31].

In the remainder of this section some specific HMTSs are presented. The notation for a particular HMTS is based on the notation of the STBC used by the specific scheme (e.g., alamouti space–time block code (STBC) (G2) or 3 transmitter antenna STBC (G3)), while each uncoded stream following the VBLAST scheme is denoted with an additional label for the hybrid according to +1. For example, the scheme designed for three transmitter antennas consisting of two layers, one space–time coded with the G2 scheme and another uncoded layer following the VBLAST scheme, is denoted G2+1.

10.3.2.1 Hybrid Scheme Designed for Three Transmitter Antennas

This HMTS, whose structure is shown in Fig. 10.7(a), employs three transmitter antenna elements with two spatial multiplexing layers. A standard G2 (Alamouti's) space–time block code is used for the first layer; the other layer is not space–time coded, similar to the VBLAST approach. In the G2+1 scheme, the transmitted signals can be organized in the equivalent space–time coding matrix:

$$\mathbf{S}_{\mathrm{G2+1}}[k,k+1] = \begin{bmatrix} s_1 & s_2 & s_3 \\ -s_2^* & s_1^* & s_4 \end{bmatrix}, \tag{10.20}$$

where the spatial dimension varies column-wise and the temporal dimension row-wise.

From (10.20), it can be seen that $K=4$ information symbols (two from each multiplexing layer) are transmitted in $T=2$ consecutive time slots. Thus, the effective spectral efficiency of this scheme is equal to $\eta = 2 \cdot \log_2 M$ bps/Hz, where M is the modulation order.

10.3.2.2 Hybrid Schemes Designed for Four Transmitter Antennas

The second HMTS, called G2+G2, is shown in Fig. 10.7(b). It employs four transmitter antenna elements with two vertically layered G2 block code schemes. Observe that the four transmitter antennas are divided into two space–time coding groups of two antennas each. The transmitted signals can be organized in an equivalent space–time coding matrix given by

$$\mathbf{S}_{\mathrm{G2+G2}}[k,k+1] = \begin{bmatrix} s_1 & s_2 & s_3 & s_4 \\ -s_2^* & s_1^* & -s_3^* & s_4^* \end{bmatrix}. \tag{10.21}$$

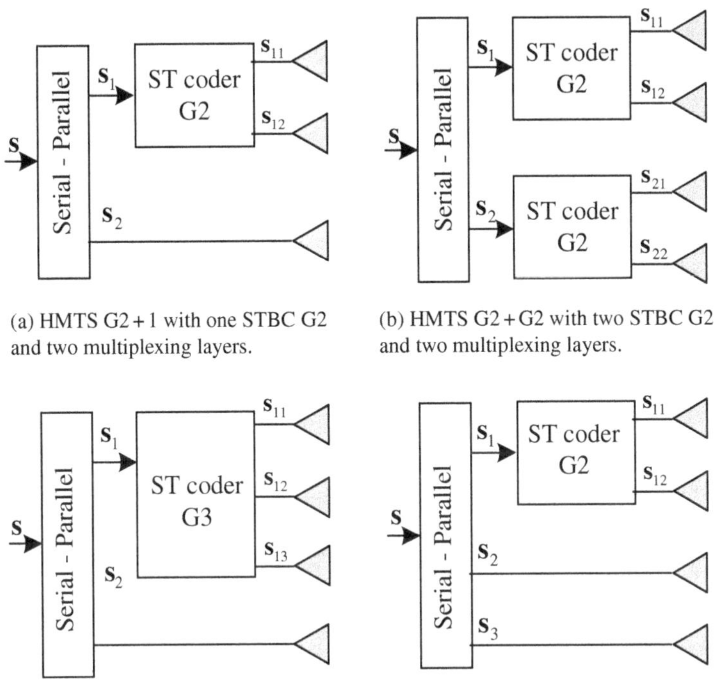

(a) HMTS G2 + 1 with one STBC G2 and two multiplexing layers.

(b) HMTS G2 + G2 with two STBC G2 and two multiplexing layers.

(c) HMTS G3 + 1 with one STBC G3 and two multiplexing layers.

(d) HMTS G2 + 1 + 1with one STBC G2 and three multiplexing layers.

Fig. 10.7 Architecture of the HMTS transmitters.

From (10.21), it can be shown that $K = 4$ information symbols (two from each multiplexing layer) are transmitted in $T = 2$ consecutive time slots. Thus, the effective spectral efficiency of this scheme is equal to $\eta = 2 \cdot \log_2 M$ bps/Hz. Compared to space–time code for four transmitter antenna elements (which has a code-rate of 1/2), the G2+G2 scheme achieves four times the data symbol rate.

Figure 10.7(c) depicts the third HMTS considered in this work. The four transmitter antennas are now divided into two multiplexing layers, where the first one consists of three antennas that are space–time coded using a G3 code [30]. The equivalent space–time coding matrix for this hybrid scheme is given by

$$\mathbf{S}_{\mathrm{G3+1}}[k,\ldots,k+7] = \begin{bmatrix} s_1 & s_2 & s_3 & s_5 \\ -s_2 & s_1 & -s_4 & s_6 \\ -s_3 & s_4 & s_1 & s_7 \\ -s_4 & -s_3 & s_2 & s_8 \\ s_1^* & s_2^* & s_3^* & s_9 \\ -s_2^* & s_1^* & -s_4^* & s_{10} \\ -s_3^* & s_4^* & s_1^* & s_{11} \\ -s_4^* & -s_3^* & s_2^* & s_{12} \end{bmatrix}. \qquad (10.22)$$

From (10.22), $K = 12$ information symbols (four from the first layer and eight from the second one) are transmitted in $T = 8$ consecutive time slots. Thus, the effective spectral efficiency of this scheme is equal to $\eta = 1.5 \cdot \log_2 \mathscr{M}$ bps/Hz. This represents three times the spectral efficiency of G4.

The fourth HMTS scheme is called G2+1+1 and is depicted in Fig. 10.7(d). Again, four transmitter antennas are employed. As it can be seen from the figure, this scheme consists of three spatial multiplexing layers; the first layer is space–time coded using G2, and the remaining ones are transmitted using VBLAST. The equivalent space–time coding matrix for the G2+1+1 scheme is given by

$$\mathbf{S}_{\mathrm{G2+1+1}}[k, k+1] = \begin{bmatrix} s_1 & s_2 & s_3 & s_4 \\ -s_2^* & s_1^* & s_5 & s_6 \end{bmatrix}. \tag{10.23}$$

In this HMTS, $K = 6$ information symbols (two from the first layer and four from the uncoded ones) are transmitted in $T = 2$ consecutive time slots. Thus, the effective spectral efficiency of this scheme is equal to $\eta = 3 \cdot \log_2 \mathscr{M}$ bps/Hz. Compared to a space–time code with four transmitter antennas (which has a code-rate of 1/2), this hybrid scheme achieves six times the data rate. Furthermore, the G2+1+1 scheme offers a 50% increase in spectral efficiency compared to the G2+G2 scheme.

In Fig. 10.8(a) and (b), the performance of BER versus SNR among the VBLAST and HMTS MIMO schemes for three (Fig. 10.8(a)) and four transmitter antennas (Fig. 10.8(b)) in a Rayleigh MIMO channel model, respectively, are illustrated. All schemes consider binary phase-shift keying (BPSK) modulation. In the receiver STBC uses maximum likelihood (ML) detection and the hybrids and VBLAST schemes use the nonlinear detector OSIC. Since the STBC presents hardly any bit errors in this scenario, the corresponding curves are not shown in Fig. 10.8(a) and 10.8(b).

According to these figures, the performance of the HMTS is between the two extremes VBLAST and STBC. In general, HMTS outperforms VBLAST with respect to robustness and outperforms STBC with respect to multiplexing gain, thus generating more diverse opportunities for considering an adaptive MIMO scheme instead of just selecting between pure diversity and multiplexing schemes.

Table 10.1 summarizes the multiplexing and diversity orders of the MIMO transceivers. The diversity order is shown for each layer under both linear and SIC detectors. When considering a STBC that has just one layer, the diversity order is based on the ML detection. Looking at the table we can clearly identify the trade-off between diversity and capacity. For example, the STBC G4 can achieve a diversity order of $4M_r$, but only achieves a symbol rate of 1/2 symbol per channel use. On the other hand, VBLAST achieves a rate of 4 symbols per channel use, but with low diversity order. The HMTS resides between the two extremes, maximal diversity order (e.g., STBC designed for four transmitter antennas) and maximal multiplexing order (e.g., VBLAST). Clearly, HMTSs are inherently flexible structures which can be adapted to the channel conditions, providing more diversity if the channel is in deep fade or more rate if the channel is experiencing good conditions for multiplexed transmission.

Table 10.1 Summary of MIMO transmission schemes.

Scheme	Achievable diversity order per layer Linear detection				SIC				Spectral efficiency η (bps/Hz)
VBLAST (4Tx-M_rRx)	M_r-3	M_r-3	M_r-3	M_r-3	M_r-3	M_r-2	M_r-1	M_r	$\min(M_t,M_r)\log_2 M$
G2+1+1 (4Tx$-M_r$Rx)	$2(M_r-2)$	M_r-3	M_r-3	–	$2(M_r-2)$	M_r-1	M_r	–	$3\cdot\log_2 M$
G2+1 (3Tx$-M_r$Rx)	$2(M_r-1)$	M_r-2	–	–	$2(M_r-2)$	M_r	–	–	$2\cdot\log_2 M$
G2+G2 (3Tx$-M_r$Rx)	$2(M_r-2)$	$2(M_r-2)$	–	–	$2(M_r-2)$	$2M_r$	–	–	$2\cdot\log_2 M$
G3+1 (4Tx$-M_r$Rx)	$3(M_r-1)$	M_r-3	–	–	$3(M_r-1)$	M_r	–	–	$1.5\cdot\log_2 M$
H3 (3Tx$-M_r$Rx)	$3M_r$	–	–	–	–	–	–	–	$3/4\cdot\log_2 M$
G3 (3Tx$-M_r$Rx)	$3M_r$	–	–	–	–	–	–	–	$0.5\cdot\log_2 M$
G4 (3Tx$-M_r$Rx)	$4M_r$	–	–	–	–	–	–	–	$0.5\cdot\log_2 M$

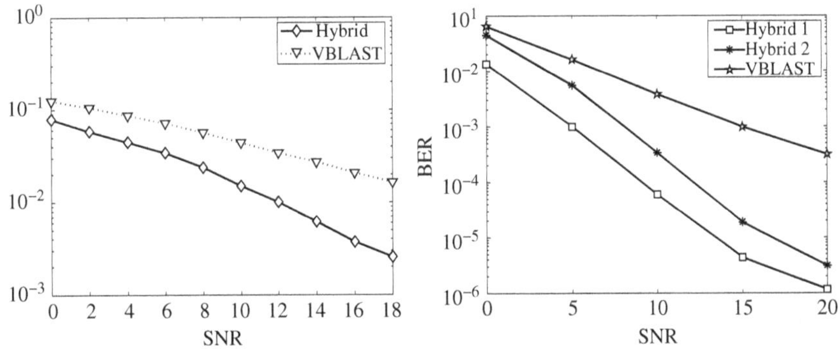

Fig. 10.8 Performance comparison of MIMO schemes with (a) three and (b) four transmitter antennas.

10.3.3 An Added Degree of Freedom

Once the operation of a MIMO system is adapted to provide gains of both diversity and multiplexing, the number of degrees of freedom is higher compared to classical MIMO systems.

In this sense, if the possibility of exploiting a layered approach to achieve both diversity and multiplexing gains is considered, a MIMO system with M_t transmitter antennas can be divided into several combinations, with each such combination being dependent on the criterion to be optimized and the conditions of the MIMO channel experienced by the considered user.

This approach thereby offers an additional degree of freedom, namely the ability to adapt between different structures. Clearly, the higher the number of transmitter antennas, the higher the flexibility of choosing different structures or the possible combinations of them.

This new free parameter can play a very important role when we consider the channel information to perform transceiver optimization. This will be discussed in detail in the following section.

10.4 Link Adaptation in Multiple Signal Dimensions

In Section 10.4.1 the basic idea of link adaptation is presented, while the inclusion of the spatial dimension is described in Section 10.4.2. Section 10.4.3 presents aspects related to OFDM and Section 10.4.4 includes the multiuser aspect to the problem of link adaptation.

10.4.1 Fundamentals of Link Adaptation: Modulation and Coding Schemes

Due to the channel variation inherent to wireless systems, changing the transmission parameters to match the current channel condition promotes a more efficient use of the available resources than just designing the system to function in a worst-case scenario. The collection of techniques that try to solve the problem of making efficient use of (radio) resources is referred to as *link adaptation (LA)*. As a more explicit example, the set of algorithms and protocols governing adaptive modulation and coding is often referred to as LA.

The main idea is that when the radio link is in a deep fade the system should adapt to a set of transmission parameters that increase reliability. On the other hand, when the channel condition is favorable the system should select a set of transmission parameters that increase the data rate and therefore result in a higher spectral efficiency. Alternatively, in cases where power is a more important resource, the power could be decreased when the channel condition is favorable and vice versa, while keeping the data rate constant.

10.4.1.1 Adaptive Modulation and Coding (AMC)

As an example of LA, adaptive modulation and coding (AMC) has been widely investigated in the literature [9, 15, 17]. In fact, sometimes the terms "link adaptation" and "adaptive modulation and coding" are used interchangeably, but any parameter that is limited and whose value influences system performance according to channel condition can be considered for adaptation. Furthermore, it is even possible to adapt different sets of parameters over different time scales.[10.7]

As an illustration, in Fig. 10.9 the various modulation and coding schemes (MCSs) for an enhanced data rate for GSM evolution (EDGE) system [2] are illustrated. The link adaptation is performed by changing the MCS according to the carrier-to-interference ratio (CIR) so that the system "stays on the LA curve" to maximize the throughput. Likewise, with the increased interest in MIMO strategies, the antenna elements used for MIMO transmission become important to adapt [12, 14] in such a way that a joint adaptation is performed over the modulation, coding and antenna scheme parameters.

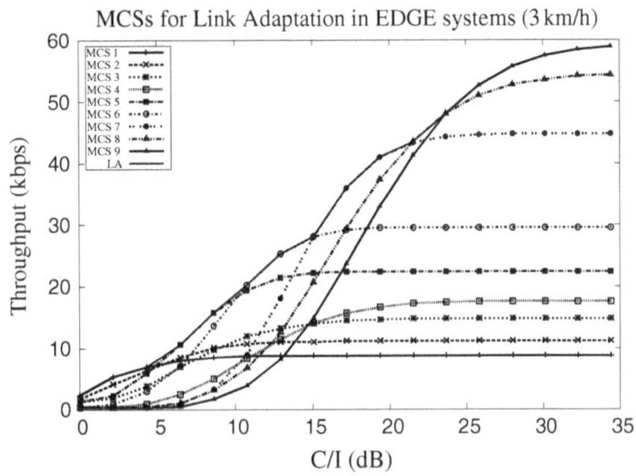

Fig. 10.9 MCSs for an EDGE system.

For a good performance of this LA process, it is necessary that the (estimated) rate of channel variation is known by the transmitter and this rate of variation in the time and frequency domains will dictate how often the estimated information must be updated. Hence, if the channel is changing faster than it can be reliably estimated and fed back to the transmitter, the adaptation will perform poorly [16]. While the channel variation is not a problem for systems with low mobility, it can become a critical factor in systems with high mobility [26]. As an example, in HSPA systems [10] the link adaptation takes place every 2 ms and, therefore, a channel quality indicator (CQI) must be fed back to the transmitter at least 500 times per second.

10.4.1.2 Link Adaptation Criteria

Two important aspects may be highlighted in LA:

- The set of parameters that can be adapted
- The optimization criteria and channel quality metrics used

[10.7] For instance, some parameters could be adapted according to path loss and shadowing while others could be adapted according to fast fading.

Both of them vary according to the application, but the optimization criteria usually reflect a desire to increase the spectral efficiency.

As for metrics, the quality of the channel may be measured by the estimated SNR in the receiver, the idea is making the choice of the transmission mode based on the estimated SNR and the transmitter is assumed to have at its disposal a pre-defined table containing the best transmission modes for each SNR interval. While the idea is simple, it has the practical limitation of requiring a good estimation of the SNR and good approximations of the BER for each scheme, which is not always a simple task, especially in scenarios with interference.

Regarding the pre-defined table, the switching points from one mode to another are the crossing points in curves such as "BER versus SNR", "Spectral Efficiency versus SNR", etc., as depicted in Fig. 10.10. That is, "choose the most spectrally efficient mode if it has a BER lower than a specified threshold".

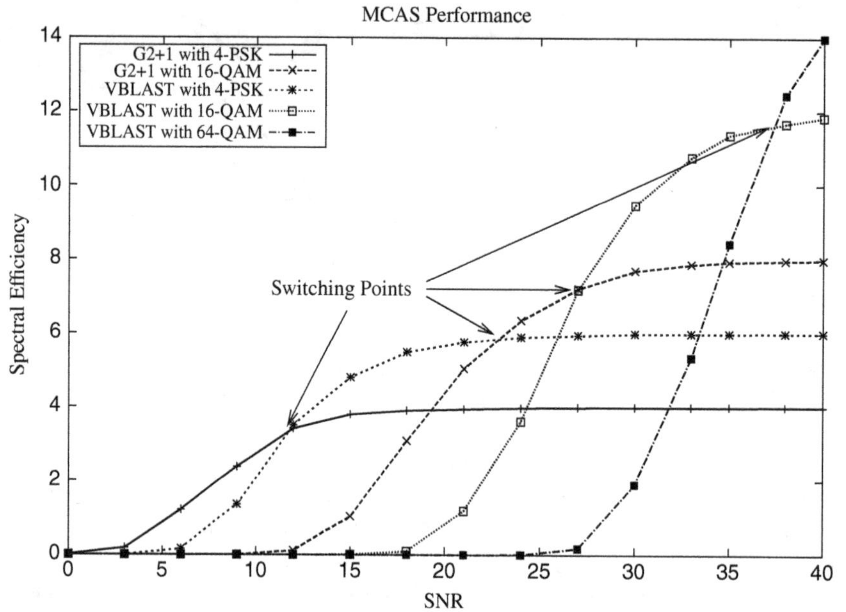

Fig. 10.10 An example of a spectral efficiency versus average SNR curve for different modulations and MIMO schemes.

Another metric that can be used for LA is based on the consecutive success or failure of transmissions. After a specific number of consecutive successful transmissions, the transmission mode is increased.[10.8] Besides the consecutive success transmission counter, the mode increase may also be triggered by a time counter. This time counter is used for increasing the mode when a specified timeout

[10.8] The transmission mode is increased in the sense that it is changed to a more spectral efficient one, while when it is decreased it is changed to a more robust one.

period has expired even if the success counter did not reach a specified threshold. This has been demonstrated to increase the system performance in a system with high load [6], since in this case the difficulties to reach the "increase mode threshold" are due to collisions instead of bad channel quality. That is, increasing the transmission mode does in this case actually result in more successful transmissions by reducing the probability of collisions due to the lowered transmission time.

Similarly, after a specific number of consecutive failed transmissions is reached the transmission mode is decreased. This method of LA is called *automatic repeat fall-back*.

In both of the previous cases, the optimized criterion is the capacity or, equivalently, the spectral efficiency. As a restriction, a minimum robustness is necessary to allow the usage of a given transmission mode. An alternative optimization criterion can be the transmit power, that is, choosing the transmission mode that requires the lowest transmit power for the channel conditions, usually for a minimum throughput restriction.

10.4.2 Adaptation Between Multiple MIMO Schemes

When the spatial dimension is added to the problem of LA the main idea of choosing the best parameters is the same but the metric is different. To clarify this aspect, note that in Fig. 10.9 the modulation and coding rate were adapted according to the metric of C/I. With MIMO, using a metric such as SNR is not clear since for the same channel condition the resultant SNR seen by the receiver (after decoding the information) is different according to which MIMO scheme was used.

In Fig. 10.10 the spectral efficiency versus SNR for different modulation and MIMO schemes in a curve that resembles Fig. 10.9 is shown. Note, however, that while it seems that the situation "hasn't changed"[10.9] from the SISO case, many important aspects of a MIMO system cannot be shown in this simple example in Fig. 10.10. Other factors such as correlation among the transmitter and/or receiver antennas must be accounted for by a channel quality metric, since the SNR alone does not capture any ill-conditioning of the channel matrix.

An important metric that should be considered in the MIMO case is the condition number of the correlation matrix of the channel **H**, which gives an insight into the performance potential of the considered MIMO channel. For instance, when the channel has a low rank, which usually represents a line-of-sight (LOS) scenario resulting in a high condition number, the user would be starved of diversity (multiplex) gain and an STBC (VBLAST) scheme is not appropriated, even if the SNR is low (high). In fact, it may even be better to fall back to the SIMO case to avoid dividing the power among the transmitter antennas, instead of trying to extract a diversity (multiplex) gain that the correlated channel cannot provide.

[10.9] The resulting throughput or spectral efficiency is plotted against a metric of C/I or SNR for different parameter configurations and the one with highest spectral efficiency is chosen.

From [23, Chapter 4], the capacity of a MIMO channel in the absence of channel knowledge at the transmitter is given by

$$C = \log_2 \det\left(\mathbf{I}_{M_r} + \frac{E_s}{M_t N_0}\mathbf{H}\mathbf{H}^H\right), \tag{10.24}$$

where E_s is the transmit power and N_0 is the noise power spectral density. Equation (10.24) may then be used as a metric of the channel quality with the advantage that the SNR and the condition number are taken into account implicitly. However, the capacity presents an upper bound on the throughput of the channel and does not cover any aspect of suboptimal decoding or STBC schemes that do not achieve the capacity bound in (10.24) [26].

The time variation of the channel is another important issue to be considered. Each MIMO scheme has some assumption of time period for which the channel is considered constant. Therefore, if the Doppler frequency is high a MIMO scheme offering the potential for a high diversity gain may perform worse than a scheme offering lower diversity gain, simply due to time variation of the channel.

Note also that while switching among different MCASs can provide extensive performance gains, the necessary rate of feedback information is larger than what is required when only MCSs are adapted. Therefore, the impact of this feedback must be observed on the overall system [18] and research on limited feedback strategies is becoming even more relevant as discussed in Chapter 12.

10.4.3 Frequency Diversity: Link Adaptation for OFDM

OFDM is an important technique to transform a wideband frequency-selective channel into several narrowband flat fading channels. In this sense, the spectrum is divided into N smaller portions called subcarriers. Again, we can profit from this parameter to select the best way (according to some criterion) to use the system resources. Hence, the link adaptation can be done separately for each subcarrier [19, 28, 29, 37] or for blocks of subcarriers. Each block is then either a group composed by consecutively located subcarriers or non-consecutively located subcarriers. This choice will depend on the process of allocation/assignment of the subcarriers for the user.

In Fig. 10.11, a scheme with five subcarriers from the same user is illustrated where we can see the amplitude of subcarriers in time and frequency domains. The transmit parameters can then be adapted according to the channel variations in both domains.

While the use of OFDM increases the flexibility of LA, some drawbacks arise such as an increase in the amount of feedback information needed to perform the required tasks for LA. If the link adaptation is executed per block of subcarriers, the system must consider metrics that represent the channel quality for all subcarriers in a block. Obviously, link adaptation based on blocks requires less feedback information than link adaptation based on individual subcarriers.

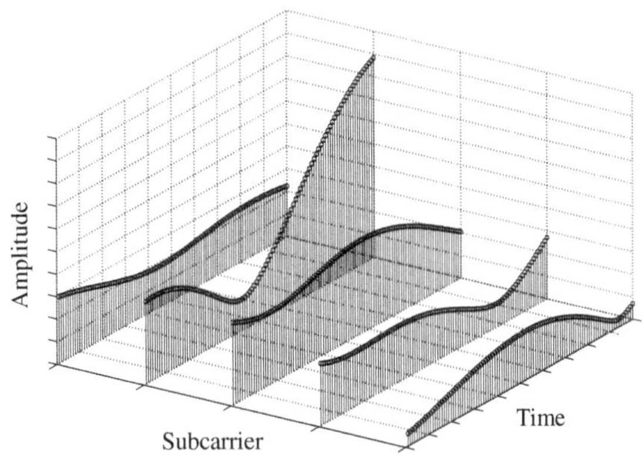

Fig. 10.11 An example of amplitude variation in terms of time and subcarriers.

10.4.4 Multiuser Diversity: Channel-Aware Subcarrier Assignment

The overarching goal for link adaptation is to select the best set of transmission parameters for a given user considering its channel conditions. Similar to that goal, channel-aware subcarrier assignments can be seen as selecting the best set of users to allocate subcarriers according to their channel state information.

From the discussion in Section 10.4.3, the whole bandwidth may be allocated to only one user, but in the present subcarrier assignment scenario, the whole bandwidth will be shared among all users. Each subcarrier (or each block of subcarriers) is allocated to only one user and since each user undergoes different fading conditions, the system may exploit this difference to obtain to a multiuser diversity by allocating the subcarrier to the "best user". Following the idea of LA, the "best user" corresponds to the user with the best channel. In Fig. 10.12 the concept of multiuser diversity is illustrated where three different users experience different channel conditions and the "best user" choice is based on channel gain amplitude. For this specific example, the system throughput is maximized and it has a special feature where the system data rate is derived from the user's best channel state rather than average one.

However, if the subcarrier assignment is just a part of a global resource allocation other metrics such as fairness must also be taken into account. Some cost functions and optimization problems have been proposed in the literature [19, 28, 29, 37] in order to assign subcarriers and to adapt link parameters.

In [28, 29], the main goal is to maximize system throughput while maintaining an acceptable BER. In these papers, adaptive modulation is used as a parameter in the cost function thus taking into account user fairness and the type of used service (best effort).

As the subcarrier assignment is dependent on the current CSI, the amount of required CSI is proportional to the number of users and subcarriers. When the uplink and downlink channels use different frequency bands, like in FDD systems, the CSI must be reported to the transmitter and this feedback information becomes another

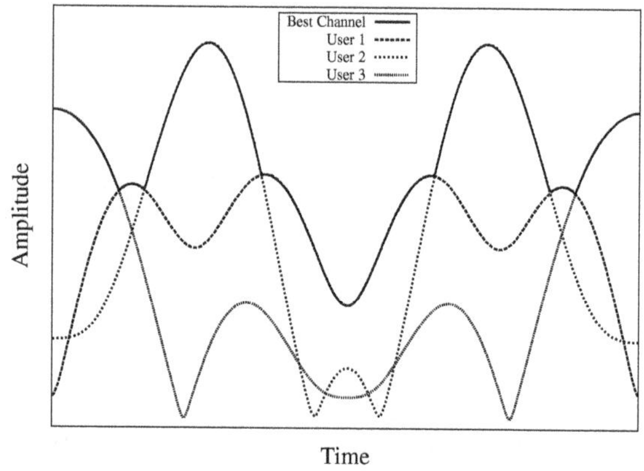

Fig. 10.12 An example of a link adaptation using the multiuser diversity.

important topic of study [5, 37]. In TDD systems, having calibrated receive and transmit RF branches, feedback information can be significantly reduced, as the base station can predict the CSI from the uplink measures. In [37] a channel-aware ALOHA-based assignment is proposed, where users send their CSI when they are above a pre-defined threshold. Similarly, in [5], thresholds are defined and users which are above these thresholds are allowed to send CSI, where the main difference compared to [37] is that in [5] thresholds are established for the employed resource allocation algorithms.

10.5 Summary

Link adaptation is essentially a very useful feature to facilitate adaptation of the system parameters to the channel variations. Although modulation and coding are the classical system parameters involved in link adaptation, the adaptation of the spatial dimension parameters resulting from the employment of multiple antenna elements for transmission and/or reception is a very active research area, the results from which are being exploited in the current and future generations of wireless systems. The adaptation possibility of the frequency dimension parameters by means of subcarrier assignment is another research area that continues to be developed.

In this chapter, the main solutions for the problem of link adaptation were discussed. Both spatial and frequency dimensions were presented with their respective benefits and drawbacks. In Section 10.2.1, the discussions centered around the two main gains provided with spatial dimension: spatial diversity, which improves link reliability, and spatial multiplexing which increases the system spectral efficiency. The associated increase of the feedback signaling and the trade-off between diversity and multiplexing were also discussed.

Link adaptation using the frequency dimension was presented and a solution taking advantage of the offered potential for multiuser diversity was described that used subcarrier assignments on a per-user basis where the subcarriers were assigned according to the channel gain distribution over the population of users.

Future research efforts, relevant for the problem of multi-antenna link adaptation, are needed within the areas outlined below:

- Different antenna schemes present different resulting signal-to-noise-plus-interference ratio (SNIR), thus such a metric is not a unique option for the selection of the transmission modes to be chosen. Different criteria/metrics for switching among the modes are then an open problem for LA when MIMO is considered.
- Regarding the selection of the transmission modes, it is still not clear how much, or what kind of, information is needed for performing a correct choice. This issue points to the problem of limited feedback information.
- When considering multiple dimensions, the granularity of the set of parameters allows the use of a high number of possibilities. However, the adaptation of parameters may benefit from updates using different time scales for different parameters. A possible solution would then define "fast" and "slow" adaptation procedures for different sets of parameters. How to define those modes is still a question.
- The MIMO-OFDM system is a very rich environment due to the frequency and space domains. Those domains can be employed to better exploit the conditions of the channel when considering frequency and space diversities. This fact may lead us to a configuration of parameters which is different from one subcarrier to another. For instance, one subcarrier could better exploit the channel (more suitable for transmission) using two antennas and another subcarrier using three antennas. How can this scenario be managed?
- The increasing interest of distributed antenna systems captures also the attention of resource allocation and LA. However, transmission modes, signaling, and what kind of metrics to be used are still not defined. This is also a foreseen research direction.

References

1. 3GPP: Performance evaluation of codebook-based precoding. Tech. rep., 3GPP, TSG RAN WG1 #46 meeting R1-062208 (2006). URL `http://www.3gpp.org`
2. 3GPP: Radio link control/medium access control (RLC/MAC) protocol. Tech. rep., 3GPP, TS 44.060 V8.1.0 (2008). URL `http://www.3gpp.org`
3. 3GPP: Requirements for further advancements for E-UTRA (LTE-advanced) (release 8). Tech. rep., 3GPP. TSG RAN (2008). URL `http://www.3gpp.org`
4. Alamouti, S.: A simple transmit diversity technique for wireless communications. IEEE Journal on Selected Areas in Communications **16**(8), 1451–1458 (1998)
5. de Araújo, C.A., Cavalcante, C.C., Freitas Jr., W.C.: Pre-processing effects for limited CSI feedback in scheduling algorithms using cross-layer issues. In: Proceedings of the XXV Brazilian Telecommunications Symposium (SBrT2007). Recife, Brazil, vol. 1 (2007)
6. Bazzi, A., Diolaiti, M., Pasolini, G.: Link adaptation algorithms over IEEE8o2. 11 WLANs in collision prone channels. In: IEEE 63rd Vehicular Technology Conference

(VTC 2006-Spring), vol. 3, pp. 1176–1181. Melbourne, Vic. (2006). DOI 10.1109/VETECS. 2006.1683020
7. Belrose, J.S.: A radioscientist's reaction to Marconi's first transatlantic experiment – revisited. In: IEEE Antennas and Propagation Society International Symposium, vol. 1, pp. 22–25. Boston, MA, USA (2001)
8. Biglieri, E., Caire, G., Taricco, G.: Limiting performance of block-fading channels with multiple antennas. IEEE Transactions on Information Theory **47**(4), 1273–1289 (2001)
9. Catreux, S., Erceg, V., Gesbert, D., Heath Jr., R.W.: Adaptive modulation and MIMO coding for broadband wireless data networks. IEEE Communications Magazine **40**(6), 108–115 (2002). DOI 10.1109/MCOM.2002.1007416
10. Dahlman, E., Parkvall, S., Sköld, J., Berming, P.: 3G Evolution HSPA and LTE for Mobile Broadband. Elsevier, Oxford, UK (2007)
11. El-Atty, S.M.A., Skoutas, D.N., Rouskas, A.N.: Reducing CQI signalling overhead in HSPA. Research Letters in Communications, vol. 2008, Article ID 982805, 5 pages (2008). DOI 10.1155/2008/982805
12. Forenza, A., Pandharipande, A., Kim, H., Heath Jr., R.W.: Adaptive transmission scheme selection for mimo systems. In: Wireless World Research Forum (WWRF12). Toronto, Canada (2004)
13. Foschini, G.J.: Layered space-time architecture for wireless communications in a fading environment when using multiple antennas. Bell Labs Technical Journal **1**(2), 41–59 (1996)
14. Freitas Jr., W.C., Cavalcanti, F.R.P., de Almeida, A.L.F., Lopes, R.R.: Exploiting dimensions of the MIMO wireless channel: multidimensional link adaptation. In: IEEE 61st Vehicular Technology Conference (VTC 2005-Spring). Stockholm, Sweden, vol. 2, pp. 924–928 (2005). DOI 10.1109/VETECS.2005.1543441
15. Glisic, S.G.: Advanced Wireless Communications. Wiley, West Sussex, UK (2004)
16. Goldsmith, A.: Wireless Communications. Cambridge University Press, New York, USA (2005). DOI 10.2277/0521837162
17. Hanzo, L., Münster, M., Choi, B.J., Keller, T.: OFDM and MC-CDMA for Broadband Multi-User Communications, WLANs and Broadcasting. Wiley, West Sussex, UK (2003)
18. Heath Jr., R.W., Love, D.J.: Multimode antenna selection for spatial multiplexing systems with linear receivers. IEEE Transactions on Communications **53**(6), 962–968 (2005)
19. Hottinen, A., Heikkinen, T.: Subcarrier allocation in a multiuser MIMO channel using linear programming. In: Proceedings of 14th European Signal Processing Conference (EUSIPCO2006). Florence, Italy (2006)
20. Jeon, S.Y., Cho, D.H.: An enhanced channel-quality indication (CQI) reporting scheme for HSDPA systems. IEEE Communications Letters **9**(5), 432–434 (2005)
21. Lo, T.K.Y.: Maximum ratio transmission. IEEE Transactions on Communications **47**(10), 1458–1461 (1999). DOI 10.1109/26.795811
22. Love, D.J., Heath Jr., R.W., Lau, V.K.N., Gesbert, D., Rao, B., Andrews, M.: An overview of limited feedback in wireless communication systems. IEEE Journal on Selected Areas in Communications **26**(8), 1341–1365 (2008). DOI 10.1109/JSAC.2008.081002
23. Paulraj, A., Nabar, R., Gore, D.: Introduction to Space-Time Wireless Communications. Cambridge University Press, Cambridge, UK (2003)
24. Paulraj, A.J., Kailath, T.: Increasing capacity in wireless broadcast systems using distributed transmission/directional reception (DTDR).
URL `http://www.freepatentsonline.com/5345599.html`
25. Roh, J.C., Rao, B.D.: Multiple antenna channels with partial channel state information at the transmitter. IEEE Transaction on Wireless Communications **3**(2), 677–688 (2004)
26. Sandell, M.: Link adaptation for MIMO systems using reliability values. In: Wireless Communications and Networking Conference, 2006. WCNC 2006. IEEE, vol. 3, pp. 1608–1613. Las Vegas, NV, USA (2006). DOI 10.1109/WCNC.2006.1696528
27. Shannon, C.E.: A mathematical theory of communication. The Bell System Technical Journal **27**, 379–423, 623–656 (1948)

28. Song, G., Li, Y.G.: Cross-layer optimization for OFDM wireless networks – Part I: theoretical framework. IEEE Transactions on Wireless Communications **4**(2), 614–624 (2005)
29. Song, G., Li, Y.G.: Cross-layer optimization for OFDM wireless networks – Part II: algorithm development. IEEE Transactions on Wireless Communications **4**(2), 625–634 (2005)
30. Tarokh, V., Jafarkhani, H., Calderbank, A.R.: Space-time block codes from orthogonal designs. IEEE Transactions on Information Theory **45**(5), 1456–1467 (1999). DOI 10.1109/18.771146
31. Tarokh, V., Naguib, A., Seshadri, N., Calderbank, A.R.: Combined array processing and space-time coding. IEEE Transactions on Information Theory **45**(4), 1121–1128 (1999)
32. Tarokh, V., Naguib, A., Seshadri, N., Calderbank, A.R.: Space-time codes for high data rate wireless communication: performance criteria in the presence of channel estimation errors, mobility, and multiple paths. IEEE Transactions on Communications **47**(2), 199–207 (1999). DOI 10.1109/26.752125
33. Tarokh, V., Seshadri, N., Calderbank, A.R.: Space-time codes for high data rate wireless communications: performance criterion and code construction. IEEE Transactions on Information Theory **44**(2), 744–765 (1998). DOI 10.1109/18.661517
34. Telatar, I.E.: Capacity of multi-antenna gaussian channels. European Transaction on Telecommunications **10**, 585–595 (1999). DOI 10.1002/ett. 4460100604
35. Visotsky, E., Madhow, U.: Space-time transmit pre-coding with imperfect feedback. IEEE Transactions on Information Theory **47**(6), 2632–2639 (2001)
36. Vu, M., Paulraj, A.: On the capacity of MIMO wireless channels with dynamic CSIT. IEEE Journal on Selected Areas in Communications **25**(7), 1269–1283 (2007)
37. Xue, Y., Kaiser, T., Gershman, A.B.: Channel-aware ALOHA-based OFDM subcarrier assignment in single-cell wireless communications. IEEE Transactions on Communications **55**, 953–962 (2007). DOI 10.1109/TCOMM.2007.896071
38. Zheng, L., Tse, D.N.C.: Diversity and multiplexing: A fundamental tradeoff in multiple-antenna channels. IEEE Transactions on Information Theory **49**(5), 1073–1096 (2003). DOI 10.1109/TIT.2003.810646

Chapter 11
Multiuser MIMO Systems Using Space–Time–Frequency Multiple-Access PARAFAC Tensor Modeling

A. L. F. de Almeida, G. Favier, and J. C. M. Mota

11.1 Introduction

Several existing signal processing problems in wireless communication systems with multiple transmit and/or receive antennas are modeled by means of matrix decompositions that represent the transformations on the transmitted signal from the transmitter to the receiver. At the receiver, signal processing is generally used to combat multipath fading effects, inter-symbol interference, and multiuser (co-channel) interference by means of multiple receive antennas. The usually considered signal processing dimensions are *space* and *time* dimensions [65]. This area has progressed over the past 20 years and has resulted in several powerful solutions. In order to allow for a higher spectral efficiency, numerous works have proposed blind signal processing techniques, which aim at avoiding the loss of bandwidth due to the use of training sequences. Blind receiver algorithms generally take special (problem-specific) structural properties of the transmitted signals into account such as constant modulus, finite alphabet, cyclostationarity, or statistical independence for performing multiuser signal separation, equalization, and channel estimation [20, 65, 86, 91–94]. Intensive research has been carried out, and the literature is abundant.

Wireless communication systems employing multiple antennas at both ends of the link, commonly known as multiple-input multiple-output (MIMO) systems, are being considered as one of the key technologies to be deployed in current and upcoming wireless communication standards [64]. MIMO systems have shown to potentially provide high spectral efficiencies by capitalizing on spatial multiplexing [35, 36, 38, 88], while considerably improving the link reliability by means of transmit spatial diversity, also known as space–time coding [3, 30, 42, 63, 87].

The integration of multiple-antenna and code-division multiple–access (CDMA) technologies has also been the subject of several studies [27, 28, 43, 44, 54, 72]. The combination of MIMO and multicarrier modulation by means of orthogonal frequency division multiplexing (OFDM) has also been the focus of a large number

F. Cavalcanti, S. Andersson (eds.), *Optimizing Wireless Communication Systems*,
DOI 10.1007/978-1-4419-0155-2_11,

of recent works and is seen as a promising basis for next-generation wireless standards [83]. In MIMO-OFDM systems, multiple transmit antennas and orthogonal subcarriers are jointly employed to achieve high data rates and to combat fading effects by means of space–time–frequency (STF) coding [2, 16, 73, 84, 85].

Despite the spectral efficiency, diversity gains, and interference rejection capabilities achieved by several existing MIMO transceivers, most of the performance figures assume perfect channel knowledge at the receiver. This assumption is too optimistic in practice. When the channel is unknown, practical receiver design is generally based on suboptimum (linear or nonlinear) equalization and signal separation structures using training sequences for channel acquisition and tracking, before decoding the transmitted data. However, practical limitations such as the receiver complexity and the training sequence overhead (which implies a reduction of the information rate) may be prohibitive in some cases. In order to cope with multiple-access and multiuser transmissions, constraints on the number of transmit and receive antennas, spreading gain, and number of subcarriers must be imposed to guarantee a satisfactory performance.

In several signal processing applications for wireless communication systems, the use of tensor decompositions has gained increased attention over the past few years. Shortly, the term *tensor* will be used here to denote a tridimensional array. As a particular case, a matrix can be interpreted as a second-order tensor. In the wireless communication context, the fact that the received signal is a third-order tensor means that each received signal sample is associated with a three-dimensional space and is represented by three indices, each one associated with a particular type of systematic variation of the received signal. In such a three-dimensional space, each dimension of the *received signal tensor* can be interpreted as a particular form of signal "diversity". In most of the cases, two of these three axes account for *space* and *time* dimensions. The *space* dimension generally corresponds to the number of receive antennas while the *time* dimension corresponds to the length of the data block to be processed at the receiver. The third dimension of the third-order tensor depends on the particular wireless communication system. This dimension is generally linked to the type of processing that is done at the transmitter and/or at the receiver. For instance, in a direct-sequence code division multiple access (DS-CDMA) system [68], the third dimension is the *code* dimension which appears due to the use of a direct sequence spreading at the transmitter. The use of multicarrier modulation at the transmitter also creates a third dimension to the received signal, that is, a *frequency* dimension. Figure 11.1 provides an illustration of the role played by tensor modeling in the wireless communication chain.

The practical motivation for a tensor modeling comes from the fact that one can simultaneously benefit from multiple (more than two) forms of diversity to perform multiuser signal separation, equalization, and channel estimation under more relaxed constraints on the system parameters than with conventional receivers that rely on matrix-based models. In this chapter, we show that tensor models have powerful uniqueness properties leading to blind receiver processing [4, 74, 78].

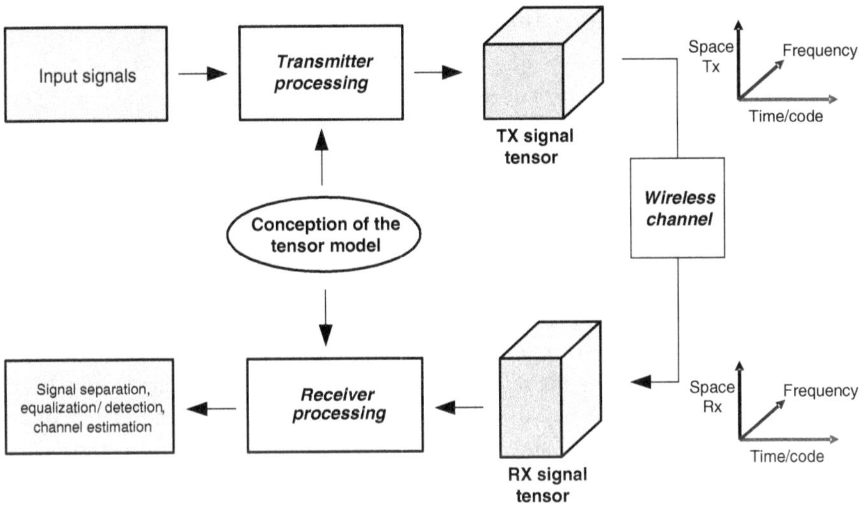

Fig. 11.1 Illustration of tensor modeling in the wireless communication chain.

Tensor models are also mathematically elegant and allow a new algebraic interpretation of the transmitter-channel-receiver transformations over the transmitted signal.

This chapter is organized as follows. Section 11.2 provides an overview of the state-of-the-art concerning the application of tensor modeling in MIMO wireless communication systems. In Section 11.3, the parallel factor (PARAFAC) tensor decomposition is briefly formulated. The fundamental issue of uniqueness enjoyed by the PARAFAC decomposition is also discussed. This section provides a background for later sections of the chapter. Section 11.4 presents the space–time–frequency multiple-access (STFMA) system, where the main building blocks of the transmitter are detailed. This section also presents a design constraint to obtain full space–frequency diversity. The generalization of the signal model to the multiuser case is also presented in this section. Section 11.5 provides illustrative simulation results for performance evaluation of the STFMA system under different STF transmit settings using a zero-forcing (ZF)-based receiver with perfect channel knowledge. The subsequent sections address the tensor modeling of the STFMA system. In Section 11.6 both the transmitted and the received signals are modeled using the tensor formalism by means of a PARAFAC modeling. Examples of special cases covered by the PARAFAC modeling are also given in this section. Section 11.7 is dedicated to the problem of blind detection in the context of the PARAFAC-based STFMA model. The refereed section capitalizes on the fundamental uniqueness results of the decomposition to study the joint blind symbol-code-channel recovery. A blind receiver based on the alternating least squares algorithm is also presented in this section. In Section 11.8, the performance of the STFMA system with PARAFAC-based blind receiver is evaluated by means of computer simulations. The chapter is concluded in Section 11.9, where some perspectives for future research are drawn.

11.2 Tensor Decompositions: A New Signal Processing Tool

The seminal works using tensor decompositions in wireless communications are due to Sidiropoulos et al. In [78], the authors show that a mixture of DS-CDMA signals received at an uniform linear array of antennas can be interpreted as a third-order tensor admitting a PARAFAC decomposition. In [75], the same authors established an interesting conceptual link between the PARAFAC decomposition and the problem of multiple invariance sensor array processing. Following these works, several works proposed applications of PARAFAC to blind multiuser detection in wideband code division multiple-access (WCDMA) systems [77], OFDM systems [45], blind beamforming [80], multiple-antenna space–time coding [76], and blind spatial signature estimation [71] (see the reference list of [74] for further related works). The PARAFAC decomposition has also been exploited for the blind identification of undetermined mixtures [22, 70] and for the blind separation of DS-CDMA signals [26] using higher-order statistics.

Generalized tensor decompositions have been proposed in [5, 9, 14, 60] to handle frequency-selective channels under different assumptions concerning the multipath propagation structure. Tensor decompositions have also been exploited recently for the blind identification and equalization of linear and nonlinear channels [32–34, 49–51] and for kernel complexity reduction of third-order Volterra models [47, 48].

In the context of MIMO antenna systems, the use of tensor modeling has first appeared in [76], where a space–time coding model with blind detection has been proposed. This multiple-antenna scheme allows to build a third-order PARAFAC model for the received signal thanks to a temporal spreading of the data streams at each transmit antenna as in a conventional CDMA system. In [15], a tensor model is proposed for a MIMO-CDMA system with multiuser spatial multiplexing, but no spreading across the transmit antennas is permitted. In more recent works [6, 7, 13], a generalization of [76] and [15] has been proposed, by covering multiple-antenna transmission systems with partial or full spatial spreading of each data stream across sets of transmit antennas. This idea was further generalized by the authors in subsequent works [8, 10–12] using the CONstrained FACtor (CONFAC) decomposition. They provide extensions of [6] and [13] by allowing to use multiple transmit antennas and spreading codes per data stream.

For the above-mentioned applications, the key characteristics of tensor-based signal processing, not covered by matrix-based signal processing, are the following:

- It does not require the use of training sequences, nor the knowledge of channel impulse responses and antenna array responses.
- It does not rely on statistical independence between the transmitted signals.
- It works on blocks of data (instead of using sample-by-sample processing) by means of joint detection with close-to-optimum performance.

The existing contributions in this growing area of research are shared between transmitter and receiver processing. Some of them focus primarily on receiver signal processing (multiuser signal separation, equalization, decoding, and channel estimation). Others emphasize the transmitter signal processing (e.g., space–time multiplexing and spreading and space–time–frequency multiple access), although these also affect the receiver processing. Figures 11.2 and 11.3 link the use of tensor modeling to the signal processing purpose at both ends of the communication chain and highlight the three signal dimensions that generally appear in each case.

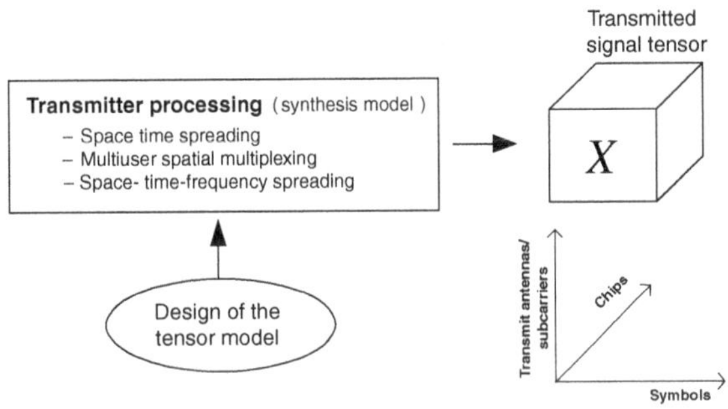

Fig. 11.2 Relationship of tensor modeling to *transmitter* signal processing.

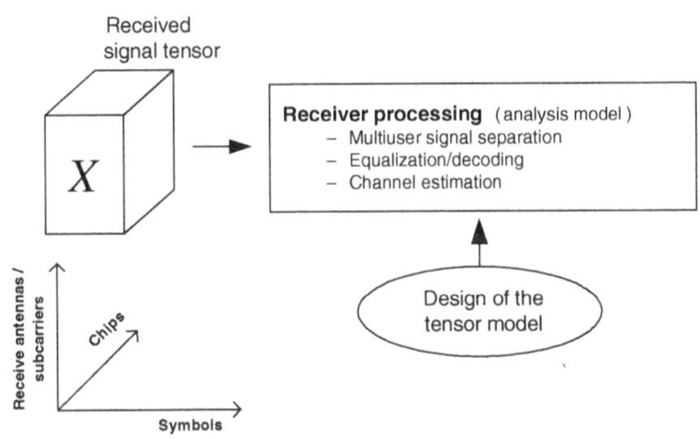

Fig. 11.3 Relationship of tensor modeling to *receiver* signal processing.

This chapter shows that the PARAFAC tensor decomposition is useful for modeling the received signal in a MIMO wireless communication system with space-time-frequency signaling. In addition to dealing with signal modeling itself, this chapter highlights the practical benefits and trade-offs of a PARAFAC modeling in MIMO transceiver design with blind detection. First, a concise background on the PARAFAC tensor decomposition will be described in the following section.

11.3 Background on the PARAFAC Tensor Decomposition

One of the most popular tensor decompositions is the PARAFAC decomposition, independently proposed by Harshman [40] and Carroll and Chang [21]. The PARAFAC decomposition can be seen as an extension of matrix (bilinear) decompositions to higher orders (a matrix is a tensor of order two). This tensor decomposition has been used as a data analysis tool in psychometrics, phonetics, statistics, arithmetic complexity, and other fields and disciplines. Intensive research on PARAFAC analysis has been conducted in the context of chemometrics in the food industry, where it is used for spectrophotometric, chromatographic, and flow injection analysis [17, 18, 81]. The attractive feature of the PARAFAC decomposition is its intrinsic uniqueness. In contrast to matrix (bilinear) decompositions, where there is the well-known problem of rotational freedom, the PARAFAC decomposition of higher-order tensors is essentially unique, up to scaling and permutation indeterminacies [52, 82].

For a third-order tensor, the PARAFAC decomposition is the factorization in a sum of *triple products*, i.e., rank-1 tensors, also called *triads*. The PARAFAC decomposition of a tensor $\mathscr{X} \in \mathbb{C}^{I_1 \times I_2 \times I_3}$ has the following scalar form:

$$x_{i_1,i_2,i_3} = \sum_{q=1}^{Q} a^{(1)}_{i_1,q} a^{(2)}_{i_2,q} a^{(3)}_{i_3,q}, \qquad (11.1)$$

where $a^{(1)}_{i_1,q} = [\mathbf{A}^{(1)}]_{i_1,q}$, $a^{(2)}_{i_2,q} = [\mathbf{A}^{(2)}]_{i_2,q}$, and $a^{(3)}_{i_3,q} = [\mathbf{A}^{(3)}]_{i_3,q}$ are entries of factor matrices $\mathbf{A}^{(j)} \in \mathbb{C}^{I_j \times Q}$, $j = 1,2,3$. Q is the number of factors, also known as the *tensor rank*.

In Fig. 11.4, a third-order PARAFAC decomposition is visualized as a sum of Q rank-1 tensors.

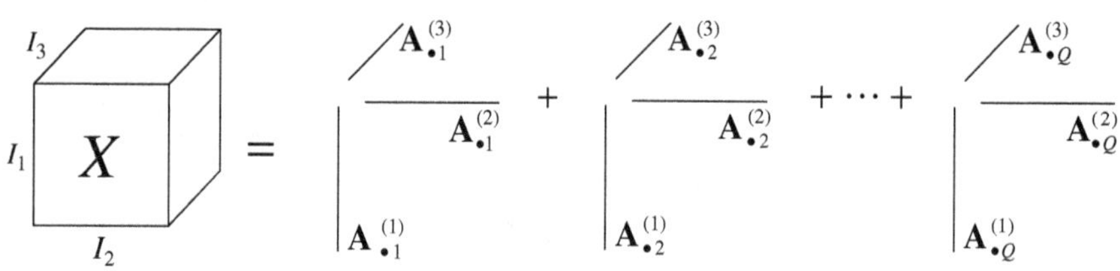

Fig. 11.4 Visualization of the third-order PARAFAC decomposition.

Alternatively, the PARAFAC decomposition can be stated using a *matrix-slice notation*. This notation characterizes the tensor by a set of parallel matrix-slices that are obtained by "slicing" the tensor in a given "direction". Each matrix-slice is obtained by fixing one index of a given *mode* and varying the two indices of the other two modes. For a third-order tensor, there are three possible slicing directions. $\mathbf{X}_{i_1\cdot\cdot} \in \mathbb{C}^{I_2 \times I_3}$ is the i_1th *first-mode slice*, $\mathbf{X}_{\cdot i_2 \cdot} \in \mathbb{C}^{I_3 \times I_1}$ is the i_2th *second-mode slice*, and $\mathbf{X}_{\cdot\cdot i_3} \in \mathbb{C}^{I_1 \times I_2}$ is the i_3th *third-mode slice*. The matrix-slice factorizations of the PARAFAC decomposition (11.1) are given by

$$\mathbf{X}_{i_1\cdot\cdot} = \mathbf{A}^{(2)} D_{i_1}(\mathbf{A}^{(1)}) \mathbf{A}^{(3)T}, \qquad (11.2)$$
$$\mathbf{X}_{\cdot i_2 \cdot} = \mathbf{A}^{(3)} D_{i_2}(\mathbf{A}^{(2)}) \mathbf{A}^{(1)T}, \qquad (11.3)$$
$$\mathbf{X}_{\cdot\cdot i_3} = \mathbf{A}^{(1)} D_{i_3}(\mathbf{A}^{(3)}) \mathbf{A}^{(2)T}, \qquad (11.4)$$

where $D_{i_j}(\mathbf{A}^{(j)})$, $j = 1,2,3$, forms a diagonal matrix holding the i_jth row of $\mathbf{A}^{(j)} \in \mathbb{C}^{I_j \times Q}$ on its main diagonal.

By stacking row-wise the second-, third-, and first-mode matrix-slices, a new representation is obtained, respectively,

$$\mathbf{X}_1 = \begin{bmatrix} \mathbf{X}_{\cdot 1 \cdot} \\ \vdots \\ \mathbf{X}_{\cdot I_2 \cdot} \end{bmatrix} = \begin{bmatrix} \mathbf{A}^{(3)} D_1(\mathbf{A}^{(2)}) \\ \vdots \\ \mathbf{A}^{(3)} D_{I_2}(\mathbf{A}^{(2)}) \end{bmatrix} \mathbf{A}^{(1)^T} = (\mathbf{A}^{(2)} \diamond \mathbf{A}^{(3)}) \mathbf{A}^{(1)^T},$$

$$\mathbf{X}_2 = \begin{bmatrix} \mathbf{X}_{\cdot\cdot 1} \\ \vdots \\ \mathbf{X}_{\cdot\cdot I_3} \end{bmatrix} = \begin{bmatrix} \mathbf{A}^{(1)} D_1(\mathbf{A}^{(3)}) \\ \vdots \\ \mathbf{A}^{(1)} D_{I_3}(\mathbf{A}^{(3)}) \end{bmatrix} \mathbf{A}^{(2)^T} = (\mathbf{A}^{(3)} \diamond \mathbf{A}^{(1)}) \mathbf{A}^{(2)^T}, \qquad (11.5)$$

$$\mathbf{X}_3 = \begin{bmatrix} \mathbf{X}_{1 \cdot\cdot} \\ \vdots \\ \mathbf{X}_{I_1 \cdot\cdot} \end{bmatrix} = \begin{bmatrix} \mathbf{A}^{(2)} D_1(\mathbf{A}^{(1)}) \\ \vdots \\ \mathbf{A}^{(2)} D_{I_1}(\mathbf{A}^{(1)}) \end{bmatrix} \mathbf{A}^{(3)^T} = (\mathbf{A}^{(1)} \diamond \mathbf{A}^{(2)}) \mathbf{A}^{(3)^T},$$

where $\diamond$ denotes the Khatri–Rao (column-wise Kronecker) product, i.e.,

$$\mathbf{A} \diamond \mathbf{B} = \left[\mathbf{A}_{\cdot 1} \otimes \mathbf{B}_{\cdot 1}, \ldots, \mathbf{A}_{\cdot Q} \otimes \mathbf{B}_{\cdot Q}\right] \in \mathbb{C}^{IJ \times Q},$$

where $\mathbf{A}_{\cdot q}$ and $\mathbf{B}_{\cdot q}$ represent the qth column of $\mathbf{A}$ and $\mathbf{B}$, respectively.

One of the most interesting properties of PARAFAC is its uniqueness. Contrary to bilinear (matrix) decompositions, which are in general not unique for ranks greater than one (rank-1 matrices are unique up to a scalar factor), the PARAFAC decomposition of tensors of rank greater than one can be unique up to scaling and permutation of factors. As will be discussed later, uniqueness is important to the context of this chapter since it will ensure a blind detection when the received signal is modeled using the PARAFAC approach.

The study of the PARAFAC uniqueness condition is based on the concept of *Kruskal*-rank, also known as k-rank, which is more restricted than the usual concept of matrix rank. The k-rank was introduced by Kruskal in his seminal paper [52], although the term "*Kruskal*-rank" was first used by Harshman and Lundy [41]. The k-rank has been extensively used as a key concept for stating PARAFAC uniqueness.

Definition 11.1 (k-rank [52]). The rank of $\mathbf{A} \in \mathbb{C}^{I \times Q}$, denoted by $rank(\mathbf{A})$, is equal to r *iff* $\mathbf{A}$ contains *at least* a set of r linearly independent columns but no set of $r+1$ linearly independent columns. The k-rank of $\mathbf{A}$, denoted by $k_{\mathbf{A}}$, is the greatest integer k such that *every* set of k columns of $\mathbf{A}$ is linearly independent. Note that the k-rank is always less than or equal to the rank, and $k_{\mathbf{A}} \leq rank(\mathbf{A}) \leq min(I,Q)$.

Theorem 11.1 (Necessary least squares (LS) identifiability condition [74, 79]). *It is assumed that none of the three factor matrices has a pair of proportional columns. A necessary condition for identifiability in the LS sense is*

$$rank(\mathbf{A}^{(2)} \diamond \mathbf{A}^{(3)}) = rank(\mathbf{A}^{(3)} \diamond \mathbf{A}^{(1)}) = rank(\mathbf{A}^{(1)} \diamond \mathbf{A}^{(2)}) = Q. \qquad (11.6)$$

Otherwise stated, LS identifiability requires that $\mathbf{A}^{(2)} \diamond \mathbf{A}^{(3)}$, $\mathbf{A}^{(3)} \diamond \mathbf{A}^{(1)}$, and $\mathbf{A}^{(1)} \diamond \mathbf{A}^{(2)}$ be full column-rank to be left-invertible, which is a necessary condition for estimating, respectively, $\mathbf{A}^{(1)}$, $\mathbf{A}^{(2)}$, and $\mathbf{A}^{(3)}$ in the LS sense, using the three unfolded matrix representations in (11.5).

Theorem 11.2 (Sufficient uniqueness condition [52, 78]). *Consider a Q-factor PARAFAC decomposition of a third-order tensor with matrix factors $\mathbf{A}^{(1)}$, $\mathbf{A}^{(2)}$, and $\mathbf{A}^{(3)}$. If*

$$k_{\mathbf{A}^{(1)}} + k_{\mathbf{A}^{(2)}} + k_{\mathbf{A}^{(3)}} \geq 2Q + 2, \tag{11.7}$$

then $\mathbf{A}^{(1)}$, $\mathbf{A}^{(2)}$, and $\mathbf{A}^{(3)}$ are unique up to column permutation and (complex) scaling. This means that any matrices $\tilde{\mathbf{A}}^{(1)}$, $\tilde{\mathbf{A}}^{(2)}$, and $\tilde{\mathbf{A}}^{(3)}$ satisfying (11.1) are linked to $\mathbf{A}^{(1)}$, $\mathbf{A}^{(2)}$, and $\mathbf{A}^{(3)}$ by $\tilde{\mathbf{A}}^{(1)} = \mathbf{A}^{(1)}\boldsymbol{\Pi}\boldsymbol{\Delta}_1$, $\tilde{\mathbf{A}}^{(2)} = \mathbf{A}^{(2)}\boldsymbol{\Pi}\boldsymbol{\Delta}_2$, and $\tilde{\mathbf{A}}^{(3)} = \mathbf{A}^{(3)}\boldsymbol{\Pi}\boldsymbol{\Delta}_3$, where $\boldsymbol{\Pi}$ is a permutation matrix and $\boldsymbol{\Delta}_1$, $\boldsymbol{\Delta}_2$, and $\boldsymbol{\Delta}_3$ are diagonal matrices satisfying the condition $\boldsymbol{\Delta}_1\boldsymbol{\Delta}_2\boldsymbol{\Delta}_3 = \mathbf{I}_Q$.

It is worth noting that if the elements of $\mathbf{A}^{(1)}$, $\mathbf{A}^{(2)}$, and $\mathbf{A}^{(3)}$ are randomly drawn from an absolutely continuous distribution, then they are full-rank with probability one. Moreover, their k-rank is equal to their rank, so that (11.7) can be equivalently stated as

$$min(I_1, Q) + min(I_2, Q) + min(I_3, Q) \geq 2Q + 2. \tag{11.8}$$

An equivalent and easy-to-check necessary and sufficient condition is proposed in [46]. A more accessible proof of uniqueness is provided in [82] using conventional linear algebra. In [25], a new uniqueness bound that is more relaxed than Kruskal bound is derived from a link between the PARAFAC decomposition and the simultaneous matrix decompositions.

The basic material presented in this section will be exploited later for a tensor modeling of the received signal in MIMO systems with space–time–frequency multiple-access transmission as well as for the conception of a blind PARAFAC-based receiver.

11.4 Space–Time–Frequency Multiple-Access MIMO System

Combinations of OFDM and CDMA technologies have been proposed in a number of different works [67]. Multicarrier code division multiple access (MC-CDMA) performs spreading of the information symbols across the different subcarriers but suffers from limited frequency diversity gains like conventional CDMA [31, 99]. Multi-carrier direct-sequence code division multiple access (MCDS-CDMA) differs from MC-CDMA by performing the spreading operation in the time-domain at each subcarrier [24]. For combating frequency-selective fading, MCDS CDMA requires forward error-correction coding and frequency-domain interleaving which implies a loss of useful bandwidth. By performing spreading after the OFDM modulation, the so-called multi-tone direct sequence (MTDS-CDMA) system [90] does not ensure orthogonality among the subcarriers. Consequently, its performance is limited by inter-symbol interference and inter-tone interference, in addition to multi-user interference (MUI).

In [37], a hybrid of MC-CDMA and OFDM systems enabling orthogonal multiple access in the frequency domain is proposed which ensures MUI-free transmission/reception regardless of the multipath channel profile. A related approach, called multi-carrier block-spread code division multiple access (MCBS-CDMA), was introduced in [66] by capitalizing on redundant block spreading and frequency-domain linear precoding to preserve orthogonal multiple accessing and to enable full multipath diversity gains. The receiver is based on low-complexity single-user equalization.

By exploiting the spatial dimension at the transmitter, in addition to time and frequency dimensions, a number of different STF transceivers were proposed to enable orthogonal multiple access in multiuser MIMO systems combining OFDM and CDMA principles. The work [29] proposed space–frequency-spreading codes for the downlink of a multiuser MIMO-OFDM system. The transmission is designed to support more multiplexed signals than transmit antennas and to provide space–frequency diversity for each multiplexed signal. Another spread spectrum-based STF transmission framework was proposed in [57] for multi-carrier spread space spectrum multiple access (MC-SSSMA), with the idea of fully spreading each user symbol over space, time, and frequency. MC-SSSMA is a generalization of its single-carrier counterpart proposed in [58, 59]. Despite the achieved spectral efficiency gains, the design of [57] was restricted to the case where the number of transmit and receive antennas is equal to the spreading gain. In [97], STF spreading was proposed for MC-CDMA based on the concatenation of a space–time spreading code with a frequency-domain spreading code. A common characteristic of all these works is the assumption of perfect channel knowledge at the receiver.

In this section, a new STFMA transceiver for MIMO wireless communication systems using PARAFAC tensor modeling is presented. The STFMA transceiver combines space, along with a time-domain block-spreading strategy by means of linear precoding. On the one hand, the use of linear precoding across space (transmit antennas) and frequency (subcarriers) potentially provides robustness against frequency-selective fading and channel ill-conditioning, while providing full transmit diversity gains. On the other hand, block spreading enables multiple accessing by preserving orthogonality among the transmitted data streams. At the receiver, by casting the received signal processing into a PARAFAC modeling approach, it is able to perform a joint blind symbol detection and channel estimation without the need to perform block despreading for interference elimination.

It is worth noting that the transmitter model is independent of the specific signal processing used at the receiver and can be associated with any other appropriate receiver strategy such as linear ZF or minimum-mean-square-error (MMSE) receivers or nonlinear decision feedback receivers. However, as will be shown later, a blind PARAFAC-based receiver exploiting the tensor structure of received signal leads to interesting trade-offs involving space, frequency, and code diversities at the receiver. These trade-offs are of practical relevance.

The proposed STFMA system is close to that of [66] in the sense that both time-domain block spreading and frequency spreading are used. The main difference is that our approach introduces the space dimension across which the transmitted signals are spread for achieving transmit spatial diversity gains. At the same time, the STFMA system induces a PARAFAC tensor structure on the transmitted and received signals, which is beneficial for blind processing at the receiver.

11.4.1 Transmission Model

In this section, the uplink of a single cell of a multicarrier multiple-access MIMO system with Q active co-channel users transmitting data using the same F subcarriers is considered. Each user terminal is equipped with M_t transmit antennas while the base station is equipped with M_r receive antennas. The transmission is composed of three main operations: (i) space spreading, (ii) frequency spreading, and (iii) time-domain block spreading. For notational simplicity, first, a single-user transmission model will be described in order to facilitate the explanation. Later on, it will be shown that the multiuser signal model is readily obtained with minor changes in notation.

After being modulated, the input sequence is serial-to-parallel (S/P) converted into R data streams, each one being constituted by N symbols. For the nth symbol period, the symbol vector is defined as

$$\mathbf{s}(n) = \left[s_1(n), \ldots, s_r(n), \ldots, s_R(n)\right]^T \in \mathbb{C}^R, \tag{11.9}$$

where $s_r(n) \doteq s\big((n-1)R + r\big)$ denotes the nth symbol associated with the rth data stream.

11.4.1.1 Space-Domain Spreading

Figure 11.5 depicts the block diagram of the transmission system by focusing on the transmission of the nth symbol of the rth data stream. The first operation is the space spreading, which consists in spreading each data stream on the M_t transmit antennas using a different code. Let $\boldsymbol{\Omega} \doteq [\boldsymbol{\Omega}_{\cdot 1}, \ldots, \boldsymbol{\Omega}_{\cdot r}, \ldots, \boldsymbol{\Omega}_{\cdot R}] \in \mathbb{C}^{M_t \times R}$ be the matrix collecting the code vectors of the R data streams. The space-domain precoded signal associated with the rth data stream is defined as the following $M_t \times 1$ vector:

$$\bar{\mathbf{s}}_r(n) = \begin{bmatrix} \bar{s}_{r,1}(n) \\ \vdots \\ \bar{s}_{r,M_t}(n) \end{bmatrix} = \boldsymbol{\Omega}_{\cdot r} s_r(n) \in \mathbb{C}^{M_t} \tag{11.10}$$

The code matrix $\boldsymbol{\Omega}$ can be any semi-unitary matrix satisfying $\boldsymbol{\Omega}^H \boldsymbol{\Omega} = \mathbf{I}_R$. This code structure adds robustness to the transmission in the space-domain by providing a

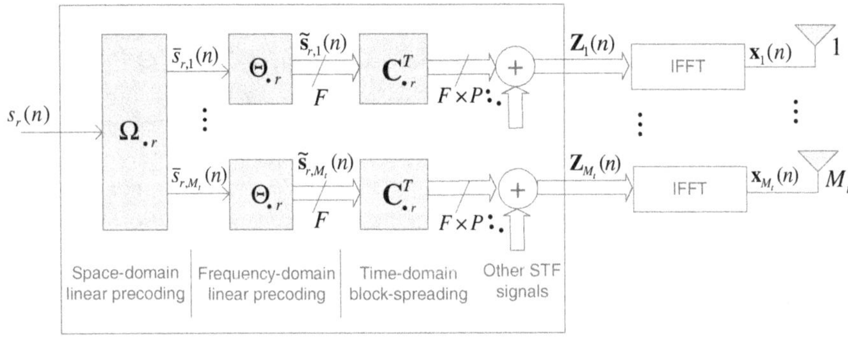

Fig. 11.5 STFMA transmission system.

diversity gain while avoiding performance loss due to ill-conditioned/rank-deficient MIMO channels [55]. In this work, the Vandermonde (or truncated fast fourier transform (FFT)) design was chosen, where $\boldsymbol{\Omega}$ is a Vandermonde matrix with complex generators $\rho_{m_t} = e^{-\mathrm{i}2\pi(m_t-1)/M_t}$, $m_t = 1,\ldots,M_t$, i.e.,

$$\boldsymbol{\Omega}(\rho_1,\ldots,\rho_{M_t}) \doteq \frac{1}{\sqrt{M_t}} \begin{bmatrix} 1 & 1 & \cdots & 1 \\ 1 & \rho_2 & \cdots & \rho_2^{R-1} \\ \vdots & \vdots & \cdots & \vdots \\ 1 & \rho_{M_t} & \cdots & \rho_{M_t}^{R-1} \end{bmatrix}. \tag{11.11}$$

11.4.1.2 Frequency-Domain Spreading

The second operation consists in spreading each component $\bar{s}_{r,m_t}(n)$, $m_t = 1,\ldots,M_t$, of the symbol vector $\bar{\mathbf{s}}_r(n)$ in the frequency-domain. Let $\boldsymbol{\Theta} \doteq [\boldsymbol{\Theta}_{\cdot 1},\ldots,\boldsymbol{\Theta}_{\cdot r},\ldots,\boldsymbol{\Theta}_{\cdot R}] \in \mathbb{C}^{F\times R}$ be the frequency-spreading matrix. The output of this linear precoder is an $M_tF \times 1$ vector given by

$$\tilde{\mathbf{s}}_r(n) = \begin{bmatrix} \tilde{\mathbf{s}}_{r,1}(n) \\ \vdots \\ \tilde{\mathbf{s}}_{r,M_t}(n) \end{bmatrix} = \begin{bmatrix} \boldsymbol{\Theta}_{\cdot r}\bar{s}_{r,1}(n) \\ \vdots \\ \boldsymbol{\Theta}_{\cdot r}\bar{s}_{r,M_t}(n) \end{bmatrix} \in \mathbb{C}^{M_tF}. \tag{11.12}$$

Using (11.10), (11.12) can be rewritten in terms of both space- and frequency-domain codes as

$$\tilde{\mathbf{s}}_r(n) = (\boldsymbol{\Omega}_{\cdot r} \otimes \boldsymbol{\Theta}_{\cdot r})s_r(n) = \mathbf{U}_{\cdot r}s_r(n), \tag{11.13}$$

where

$$\mathbf{U}_{\cdot r} = \boldsymbol{\Omega}_{\cdot r} \otimes \boldsymbol{\Theta}_{\cdot r} \in \mathbb{C}^{M_tF} \tag{11.14}$$

is the joint space–frequency-spreading vector associated with the rth transmitted data stream to be transmitted.

The code matrix $\boldsymbol{\Theta}$ is also a semi-unitary matrix satisfying $\boldsymbol{\Theta}^H\boldsymbol{\Theta} = \mathbf{I}_R$. Such a precoding is introduced to combat frequency-selective fading and thus guarantee symbol detection even in the presence of deep channel fades over one or more subcarrier channels. At the same time, it affords both frequency diversity and coding gains. The frequency spreading can be redundant ($F > R$) or non-redundant ($F = R$). Some designs for $\boldsymbol{\Theta}$ have been reported in the literature (see e.g., [96]). As for the space precoder, here we also choose the Vandermonde design and $\boldsymbol{\Theta}$ as a Vandermonde matrix with complex generators $\xi_f = e^{-i2\pi(f-1)/F}$, $f = 1,\ldots,F$, i.e.,

$$\boldsymbol{\Theta}(\xi_1,\ldots,\xi_F) \doteq \frac{1}{\sqrt{F}}\begin{bmatrix} 1 & 1 & \cdots & 1 \\ 1 & \xi_2 & \cdots & \xi_2^{R-1} \\ \vdots & \vdots & \cdots & \vdots \\ 1 & \xi_F & \cdots & \xi_F^{R-1} \end{bmatrix}. \tag{11.15}$$

Note that spreading in the space-domain consists in multiplying the symbol $s_r(n)$ by a complex code that depends on the transmit antenna number m_t while spreading in the frequency-domain results in a multiplication of the same symbol by a complex code that depends on the frequency number f.

11.4.1.3 Time-domain Block Spreading

The third operation of our transmitter consists in a time-domain spreading of the space–frequency precoded sequence $\tilde{\mathbf{s}}_r(n)$. Instead of performing a *symbol* spreading, as in a classical CDMA system, in this work a block-spreading approach is considered, as suggested in [66]. In this context, the precoded symbols $\tilde{\mathbf{s}}_1(n),\ldots,\tilde{\mathbf{s}}_r(n),\ldots,\tilde{\mathbf{s}}_R(n)$ are spread by a factor P using time-domain spreading codes. Let $\mathbf{C} \doteq [\mathbf{C}_{\cdot 1},\ldots,\mathbf{C}_{\cdot r},\ldots,\mathbf{C}_{\cdot R}] \in \mathbb{R}^{P\times R}$ be the spreading code matrix, the columns/rows of which belong to a (possibly truncated) walsh–hadamard (WH) code matrix. Two situations shall be considered. When $P \geq R$, $\mathbf{C}$ is formed by selecting the R first columns of a $P\times P$ WH matrix. On the other hand, when $P \leq R$, $\mathbf{C}$ is formed by selecting its P first rows.

The space–frequency precoded sequence $\tilde{\mathbf{s}}_r(n)$ is repeated P times and multiplied by the rth spreading code with period $T_c = T/P$, where T_c corresponds to the chip period and T corresponds to the duration of a data block (i.e., an OFDM symbol). The R block-spread signals are summed up to form a multi-stream STF signal, yielding the following expression:

$$\mathbf{Z}(n) = \begin{bmatrix} \mathbf{Z}_1(n) \\ \vdots \\ \mathbf{Z}_{M_t}(n) \end{bmatrix} = \sum_{r=1}^{R} \tilde{\mathbf{s}}_r(n)\mathbf{C}_{\cdot r}^T \in \mathbb{C}^{M_tF\times P}, \tag{11.16}$$

with

$$\mathbf{Z}_{m_t}(n) = \sum_{r=1}^{R} \tilde{\mathbf{s}}_{r,m_t}(n)\mathbf{C}_{\cdot r}^T \in \mathbb{C}^{F\times P}. \tag{11.17}$$

As shown in (11.16), the STF signal $\mathbf{Z}(n) \in \mathbb{C}^{M_t F\times P}$ is treated as the concatenation of M_t matrix blocks. Substituting (11.13) into (11.16), we obtain

$$\mathbf{Z}(n) = \sum_{r=1}^{R} \mathbf{U}_{\cdot r} s_r(n)\mathbf{C}_{\cdot r}^T = \left[\mathbf{U}_{\cdot 1},\ldots,\mathbf{U}_{\cdot R}\right] \begin{bmatrix} s_1(n) & & \\ & \ddots & \\ & & s_R(n) \end{bmatrix} \begin{bmatrix} \mathbf{C}_{\cdot 1}^T \\ \vdots \\ \mathbf{C}_{\cdot R}^T \end{bmatrix}, \tag{11.18}$$

i.e.,

$$\mathbf{Z}(n) = \mathbf{U}\bar{\mathbf{S}}(n)\mathbf{C}^T, \tag{11.19}$$

where $\mathbf{U} \in \mathbb{C}^{M_t F\times R}$ represents the combined space- and frequency-spreading structure, which can be viewed as a joint space–frequency linear precoder,

$$\bar{\mathbf{S}}(n) = diag\big(\mathbf{s}(n)\big) \in \mathbb{C}^{R\times R}, \tag{11.20}$$

and $\mathbf{s}(n)$ is defined in (11.9). Using (11.14), $\mathbf{U}$ can be factorized as the Khatri–Rao product of space- and frequency-spreading matrices:

$$\mathbf{U} = \left[\boldsymbol{\Omega}_{\cdot 1}\otimes\boldsymbol{\Theta}_{\cdot 1},\ldots,\boldsymbol{\Omega}_{\cdot R}\otimes\boldsymbol{\Theta}_{\cdot R}\right] = \boldsymbol{\Omega}\diamond\boldsymbol{\Theta}. \tag{11.21}$$

Remark 1: It is worth noting that block spreading will preserve the orthogonality between the transmitted data streams provided that $\mathbf{C}$ has orthonormal columns. This condition is generally assumed in downlink transmissions. In this case, single-stream/single-user detection can be performed at the receiver by exploiting the orthogonality of the spreading codes. In uplink transmissions, however, such an orthogonality condition does not hold in the presence of unknown interference caused by out-of-cell users [100], and some of the spreading codes (i.e., some columns of $\mathbf{C}$) are *nonorthogonal* (or even *unknown*) at the receiver. In this situation, serious performance degradation is expected using receiver techniques based on despreading and approaches based on multiuser detection are preferable. By exploiting the PARAFAC tensor structure of the received signal, *blind detection* is possible under more challenging situations encountered in practice (for instance, when code matrix $\mathbf{C}$ is nonorthogonal and/or unknown at the receiver).

11.4.1.4 OFDM Modulation

Before being transmitted, the STF block-spread signal passes through the OFDM modulator. An inverse fast fourier transform (IFFT) is applied to $\mathbf{Z}_{m_t}(n), m_t, 1,\ldots,M_t$, and a transmit redundancy in the form of a cyclic prefix (CP) of N_{cp} chips is appended to each length F chip sequence at each transmit antenna, resulting in the following time-domain signal:

$$\mathbf{X}_{m_t}(n) = \mathbf{T}_{cp}\mathbf{F}^H\mathbf{Z}_{m_t}(n) \in \mathbb{C}^{\bar{F}\times P}, \tag{11.22}$$

where $\bar{F} = F + N_{cp}$, $\mathbf{F}^H \in \mathbb{C}^{F\times F}$, represents the IFFT matrix with $[\mathbf{F}]_{i,j} = e^{-i2\pi(i-1)(j-1)/F}$, $\mathbf{T}_{cp} = [\mathbf{I}_{cp}^T, \mathbf{I}_F]^T \in \mathbb{C}^{\bar{F}\times F}$ represents the CP-adding matrix, and $\mathbf{I}_{cp}$ is a matrix formed from the N_{cp} last rows of $\mathbf{I}_F$.

Note that the length N_{cp} of the CP is chosen as a known upper bound to the order of the chip-sampled finite impulse response (FIR) of the channel linking each transmit and receive antenna. It is used to avoid the interference between two adjacent chip sequences due to the time-dispersive nature of the channel. In practice, the minimum channel order L_{min} can be approximated as $L_{min} \approx \lceil \tau_{\max}/T_c \rceil$, where $\tau_{\max}$ is the maximum excess delay within the multipath propagation channel. Therefore, N_{cp} must be greater than L_{min}. For further details on the OFDM modulation, see [56, 67].

Before being transmitted, the time-domain signal matrices $\mathbf{X}_1(n),\ldots,\mathbf{X}_{M_t}(n)$ are serialized resulting in the chip sequences $\mathbf{x}_1(n),\ldots,\mathbf{x}_{M_t}(n)$ of length $\bar{F}P$ samples: $\mathbf{x}_{m_t}(n) = vec\left[\mathbf{X}_{M_t}(n)\right] \in \mathbb{C}^{\bar{F}P}$. Each one of these sequences are pulse shaped, upconverted, and then launched into the wireless channel. In terms of bandwidth efficiency, note that R data streams are transmitted during $\bar{F}$ chip periods at the nth symbol block. Therefore, the bandwidth efficiency of the proposed STFMA system is given by

$$\varepsilon_{\text{STFMA}} = \frac{R}{\bar{F}P}. \tag{11.23}$$

11.4.2 Received Signal Model

In this section a discrete-time baseband equivalent model for the received signal is adopted. The block diagram of the receiver is depicted in Fig. 11.6. It is considered a perfect chip- and symbol-level synchronization at the receiver. The M_r received sequences are denoted by $\mathbf{y}_1(n),\ldots,\mathbf{y}_{M_r}(n)$, each one of length $\bar{F}P$ samples. Let $\bar{\mathbf{Y}}_{m_r}(n) = unvec\left[\mathbf{y}_{m_t}(n)\right] \in \mathbb{C}^{\bar{F}\times P}$, $m_r = 1,\ldots,M_r$, be the matrices collecting the S/P converted sequences. Ignoring the additive noise term, $\bar{\mathbf{Y}}_{m_r}(n)$ can be written as

$$\bar{\mathbf{Y}}_{m_r}(n) = \sum_{m_t=1}^{M_t} \dot{\mathbf{H}}_{m_r,m_t}\mathbf{X}_{m_t}(n) \in \mathbb{C}^{\bar{F}\times P}, \tag{11.24}$$

where $\dot{\mathbf{H}}_{m_r,m_t} \in \mathbb{C}^{\bar{F}\times\bar{F}}$ is a lower triangular Toeplitz matrix constructed from the chip-sampled FIR channel vector $\mathbf{h}_{m_r,m_t} = [h_{m_r,m_t}(1),\ldots,h_{m_r,m_t}(L_c)]^T \in \mathbb{C}^{L_c}$ modeling the chip-sampled frequency-selective channel between the m_tth transmit antenna and the m_rth receive antenna, including the effect of transmit and receive filters. Note that $[\dot{\mathbf{H}}_{m_r,m_t}]_{i,j} = h(i-j)$. After removal of the CP and the application of the FFT, $\mathbf{Y}_{m_r}(n)$ can be written as

$$\mathbf{Y}_{m_r}(n) = \mathbf{F}\mathbf{R}_{cp}\bar{\mathbf{Y}}_{m_r}(n) \in \mathbb{C}^{F\times P}, \tag{11.25}$$

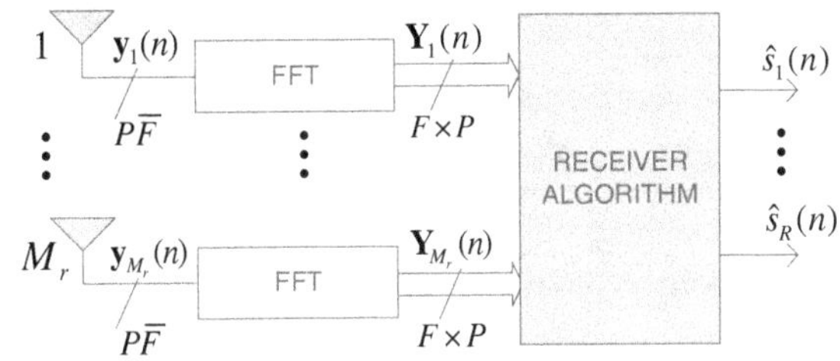

Fig. 11.6 Receiver block diagram.

where $\mathbf{F} \in \mathbb{C}^{F \times F}$ represents the FFT matrix and $\mathbf{R}_{cp} = [\mathbf{0}_{F \times N_{cp}}, \mathbf{I}_F] \in \mathbb{C}^{F \times \bar{F}}$ represents the CP-removal matrix. Combining (11.22) and (11.24), (11.25) can be rewritten as

$$\mathbf{Y}_{m_r}(n) = \sum_{m_t=1}^{M_t} \left[\mathbf{F}\left(\mathbf{R}_{cp}\dot{\mathbf{H}}_{m_r,m_t}\mathbf{T}_{cp}\right)\mathbf{F}^H\right]\mathbf{Z}_{m_t}(n), \tag{11.26}$$

or, alternatively,

$$\mathbf{Y}_{m_r}(n) = \sum_{m_t=1}^{M_t} \left[\mathbf{F}\tilde{\mathbf{H}}_{m_r,m_t}\mathbf{F}^H\right]\mathbf{Z}_{m_t}(n), \tag{11.27}$$

where $\tilde{\mathbf{H}}_{m_r,m_t} = \mathbf{R}_{cp}\dot{\mathbf{H}}_{m_r,m_t}\mathbf{T}_{cp}$ is a *circulant* channel matrix. Using the fact that circulant matrices are diagonalized by Fourier transformations [39],

$$\mathbf{F}\tilde{\mathbf{H}}_{m_r,m_t}\mathbf{F}^H = diag(\bar{\mathbf{h}}_{m_r,m_t}), \tag{11.28}$$

where

$$\bar{\mathbf{h}}_{m_r,m_t} = \left[H_{m_r,m_t}(e^{i0}), H_{m_r,m_t}(e^{i2\pi/F}), \ldots, H_{m_r,m_t}(e^{i2\pi(F-1)/F})\right]^T \in \mathbb{C}^F \tag{11.29}$$

is the frequency-domain channel impulse response and $H_{m_r,m_t}(z) = \sum_{l=1}^{L_c} h_{m_r,m_t}(l)z^{-l}$ is the z-transform of $h_{m_r,m_t}(l)$. This allows to rewrite (11.27) as the following frequency-domain input–output model [56, 67]:

$$\mathbf{Y}_{m_r}(n) = \sum_{m_t=1}^{M_t} diag(\bar{\mathbf{h}}_{m_r,m_t})\mathbf{Z}_{m_t}(n) \in \mathbb{C}^{F \times P}. \tag{11.30}$$

Concatenating the received signal of the M_r receive antennas and eliminating the summation, the following compact expression is obtained:

$$\mathbf{Y}(n) = \mathbf{H}\mathbf{Z}(n) \in \mathbb{C}^{M_rF \times P}, \tag{11.31}$$

where

$$\mathbf{Y}(n) = \begin{bmatrix} \mathbf{Y}_1(n) \\ \vdots \\ \mathbf{Y}_{M_r}(n) \end{bmatrix} \quad \text{and} \quad \mathbf{H} = \begin{bmatrix} diag(\bar{\mathbf{h}}_{1,1}) & \cdots & diag(\bar{\mathbf{h}}_{1,M_t}) \\ \vdots & \vdots & \vdots \\ diag(\bar{\mathbf{h}}_{M_r,1}) & \cdots & diag(\bar{\mathbf{h}}_{M_r,M_t}) \end{bmatrix} \in \mathbb{C}^{M_rF \times M_tF}. \tag{11.32}$$

Using (11.16), (11.31) can be rewritten as

$$\mathbf{Y}(n) = \mathbf{H}\left(\sum_{r=1}^{R} \tilde{\mathbf{s}}_r(n)\mathbf{C}_{\cdot r}^T\right) = \mathbf{H}\tilde{\mathbf{S}}(n)\mathbf{C}^T \in \mathbb{C}^{M_r F \times P}, \tag{11.33}$$

where $\tilde{\mathbf{S}}(n) = [\tilde{\mathbf{s}}_1(n), \ldots, \tilde{\mathbf{s}}_R(n)] \in \mathbb{C}^{M_t F \times R}$.

11.4.3 Multiuser Signal Model

The extension of the transmitted and received signal models to the multiuser case is straightforward. Let us assume that Q users are transmitting to the base station (uplink transmission) and that all users have the same number M_t of transmit antennas, M_r denoting the number of receive antennas at the base station. The multiuser signal model follows that of the single-user case by working with a block-partitioned matrix notation. In the multiuser case, R denotes the total number of transmitted symbols *summed over all the users*, i.e.,

$$R = R^{(1)} + \cdots + R^{(Q)},$$

where $R^{(q)}$ denotes the number of STF-spread data streams transmitted by the qth user. With these definitions, the transmitted signal model (11.19) becomes a column-wise concatenation of Q blocks:

$$\mathbf{Z}(n) = \begin{bmatrix} \mathbf{Z}_1(n) \\ \vdots \\ \mathbf{Z}_Q(n) \end{bmatrix} = \begin{bmatrix} \mathbf{U}_1\bar{\mathbf{S}}_1(n)\mathbf{C}_1^T \\ \vdots \\ \mathbf{U}_Q\bar{\mathbf{S}}_Q(n)\mathbf{C}_Q^T \end{bmatrix} = \begin{bmatrix} \mathbf{U}_1 & & \\ & \ddots & \\ & & \mathbf{U}_Q \end{bmatrix} \begin{bmatrix} \bar{\mathbf{S}}_1(n) & & \\ & \ddots & \\ & & \bar{\mathbf{S}}_Q(n) \end{bmatrix} \begin{bmatrix} c\mathbf{C}_1^T \\ \vdots \\ \mathbf{C}_Q^T \end{bmatrix}, \tag{11.34}$$

where $\mathbf{U}_q \in \mathbb{C}^{M_t F \times R^{(q)}}$, $\bar{\mathbf{S}}_q(n) \in \mathbb{C}^{R^{(q)} \times R^{(q)}}$, $\mathbf{C}_q \in \mathbb{C}^{P \times R^{(q)}}$, and, compactly,

$$\mathbf{Z}(n) = \mathbf{U}\bar{\mathbf{S}}(n)\mathbf{C}^T, \tag{11.35}$$

where $\bar{\mathbf{S}}_q(n) = diag(\mathbf{s}_q(n))$ and $\mathbf{s}_q(n) = [s_1^{(q)}(n), \ldots, s_R^{(q)}(n)]^T \in \mathbb{C}^{R^{(q)}}$, $\mathbf{U} \in \mathbb{C}^{QM_t F \times R}$ and $\bar{\mathbf{S}}(n) \in \mathbb{C}^{R \times R}$ are block-diagonal matrices, and $\mathbf{C}$ is a column-wise partitioned matrix composed of Q blocks. In this case, using (11.21), $\mathbf{U}$ is given by

$$\mathbf{U} = \begin{bmatrix} \boldsymbol{\Omega}_1 \diamond \boldsymbol{\Theta}_1 & & \\ & \ddots & \\ & & \boldsymbol{\Omega}_Q \diamond \boldsymbol{\Theta}_Q \end{bmatrix}, \tag{11.36}$$

where $\boldsymbol{\Omega}_q$ and $\boldsymbol{\Theta}_q$ are the qth user space- and frequency-spreading matrices.

The overall received signal model also follows the single-user model in (11.31) and can be written as

$$\mathbf{Y}(n) = \left[\mathbf{H}_1, \ldots, \mathbf{H}_Q\right] \begin{bmatrix} \mathbf{Z}_1(n) \\ \vdots \\ \mathbf{Z}_Q(n) \end{bmatrix} = \mathbf{HZ}(n) \in \mathbb{C}^{M_r F \times P}, \tag{11.37}$$

where

$$\mathbf{H} = [\mathbf{H}_1, \ldots, \mathbf{H}_Q] \in \mathbb{C}^{M_r F \times Q M_t F} \tag{11.38}$$

is the multiuser channel matrix.

11.4.4 Subcarrier Grouping

Up to this point, it was assumed that frequency spreading of each transmitted user signal is performed across all the F subcarriers. However, this implies a high decoding complexity due to the large number of subcarriers used in practical systems. Receiver complexity can become even prohibitive when dealing with multiuser detection receivers. However, spreading over L subcarriers, where L is the number of independent multipaths, is sufficient to obtain the best diversity performance while significantly reducing the receiver complexity [95].

Similar to the methodology used in some recent works [53, 89, 95], we propose to divide the set of F subcarriers into J nonintersecting subsets of $K \geq L$ equispaced subcarriers. The same frequency spreading is applied within each group of K subcarriers, while different subcarrier groups transmit different information. In this case, the bandwidth efficiency of the STFMA system is given by

$$\varepsilon_{\text{STFMA}} = \frac{RJ}{\bar{F}P}. \tag{11.39}$$

Note that, by comparing (11.39) with (11.23), a J-fold increase in the bandwidth efficiency is obtained over a system without subcarrier grouping, since now each data block contains RJ information symbols.

Since both F and K are system design parameters, they can be properly chosen so that $J = F/K$ is an integer. For instance, let us suppose a system using a total of $F = 64$ subcarriers. If the channel has $L = 6$ paths, a possible subcarrier grouping strategy would consist in dividing the subcarriers into $J = 8$ groups of $K = 8$ subcarriers. Supposing that each group transmits R data streams, this would imply a total of $8R$ data streams over the whole set of 64 subcarriers. Moreover, under the assumption of subcarrier orthogonality, information recovery can be carried out independently within each subcarrier group carrying R data streams.

Remark 2: Note that the subcarrier grouping from the previously described STFMA signal model has been precluded in order to avoid unnecessary complication in mathematical notation.

Throughout the rest of the chapter, the subcarrier grouping will be taken into account in our signal model by replacing F by K, i.e., K will denote the number of subcarriers across which a given data stream is spread while F will denote the total number of subcarriers. It is worth noting that receiver processing will be performed in a group-wise fashion by means of J parallel detection layers under the assumption of subcarrier orthogonality. For notational simplicity, our developments will now focus on a particular subcarrier group of K subcarriers.

Example 11.1 (System design example). The STFMA system with subcarrier grouping has additional flexibility to cover different space–time–frequency transmit schemes with different trade-offs involving diversity and bandwidth efficiency. In the following, a system design example is provided for illustration purpose.

Assume an STFMA system using $F = 64$ subcarriers and a spreading factor $P = 8$. The wireless channel is frequency selective and characterized by $L = 2$ paths. A CP of length $N_{cp} = 3$ is used to ensure inter-block interference-free transmission. In order to add robustness to transmission as well as to benefit from spatial and frequency diversities, both space and frequency spreading are performed using, for instance, $M_t = 3$ transmit antennas and $K = 2$ subcarriers, respectively. Note that, in this case, each data stream can potentially achieve a transmit spatial diversity of order 3 and a multipath diversity of order 2. The joint space–frequency diversity is therefore of order $M_tK = 6$. By fixing K, the number of parallel transmission groups $J = F/K = 32$ is determined. Supposing that each group transmits $R = 8$ data streams, we have a total of $RJ = 256$ transmitted data streams. Using (11.39), the bandwidth efficiency is approximately equal to 0.48.

If we are interested in adding resilience to transmitted information against deep fades across the subcarriers, the number of subcarriers per group can be increased, so that the same data stream will now be spread over a higher bandwidth. The price to pay is, of course, a decrease in the bandwidth efficiency. Supposing, for example, $K = 4$ subcarriers per group, the bandwidth efficiency falls to 0.24, i.e., is reduced by a factor of two. Suppose now that the system operates in a flat-fading propagation channel. In this case, $F = 1$ and $K = J = 1$. Consequently, the diversity gain is reduced by a factor of $K = 2$, since frequency spreading is no more performed and only spatial diversity is obtained. Such a diversity loss comes, however, with a two-fold increase in the bandwidth efficiency which is now equal to $R/P = 1$.

11.4.5 Design Constraint: Space and Frequency Diversity Trade-Off

Choosing a Vandermonde matrix with arbitrary dimensions for the space- and frequency-spreading matrices is not sufficient for symbol detectability. To be specific, not all combinations of $\boldsymbol{\Omega}$ and $\boldsymbol{\Theta}$ lead to a full-rank $\mathbf{U} = \boldsymbol{\Omega} \diamond \boldsymbol{\Theta}$, which is necessary for obtaining symbol detectability. In the following a simple design constraint on the number M_t of transmit antennas and subcarriers K is derived. Recall that $\mathbf{U}$ defined in (11.36) for the multiuser model depends on the Khatri–Rao product of space - and frequency-spreading matrices. For convenience, let us recall this Khatri–Rao factorization:

$$\mathbf{U} = \begin{bmatrix} \boldsymbol{\Omega}_1 \diamond \boldsymbol{\Theta}_1 & & \\ & \ddots & \\ & & \boldsymbol{\Omega}_Q \diamond \boldsymbol{\Theta}_Q \end{bmatrix} \in \mathbb{C}^{QM_tK \times R}, \quad R = \sum_{q=1}^{Q} R^{(q)}.$$

The problem consists in choosing M_t and K such that every diagonal block of $\mathbf{U}$ is full column-rank. This ensures detectability of the $R^{(1)}, \ldots, R^{(Q)}$ data streams. Therefore, the rank of $\boldsymbol{\Omega} \diamond \boldsymbol{\Theta}$ has to be evaluated.

An upper bound on the rank of the Khatri–Rao product of two matrices has been derived in [80]. Hereafter, these results are briefly recalled:

Lemma 11.1. *If neither* $\mathbf{A} \in \mathbb{C}^{I \times M}$ *nor* $\mathbf{B} \in \mathbb{C}^{J \times M}$ *contains a zero column, then*

$$rank(\mathbf{A} \diamond \mathbf{B}) \geq min\big(rank(\mathbf{A}) + rank(\mathbf{B}) - 1, M\big). \tag{11.40}$$

Assuming that both $\mathbf{A}$ and $\mathbf{B}$ are full row-rank (i.e., $I \leq M$ and $J \leq M$), this lemma implies that $\mathbf{A} \diamond \mathbf{B}$ is full column-rank if $rank(\mathbf{A}) + rank(\mathbf{B}) \geq M + 1$, i.e.,

$$I + J \geq M + 1.$$

Applying this result to our context with $\big(\mathbf{A}, \mathbf{B}, I, J, M\big) \to \big(\boldsymbol{\Omega}_q, \boldsymbol{\Theta}_q, M_t, K, R^{(q)}\big)$, the following design constraint is deduced:

Design constraint: For $M_t \leq R^{(q)}$ and $K \leq R^{(q)}$, symbol detectability for the qth user requires

$$M_t + K \geq R^{(q)} + 1, \tag{11.41}$$

if $\boldsymbol{\Omega}_q$ and $\boldsymbol{\Theta}_q$ are chosen full-rank.

This design constraint shows the symmetry in the roles of M_t and K that arises when combining space and frequency spreadings to achieve symbol recovery. Otherwise stated, this condition clearly indicates the existing trade-off between space and frequency spreadings that is inherent to the STFMA system concept.

11.5 STFMA Performance with Perfect Channel Knowledge

This section presents a set of preliminary computer simulation results to access the bit error rate (BER) performance of the STFMA system under a variety of transmit configurations. Specifically, these illustrative simulation results allow to study the influence of the main transmit parameters of interest (number K of subcarriers per group, number M_t of transmit antennas and spreading factor P) on the BER performance. All the simulations are carried out using a ZF-based receiver.

11.5.1 Joint-ZF Receiver Without Despreading

As previously mentioned, the orthogonality among the R transmitted signals at the receiver cannot always be guaranteed in uplink transmissions, as a consequence, for instance, of unknown interference caused by out-of-cell users [100]. Since $\mathbf{C}$ is no more orthogonal at the receiver, it is not possible to rely on a despreading operation prior to the ZF receiver. Otherwise, performance degradation is expected. Instead of decoupling the despreading operation and ZF equalization in two consecutive stages, a joint-ZF receiver that directly and simultaneously estimates all the R transmitted signals without despreading is proposed.

Substituting (11.35) into (11.37), $\mathbf{Y}(n)$ is obtained by

$$\mathbf{Y}(n) = \mathbf{HU}\bar{\mathbf{S}}(n)\mathbf{C}^T. \tag{11.42}$$

Now, consider the following property:

$$vec\left[\mathbf{A}\ \text{diag}(\mathbf{x})\mathbf{B}^T\right] = (\mathbf{B} \diamond \mathbf{A})\mathbf{x}, \tag{11.43}$$

with $\mathbf{A} \in \mathbb{C}^{I\times R}, \mathbf{B} \in \mathbb{C}^{J\times R}$, and $\mathbf{x} \in \mathbb{C}^R$. Applying this property, (11.42) can be rewritten as

$$\mathbf{y}(n) = vec[\mathbf{Y}(n)] = \big(\mathbf{C} \diamond (\mathbf{HU})\big)\mathbf{s}(n) \in \mathbb{C}^{PM_rK}, \tag{11.44}$$

where

$$\mathbf{s}(n) = [\mathbf{s}_1^T(n), \ldots, \mathbf{s}_Q^T(n)]^T \in \mathbb{C}^R \tag{11.45}$$

is a vector formed from the diagonal of $\bar{\mathbf{S}}(n)$.

By minimizing $\|\mathbf{y}(n) - \big(\mathbf{C} \diamond (\mathbf{HU})\big)\mathbf{s}(n)\|^2$ in the least squares sense, the joint-ZF solution is given by

$$\mathbf{W} = \big(\mathbf{C} \diamond (\mathbf{HU})\big)^\dagger \in \mathbb{C}^{R\times PM_rK}.$$

A simultaneous estimate of the R transmitted data streams is then found as

$$\widehat{\mathbf{s}}(n) = \mathbf{W}\mathbf{y}(n) \in \mathbb{C}^R,\ n = 1, \ldots, N. \tag{11.46}$$

From the structure of (11.5.1), note that the joint-ZF receiver takes the correlation of the spreading codes into account. It is important noting that the joint-ZF receiver, similar to the two previous ones, assumes the knowledge, or estimation, of the code matrix $\mathbf{C}$. As will be shown later, by exploiting the PARAFAC tensor structure of the received signal, a *code-blind detection* is still possible in these more challenging situations, where the code matrix $\mathbf{C}$ is unknown and/or nonorthogonal. Since $\mathbf{C} \diamond (\mathbf{HU}) \in \mathbb{C}^{PM_rK\times R}$ must be full column-rank, the joint-ZF receiver requires that $PM_rK \geq R$.

11.5.2 Simulation Results

The simulated STFMA system operates at a transmission rate of $R_c = 1/T_c = 4.096$ Mcps, using a total of $F = 64$ subcarriers divided into J groups of K subcarriers each. Note that $F = 64$ is a fixed parameter, while K is a transmission design parameter that will be varied in our simulations. Due to subcarrier grouping, each data block contains RJ information symbols. The transmission of $N = 10$ data blocks is assumed. At each run, the transmitted symbols are drawn from a pseudo-random quaternary phase shift keying (QPSK) sequence. The channel is assumed to be quasi-static, which means the channel responses do not change during the transmission of a data block. Perfect time and frequency synchronization is assumed. Table 11.1 summarizes the STFMA system parameters.

Table 11.1 STFMA system parameters.

Chip rate	4.096 Mcps
Number of subcarriers (F)	64
Number of subcarriers per group (K)	2 or 4
Number of subcarrier groups (J)	32 or 16
CP length (N_{cp})	5 (channel A)/20 (channel B)
Number of transmitted data blocks (N)	10
Modulation format	QPSK

Each plotted BER curve is on average over 1000 independent Monte Carlo runs and is shown as a function of an overall effective signal-to-noise ratio (SNR) measure, given by

$$\text{SNR} = 10\log_{10}\left(\sum_{n=1}^{N} \frac{\|\mathbf{Y}(n)\|_F^2}{\|\mathbf{V}(n)\|_F^2}\right),$$

where $\mathbf{V}(n) \in \mathbb{C}^{M_r F \times P}$, $n = 1,\ldots,N$, is the noise matrix, the elements of which are circularly symmetric complex Gaussian random variables. Note that this SNR measure takes all the received signal dimensions into account, i.e., the number N of data blocks, the spreading factor P, and the number M_r of receive antennas. At each run, the additive noise power is generated according to this SNR measure. The BER curves represent the performance averaged over the RJ transmitted signals.

Two frequency-selective channel models are adopted for modeling the channel between each pair of transmit and receive antenna. Both are ITU's outdoor-to-indoor models and are valid for typical urban propagation environments: (i) the 4-ray pedestrian channel A and (ii) the 6-ray pedestrian channel B [1]. Note that, for channel A, the maximum multipath delay is $\tau_{max} = 410$ ns, and the channel impulse response has $L_c = \lceil \tau_{max}/T_c \rceil = 2$ chip-sampled coefficients. A CP length of $N_{cp} = 5$ chips is chosen when considering channel A. For channel B, the maximum multipath delay is $\tau_{max} = 3700$ ns so that the channel impulse response has $L_c = \lceil \tau_{max}/T_c \rceil = 16$ chip-sampled coefficients. A CP length of $N_{cp} = 20$ chips is chosen when channel B is simulated.

In the following results, the BER performance of the STFMA system is evaluated for different transmission settings. The objective of the following simulations is to study the impact of the different building blocks of the transmitter on the performance and, in particular, to evaluate the diversity and coding gains that can be afforded in a single-user transmission. First, single-user transmissions ($Q = 1$) will be described, while the multiuser case will be considered later. In the following simulations, the joint-ZF receiver is selected as it offers the best achievable performance with perfect channel knowledge. The pedestrian channel B is used in all cases.

11.5.2.1 Impact of Frequency Spreading

As previously discussed, spreading the transmitted signals across different subcarriers allows to benefit from the frequency (multipath) diversity gain. Coding gain is also obtained when $K > L$. Figure 11.7 depicts the performance of a system with $M_t = 3$ and $M_r = 2$, $P = 4$, using (i) $K = 2$ and $R = 3$ and (ii) $K = 4$ and $R = 6$. Note that both configurations have the same ratio R/K which leads to the same bandwidth efficiency according to (11.39). The system is simulated over ITU channels A and B. It is worth noting that a remarkable performance improvement is obtained under channel B when more subcarriers are used for frequency spreading. This is not the case for channel A where less multipath diversity is available.

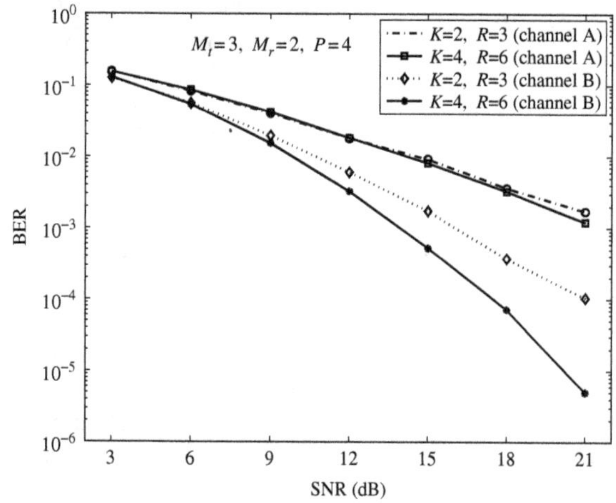

Fig. 11.7 Impact of frequency spreading with ITU channels A and B.

11.5.2.2 Impact of Space Spreading

The next experiment evaluates the impact of space spreading in the STFMA system performance. From now on, the ITU channel B is assumed in all simulations. $M_r = 2$ receive antennas and $R = 8$ transmitted data streams are considered. Frequency spreading uses $K = 4$ subcarriers. Space spreading uses $M_t = 2$ and 4 transmit antennas. Figure 11.8 shows that for $P = 2$, an irreducible BER floor exists when using $M_t = 2$ transmit antennas. The BER performance, however, improves significantly when spreading the transmitted data across $M_t = 4$ transmit antennas. For $P = 4$, a higher coding gain is present so that the gap between the two

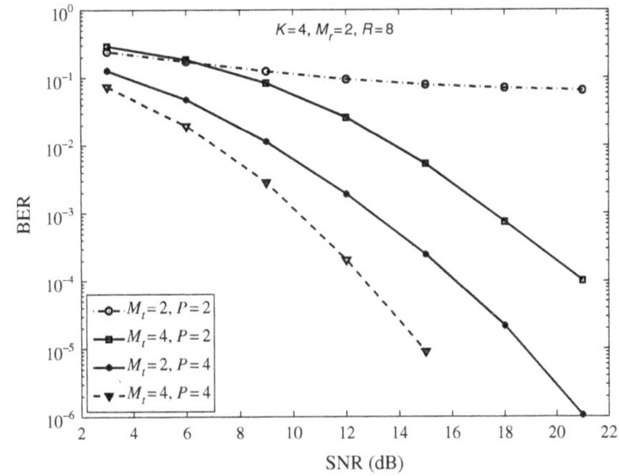

Fig. 11.8 Impact of space spreading with ITU channel B.

space-spreading configurations is reduced. In any case, these results confirm that use of the space dimension is important and even complementary to the frequency dimension, in particular when using small spreading factors.

11.5.2.3 Impact of the System Load

In order to evaluate the impact of the number of users, a multiuser transmission with $Q = 8, 16$ and 32 users is considered. The system uses $M_t = M_r = 2$, $K = 2$, and $R = Q$ (which means that each user transmits a single data stream). The results are depicted in Fig. 11.9. Despite the BER performance degradation as more users are present in the system (as in classical DS-CDMA systems), it is worth noting that symbol detectability is still possible with less spreading than users (which is the case when $Q = 16$ and 32).

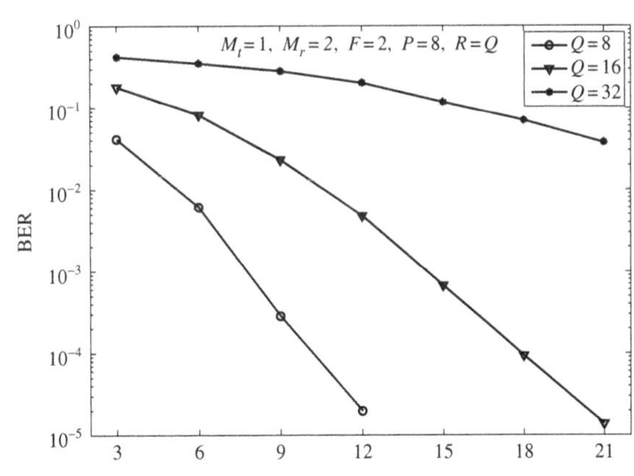

Fig. 11.9 Impact of the system load ($Q = 8$, 16, and 32).

11.6 PARAFAC Tensor Modeling for the STFMA System

This section shows that the PARAFAC decomposition is useful for a tensor modeling of the received signal in the STFMA system. It is shown that different multiple-antenna signaling schemes are easily obtained from this tensor model by making some assumptions and simplifications. It is important to recall that the main motivation for using a PARAFAC modeling at both transmitter and receiver comes from its powerful identifiability properties, affording a blind multiuser detection even in more challenging situations (e.g., unknown spreading codes or multipath signatures).

Let us recall the multiuser transmitted signal model (11.35) and (11.36) in the following form:

$$\mathbf{Z}(n) = \mathbf{U}\bar{\mathbf{S}}(n)\mathbf{C}^T, \tag{11.47}$$

where $\mathbf{U} = blockdiag\big(\boldsymbol{\Omega}_1 \diamond \boldsymbol{\Theta}_1, \ldots, \boldsymbol{\Omega}_Q \diamond \boldsymbol{\Theta}_Q\big)$. Substituting (11.47) and using (11.38), (11.37) is rewritten as

$$\mathbf{Y}(n) = \mathbf{G}\bar{\mathbf{S}}(n)\mathbf{C}^T \in \mathbb{C}^{M_rK \times P}, \tag{11.48}$$

where $\mathbf{G} = blockdiag\big(\mathbf{H}_1(\boldsymbol{\Omega}_1 \diamond \boldsymbol{\Theta}_1), \ldots, \mathbf{H}_Q(\boldsymbol{\Omega}_Q \diamond \boldsymbol{\Theta}_Q)\big) \in \mathbb{C}^{M_rK \times R}$ represents the *effective* channel between the transmitter and the receiver, linking the R multiplexed data streams at the transmitter to the M_rK equivalent subchannel outputs.

Using property (11.43), (11.48) can be rewritten as

$$\mathbf{y}(n) = vec[\mathbf{Y}(n)] = (\mathbf{C} \diamond \mathbf{G})\mathbf{s}(n) \in \mathbb{C}^{PM_rK}, \tag{11.49}$$

where $\mathbf{s}(n)$ is defined in (11.45). Collecting N received signal vectors $\mathbf{y}(n)$ in the matrix $\overline{\mathbf{Y}}_1 = [\mathbf{y}(1), \ldots, \mathbf{y}(N)] \in \mathbb{C}^{PM_rK \times N}$ yields

$$\overline{\mathbf{Y}}_1 = (\mathbf{C} \diamond \mathbf{G})\mathbf{S}^T, \tag{11.50}$$

where

$$\mathbf{S} = [\mathbf{s}(1), \ldots, \mathbf{s}(N)]^T \in \mathbb{C}^{N \times R}. \tag{11.51}$$

By comparing (11.50) with the first expression in (11.5), we recognize a trilinear PARAFAC of rank R for the received signal, and we can deduce the following correspondences:

$$\begin{aligned} &(I_1, I_2, I_3, Q) \leftrightarrow (N, P, M_rK, R), \\ &(\mathbf{A}^{(1)}, \mathbf{A}^{(2)}, \mathbf{A}^{(3)}) \leftrightarrow (\mathbf{S}, \mathbf{C}, \mathbf{G}). \end{aligned} \tag{11.52}$$

By analogy with the second and third expressions in (11.5), the two other matrix representations are obtained by

$$\overline{\mathbf{Y}}_2 = (\mathbf{G} \diamond \mathbf{S})\mathbf{C}^T \in \mathbb{C}^{M_rKN \times P}, \tag{11.53}$$

$$\overline{\mathbf{Y}}_3 = (\mathbf{S} \diamond \mathbf{C})\mathbf{G}^T \in \mathbb{C}^{NP \times M_rK}. \tag{11.54}$$

In the following, the received signal model is rewritten in a scalar form by means of the PARAFAC decomposition. Let $s_{n,r}$, $c_{p,r}$, and $g_{i,r}$ be the entries of $\mathbf{S} \in \mathbb{C}^{N\times R}$, $\mathbf{C} \in \mathbb{C}^{P\times R}$, and $\mathbf{G} \in \mathbb{C}^{M_r K\times R}$, respectively. Using the correspondences (11.52) in (11.1), the scalar notation for the PARAFAC decomposition of the received signal tensor (in the absence of noise) is given by

$$y_{n,p,i} = \sum_{r=1}^{R} s_{n,r} c_{p,r} g_{i,r}, \tag{11.55}$$

where $y_{n,p,i}$ is an entry of the third-order tensor $\mathscr{Y} \in \mathbb{C}^{N\times P\times I}$ representing the received signal, and let $i \doteq (m_r-1)K+k$, $i=1,\ldots,I$, $k=1,\ldots,K$, and $I=M_rK$.

The adaptation of this PARAFAC model to the multiuser case is obtained by modeling the symbol and effective channel matrices as block matrices partitioned into Q blocks, and we have

$$\mathbf{S} = [\mathbf{S}^{(1)},\ldots,\mathbf{S}^{(Q)}] \in \mathbb{C}^{N\times R}, \quad \mathbf{S}^{(q)} \in \mathbb{C}^{N\times R^{(q)}}, \tag{11.56}$$

$$\mathbf{C} = [\mathbf{C}^{(1)},\ldots,\mathbf{C}^{(Q)}] \in \mathbb{C}^{P\times R}, \quad \mathbf{C}^{(q)} \in \mathbb{C}^{P\times R^{(q)}}, \tag{11.57}$$

$$\mathbf{G} = [\mathbf{G}^{(1)},\ldots,\mathbf{G}^{(Q)}] \in \mathbb{C}^{M_r K\times R}, \quad \mathbf{G}^{(q)} \in \mathbb{C}^{M_r K\times R^{(q)}}. \tag{11.58}$$

In this case, (11.55) can be rewritten as

$$y_{n,p,i} = \sum_{q=1}^{Q} \sum_{r^{(q)}=1}^{R^{(q)}} s^{(q)}_{n,r^{(q)}} c^{(q)}_{p,r^{(q)}} g^{(q)}_{i,r^{(q)}}. \tag{11.59}$$

Note that (11.59) is simply a partitioned version of (11.55).

11.6.1 Examples of Special Cases

The PARAFAC model (11.55) is general in the sense that it incorporates several existing multiple-access/multiple-antenna signaling schemes/models. By making appropriate assumptions/simplifications on the model, the structure of (11.55) can be gradually simplified, and different transmission models are obtained as special cases:

- **Space–time spreading**: For $F=K=1$, which corresponds to single-carrier transmission over a flat-fading channel, we can abandon the frequency-dependent index and eliminate the frequency-spreading matrix $\boldsymbol{\Theta}$, so that $\mathbf{G}=\mathbf{H}\boldsymbol{\Omega}$. Thus, the trilinear model (11.50) reduces to classical space–time spreading using multiple spreading codes and can be written as

$$\overline{\mathbf{Y}}_1 = (\mathbf{C} \diamond \mathbf{H}\boldsymbol{\Omega})\mathbf{S}^T \in \mathbb{C}^{PM_r\times N}. \tag{11.60}$$

  This model is valid for modeling the multiple-antenna transmission systems proposed in [23, 55].

- **Spatial multiplexing CDMA**: In spatial multiplexing CDMA systems, the space-spreading operation (which is responsible for spreading the R data streams across the M_t transmit antennas) is eliminated. In other words, each data stream is transmitted by a different transmit antenna. Still considering $F = K = 1$, in this case $R^{(q)} = M_t$, $\boldsymbol{\Omega}^{(q)} = \mathbf{I}_{M_t}$, and $\boldsymbol{\Theta}^{(q)} = \mathbf{1}_K^T$, $q = 1,\ldots,Q$, which implies $\mathbf{G} = \mathbf{H}$, and model (11.50) becomes

$$\overline{\mathbf{Y}}_1 = (\mathbf{C} \diamond \mathbf{H})\mathbf{S}^T \in \mathbb{C}^{PM_r \times N}. \tag{11.61}$$

  This model covers a spatial multiplexing/multiple-access CDMA system using a different spreading code per transmit antenna [44] and is the same as the PARAFAC-CDMA model proposed in the seminal paper [78]. It also coincides with the Khatri-Rao space-time (KRST) coding model of [76].
- **Multicarrier CDMA systems (MCBS-CDMA/MCDS-CDMA/MC-CDMA)**: The transmission model of a MCDS-CDMA system where both time and frequency spreadings take place is considered (e.g., see [98, 100]). This is a single-input single-output antenna system ($M_r = M_t = 1$), which means that the channel matrix in (11.32) reduces to a diagonal matrix $\mathbf{H} \in \mathbb{C}^{K \times K}$ and the space-spreading matrix can be eliminated so that $\mathbf{G} = \mathbf{H}\boldsymbol{\Theta} \in \mathbb{C}^{K \times R}$. Consequently, the general PARAFAC model (11.50) becomes

$$\overline{\mathbf{Y}}_1 = (\mathbf{C} \diamond \mathbf{H}\boldsymbol{\Theta})\mathbf{S}^T \in \mathbb{C}^{PK \times N}. \tag{11.62}$$

  It is worth noting that this special model can be interpreted as the tensorial formulation of the MCBS-CDMA system proposed in [66]. In particular, if frequency spreading is not used, $R^{(q)} = 1$, $q = 1,\ldots,Q$ (i.e., $R = Q$), and $K = Q$ so that $\boldsymbol{\Theta} = \mathbf{I}_Q$. In this case, (11.62) reduces to a PARAFAC model for an MCDS-CDMA system [24].
  Finally, if time-domain spreading is not used ($P = 1$ and $\mathbf{C} = \mathbf{1}_R^T$), then (11.62) reduces to a (matrix-based) MC-CDMA model, given by

$$\overline{\mathbf{Y}}_1 = \mathbf{H}\boldsymbol{\Theta}\mathbf{S}^T \in \mathbb{C}^{K \times N}. \tag{11.63}$$

- **Classical spatial multiplexing**: This is the well-known single-user single-carrier MIMO system with spatial multiplexing, without time-domain block spreading (such as the V-BLAST system of [35]). Therefore, $Q = 1$, $K = P = 1$, $R = M_t$, and $\mathbf{C} = \mathbf{1}_R^T$, $\boldsymbol{\Omega} = \mathbf{I}_{M_t}$, $\boldsymbol{\Theta} = \mathbf{1}_R^T$. In this case, the general PARAFAC model (11.50) simplifies to a conventional matrix-based model:

$$\overline{\mathbf{Y}}_1 = \mathbf{H}\mathbf{S}^T \in \mathbb{C}^{M_r \times N}. \tag{11.64}$$

11.7 Blind Detection

As far as blind symbol recovery/multiuser detection is concerned, the goal of the base station receiver is to separate the co-channel transmissions while recovering the data transmitted by each user without relying on training sequences in order to increase the transmitted information rate. In our system model, the co-channel transmissions are represented by the R STF-domain signals simultaneously accessing the space, time, and frequency channel resources. We are interested in a blind receiver processing without relying on the knowledge of channel impulse responses and antenna array responses. Moreover, statistical independence between the transmitted signals is not required. These are distinguishing features of the PARAFAC-based approach and constitute the main motivation for using the previously developed PARAFAC tensor model.

11.7.1 Joint Symbol-Code-Channel recovery: Identifiability Issues

This section applies the fundamental results of Section 11.3 for studying the joint blind symbol-code-channel recovery based on the PARAFAC model (11.55). These fundamental results yield several practical corollaries, which provide lower bounds on the required number of transmit/receive antennas, subcarriers, spreading factor, and data block length for ensuring a blind symbol-code-channel recovery. They also clearly illustrate the existing trade-offs involving space, frequency, and code diversities.

Remark 3: Recall that, when subcarrier grouping is used, receiver processing is parallelized into J independent detection "layers", each one associated with $K = F/J$ subcarriers. For this reason, identifiability can be studied group-wise (i.e., what matters for identifiability is K and *not* F) since the results obtained for a given subcarrier group are equally valid for all the other groups. For convenience, the identifiability issue is studied for a given subcarrier group by avoiding the use of subcarrier group indexing in the PARAFAC signal model. Therefore, let us rewrite the three unfolded matrices of the received signal in (11.50), (11.53), and (11.54) in the following manner:

$$\overline{\mathbf{Y}}_1 = \mathbf{Z}_{(\mathbf{c},\mathbf{g})}\mathbf{S}^T, \quad \overline{\mathbf{Y}}_2 = \mathbf{Z}_{(\mathbf{g},\mathbf{s})}\mathbf{C}^T, \quad \overline{\mathbf{Y}}_3 = \mathbf{Z}_{(\mathbf{s},\mathbf{c})}\mathbf{G}^T, \tag{11.65}$$

where $\mathbf{Z}_{(\mathbf{c},\mathbf{g})} = \mathbf{C} \diamond \mathbf{G} \in \mathbb{C}^{PM_rK \times R}$, $\mathbf{Z}_{(\mathbf{g},\mathbf{s})} = \mathbf{G} \diamond \mathbf{S} \in \mathbb{C}^{M_rKN \times R}$, and $\mathbf{Z}_{(\mathbf{s},\mathbf{c})} = \mathbf{S} \diamond \mathbf{C} \in \mathbb{C}^{NP \times R}$. Applying condition (11.6) of Theorem 1 (see Section 11.3) to our context, it follows that a joint symbol-code-channel recovery in the LS sense from (11.65) requires that $\mathbf{Z}_{(\mathbf{c},\mathbf{g})}$, $\mathbf{Z}_{(\mathbf{g},\mathbf{s})}$, and $\mathbf{Z}_{(\mathbf{s},\mathbf{c})}$ be full column-rank, which implies

$$min(PM_rK, M_rKN, NP) \geq R. \tag{11.66}$$

This condition is useful when one is interested in eliminating system configurations leading to a nonidentifiable model. It is important to emphasize that (11.66) does not imply joint symbol-code-channel recovery since it is *not* a sufficient condition for PARAFAC model uniqueness. However, using condition (11.7) and the correspondences (11.52), joint symbol-code-channel recovery is guaranteed if

$$k_{\mathbf{S}} + k_{\mathbf{G}} + k_{\mathbf{C}} \geq 2R + 2. \tag{11.67}$$

Under the assumption that **S**, **G**, and **C** are full-rank with no proportional columns, condition (11.67) is equivalent to the following one:

$$min(N,R) + min(M_r K,R) + min(P,R) \geq 2R + 2. \tag{11.68}$$

Remark 4: Practical assumptions about the structure of **S**, **G**, and **C** can ensure that they are full-rank with no proportional columns (i.e., full k-rank). First, **S** is full k-rank almost surely if N is big enough so that $k_{\mathbf{S}} = min(N,R)$. **C** is also full k-rank if its columns are the columns of a (possibly truncated) Walsh–Hadamard matrix, so that $k_{\mathbf{C}} = min(P,R)$. Moreover, space and frequency spreading can be designed to ensure that **G** is full k-rank so that $k_{\mathbf{G}} = min(M_r K,R)$.

In the following, it is assumed that $\mathbf{G} = \mathbf{HU}$, with **U** given in (11.36) and consider particular cases leading to simplifications of (11.67) which are of practical relevance. Interesting trade-offs for blind multiuser detection can be explicitly obtained.

11.7.1.1 Single-Carrier Transmission ($F = K = 1$)

1. $\underline{M_r \geq M_t}$. Note that in this case $\mathbf{G} = \mathbf{H}\boldsymbol{\Omega}$. Assuming that **H** is full-rank, which means that fading is spatially uncorrelated at the transmitter and receiver, it follows that $k_{\mathbf{G}} = rank(\mathbf{G}) = rank(\boldsymbol{\Omega}) = min(M_t,R)$, which implies

$$min(N,R) + min(M_t,R) + min(P,R) \geq 2R + 2. \tag{11.69}$$

2. $\underline{R \geq M_t}$. In this case $\boldsymbol{\Omega}$ is full row-rank by definition, since $\boldsymbol{\Omega}$ is a Vandermonde matrix with distinct generators. It thus follows that $k_{\mathbf{G}} = rank(\mathbf{G}) = rank(\mathbf{H}) = min(M_r,M_t)$, which implies

$$min(N,R) + min(M_r,M_t) + min(P,R) \geq 2R + 2. \tag{11.70}$$

These two conditions have interesting practical corollaries. Assuming that $N \geq R$ and $P \geq R$ (this situation is generally verified in practical systems), conditions (11.69) and (11.70) become, respectively,

$$min(M_t,R) \geq 2, \quad (M_r \geq M_t) \tag{11.71}$$

and

$$min(M_r,M_t) \geq 2, \quad (R \geq M_t) \tag{11.72}$$

and can be, respectively, interpreted in the following manner:

- For $M_r \geq M_t$, spreading/precoding across $M_t = 2$ transmit antennas are enough for a joint blind symbol-code-channel recovery, regardless of the number $R \geq 2$ of data streams.
- For $R \geq M_t$, $M_r = 2$ receive antennas are enough for a joint blind symbol-code-channel recovery, regardless of the number $M_t \geq 2$ of transmit antennas.

11.7.1.2 Single-Antenna Transmission ($M_t = 1$)

In this case, $\mathbf{G} = \mathbf{H}\boldsymbol{\Theta}$. If $\mathbf{H}$ is full column-rank, it follows that $k_{\mathbf{G}} = rank(\mathbf{G}) = rank(\boldsymbol{\Theta}) = min(K,R)$, which implies

$$min(N,R) + min(K,R) + min(P,R) \geq 2R + 2. \tag{11.73}$$

Assuming that $N \geq R$ and $P \geq R$, condition (11.73) is equivalent to

$$min(K,R) \geq 2, \tag{11.74}$$

and the following important corollary is obtained:

- For $M_t = 1$, spreading across $K = 2$ subcarriers per group is enough for a joint blind symbol-code-channel recovery, regardless of the number $R \geq 2$ of data streams.

Note that this condition is independent of the number M_r of receive antennas, which means that joint blind symbol-code-channel recovery is achieved even with a single receive antenna. This clearly illustrates the trade-off between frequency diversity and space diversity at the receiver, which is inherent to the proposed STFMA PARAFAC model.

11.7.1.3 Small Spreading Factors ($P < R$)

A different interpretation of (11.69) and (11.70) arises if $N \geq R$ but $P < R$, i.e., the spreading factor is smaller than the number R of data streams. This is a challenging situation, since most of the multiuser detection receivers (as well as the single-user ones) need $P \geq R$ in order to achieve multiuser interference rejection or despreading. In this case, for single-carrier transmissions ($F = K = 1$), conditions (11.69) and (11.70) reduce, respectively, to the following ones:

$$min(M_t,R) + P \geq R + 2, \tag{11.75}$$

and

$$min(M_r,M_t) + P \geq R + 2. \tag{11.76}$$

These two simplified conditions can be respectively interpreted as

- For $M_r \geq M_t \geq R$, spreading across $P = 2$ chips is enough for a joint blind symbol-code-channel recovery, regardless of the number $R \geq 2$ of data streams and receive antennas.

In other words, this condition establishes a trade-off between spreading diversity and space diversity afforded by the STFMA PARAFAC modeling.

11.7.2 Receiver Algorithm: alternating least squares (ALS)

The blind symbol-code-channel recovery is carried out by estimating each one of the three matrix factors **S**, **C**, and **G** of the trilinear PARAFAC model through minimization of the following nonlinear quadratic cost function:

$$f(\mathbf{S},\mathbf{C},\mathbf{G}) = \sum_{n=1}^{N}\sum_{p=1}^{P}\sum_{i=1}^{M_rF}\left|y_{n,p,i} - \sum_{r=1}^{R} s_{n,r}c_{p,r}g_{i,r}\right|^2. \tag{11.77}$$

The ALS algorithm is the classical solution to minimize this cost function [17, 78, 81]. It is an iterative algorithm that alternates among the estimation of **S**, **C**, and **G**. In other words the ALS algorithm converts a nonlinear optimization problem into three independent linear LS problems. Also, each iteration is composed of three LS estimation steps. At each step, one factor matrix is updated while the other two are fixed to their values obtained in previous estimation steps. The ALS algorithm exploits the Khatri–Rao factorizations of the received signal $\overline{\mathbf{Y}}_1$, $\overline{\mathbf{Y}}_2$, and $\overline{\mathbf{Y}}_3$ given in (11.65).

11.7.2.1 Summary of the ALS Algorithm

Define $\widetilde{\mathbf{Y}}_i = \overline{\mathbf{Y}}_i + \mathbf{V}_i$, $i = 1,2,3$, as the noisy versions of $\overline{\mathbf{Y}}_i$, where $\mathbf{V}_i$ is an additive complex-valued white Gaussian noise matrix. The steps of the ALS algorithm are summarized in Algorithm 11.1:

The convergence at the tth iteration is declared when the error between the true tensor and its reconstructed version from the estimated matrix factors does not significantly change between iterations t and $t+1$. An error measure at the end of the tth iteration can be calculated from the following formula:

$$e_{(t)} = \left\|\widetilde{\mathbf{Y}}_1 - (\widehat{\mathbf{C}}_{(t)} \diamond \widehat{\mathbf{G}}_{(t)})\widehat{\mathbf{S}}^T_{(t)}\right\|_F.$$

The convergence at the tth iteration is declared when $\|e_{(t+1)} - e_{(t)}\| < \delta$, where δ is a prescribed threshold value (e.g., $\delta = 10^{-6}$). If convergence is declared, then we have $\widehat{\mathbf{S}}_{(conv)} = \widehat{\mathbf{S}}_{(t)}$, $\widehat{\mathbf{C}}_{(conv)} = \widehat{\mathbf{C}}_{(t)}$, and $\widehat{\mathbf{G}}_{(conv)} = \widehat{\mathbf{G}}_{(t)}$.

Although the final goal of the blind receiver is to detect the transmitted symbols, we can rely on the knowledge of the space- and frequency-spreading matrices to obtain a final estimate of the original channel **H** from the estimate of the effective channel after convergence $\widehat{\mathbf{G}}_{(conv)}$. If $\boldsymbol{\Omega} \diamond \boldsymbol{\Theta}$ is full row-rank, (11.38) can be used to calculate an LS channel estimate as

$$\widehat{\mathbf{H}} = \widehat{\mathbf{G}}_{(conv)}[\boldsymbol{\Omega} \diamond \boldsymbol{\Theta}]^{\dagger} \in \mathbb{C}^{M_rK \times M_tK}. \tag{11.78}$$

Algorithm 11.1 Alternating LS updates

Initialization: Set $t = 0$; Randomly initialize $\widehat{\mathbf{C}}_{(t=0)}$ and $\widehat{\mathbf{G}}_{(t=0)}$
for all $t > 0$ **do**
repeat
From $\widetilde{\mathbf{Y}}_1$ and using $\widehat{\mathbf{C}}_{(t-1)}$ and $\widehat{\mathbf{G}}_{(t-1)}$, find an estimate of $\mathbf{S}$ by solving the following LS problem:

$$\widehat{\mathbf{S}}_{(t)} = \underset{\mathbf{S}}{\operatorname{argmin}} \left\| \widetilde{\mathbf{Y}}_1 - \widehat{\mathbf{Z}}_{(\mathbf{c},\mathbf{g})} \mathbf{S}^T \right\|_F^2, \quad \widehat{\mathbf{Z}}_{(\mathbf{c},\mathbf{g})} = \widehat{\mathbf{C}}_{(t-1)} \diamond \widehat{\mathbf{G}}_{(t-1)},$$

the solution of which is given by

$$\widehat{\mathbf{S}}^T_{(t)} = \left[\widehat{\mathbf{Z}}_{(\mathbf{c},\mathbf{g})}\right]^{\dagger} \widetilde{\mathbf{Y}}_1.$$

From $\widetilde{\mathbf{Y}}_2$ and using $\widehat{\mathbf{G}}_{(t-1)}$ and $\widehat{\mathbf{S}}_{(t)}$, find an estimate of $\mathbf{C}$ by solving the following LS problem:

$$\widehat{\mathbf{C}}_{(t)} = \underset{\mathbf{C}}{\operatorname{argmin}} \left\| \widetilde{\mathbf{Y}}_2 - \widehat{\mathbf{Z}}_{(\mathbf{g},\mathbf{s})} \mathbf{C}^T \right\|_F^2, \quad \widehat{\mathbf{Z}}_{(\mathbf{g},\mathbf{s})} = \widehat{\mathbf{G}}_{(t-1)} \diamond \widehat{\mathbf{S}}_{(t)},$$

the solution of which is given by

$$\widehat{\mathbf{C}}^T_{(t)} = \left[\widehat{\mathbf{Z}}_{(\mathbf{g},\mathbf{s})}\right]^{\dagger} \widetilde{\mathbf{Y}}_2.$$

From $\widetilde{\mathbf{Y}}_3$ and using $\widehat{\mathbf{S}}_{(t)}$ and $\widehat{\mathbf{C}}_{(t)}$, find an estimate of $\mathbf{G}$ by solving the following LS problem:

$$\widehat{\mathbf{G}}_{(t)} = \underset{\mathbf{G}}{\operatorname{argmin}} \left\| \widetilde{\mathbf{Y}}_3 - \widehat{\mathbf{Z}}_{(\mathbf{s},\mathbf{c})} \mathbf{G}^T \right\|_F^2, \quad \widehat{\mathbf{Z}}_{(\mathbf{s},\mathbf{c})} = \widehat{\mathbf{S}}_{(t)} \diamond \widehat{\mathbf{C}}_{(t)},$$

the solution of which is given by

$$\widehat{\mathbf{G}}^T_{(t)} = \left[\widehat{\mathbf{Z}}_{(\mathbf{s},\mathbf{c})}\right]^{\dagger} \widetilde{\mathbf{Y}}_3$$

until convergence.
end for

In the ALS algorithm, the conditional update of any given matrix may either improve or maintain but cannot worsen the current fit. The algorithm always monotonically converges to (at least) a local minimum. However, the ALS algorithm is dependent on the initialization, and convergence to the global minimum can sometimes be slow if all the matrix factors **S**, **C**, and **G** are unknown. Several alternative algorithms have been proposed in the literature to alleviate the slow convergence problems caused by a random initialization of the algorithm. For instance, an eigenanalysis solution based on compression of the tensor dimensions can be used [78]. The work [25] proposes a generalization of eigenanalysis solutions by means of simultaneous matrix diagonalization. The convergence of the ALS algorithm can also be improved by means of enhanced line search (ELS) [62, 69] or using nonlinear optimization algorithm such as the Levenberg–Marquardt algorithm [61]. In fact, the ALS algorithm rapidly converges when one of the three matrix factors of the model is known. This is typically the case when the spreading code matrix **C** is known.

Remark 5: After convergence of the ALS algorithm, the three estimated matrix factors $\widehat{\mathbf{S}}_{(conv)}$, $\widehat{\mathbf{C}}_{(conv)}$, and $\widehat{\mathbf{G}}_{(conv)}$ are affected by unknown column scalings. In order to eliminate the scaling ambiguity from the columns of the estimated symbol matrix $\widehat{\mathbf{S}}_{(conv)}$, thus leading to an unambiguous symbol recovery, it is assumed that "all ones" symbols are introduced at the beginning of the transmission, i.e., at the first symbol block. Mathematically speaking, this means that the first row of the symbol matrix is a row vector of ones: $\mathbf{S}_{1\cdot} = [1\ 1 \cdots 1] \in \mathbb{C}^{1\times R}$. Under this assumption, a final estimate of the symbol matrix can therefore be obtained in the following manner:

$$\widehat{\mathbf{S}}_{(final)} = \widehat{\mathbf{S}}_{(conv)} \left[D_1(\widehat{\mathbf{S}}_{(conv)})\right]^{-1},$$

where $D_1(\widehat{\mathbf{S}}_{(conv)})$ is the diagonal matrix formed from the first row of $\widehat{\mathbf{S}}_{(conv)}$.

11.8 Simulation Results with Blind Detection

The following simulation results study the performance of the STFMA system using the PARAFAC-based blind ALS receiver described in the previous section. These results aim at illustrating the potential of PARAFAC modeling when applied to blind multiuser detection in the STFMA system. The main objectives are the following:

1. to compare the performance of the channel-blind ALS receiver with that of the nonblind joint-ZF receiver, which assumes perfect channel knowledge;
2. to compare STFMA with other CDMA-based systems when blind ALS detection is used;
3. to evaluate the capability of the blind ALS receiver to perform blind symbol recovery without full knowledge of the channel and code matrices; and
4. to evaluate the channel estimation accuracy.

All the simulations are performed assuming $F = 64$ subcarriers divided into groups of $K = 2$ or $K = 4$ subcarriers. If not explicitly mentioned, the pedestrian channel B is considered in the simulations. When using code-blind detection in the ALS receiver, the spreading code matrix is initialized as a noisy version of the original code matrix. The noise term follows a zero-mean complex Gaussian distribution with variance 0.01.

11.8.1 Blind ALS vs. Nonblind Joint-ZF Receivers

As a reference for comparison, this section evaluates the performance of the nonblind joint-ZF receiver presented in Section 11.5.1. Our aim is to evaluate the performance loss due to blind detection. Contrary to the blind ALS receiver (which is both channel and code blind), the joint-ZF receiver assumes perfect channel and code knowledge. Let $M_t = M_r = 2$, $K = 2$, $P = 8$, $N = 10$, and $Q = 4$ with $R^{(q)} = 2$, $q = 1,\ldots,4$. It can be observed in Fig. 11.10 that the blind ALS receiver is not far from the joint-ZF receiver. In particular, the slope of the BER curves is the same, which means that the ALS receiver presents the same BER improvement as the joint-ZF receiver according to SNR. As expected, both receivers perform better in channel B due to the increased multipath diversity. Note that, for both scenarios, the SNR gap between ALS and ZF is 3 dB for a target BER of 10^{-2}.

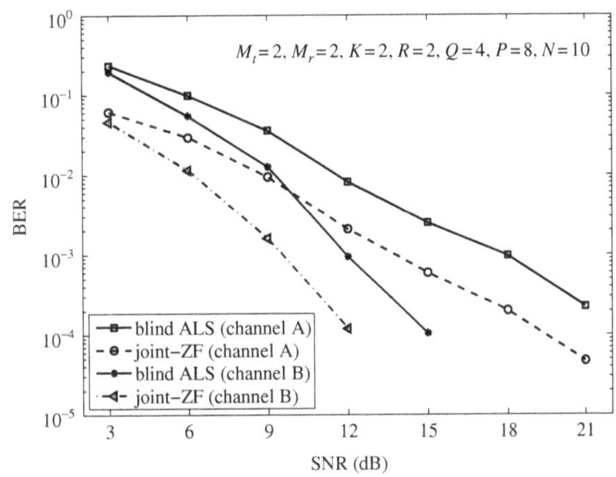

Fig. 11.10 Comparison between blind ALS and ZF receivers.

11.8.2 Performance for Different System Loads

The following results illustrate the performance of the blind ALS receiver for different system loads. Let $M_t = 2$, $K = 2$, $P = 16$, and $N = 20$ while the number of users is varied ($Q = 4$, 6, and 8). Each user transmits two data streams ($R^{(q)} = 2$, $q = 1,\ldots,Q$). Both channel- and code-blind detection and $M_r = 1$ or 2 are assumed. Note that these are challenging configurations in terms of receiver spatial diversity, since M_r is always smaller than Q. The aim of this section is, therefore, to show that blind multiuser detection is still possible in this situation thanks to the joint use of STFMA and PARAFAC modeling. Note that the sufficient uniqueness condition (11.68) is satisfied in the chosen configuration. In fact, as can be observed from Fig. 11.11, blind multiuser detection is achieved even when $M_r = 1$. For instance, with $M_r = 2$ receive antennas, increasing the number of users from $Q = 4$ to $Q = 6$, or from $Q = 4$ to $Q = 8$, implies nearly a 2 dB increase in the required SNR for a target BER of 10^{-2}. We can also note that the BER performance is more sensitive to a variation in the system load when $M_r = 2$ receive antennas are used.

Fig. 11.11 Performance of STFMA with blind detection under different system loads.

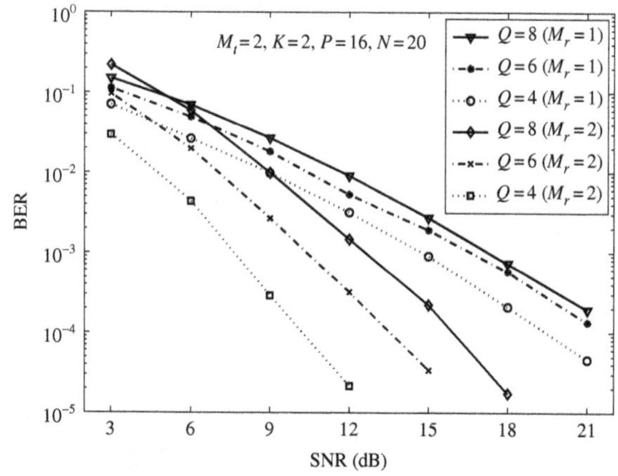

11.8.3 STFMA vs. MCDS-CDMA

The MCDS-CDMA system is a multicarrier extension of the classical DS-CDMA to frequency-selective channels, by performing the spreading operation in the time-domain at each subcarrier [24]. As shown in Section 11.6.1, the PARAFAC modeling is also valid to model the MCDS-CDMA system, which is a special case of the STFMA system without space and frequency spreading (i.e., $M_t = 1$ and $K = 1$). This section compares the performance of both systems using the same PARAFAC-based ALS receiver with spreading code knowledge. The joint-ZF receiver with perfect channel knowledge was also simulated as a reference for comparisons. By comparing STFMA with MCDS-CDMA, the impact of space and frequency spreading as the distinguishing feature of the STFMA system can be verified. Here, it is considered that $M_r = 2$, $P = 8$, $N = 50$, and $Q = 8$ users, each one transmitting a single data stream ($R^{(q)} = 1, q = 1, \ldots, 8$). In Fig. 11.12 the substantial performance gains obtained with the STFMA system are illustrated, which corroborates the advantages of space and frequency spreading. It is worth noting that the gap between ALS and ZF receivers is smaller when STFMA is used.

11.8.4 Channel Estimation Performance

As previously mentioned (see (11.78)), a byproduct of the proposed blind ALS receiver is the estimation of the MIMO channel using the knowledge of the space- and frequency-spreading matrices. The channel estimation accuracy is now evaluated from a root mean square error (RMSE) measure obtained from 100 independent Monte Carlo runs. The overall RMSE is calculated using the following formula:

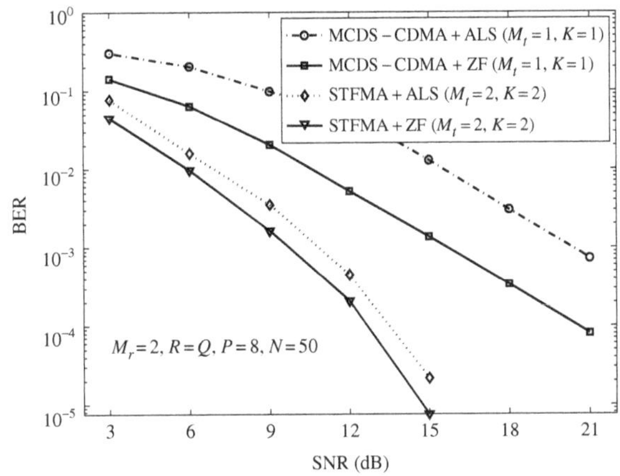

Fig. 11.12 STFMA vs. MCDS-CDMA (ALS and ZF receivers).

$$\text{RMSE} = \sqrt{\frac{1}{100 M_t M_r} \sum_{i=1}^{100} \left\| \widehat{\mathbf{H}}_{(i)} - \mathbf{H} \right\|_F^2}.$$

The following system configuration is considered: $M_t = K = 2$, $P = 2$, $R = 4$, $N = 10$, and $M_r = 1$ or 2. In Fig. 11.13 it can be observed that the root mean square error (RMSE) has a linear decrease as a function of the SNR in both cases. Using $M_r = 2$ antennas provides a performance gain of 3 dB over the single receive antenna case. Such a gain obviously comes from the increased receiver spatial diversity that helps the separation of the data streams, despite the larger number of parameters to estimate.

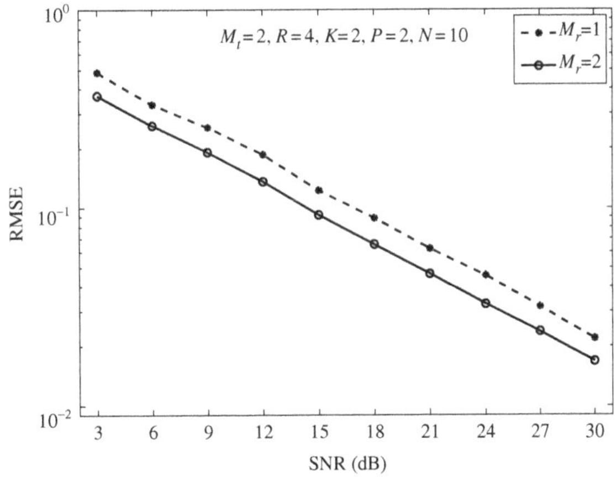

Fig. 11.13 Root mean square error (RMSE) of the estimated channel.

11.9 Conclusions and Research Directions

This chapter is a connection between MIMO transceiver design and PARAFAC tensor modeling. It starts with an overview of the state-of-the-art tensor decompositions and its applications to wireless communication, followed by an illustration of the role of tensor modeling in the wireless communication chain. Some background on the PARAFAC tensor decomposition and its classical identifiability/uniqueness results are provided. Then, a new transceiver architecture for multiple-access multiple-antenna MIMO transmissions was introduced. The distinguishing feature of the so-called STFMA system is that it amalgamates different MIMO techniques into a single transceiver architecture by combining multi-stream space and frequency spreading along with classical time-domain block spreading. It merges the advantages of OFDM and CDMA to minimize multiuser interference effects while providing space and frequency diversity gains. Under the assumption of perfect channel knowledge, the performance of the STFMA system was benchmarked under different transmit configurations using a ZF-based receiver.

The second part of this chapter is dedicated to the application of PARAFAC modeling to model the received signal in the STFMA system in tensorial form. The implications of the identifiability/uniqueness conditions of this PARAFAC model to blind multiuser detection are discussed. Specifically, these conditions help the understanding of the existing trade-offs involving space, frequency, and code diversities that are inherent to the STFMA system. The BER performance of this system was tested using PARAFAC modeling in a variety of system configurations. The results confirm that PARAFAC-based blind multiuser detection is powerful by supporting more users/data streams than transmit antennas. Channel-blind symbol detection is also possible with small spreading factors and, in particular, with only one receive antenna in some cases. Moreover, different from classical multiuser detection receivers, perfect knowledge of the spreading codes is not required by the receiver (although code knowledge can lead to improved performance). These are the practical advantages of PARAFAC modeling when combined with the STFMA system.

Some research directions can be raised from this chapter:

- The optimal design of the space- and frequency-spreading matrices from the viewpoint of diversity and coding gains is an issue to be investigated. The chosen Vandermonde structure is flexible in the sense that a variable degree of spreading/multiplexing can be achieved by simple row/column truncation. However, this structure is not optimized for a performance-oriented transmission. Optimization of the subcarrier grouping is another issue to be considered in future works (see [19, 89]).
- An interesting extension of the STFMA transmission architecture would consist in introducing a joint allocation/scheduling of the data streams in space (transmit antenna) and frequency (subcarrier) dimensions by exploiting some form of channel knowledge at the transmitter.

- The use of a more efficient receiver algorithm (other than ALS) is desirable. The standard ALS algorithm was chosen since it is the classical algorithm for estimating the parameters of a PARAFAC model. However, this algorithm may exhibit a slow convergence for a large number of users/data streams. In order to boost the applicability of PARAFAC modeling in MIMO transceivers, fast-convergent algorithms are preferable. Some recently proposed solutions could be equally applied here [25, 61, 62, 69].

It should also be mentioned that this chapter was restricted to the PARAFAC decomposition/modeling as it completely describes received signal structure in the STFMA system. However, when more complicated signal transformations are present in the transmitter precoding, the PARAFAC modeling can be too restrictive to be used as a signal modeling tool. As detailed in [4], several MIMO transmission schemes need to resort to generalized tensor decompositions offering additional modeling flexibility. Among them, the CONstrained FACtor decomposition (CONFAC) [11] is highlighted. When applied in MIMO transceiver design, this decomposition introduces a set of constraint matrices for modeling different couplings, or combinations, involving data streams, transmit antennas, and spreading codes. Constrained factor (CONFAC) modeling can be viewed as a generalization of PARAFAC modeling and will be exploited in the next chapter as a tool for modeling space–time precoding structures for MIMO systems with limited-feedback transmission.

References

1. ITU-R recommendation M.1225. Guidelines for Evaluation of Radio Transmission Technologies for IMT-2000 (1997)
2. Agrawal, D., Tarokh, V., Naguib, A., Seshadri, N.: Space-time coded OFDM for high data-rate wireless communications over wideband channels. In: Proc. of Vehic. Tech. Conf., pp. 2232–2236. Ottawa, Canada (1998)
3. Alamouti, S.: A simple transmit diversity technique for wireless communications. IEEE J. Sel. Areas Commun. **16**(8), 1451–1458 (1998)
4. de Almeida, A.L.F.: Tensor modeling and signal processing for wireless communication systems. Ph.D. thesis, University of Nice, Sophia-Antipolis (2008)
5. de Almeida, A.L.F., Favier, G., Mota, J.C.M.: PARAFAC models for wireless communication systems. In: Int. Conf. on Physics in Signal and Image processing (PSIP). Toulouse, France (2005)
6. de Almeida, A.L.F., Favier, G., Mota, J.C.M.: Space-time multiplexing codes: A tensor modeling approach. In: IEEE 7th Workshop on Sig. Proc. Advances in Wireless Commun. (SPAWC). Cannes, France (2006)
7. de Almeida, A.L.F., Favier, G., Mota, J.C.M.: Tensor-based space-time multiplexing codes for MIMO-OFDM systems with blind detection. In: Proc. IEEE Symp. Pers. Ind. Mob. Radio Commun. (PIMRC). Helsinki, Finland (2006)
8. de Almeida, A.L.F., Favier, G., Mota, J.C.M.: Constrained space-time spreading for MIMO-CDMA systems: Tensor modeling and blind detection. In: European Signal Process. Conf. (EUSIPCO). Poznan, Poland (2007)
9. de Almeida, A.L.F., Favier, G., Mota, J.C.M.: PARAFAC-based unified tensor modeling for wireless communication systems with application to blind multiuser equalization. Signal Processing **87**(2), 337–351 (2007)

10. de Almeida, A.L.F., Favier, G., Mota, J.C.M.: Space-time spreading MIMO system using canonical precoding tensor model. In: Asilomar Conference Sig. Syst. Comp. Pacific Grove, CA (2007)
11. de Almeida, A.L.F., Favier, G., Mota, J.C.M.: A constrained factor decomposition with application to MIMO antenna systems. IEEE Trans. on Signal Process. **56**(6), 2429–2442 (2008)
12. de Almeida, A.L.F., Favier, G., Mota, J.C.M.: Constrained tensor modeling approach to blind multiple-antenna CDMA schemes. IEEE Trans. on Signal Process. **56**(6), 2417–2428 (2008)
13. de Almeida, A.L.F., Favier, G., Mota, J.C.M.: Multiuser MIMO system using block space-time spreading and tensor modeling. Elsevier Signal Processing **88**(10), 2388–2402 (2008)
14. de Baynast, A., De Lathauwer, L.: Détection autodidacte pour des systèmes à accès multiple basée sur l'analyse PARAFAC. In: Proc. of XIX GRETSI Symp. Sig. Image Proc. Paris, France (2003)
15. de Baynast, A., De Lathauwer, L., Aazhang, B.: Blind PARAFAC receivers for multiple access-multiple antenna systems. In: Proc. VTC Fall. Orlando, USA (2003)
16. Bolcskei, H., Paulraj, A.: Space-frequency coded broadband OFDM systems. In: Proc. of Wirel. Comm. Networking Conf., pp. 1–6. Chicago, IL (2000)
17. Bro, R.: PARAFAC: Tutorial and applications. Chemometrics Intell. Lab. Syst. **38**, 149–171 (1997)
18. Bro, R.: Multi-way analysis in the food industry: Models, algorithms and applications. Ph.D. thesis, University of Amsterdam, Amsterdam (1998)
19. Cai, X., Zhou, S., Giannakis, G.B.: Group-orthogonal multicarrier CDMA. IEEE Trans. on Commun. **52**, 90–99 (2004)
20. Cardoso, J.F.: Blind signal separation: Statistical principles. In: Proc. of the IEEE, vol. 9, pp. 2009–2025 (1998)
21. Carroll, J.D., Chang, J.: Analysis of individual differences in multidimensional scaling via an N-way generalization of "Eckart–Young" decomposition. Psychometrika **35**(3), 283–319 (1970)
22. Comon, P.: Blind identification and source separation in 2x3 under-determined mixtures. IEEE Trans. on Signal Process. **52**(1), 11–22 (2004)
23. Damen, M.O., Abed-Meraim, K., Safavi, A.: On CDMA with space-time codes over multi-path fading channels. IEEE Trans. on Wireless Commun. **2**(1), 11–19 (2003)
24. DaSilva, V.M., Sousa, E.S.: Multicarrier orthogonal CDMA signals for quasi-synchronous communication systems. IEEE J. Sel. Areas Commun. **12**(5), 842–852 (1994)
25. De Lathauwer, L.: A link between the Canonical Decomposition in multilinear algebra and simultaneous matrix diagonalization. SIAM J. Matrix Anal. Appl. **28**(3), 642–666 (2006)
26. De Lathauwer, L., Castaing, J.: Tensor-based techniques for the blind separation of DS-CDMA signals. Signal Processing **87**(2), 322–336 (2007)
27. Doostnejad, R., Lim, T.J., Sousa, E.: Space-time spreading codes for a multiuser MIMO system. In: Proc. of 36th Asilomar Conf. Signals, Syst. Comp., pp. 1374–1378. Pacific Grove, USA (2002)
28. Doostnejad, R., Lim, T.J., Sousa, E.: Space-time multiplexing for MIMO multiuser downlink channels. IEEE Trans. on Wireless Commun. **5**(7), 1726–1734 (2006)
29. Doostnejad, R., Lim, T.J., Sousa, E.S.: On spreading codes for the down-link in a multiuser MIMO-OFDM system. In: Proc. of IEEE Vehic. Tech. Conf., vol. 1, pp. 498–502. Orlando, FL (2003)
30. El Gamal, H., Damen, M.O.: Universal space-time coding. IEEE Trans. on Inf. Theory **49**(5), 1097–1119 (2003)
31. Fazel, K., Kaiser, S., Schnell, M.: A flexible and high performance cellular mobile communications system based on orthogonal multi-carrier SSMA. Wireless Pers. Commun. **2**(1/2), 121–144 (1995)
32. Fernandes, C.A.R., Favier, G., Mota, J.C.M.: Blind tensor-based identification of memoryless multiuser Volterra channels using SOS and modulation codes. In: European Signal Process. Conf. (EUSIPCO). Poznan, Poland (2007)

33. Fernandes, C.E.R., Favier, G., Mota, J.C.M.: Tensor-based blind channel identification. In: IEEE Int. Conf. on Commun. (ICC). Glasgow, UK (2007)
34. Fernandes, C.E.R., Favier, G., Mota, J.C.M.: Blind channel identification algorithms based on the Parafac decomposition of cumulant tensors: The single and multiuser cases. Elsevier Signal Processing **88**, 1382–1401 (2008)
35. Foschini, G.J.: Layered space-time architecture for wireless communications in a fading environment when using multiple antennas. Bell Labs Tech. J. **1**(2), 41–59 (1996)
36. Foschini, G.J., Gans, M.J.: On limits of wireless communications when using multiple antennas. Wireless Pers. Commun. **6**(3), 311–335 (1998)
37. Giannakis, G.B., Wang, Z., Scaglione, A., Barbarossa, S.: AMOUR-generalized multicarrier transceivers for blind CDMA regardless of multipath. IEEE Trans. on Communications **48**(12), 2064–2076 (2000)
38. Goldsmith, A., Jafar, S.A., Jindal, N., Vishwanath, S.: Capacity limits of MIMO channels. IEEE J. Sel. Areas Commun. **21**(5), 684–702 (2003)
39. Golub, G., van Loan, C.: Matrix Computations. The Johns Hopkins University Press, London, UK (1996)
40. Harshman, R.A.: Foundations of the PARAFAC procedure: Model and conditions for an "explanatory" multi-mode factor analysis. UCLA Working Papers in Phonetics **16**, 1–84 (1970)
41. Harshman, R.A., Lundy, M.E.: The PARAFAC Model for Three-Way Factor Analysis and Multidimensional Scaling. H. G. Law, C. W. Snyder Jr., J. Hattie, and R. P. McDonald, Eds. Praeger, New York (1984)
42. Hassibi, B., Hochwald, B.M.: High-rate codes that are linear in space and time. IEEE Trans. on Inf. Theory **48**(7), 1804–1824 (2002)
43. Hochwald, B., Marzetta, T.L., Papadias, C.B.: A transmitter diversity scheme for wideband CDMA systems based on space-time spreading. IEEE J. Sel. Areas Commun. **19**(1), 48–60 (2001)
44. Huang, H., Viswanathan, H., Foschini, G.J.: Multiple antennas in cellular CDMA systems: Transmission, detection, and spectral efficiency. IEEE Trans. on Wireless Commun. **1**(3), 383–392 (2002)
45. Jiang, T., Sidiropoulos, N.D.: A direct semi-blind receiver for SIMO and MIMO OFDM systems subject to frequency offset. In: Proc. SPAWC. Rome, Italy (2003)
46. Jiang, T., Sidiropoulos, N.D.: Kruskal's permutation lemma and the identification of CANDECOMP/PARAFAC and bilinear models with constant modulus constraints. IEEE Trans. on Signal Process. **52**, 2625–2636 (2004)
47. Khouaja, A., Favier, G.: Identification of PARAFAC-Volterra cubic models using an alternating recursive least squares algorithm. In: European Signal Process. Conf. (EUSIPCO). Vienna, Austria (2004)
48. Khouaja, A., Kibangou, A.Y., Favier, G.: Third-order Volterra kernels complexity reduction using PARAFAC. In: First Int. Symp. on Control, Commun. and Signal Process. (ISCCSP), pp. 857–860. Hammamet, Tunisia (2004)
49. Kibangou, A., Favier, G.: Identification aveugle de canaux de communication non linéaires basée sur la décomposition PARAFAC. In: GRETSI Symp. on Signal and Image Process. Troyes, France (2007)
50. Kibangou, A., Favier, G.: Blind equalization of nonlinear channels using a tensor decomposition with code/space/time diversities. Elsevier Signal Processing. **89**(2), 133–143 (2008)
51. Kibangou, A., Favier, G., Hassani, M.M.: Récepteur aveugle basé sur la décomposition PARAFAC pour des canaux de communication non-linéaires. In: Proc. GRETSI, pp. 177–180. Louvain-la-neuve, Belgium (2005)
52. Kruskal, J.B.: Three-way arrays: Rank and uniqueness of trilinear decompositions, with applications to arithmetic complexity and statistics. Linear Algebra Appl. **18**, 95–138 (1977)
53. Liu, Z., Xin, Y., Giannakis, G.: Linear constellation precoding for ofdm with maximum multipath diversity and coding gains. IEEE Trans. on Commun. **51**(3), 416–427 (2003). DOI 10.1109/TCOMM.2003.809791

54. Mailaender, L.: Linear MIMO equalization for CDMA downlink signals with code reuse. IEEE Trans. on Wireless Commun. **4**(5), 2423–2434 (2005)
55. Mudulodu, S., Paulraj, A.J.: A simple multiplexing scheme for MIMO systems using multiple spreading codes. In: Proc. 34th ASILOMAR Conf. on Signals, Systems and Computers, vol. 1, pp. 769–774. Pacific Grove, USA (2000)
56. van Nee, R.D., Prasad, R.: OFDM for Wireless Multimedia Communications. Artech House. London, UK (2000)
57. Ng, B.K., Sousa, E.S.: Multicarrier spread space-spectrum multiple access for the MIMO forward link transmission. In: Proc. IEEE Int. Symp. Pers. Ind. Mob. Radio Commun. (PIMRC). Lisbon, Portugal (2002)
58. Ng, B.K., Sousa, E.S.: A novel spread space-spectrum multiple access scheme for the forward link. In: Proc. IEEE WCNC'02. Orlando, FL, USA (2002)
59. Ng, B.K., Sousa, E.S.: SSSMA for multiuser MIMO systems. IEEE Microwave Magazine pp. 1527–3342 (2004)
60. Nion, D., De Lathauwer, L.: A block factor analysis based receiver for blind multi-user access in wireless communications. In: Proc. ICASSP. Toulouse, France (2006)
61. Nion, D., De Lathauwer, L.: Levenberg–Marquadt computation of the block factor model for blind multi-user access in wireless communications. In: European Signal Process. Conf. (EUSIPCO). Florence, Italy (2006)
62. Nion, D., De Lathauwer, L.: Line search computation of the block factor model for blind multi-user access in wireless communications. In: IEEE 7th Workshop on Sig. Proc. Advances in Wireless Commun. (SPAWC). Cannes, France (2006)
63. Paulraj, A., Nabar, R., Gore, D.: Introduction to Space-Time Wireless Communications. Cambridge University Press, Cambridge, UK (2003)
64. Paulraj, A.J., Gore, D.A., Nabar, R.U., Bolcskei, H.: An overview of MIMO communications: A key to gigabit wireless. Proc. of IEEE **92**(2), 198–218 (2004)
65. Paulraj, A.J., Papadias, C.B.: Space-time processing for wireless communications. IEEE Signal Process. Magazine **14**(6), 49–83 (1997)
66. Petré, F., Leus, G., Moonen, M., De Man, H.: Multi-carrier block-spread CDMA for broadband cellular downlink. EURASIP J. Appl. Signal Process. (10), 1568–1584 (2004)
67. Prasad, R., Hara, S.: Overview of multicarrier CDMA. IEEE Commun. Magazine pp. 126–133 (1997)
68. Proakis, J.G.: Digital Communications. McGraw-Hill, New York (2001)
69. Rajih, M., Comon, P.: Enhanced line search: A novel method to accelerate PARAFAC. In: Proc. EUSIPCO. Antalya, Turkey (2005)
70. Rajih, M., Comon, P.: Alternating least squares identification of under-determined mixtures based on the characteristic function. In: ICASSP'06. Toulouse, France (2006)
71. Rong, Y., Vorobyov, S.A., Gershman, A.D., Sidiropoulos, N.D.: Blind spatial signature estimation via time-varying user power loading and parallel factor analysis. IEEE Trans. on Signal Process. **53**(5), 1697–1710 (2005)
72. Sfar, S., Murch, R.D., Letaief, K.B.: Layered space-time multiuser detection over wireless uplink systems. IEEE Trans. on Wireless Commun. **2**(4), 653–668 (2003)
73. Shao, L., Roy, S.: Rate-one space-frequency block codes with maximum diversity for MIMO-OFDM. IEEE Trans. on Wireless Commun. **4**(4), 1674–1686 (2005)
74. Sidiropoulos, N.D.: Low-rank decomposition of multi-way arrays: A signal processing perspective. In: Invited Plenary Lecture at IEEE SAM. Barcelona, Spain (2004)
75. Sidiropoulos, N.D., Bro, R., Giannakis, G.B.: Parallel factor analysis in sensor array processing. IEEE Trans. on Signal Process. **48**(8), 2377–2388 (2000)
76. Sidiropoulos, N.D., Budampati, R.: Khatri–Rao space-time codes. IEEE Trans. on Signal Process. **50**(10), 2377–2388 (2002)
77. Sidiropoulos, N.D., Dimic, G.Z.: Blind multiuser detection in WCDMA systems with large delay spread. IEEE Signal Process. Lett. **8**(3), 87–89 (2001)
78. Sidiropoulos, N.D., Giannakis, G.B., Bro, R.: Blind PARAFAC receivers for DS-CDMA systems. IEEE Trans. on Signal Process. **48**(3), 810–822 (2000)

79. Sidiropoulos, N.D., Liu, X.: Cramer–Rao bounds for low-rank decomposition of multidimensional arrays. IEEE Trans. on Signal Process. **49**(9), 2074–2086 (2001)
80. Sidiropoulos, N.D., Liu, X.: Identifiability results for blind beamforming in incoherent multipath with small delay spread. IEEE Trans. on Signal Process. **49**(1), 228–236 (2001)
81. Smilde, A., Bro, R., Geladi, P.: Multi-Way Analysis. Applications in the Chemical Sciences. John Wiley and Sons, Chichester, UK (2004)
82. Stegeman, A., Sidiropoulos, N.D.: On Kruskal's uniqueness condition for the Candecomp/Parafac decomposition. Lin. Alg. Appl. **420**, 540–552 (2007)
83. Stuber, G.R., Barry, J.R., Mclaughlin, S.W., Li, Y., Ingram, M.A., Pratt, T.G.: Broadband MIMO-OFDM wireless communications. Proc. of the IEEE **92**(2), 271–294 (2004)
84. Su, W., Safar, Z., Liu, K.J.R.: Full-rate full-diversity space-frequency codes with optimum coding advantage. IEEE Trans. Inf. Theory **51**, 229–249 (2005)
85. Su, W., Safar, Z., Liu, K.J.R.: Towards maximum achievable diversity in space, time and frequency: Performance analysis and code design. IEEE Trans. Commun. **4**(4), 1847–1857 (2005)
86. Talwar, S., Viberg, M., Paulraj, A.J.: Blind separation of synchronous co-channel digital signals using an antenna array – Part I: Algorithms. IEEE Trans. on Signal Process. **44**(5), 1184–1197 (1996)
87. Tarokh, V., Seshadri, N., Calderbank, A.R.: Space-time codes for high data rate wireless communications: Performance criterion and code construction. IEEE Trans. on Inf. Theory **44**(2), 744–765 (1998)
88. Telatar, I.E.: Capacity of multi-antenna gaussian channels. European Trans. on Telecommun. **10**(6), 585–595 (1999)
89. Tran, N., Nguyen, H., Le-Ngoc, T.: Subcarrier grouping for ofdm with linear constellation precoding over multipath fading channels. IEEE Trans. on Vehic. Techn. **56**(6), 3607–3613 (2007). DOI 10.1109/TVT.2007.899970
90. Vanderdope, L.: Multitone spread spectrum multiple access communications system in a multipath rician fading channel. IEEE Trans. on Vehic. Techn. **44**(2), 327–337 (1995)
91. Vanderveen, M.C., Papadias, C.B., Paulraj, A.: Joint angle and delay estimation (JADE) for multipath signals arriving at an antenna array. IEEE Commun. Lett. **1**(1), 12–14 (1997)
92. Vanderveen, M.C., van der Veen, A.J., Paulraj, A.: Estimation of multipath parameters in wireless communications. IEEE Trans. on Signal Process. **46**(3), 682–690 (1998)
93. van der Veen, A.J.: Algebraic methods for deterministic blind beamforming. Proc. of IEEE **86**(10), 1987–2008 (1998)
94. van der Veen, A.J., Vanderveen, M.C., Paulraj, A.: Joint angle and delay estimation using shift invariance techniques. IEEE Trans. on Signal Process. **46**(2), 405–418 (1998)
95. Wang, Z., Giannakis, G.: Wireless multicarrier communications. IEEE Signal Process. Magazine **17**(3), 29–48 (2000). DOI 10.1109/79.841722
96. Wang, Z., Giannakis, G.B.: Complex-field coding for OFDM over fading wireless channels. IEEE Trans. on Information Theory **49**(3), 707–720 (2003)
97. Yang, L.L., Hanzo, L.: Broadband MC DS-CDMA using space-time and frequency-domain spreading. In: Proc. IEEE Vehic. Tech. Conf. (VTC Fall), pp. 1632–1636. Vancouver, Canada (2002)
98. Yang, L.L., Hua, W., Hanzo, L.: Multiuser detection assisted time- and frequency-domain spread multicarrier code-division multiple-access. IEEE Trans. on Vehic. Techn. **55**(1), 397–405 (2006)
99. Yee, N., Linnartz, J.P., Fettweis, G.: Multi-carrier CDMA in indoor wireless radio networks. In: Proc. IEEE Int. Symp. on Personal, Indoor, and Mobile Radio Commun. (PIMRC), pp. 109–113. Yokohama, Japan (1993)
100. You, C.W., Hong, D.S.: Multicarrier CDMA systems using time-domain and frequency-domain spreading codes. IEEE Trans. on Commun. **51**(1), 17–21 (2003)

Chapter 12
MIMO Transceiver Design for Enhanced Performance Under Limited Feedback

Ícaro L. J. da Silva, André L. F. de Almeida, Francisco R. P. Cavalcanti, and Gérard Favier

12.1 Introduction

As described in the previous chapters, multiple-input multiple-output (MIMO) wireless communication systems are able to provide large gains in capacity and link quality by capitalizing either on spatial multiplexing [7, 8, 11, 12, 40] or on space–time coding [1, 39] techniques or, yet, on hybrid combinations of both [6, 14, 44]. Conventional spatial multiplexing and transmit diversity techniques are referred to as "open-loop" since they do not rely on any form of channel information at the transmitter. Despite their practical importance, open-loop multiple-antenna techniques are very sensitive to the ill-conditioning (i.e., rank deficiency) of the channel matrix which incurs limited diversity gains and increased probability of error. Performance enhancements of MIMO systems in terms of higher spectral efficiencies and lower error rates can be achieved by using "closed-loop" transmission techniques [31]. In contrast to the open-loop schemes, these techniques exploit the knowledge of the channel conditions at the transmitter for adapting the transmission signal aiming at improving the link performance. Multiple-antenna techniques like *transmit beamforming* or *precoding* [20] work traditionally in a closed-loop mode. The set of all closed-loop techniques is also called herein *channel-adaptive MIMO*. They are mandatory in upcoming wireless standards for achieving the challenging requirements of beyond 3G systems like 3rd. Generation Partnership Project (3GPP) long-term evolution (LTE).

In the actual wireless communication system, when closed-loop transmission is implemented, some information is sent back to the transmitter from the receiver through a *feedback control channel*. This channel is usually shared among all users, being modeled as a limited-rate channel. The design of such channels and the quantization of the information to be reported back through them is defined as a *feedback problem*. In the past few years, this problem has gained increased attention due to the importance of considering channel-adaptive MIMO transmission. A first approach

F. Cavalcanti, S. Andersson (eds.), *Optimizing Wireless Communication Systems*,
DOI 10.1007/978-1-4419-0155-2_12, © Springer Science+Business Media, LLC 2009

to exploit feedback in closed-loop is to adapt the transmission based on the knowledge, or estimation, of the full channel state information (CSI) at the transmitter [33, 34]. However, although optimal, these techniques are computationally complex and often require too much feedback information which is not practical in current systems. The increase in the number of channel parameters to be estimated in MIMO systems is the main reason for not using *full channel feedback*. For instance, if that the signal adaptation for each channel realization in a 4×4 MIMO system is considered, the quantization of 32 parameters (16 channel phases and gains) has to be performed, which represents an increase in the number of feedback bits by a factor of 30 when compared to fast power control used in a single-antenna system. A second approach to design feedback-based communication links is to use *statistical feedback* by relying on mean or covariance information about the MIMO channel [32, 45, 46]. These techniques, although less complex and more feasible than those based on full channel feedback, may not be capable of tracking the rapid fluctuations of the channel.

A more practical approach to MIMO transceiver design operating in closed-loop relies on the use of "limited" feedback information. The so-called limited feedback (LF) techniques are based on the quantization of the instantaneous channel information at the receiver followed by its conveyance to the transmitter using a low-rate feedback channel [15, 22–24, 26, 27, 36]. To handle possible channel-adaptive MIMO into a LF context (herein called MIMO-LF), first, a *transceiver framework* has to be considered, where it is possible to implement the most important MIMO techniques such as beamforming, space–time block code (STBC) or spatial multiplexing (SM) systems operating in closed loop. The reference to design such a framework are the channel-adaptive MIMO techniques considering full CSI available at the transmitter. This framework need to take into account the design of the limited-rate feedback channel, a storage scheme for the quantized CSI, and a precoding model to format the signal.

Once a transceiver framework is defined, a quantization strategy has to be chosen. This process consists of finding a finite set representation for some channel information. This problem is defined herein as *CSI quantization problem*. In this chapter, two quantization approaches are considered: channel quantization (CQ) and quantized signal adaptation (QSA). The goal of this chapter is to optimize the MIMO link when some quantized information is available. When a CQ technique is considered the channel matrix itself is represented by a finite set of vectors, where each vector is a quantized version of the channel. Then, the receiver decides what is the best quantized representation and sends it back to the transmitter, which formats the signal based on this quantized version. The QSA approach, on the other hand, does not employ a quantized channel. Instead, some "structure" which optimizes the signal is directly quantized. This structure can be a precoder and can be seen as a kind of "information" about the channel. Both CQ and QSA approaches will be discussed later.

Apart from the approaches to solve the quantization problem, another issue that is inherent to LF-based transceiver design concerns the definition of a selection criterion (also called *selection function*). Specifically, a criterion to choose a quantized version from the finite set of quantized information has to be decided. This criterion has to optimize a given performance metric such the SNR, the BER, or the channel capacity. Different channel-adaptive MIMO-LF techniques can be associated with different selection functions. For example, in Section 12.4.1 the maximization of the worst-case SNR is considered as the selection function to solve the antenna selection problem for SM systems.

The most common assumption made when assessing the performance of LF transceivers is that of perfect channel knowledge at the receiver. This assumption leads to an idealized transceiver design, which is not applicable to practical systems. In this chapter, we are also interested in overcoming the perfect channel assumption. As will be seen in the sequel, this is possible by means of a tensor modeling approach [2]. The innovation aspect of this approach is on the algebraic formulation of the precoder model leading to a new formulation of the received signal. Instead of using usual matrix modeling, a tensor modeling approach is adopted to design MIMO-LF transceivers. The distinguishing feature of this approach is the possibility of employing blind channel estimation and detection methodology. This avoids the use of bandwidth-consuming training signals for channel estimation. In addition, using the tensor modeling framework, the performance of LF methods using actual (blind) channel estimates can be evaluated.

This chapter is organized as follows. Section 12.2 presents a general framework for MIMO-LF based on linear precoding, and CSI quantization approaches are discussed. Then, these approaches are applied to channel-adaptive MIMO techniques to be able to consider them in a MIMO-LF context. In Section 12.3, the LF concept is applied to transmit beamforming. Selection diversity transmission, equal gain transmission, and Grassmannian beamforming are also presented in this section. Sections 12.4 and 12.5 discuss the main LF methods used to improve the performance of SM and STBC systems, respectively. This includes the problems of transmit antenna selection and unitary precoding. In order to evaluate the performance of the different MIMO linear precoding techniques under LF transmission, a set of computer simulation results are provided. Section 12.6 presents a new space–time precoding model based on tensor modeling. Two different precoding models are proposed and are then exploited when formulating new transmit antenna selection algorithms. The transceiver design is completed with the proposal of a blind receiver based on tensor modeling. Simulation results are also provided in this section to illustrate the performance of the tensor-based precoding approach. The chapter is concluded in Section 12.7 along with some perspectives for future research.

12.2 Background on Limited Feedback-Based MIMO Systems

Channel-adaptive MIMO brings new challenges to MIMO transceiver design due to the increase in the amount of feedback control information that must be made available to the transmitter by the receiver. This is a consequence of the added spatial dimension at the transmitter due to its use of multiple antennas. In the past few years,

MIMO-LF has gained increased attention. By relying on the use of quantized CSI sent from the receiver back to the transmitter through a low-rate feedback control channel, LF precoding methods can potentially enhance the link performance of MIMO transceivers [25].

In time division duplex (TDD) systems with radio frequency (RF) calibration, the need for feedback is alleviated as the forward link channel can be inferred from the reverse link, since it is assumed that both links are affected by the same fading realizations. It is important to assume also that the channel is not time-varying during the channel acquisition and transmit adaptation loop. Based on this channel reciprocity, several channel-adaptive methods exploiting statistical channel characteristics have been proposed in the literature based, e.g., on channel correlations (in space, time, and/or frequency) [9]. However, in more practical time-varying channel conditions, including also interference variations, these methods can present a significant performance degradation compared with those based on the instantaneous CSI, since the efficient computation of channel statistics may need a considerable amount of transmitted data. Such techniques based on channel reciprocity are not applicable to frequency division duplex (FDD) systems, where different RF paths are taken for receive and transmit branches.

This chapter focuses on MIMO-LF methods based on instantaneous CSI. Therefore, some type of information about the channel needs to be quantized and sent back to the transmitter through a feedback control channel for each channel realization. To apply channel-adaptive MIMO techniques employing LF with instantaneous quantized CSI, first, it is necessary to specify a transceiver framework. A transceiver architecture is adopted where a finite set of *quantized information* is known at the transmitter and receiver. For each channel realization an element in this set is chosen, and based on it, data are encoded before transmission. To conclude this transceiver framework, in this section, the precoding process is also defined as well as how it makes use of the quantized information.

12.2.1 General Transceiver Framework

This section provides an overview of the general framework for MIMO-LF transceiver design by means of linear precoding. The idea is to show how the MIMO link can be optimized as a function of quantized information about the channel obtained from a finite precoding *codebook*. Two quantization approaches are described for this purpose and the associated criteria for transmission optimization are presented.

12.2.1.1 Codebook-Based Scheme

This scheme represents a solution to the problem of storage of the quantized information at both sides of the link. At the same time, the problem of sending a feedback from the receiver to the transmitter is solved. A codebook can be defined as a finite set of N elements which is designed off-line and known to both the transmitter and receiver. This codebook can be represented as below:

$$\mathscr{H}_N^{(q)} = \{\mathbf{H}_1^{(q)}, \ldots, \mathbf{H}_i^{(q)}, \ldots, \mathbf{H}_N^{(q)}\}, \tag{12.1}$$

where $\mathbf{H}_i^{(q)} = q(\mathbf{H})$ is the quantized information concerning a given channel realization $\mathbf{H}$ for a quantization function $q(\cdot)$. The optimum element $\mathbf{H}_{i_{opt}}^{(q)}$ is chosen in such a way that some *performance metric function* $f\left(\mathbf{H},\ \mathbf{H}_i^{(q)}\right)$ is optimized.

Note that the metric depends on the channel realization $\mathbf{H}$ and the quantized information $\mathbf{H}_i^{(q)}$ from the codebook, hence it can be denoted by $f(\mathbf{H},\ i)$. An index *selection function* can be defined as

$$i_{opt} = \arg \underset{1 \le i \le N}{\text{opt}} f(\mathbf{H}, i). \tag{12.2}$$

Note that the elements are ordered so that once the receiver decides on an optimal element it conveys back to the transmitter just its index, which is defined as our *feedback implementation*. In our model, the feedback channel is also considered delay- and error-free. For a codebook with N elements, $B = \lceil \log_2(N) \rceil$ feedback bits are needed to the feedback channel. The operator $\lceil x \rceil$ denotes the closest integer number that is greater than x.

Example 12.1 (The Feedback Channel Implementation). As an example of selection procedure, suppose that $B = 3$ bits. Then, a codebook with $2^3 = 8$ elements can be constructed. Suppose that for a given channel realization if $\mathbf{H}_{i_{opt}}^{(q)} = \mathbf{H}_2^{(q)}$, then $i_{opt} = 2$ and the receiver sends back to the transmitter the bits 010. Note that this framework requires a cooperation between the transmitter and the receiver. Figure 12.1 summarizes this MIMO-LF general framework.

It is worth mentioning that the function $f(\cdot)$ can be determined according to different optimization criteria, such as SNR maximization, outage probability minimization, or error probability minimization. In any case, the ith optimal element is selected so that the performance metric function $f(\cdot)$ is optimized. Moreover, the form taken by the function $f(\cdot)$ is also dependent on the chosen MIMO technique.

12.2.1.2 The Linear Precoding Model

Consider a narrowband MIMO wireless communication system with M_t transmit antennas and M_r receive antennas. The received signal at the kth channel realization can be modeled by the following input–output relation:

$$\mathbf{y}[k] = \sqrt{E_s}\, \mathbf{H}\mathbf{x}[k] + \mathbf{v}[k], \tag{12.3}$$

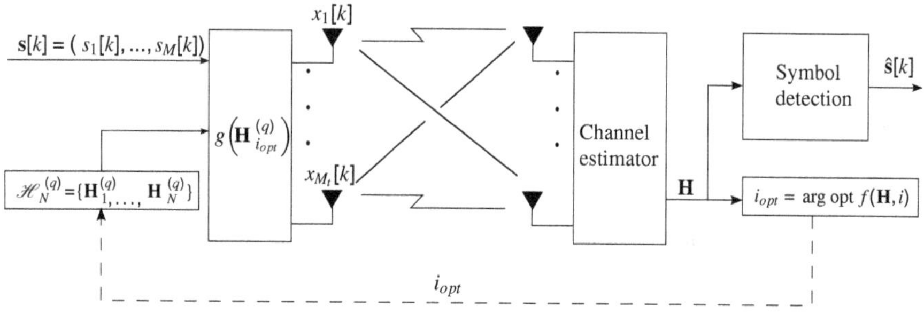

Fig. 12.1 General linear precoding framework for MIMO communications under LF.

where $\mathbf{x}[k]$ is an $M_t \times 1$ transmit vector, $\mathbf{H}$ is the $M_r \times M_t$ channel matrix composed by i.i.d. zero-mean circularly symmetric complex Gaussian with unit variance, $\mathbf{y}[k]$ and $\mathbf{v}[k]$ are M_r-dimensional vectors representing the received signal and noise signal, respectively. Additive noise is modeled as complex zero-mean Gaussian random variable with variance N_0. The total transmitted signal power satisfies the constraint $\|\mathbf{x}[k]\|_F^2 = 1$ and E_s represents the average constellation energy. The goal to be achieved is to design the transmitted signal based on some quantized information from the N-element codebook $\mathscr{H}_N^{(q)}$. The precoding of the transmit signal can be represented as follows:

$$\mathbf{x}[k] = g\left(\mathbf{H}^{(q)}, \mathbf{s}[k]\right), \tag{12.4}$$

where $\mathbf{s}[k]$ is the transmitted symbol vector and $g(\cdot)$ is called *precoding function.* From a transceiver point of view, it is necessary to search for an appropriate precoding function.

Many interesting channel-adaptive MIMO techniques can be implemented by means of linear precoding. This generally consists of applying a linear transformation to the space–time codeword before its transmission in order to add resiliency against channel ill-conditioning or to improve the link performance [34]. Then, $g(.)$ can be considered as a linear function in order to conclude the design of our general framework for MIMO-LF. Using (12.4), the linear precoding model can be stated as

$$\mathbf{x}[k] = g\left(\mathbf{H}^{(q)}, \mathbf{s}[k]\right) = g\left(\mathbf{H}^{(q)}\right)\mathbf{s}[k] = \mathbf{F}\mathbf{s}[k], \tag{12.5}$$

where $\mathbf{F}$ is a *linear precoder* which can be the quantized information $\mathbf{H}^{(q)}$ itself or can be designed using it. These two possibilities define two different approaches to solve the CSI quantization problem, which will be detailed in the sequel.

12.2.2 The Quantization Problem

The general framework based on linear precoding presented previously assumes that the MIMO link is optimized as a function of quantized information about the channel obtained from a finite codebook. This section describes the methods to solve the problem of how to quantize the information contained in $\mathbf{H}$ by means of a quantization function $q(\cdot)$. Two approaches are proposed for solving this problem. The first approach is called CQ and consists of quantizing the entire MIMO channel matrix into a finite codebook. In the second one, called QSA, a finite set of precoders is designed and stored into an ordered codebook following the proposed general framework. It will be described that when this QSA approach is considered *codebook design* and *CSI quantization* can be seen as two independent problems. While the quantization problem consists of finding codebook design criteria based on the properties of optimal adaptive structures, the former problem is related to the construction of codebooks fulfilling such criteria. In the following both approaches are discussed in more detail.

12.2.2.1 Channel Quantization (CQ)

The CQ approach is based on vector quantization (VQ) techniques, applied mostly for data compression purposes. A vector quantizer maps L-dimensional vectors in the vector space $\mathbb{C}^L$ into a finite set of vectors $\mathscr{C}_N = \{\mathbf{c}_1, \ldots, \mathbf{c}_N\}$ and can be seen as an approximation function $q(\cdot)$. Each vector $\mathbf{c}_i$ is called a code vector, or a *codeword*, and the set of all codewords is called a codebook. A nearest neighbor region, c_i, called Voronoi region [10], is associated with each codeword.

Example 12.2 (A 4-bit Bi-dimensional VQ). In this case, the concepts presented for the VQ problem are depicted in Fig. 12.2, where a codebook with 12 codewords corresponds to 12 bi-dimensional Voronoi regions.

The VQ design problem can be stated as follows. Given a vector source $\mathbf{w}_k$, with known statistical properties, and a distortion measure D_{ave}, find a codebook $\mathscr{C}_N = \{\mathbf{c}_1, \ldots, \mathbf{c}_N\}$ and a partition $C = \{C_1, \ldots, C_N\}$ resulting in the smallest average distortion. Once the codebook and the regions are defined, if a given source vector realization $\mathbf{w}_k$ lies in a region C_i, then its approximation is denoted by

$$q(\mathbf{w}_k) = \mathbf{c}_i. \tag{12.6}$$

Assuming a *mean squared error distortion measure*, the average distortion can be defined as

$$D_{ave} = \frac{1}{K} \sum_{k=1}^{K} \|\mathbf{w}_k - q(\mathbf{w}_k)\|^2 = \frac{1}{K} \sum_{k=1}^{K} \|\mathbf{w}_k - \mathbf{c}_i\|^2, \tag{12.7}$$

for K vector samples used in the training process to construct the codebook and define the regions.

Once the design process is finished, for each vector realization, it is necessary to decide what is the optimum codeword $\mathbf{c}_{i_{opt}}$. This *selection criterion* or *selection function* can then be formalized as

$$i_{opt}[k] = \arg \min_{1 \le i \le N} f(\mathbf{w}_k) \tag{12.8}$$

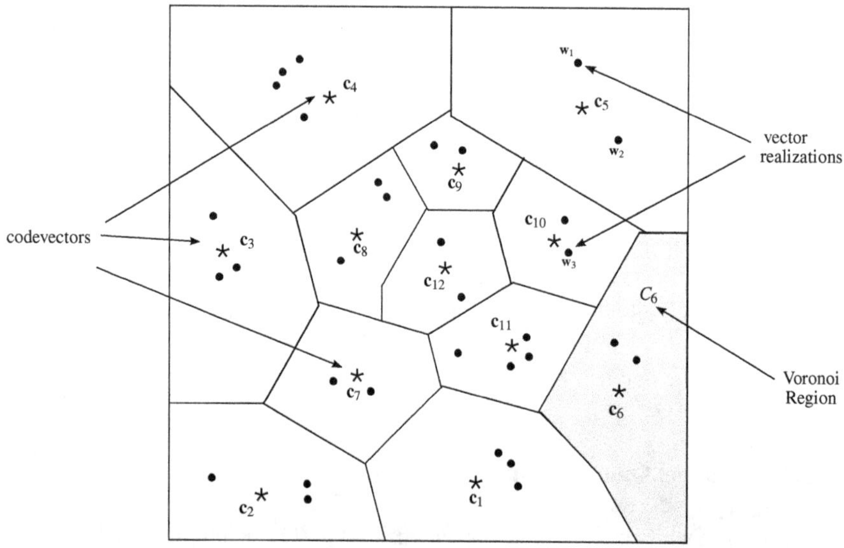

Fig. 12.2 Example of a 4-bit bi-dimensional VQ problem.

for the kth vector realization. In this case, the performance metric function to be optimized is given by

$$f(\mathbf{w}_k) = \|\mathbf{w}_k - \mathbf{c}_i\|^2. \tag{12.9}$$

It is worth mentioning that the application of this VQ concept to the CQ problem consists in quantizing the channel matrix $\mathbf{H} \in \mathbb{C}^{M_r \times M_t}$ in a vector $\mathbf{h} \in \mathbb{C}^{1 \times M_r M_t}$, which is a concatenation of the M_t columns of $\mathbf{H}$. The codebook is then composed of codewords which are quantized versions of the channel itself, as follows:

$$\mathscr{H}_N = \{\mathbf{h}_1, \ldots, \mathbf{h}_i, \ldots, \mathbf{h}_N\}. \tag{12.10}$$

Note that the notation from (12.1) will change once it is defined now that the quantized information about the channel is the channel itself, then $\mathscr{H}_N^{(q)} = \mathscr{H}_N$ and $\mathbf{H}_i^{(q)} = \mathbf{h}_i$.

The difference between the fundamental VQ concept and the CQ for MIMO-LF is that pure VQ attempts to obtain a good approximation of a given channel realization. Instead, the goal of CQ for MIMO-LF problem is rather to optimize some performance measure such as capacity, SNR, or error rate. To achieve this goal, a better idea is to consider a *distortion measure* which exploits some form of *channel invariance*. The channel invariance concept has first appeared in [30], where it was noticed that beamforming is invariant to the channel being multiplied by $exp(j\theta)$ for any θ. Based on this observation, the work [30] derived the phase-invariant distortion measure:

$$D_{ave} = \frac{1}{K} \sum_{k=1}^{K} |\mathbf{h}_k \, q(\mathbf{h}_k)^H| \tag{12.11}$$

such that the performance metric function is defined as $f(\mathbf{h}_k) = |\mathbf{h}_k\, \mathbf{h}_i^H|$ based on SNR maximization of a multiple-input single-output (MISO) beamforming system. This distortion measure then leads to the following selection function:

$$i_{opt} = \arg\max_{1\leq i\leq N} |\mathbf{h}_k\, \mathbf{h}_i^H|, \tag{12.12}$$

with $\mathbf{h}_i \in \mathscr{H}_N$ and $\mathbf{h}_k$ the kth channel realization. Finally, the index i_{opt} is conveyed back to the transmitter to identify the optimum quantized channel. Then, considering the linear precoding model from (12.5), this quantized channel $\mathbf{h}_{i_{opt}}$ is used to construct the precoding matrix $\mathbf{F}_i$. This latter mapping depends on the specific multi-antenna application under consideration.

12.2.2.2 Quantized Signal Adaptation (QSA)

The approach considered in [30] has shifted the paradigm of MIMO-LF by providing solutions to the problem of reducing the number of feedback bits without significant performance loss. This work showed that it is not necessary to quantize the entire channel for the purpose of link adaptation (LA), but only its relevant information.

A different approach derived from CQ, called QSA, consists in exploiting the linear precoding model in (12.5) to quantize directly a precoder that *adapts* the transmitted signal to the channel. Instead of quantizing the channel itself, the idea of QSA is to find design criteria for good codebooks then to construct a precoder codebook instead of a channel codebook. To find such design criteria, the structure from the optimum precoder when full CSI is available is taken into account. This optimum precoder has a particular form for each channel-adaptive MIMO technique. When QSA-based technique is considered, the notation from (12.1) in Section 12.2 is changed and the codebook is redefined as

$$\mathscr{F}_N = \{\mathbf{F}_1, \ldots, \mathbf{F}_N\}, \tag{12.13}$$

where $\mathscr{H}_N^{(q)} = \mathscr{F}_N$ and $\mathbf{H}_i^{(q)} = \mathbf{F}_i$. The feedback is also based on the index of the quantized information as in the first approach.

An example of a QSA applied to the problem of optimum beamforming in MISO systems is given by [27]. Therein, the beamforming vectors are chosen to minimize the outage probability. For a given codebook $\mathscr{F}_N = \{\mathbf{f}_1, \ldots, \mathbf{f}_N\}$, where $\mathbf{f}_i$ is the ith quantized beamformer, the following selection function is used to choose the beamforming vectors:

$$i_{opt} = \arg\max_{1\leq i\leq N} f(\mathbf{h}) = \arg\max_{1\leq i\leq N} \langle \mathbf{h}, \mathbf{f}_i \rangle, \tag{12.14}$$

where $\langle \mathbf{h}, \mathbf{f}_i \rangle = \mathbf{h}\, \mathbf{f}_i^H$ defines the inner product between the vectors $\mathbf{h}$ and $\mathbf{f}_i$.

An interesting aspect of the QSA approach concerns the codebook design. For the CQ approach, among the existing algorithms to define the Voronoi regions and the codewords, the Lloyd algorithm has been extensively considered in the literature for either classical VQ or CQ [10]. On the other hand, methods to construct precoders codebook that are not based on the Lloyd algorithm [10] have been proposed in [22, 23, 26, 27] and some metrics have been derived to compare the codebooks.

12.3 Channel-Adaptive Limited Feedback Beamforming Techniques

This section presents the most important closed-loop LF techniques relying on transmit beamforming, including the problem of selection diversity and equal gain transmission as well as Grassmannian beamforming. Codebook design issues and transmission optimization criteria are also presented. The different techniques are then compared by means of computer simulation results.

12.3.1 Transmit Beamforming Under LF

Transmit beamforming aims at concentrating the transmitted signal energy in relevant directions to maximize the received SNR. It consists of adapting the antenna weights to the channel structure in order to form spatial beams in different directions from the transmitter to the receiver. For a MISO link, the transmit beamforming model can be represented as

$$\mathbf{x}[k] = \mathbf{f}s[k], \tag{12.15}$$

where a transmit power constraint exists such that $\|\mathbf{x}[k]\|^2 = 1$. Equation (12.15) can be seen as a special case of (12.5), where the precoding matrix is replaced by a precoding vector, or beamforming vector, $\mathbf{f} \in \mathbb{C}^{M_t}$ converting the single data stream $s[k]$ into a vector $\mathbf{x}[k]$. In this case, the received signal in (12.3) can be rewritten as

$$y[k] = \sqrt{E_s}\,\mathbf{h}\mathbf{f}s[k] + v[k], \tag{12.16}$$

where $\mathbf{h}$ is an $1 \times M_t$ vector representing the MISO channel.

From (12.16) the received SNR for a given channel realization is

$$\gamma = \frac{E_s}{N_0}|\mathbf{h}\mathbf{f}|^2, \tag{12.17}$$

where the optimum beamforming vector, which maximizes γ, is given by [41]

$$\mathbf{f}_{opt} = \frac{\mathbf{h}^H}{\sqrt{\|\mathbf{h}^H\|^2}}. \tag{12.18}$$

This technique is also called maximal ratio transmission (MRT) and is possible only if perfect CSI at the transmitter is available. When we also have multiple receive antennas, a receiver combining is designed to capture the energy of the received signal from preferred directions, adapting the receive antenna weights to the estimated channel. The previous results can be extended for a MIMO link with joint transmitter beamforming and receiver combining, by means of the following model:

$$y[k] = \sqrt{E_s}\, \mathbf{z}^H \mathbf{H}\mathbf{f}s[k] + \mathbf{z}^H \mathbf{v}[k], \tag{12.19}$$

where $\mathbf{z}$ denotes the $M_r \times 1$ linear receive combiner. In this case, the received SNR can be written as

$$\gamma = \frac{E_s}{N_0} |\mathbf{z}^H \mathbf{H}\, \mathbf{f}|^2. \tag{12.20}$$

It is shown in [42] that the SNR is maximized when $\mathbf{f}$ and $\mathbf{z}$ are the input and output singular vectors, respectively, associated with the maximum singular value σ_{max} of the channel matrix $\mathbf{H}$. The technique which considers only the optimum receiver is called maximal ratio combining (MRC) [31]. Various forms of receiver combining methods can be found in [21], [35] and references therein.

In the following, it is demonstrated that beamforming vectors can be designed using partial channel knowledge only. Toward this path, the properties of the quantization approaches that were introduced in Section 12.2 are of interest. The following MIMO-LF techniques will be examined for transmit beamforming:

1. selection diversity transmission (SDT)
2. quantized equal gain transmission (QEGT)
3. Grassmannian beamforming

12.3.1.1 Selection Diversity Transmission (SDT)

The SDT method was first proposed in [37] for MISO systems and is also commonly known as *transmit antenna selection*. It represents the simplest way to exploit the linear precoding model in (12.5). Specifically, it consists of choosing the best transmit antenna to send the transmitted signal through in such a way that the received SNR is maximized. From (12.17), the received SNR when SDT is considered is given by

$$\gamma = \frac{E_s}{N_0}|\mathbf{h}\mathbf{f}|^2 = \frac{E_s}{N_0}|h_i|^2, \tag{12.21}$$

where $|h_i|$ denotes the channel gain associated with the ith transmit antenna. In this case, all the available transmit power is allocated to a single transmit antenna. Let us denote by $i_{opt}[k]$ the index of the selected transmit antenna. Then, the SDT technique optimizes the following selection function:

$$i_{opt} = \arg \max_{1 \le i \le M_t} |h_i|^2. \tag{12.22}$$

Since the transmitter must select one among M_t antennas to send the transmitted signal through, it then needs $B = \lceil \log_2(M_t) \rceil$ feedback information bits to code the index of the selected antenna. Using the same notation as for the linear precoding model (12.5), if M_t transmit antennas are employed, the following codebook of vectors can be defined as

$$\mathscr{F}_{M_t} = \{\mathbf{e}_1^{(M_t)}, \ldots, \mathbf{e}_i^{(M_t)}, \ldots, \mathbf{e}_{M_t}^{(M_t)}\}, \tag{12.23}$$

where each precoder is a canonical vector from M_t-dimensional space.[12.1]

Example 12.3 (Codebook Design for SDT). For MISO system with $M_t = 3$ transmit antennas, the SDT codebook is given by

$$\mathscr{F}_3 = \{\mathbf{e}_1^{(3)}, \mathbf{e}_2^{(3)}, \mathbf{e}_3^{(3)}\} = \left\{ \begin{bmatrix} 1 \\ 0 \\ 0 \end{bmatrix}, \begin{bmatrix} 0 \\ 1 \\ 0 \end{bmatrix}, \begin{bmatrix} 0 \\ 0 \\ 1 \end{bmatrix} \right\}. \tag{12.24}$$

Supposing that the receiver selects the second antenna as the best one for transmission, it will send back to the transmitter the antenna index $i_{opt} = 2$, which means that the second precoding vector $\mathbf{e}_2^{(3)}$ of the codebook will be selected by the transmitter. Note that, in this case, $B = \lceil log_2(3) \rceil = 2$ feedback information bits. When multiple receive antennas are available, a filter can be used at the receiver side derived from the MRC principle to improve the array gain. One alternative also employed is the dual solution of SDT called selection diversity combining (SDC) where the receive antenna which maximizes the SNR is chosen [21].

12.3.1.2 Quantized Equal Gain Transmission (QEGT)

Equal gain transmission is a beamforming method that maximizes the MISO channel capacity when there is a constraint on equal power per-antenna at the transmitter. It consists of beamforming with a uniform per-antenna power allocation. Only the phases in the beamforming vector are modified. From this assumption and using the model (12.15), the power constraint can be rewritten as

$$|x_i[k]|^2 = \frac{1}{M_t}, \tag{12.25}$$

differently from the total power constraint considered in (12.3). Per-antenna power constraints, rather than total power constraints, have a practical implication in the design of transmit beamforming vectors for MISO systems, as they impose much less dynamic range requirements on the RF power amplifiers.

Considering the per-antenna power constraint assumption, the optimum beamforming vector for equal gain transmission (EGT) can be expressed as

$$\mathbf{f}_{opt} = \frac{1}{\sqrt{M_t}}[exp(j\theta_1), \ \ldots \ , \ exp(j\theta_{M_t})], \tag{12.26}$$

12.1 A canonical vector $\mathbf{e}_i^{(M_t)}$ of the M_t-dimensional space is a unitary vector containing an element equal to 1 in its ith position and 0s elsewhere

where θ_m represents the phase of the mth optimum beamforming vector coefficient $[\mathbf{f}_{opt}]_m = 1/\sqrt{M_t}\, exp(j\theta_m)$. From (12.17), we can see that if $\mathbf{f}$ maximizes the SNR in a MISO system, then $\mathbf{f}\, exp(j\theta)$ provides the same solution, given by

$$\gamma = \frac{E_s}{N_0}\, |\mathbf{h}\,\mathbf{f}\, exp(j\theta)|^2 = \frac{E_s}{N_0}\, |\mathbf{h}\,\mathbf{f}|^2 |exp(j\theta)|^2 = \frac{E_s}{N_0}\, |\mathbf{h}\,\mathbf{f}|^2. \tag{12.27}$$

This equality demonstrates that beamforming is a phase-invariant method. It means that the information about the phase coefficients does not need to be known to maximize the SNR, but only the relative phase are needed. This implies that the vector to be reported to the transmitter can be restricted to

$$\mathbf{f}_{opt} = \frac{1}{\sqrt{M_t}}[1,\ exp(j\phi_2),\ exp(j\phi_3),\ \ldots\ ,\ exp(j\phi_{M_t})], \tag{12.28}$$

where $\phi_m = \theta_m - \theta_1$ is the relative phase between the mth and the first beamforming vector coefficient. In a practical scenario, where a limited feedback channel is considered, the problem that arises is how to quantize these relative phases.

The answer to this problem was given in [16], where a QEGT method was first proposed for a MISO system with $M_t = 2$ transmit antennas. This method has been called *partial phase combining (PPC)*. A uniform per-antenna-phase quantization has been employed. From (12.18), it is possible to observe that in a MISO system the relative phases of the optimum beamforming vector are the same as those of the MISO channel itself. Considering a Rayleigh fading distribution for each channel coefficient and knowing that the phase of a circular complex Gaussian random variable is uniformly distributed, the most reasonable choice is to use a uniform phase quantization. For $M_t = 2$, the beamforming vector $\mathbf{f} = \frac{1}{\sqrt{2}}[1,\ exp(j\phi)]$ is then quantized in a set of vectors $\mathscr{F}_N = \{\mathbf{f}_1, \ldots, \mathbf{f}_N\}$ such that

$$\mathbf{f}_i = 1/\sqrt{2}[1,\ exp(j\phi_i)], \tag{12.29}$$

where $\phi_i \in \Phi_N$ and

$$\Phi_N : \left\{\frac{2\pi i}{N}\right\}_{i=0}^{N-1}. \tag{12.30}$$

To select the relative phase which maximize the SNR, the receiver selects the precoder $\mathbf{f}_i \in \mathscr{F}_N$ such that

$$i_{opt} = \arg \max_{1 \le i \le N} \|h_1 + exp(j\phi_i)h_2\|^2 = \arg \max_{1 \le i \le N} |\mathbf{h}\,\mathbf{f}_i|, \tag{12.31}$$

where h_1 and h_2 represent the channel coefficients. Finally the ith precoder index is conveyed to the transmitter via the feedback channel. Note that the selection function in (12.31) is an attempt to align the relative phase of h_1 and h_2 as closely as possible by rotating the output of antenna two by an angle $exp(j\phi_i)$.

Some similarities between PPC and SDT exist for the case when we have a single feedback bit available. From (12.31), the receiver chooses the best phase $\phi_i \in \Phi_2$ by deciding whether $g_1 = (h_1 + h_2)/\sqrt{2}$ or $g_2 = (h_1 - h_2)/\sqrt{2}$ is greater in magnitude. Since h_1 and h_2 are complex i.i.d. Gaussian random variables, it can be shown that g_1 and g_2 have the same distribution as h_1 and h_2. Therefore, the average performance when selecting either $\max(g_1, g_2)$ or $\max(h_1, h_2)$ is identical. Consequently, PPC with a single bit and SDT are equivalent.

The extension of these results to MIMO systems without receiver combining consists in selecting the beamforming vector that optimizes the following selection criterion:

$$i_{opt} = \arg \max_{1 \le i \le N} \|\mathbf{H}\mathbf{f}_i\|^2. \tag{12.32}$$

PPC implies a low implementation complexity, although being suboptimal compared to other methods based on random CQ even for the case of i.i.d. Rayleigh fading channels [28]. Codebook design techniques based on CQ have been proposed for QEGT in [18, 43]. In [18] a random search-based algorithm is derived while in [43] the minimum value of the maximum magnitude of the inner product between any two code vectors has been used as a performance metric to derive several families of codebooks. In [21], a codebook design criterion following a QSA approach has been proposed.

The performance evaluation between various quantization approaches for quantized EGT has been carried out in [28]. Therein, a codebook design criterion based on the capacity loss using a quantized approach is also proposed jointly with a modified version of the Lloyd algorithm [10]. A performance evaluation of QEGT over feedback noisy channel is presented in [29]. QEGT has been chosen as one of the channel-adaptive beamforming techniques in the WCDMA system [19].

12.3.1.3 Grassmannian Beamforming

The EGT method for LF-based transmission represents an important application where the link optimization has a per-antenna power constraint. However, in the case of total power constraint, a more accurate method can be used to quantize the beamforming vector and consists of quantizing both the antenna phases and gains such as in [30]. The method presented in this section is based on the same principle. The Grassmannian beamforming method relies on the Grassmann manifold along with a QSA approach to design the codebook of beamforming vectors, as is proposed in [26]. The use of QSA is the key difference between Grassmannian beamforming and the other techniques presented in this section since the latter ones are generally based on CQ. Recall that QSA is not based on channel quantization but instead on precoder quantization.

Recall that, for a MISO system, from (12.18), it is possible to observe that $\mathbf{h}$ and $\mathbf{f}_{opt}$ lie in the same hypersphere $\|\mathbf{h}\|^2 = \rho$ when the precoder is not normalized. Note that ρ is a function of a random variable that depends on the channel distribution. Considering that $\mathbf{f}_{opt}$ is quantized into a codebook $\mathscr{F}_N = \{\mathbf{f}_1, \ldots, \mathbf{f}_i, \ldots, \mathbf{f}_N\}$, an angle ω_i between $\mathbf{h}$ and each quantized vector $\mathbf{f}_i$ on the hypersphere can be defined. From (12.16), the received signal from a geometrical point of view can be rewritten as

$$y = \sqrt{E_s}\,\langle \mathbf{h}, \mathbf{f}_i \rangle s + v. \tag{12.33}$$

In this case, the effective SNR is represented by

$$\gamma_{eff} = |\langle \mathbf{h}, \mathbf{f}_i \rangle|^2, \tag{12.34}$$

where the inner product between $\mathbf{h}$ and $\mathbf{f}_i$ can be written as

$$\langle \mathbf{h}, \mathbf{f}_i \rangle = |\mathbf{h}||\mathbf{f}_i|cos(\omega_i). \tag{12.35}$$

The goal to be achieved is to find a design criteria for the codebook $\mathscr{F}_N$ that optimizes some performance metric. Assuming a quasi-static channel,[12.2] an appropriate performance metric is the outage probability. The outage probability, denoted herein by $P_{out} = Prob\{\hat{s} \neq s|\mathbf{h}\}$, is the probability for which an error occurs during the estimation process for a given channel realization. Then, the estimated symbol $\hat{s}$ is not decoded correctly. In this context, it happens when the effective SNR is smaller than a given threshold called outage SNR. In [27], a geometrical approach is proposed to calculate the probability of successful transmission,[12.3] denoted herein by P_{out}^c for an outage SNR of ρ_o. A transmission is considered successful for a given channel realization $\mathbf{h}$ when the selected precoder $\mathbf{f}_i$ from $\mathscr{F}_N$ is such that the effective SNR is greater than the outage SNR. Geometrically, for $M_t = 2$, we are in outage, if the angle between $\mathbf{f}_i$ and $\mathbf{h}$ is greater than ω_{max}. This angle defines a non-outage region in the circle for a given precoder, which can also be represented as

$$|\langle \mathbf{f}_i, \mathbf{h} \rangle| \geq \rho_o. \tag{12.36}$$

This concept is depicted in Fig. 12.3 where each precoder $\mathbf{f}_i$ and ω_{max} defines a non-outage region in the circle. If a channel realization lies in that region and the ith precoder is chosen, there is a successful transmission. In the general case ($M_t > 2$), we can approximate the non-outage regions to spherical caps and the success probability is calculated from the area surface of this sphere using some approximations [27]. Note that the partitions of this sphere can be interpreted as the quantization regions where each region is associated with a precoder.

It can be shown [27] that the probability of successful transmission conditioned on the channel realization in the hypersphere of radius ρ is bounded by

$$P_{out}^c\{r_t, E_s \mid \|\mathbf{h}\|^2 = \rho\} \leq N\left(1 - \frac{\rho_o}{\rho}\right)^{(M_t-1)} \qquad \rho^* < \rho \tag{12.37}$$

for a given data rate r_t and transmit power E_s. The overlap radius ρ^* of the beamformer codebook $\mathscr{F}_N$ is defined as the minimum radius such that the spherical caps referent to the non-outage regions do not overlap. This radius is a function of the

---

[12.2] A channel is said to be quasi-static if it remains constant over the time necessary to transmit a data block, which can be, e.g., a space–time codeword.

[12.3] P_{out}^c stands for the complementor of the outage probability, i.e., $P_{out}^c = 1 - P_{out}$.

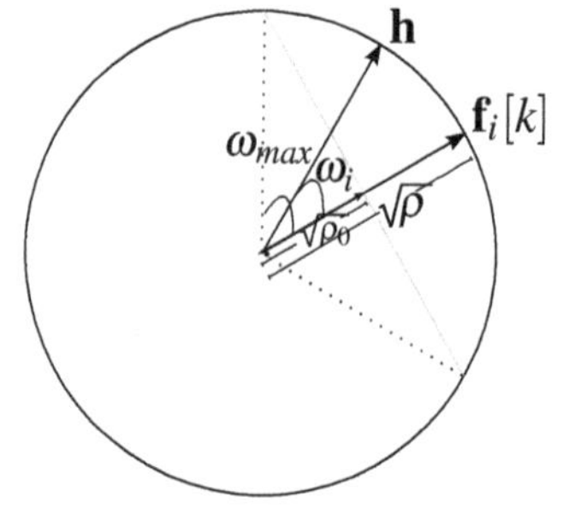

Fig. 12.3 Non-outage region for $M_t = 2$.

partition of the hypersphere $||\mathbf{h}||^2 = \rho$ in spherical caps, and consequently is a function of the codebook design.

The probability $P_{out}^{c}\{.\}$ of successful transmission is bounded by the sum of each surface area for each codebook vector. If the caps do not overlap (i.e., if $\rho < \rho^*$) the bound is achieved. Otherwise, if $\rho \geq \rho^*$ some of the spherical caps overlap each other, and the area corresponding to $P_{out}^{c}\{.\}$ is less than that of the sum of the individual spherical caps. Then, it is clear that the optimization of a beamforming codebook is closely related to the maximization of the overlap radius. This maximum overlap radius is expressed as [27]

$$[\rho^*(\mathscr{F}_N)]_{opt} = \left(\frac{2\rho_o}{1 + \max\limits_{1 \leq i \leq j \leq N} |\langle \mathbf{f}_i, \mathbf{f}_j \rangle|} \right). \tag{12.38}$$

Consequently, the design criterion of a good beamformer codebook $\mathscr{F}_N$ with M_t transmit antennas can be formulated as

$$\min_{\mathscr{F}_N \in \mathbb{F}^{M_t}} \max_{1 \leq i \leq j \leq N} |\langle \mathbf{f}_i, \mathbf{f}_j \rangle|, \tag{12.39}$$

where $\mathbb{F}^{M_t}$ is the set of N unitary vectors each one with M_t terms. Maximizing the inner product between two vectors is equivalent to maximizing the angular distance between the two closest vectors in the codebook, once $|\langle \mathbf{f}_i, \mathbf{f}_j \rangle| = |\mathbf{f}_i||\mathbf{f}_j||cos\beta_{ij}|$. This approach is similar to the design criterion of signal constellations for Gaussian channels, where the minimum Euclidian distance is maximized. The difference is that the distance between *lines* passing through the origin is going to be maximized.

An interesting similarity to the problem of packings in the Grassmannian space [5] have been observed in [27]. Grassmannian line packing is the problem of optimally packing one-dimensional subspaces [5], which is the same as finding an optimal set of one-dimensional precoders where each precoder can be interpreted as a vector which generates a subspace in $\mathbb{C}^{M_t}$. The complex Grassmann manifold $G(M_t, 1)$ is the set of all one-dimensional subspaces of the space $\mathbb{C}^{M_t}$. Ω_{M_t} is called the space of unit-norm transmit beamforming vectors to define a distance function on $G(M_t, 1)$ by letting the distance between two lines generated from unit vectors $\mathbf{f}_1$

and $\mathbf{f}_2 \in \Omega_{Mt}$ taking the sine of the angle β_{12} between the two lines. This distance is expressed as [3]

$$d_c(\mathbf{f}_1,\mathbf{f}_2) = sin(\beta_{12}) = \sqrt{1-|\langle \mathbf{f}_1,\mathbf{f}_2\rangle|^2}, \tag{12.40}$$

and Fig. 12.4 shows two lines generated from the quantized beamforming vectors $\mathbf{f}_i$ and $\mathbf{f}_j$ and the angle β_{ij} between them.

The *Grassmannian line packing problem* is the problem of finding the set, or packing, of N lines in $\mathbb{C}^{M_t}$ that has a maximal minimum distance between any pair of lines. A packing of N lines in $G(M_t,1)$ can be represented by the set $\mathscr{F}_N = \{\mathbf{f}_1 \ldots \mathbf{f}_N\}$, where $\mathbf{f}_i \in \Omega_{M_t}$ is the unit-norm vector which generates the ith line in the packing. The packing problem is only of interest where $N > M_t$. The minimum distance of a packing is the sine of the smallest angle between any pair of lines. This is written as

$$\delta(\mathscr{F}_N) = \min_{1\leq i \leq j \leq N} \sqrt{1-|\langle \mathbf{f}_i,\mathbf{f}_j\rangle|^2} = sin(\beta_{min}), \tag{12.41}$$

where β_{min} is the smallest angle between any pair of lines in the packing. The problem of finding algorithms to design packings for arbitrary M_t and N has been studied by many researchers in applied mathematics and information theory. Thus, from these results, the following criterion for designing quantized beamforming codebooks has been proposed in [26] and stated as below:

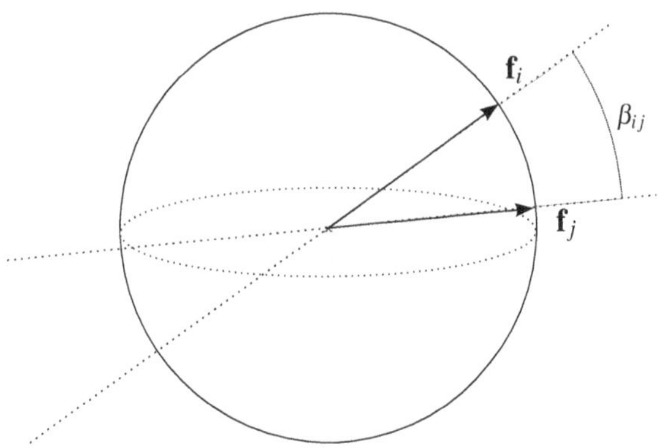

Fig. 12.4 Grassmannian packing lines in $\mathbb{R}^3$.

Codebook design criterion for Grassmannian Beamforming: Design the set of vectors $\{\mathbf{f}_i\}_{i=1}^N$ such that the corresponding codebook $\mathscr{F}_N$ maximizes $\delta(\mathscr{F}_N)$.

In [26] the authors proved another important result: the distribution of the optimal beamforming vector is independent of the number of receive antennas. Thus, the problem of finding quantized beamformers for MISO systems is the same problem as that of finding quantized beamformers for MIMO systems.

12.3.1.4 Performance Results for MIMO-LF Beamforming

The performance comparisons between some of the studied beamforming MIMO-LF methods are depicted in five plots, generated by Monte Carlo simulations. The performance comparisons are made in terms of average symbol error rate (SER). Each plotted SER curve is an average over 10^8 simulations for each SNR value. The channel is assumed quasi-static during the closed-loop adaptive transmission. The transmitted symbols are generated from a M-quadrature amplitude modulation (QAM) or M-phase-shift keying (PSK) constellation and the modulation order M is the function of the fixed data rate in bits per second per Hertz (bps/Hz).

Experiment 1 – SDT (single diversity transmission): This experiment compares 2×1 and 3×1 MISO schemes at a fixed rate of $r_c = 2$ bps/Hz constructed from a 4-QAM constellation. We observe from Fig. 12.5(a) a diversity advantage which increases with M_t over the uncoded SISO system when $M_t = 2$ and $M_t = 3$. When optimum beamforming is employed an array gain of 2 dB is provided for $M_t = 3$.

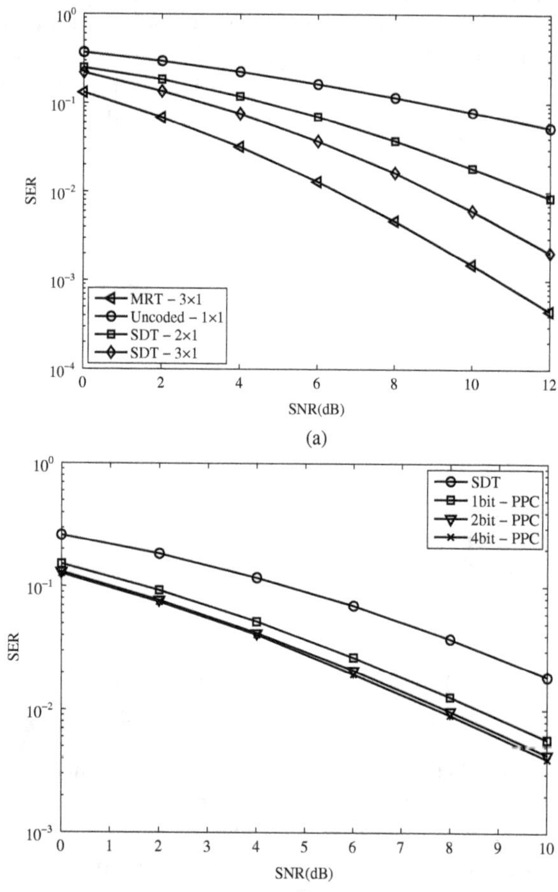

Fig. 12.5 Performance comparison of (a) SDT and MRT in MISO configurations with $M_t = 2$ and $Mt = 3$ and (b) SDT and PPC with different phase quantization degrees.

The price to pay in terms of feedback is equal to 1 bit, when considering $M_t = 3$. As described in Section 12.3.1.1, the SDT technique, which is based on single transmit antenna selection, is a good way to increase the diversity order when having multiple transmit antennas.

Experiment 2 – PPC (partial phase combining): Concerning the channel-adaptive techniques with per-antenna power constraint, PPC is compared with SDT in Fig. 12.5(b) for a 2×1 MISO system and 4-QAM constellation. As described in 12.3.1.2, the SDT and PPC with 1 feedback bit present the same diversity order (generate the same slope of the SER curve), the only difference being an array gain that is provided by PPC. Note that there is an additional coding gain of 0.5 dB when 2 feedback bits are used (instead of 1) but there is no substantial gain when the number of feedback bits is increased from 2 to 4.

Experiment 3 – QEGT (quantized equal gain transmission): Figure 12.6(a) depicts the results for QEGT, another technique based on a per-antenna power constraint. This experiment is focused on a 2×2 MIMO system with an equal gain

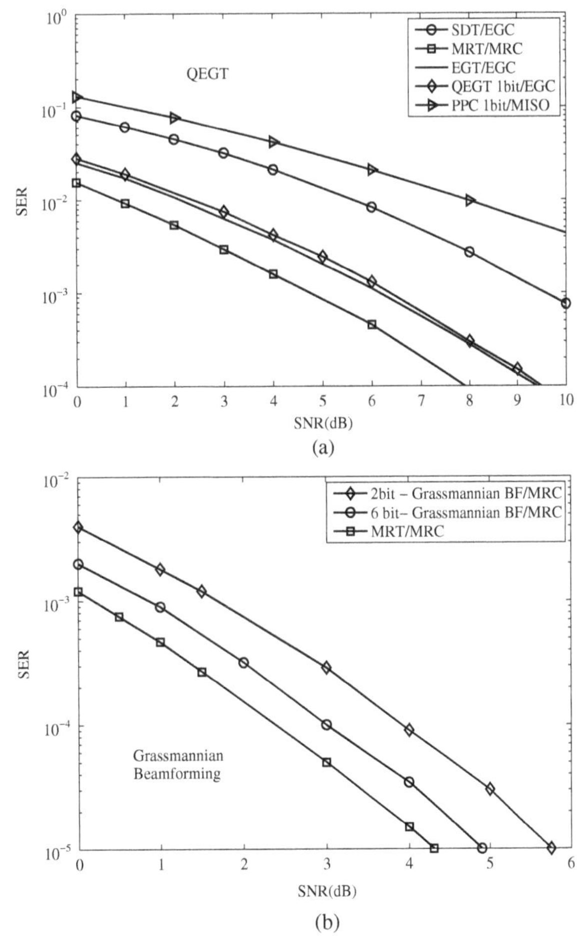

Fig. 12.6 Performance comparison of EGT, MIMO optimum BF and SDT (a) evaluated results and (b) results from [26], reproduced with permission from Love, D.J., Heath, R.W. Jr., Strohmer, T., Grassmannian beamforming for multiple-input multiple-output wireless systems. IEEE Trans. Inf. Theory, **49**(10), 2735–2747 (2003) © 2003 IEEE.

combining (EGC) receiver [21]. This EGC technique is the dual of EGT and is applied at the receiver side. According to this figure, a constant coding gain of 3 dB is obtained over SDT when 1 bit QEGT is employed. Note that 1-bit QEGT (based on QSA) outperforms PPC in all the SNR range. As QEGT increases the number of feedback bits, we approximate the optimum performance under the same power constraint.

Experiment 4 – Grassmannian beamforming: Finally, the performance of Grassmannian beamforming is evaluated with different number of available feedback bits. This is a 3×3 MIMO system with BPSK constellation. It can be seen from Fig. 12.6(b) that four additional bits for the feedback channel provides a gain of 1 dB when Grassmannian beamforming is employed. Note also that the performance approximates that of the optimum MRT/MRC beamforming/combining scheme when the feedback rate is increased.

12.4 Linear Precoding for Spatial Multiplexing Systems

As it has been shown in Chapter 10, spatial multiplexing systems employ the multiple transmit antennas for increasing the information rate and link capacity over a MIMO channel [7, 8]. The serial input data stream is divided into multiple substreams and each one being transmitted using, possibly, a subset of the available transmit antennas. At the receiver, these substreams are separated by means of spatial filtering and interference cancellation techniques [31]. Linear precoding for SM systems represents an extension of the beamforming concept applied to single stream systems. Let $\mathbf{s}[k] = [s_1[k], \ldots, s_{M_s}[k]]$ denote a vector containing M_s symbols to be transmitted at the kth symbol period. Linear precoding consists in pre-multiplying this symbol vector by a precoding matrix $\mathbf{F}$ with dimensions $(M_t \times M_s)$ chosen for each channel realization. The received signal can be modeled as

$$\mathbf{y}[k] = \sqrt{E_s}\,\mathbf{HFs}[k] + \mathbf{v}[k]. \tag{12.42}$$

Recall that $M_s > \min(M_r, M_t)$ must be assumed in order to benefit from the link capacity gains of the MIMO channel [40]. In the following, we are interested in coupling this linear precoding model with the MIMO-LF quantization approaches (c.f. Section 12.2), i.e., we need to choose the optimum $\mathbf{F}$ by solving the problems of determining (i) a selection function and (ii) a precoder codebook.

Based on the linear precoding model (12.42) as well as its special cases, the following section presents some recently proposed MIMO-LF techniques.

12.4.1 Transmit Antenna Selection for SM Systems

Transmit antenna selection can be viewed as the simplest linear precoding method exploiting the model (12.42). It consists of choosing a subset of M_s among M_t transmit antennas to transmit the M_s data substreams so that the link performance can be optimized according to some metric. The use of transmit antenna selection is of practical interest since it allows a reduction in the hardware complexity by using fewer RF chains than transmit antennas. Differently from a non-adaptive MIMO technique where all the transmit antennas are simultaneously used, an adaptive scheme based on transmit antenna selection can identify the best subset of transmit antennas for a given channel realization and switch them to the RF chains accordingly.

For a SM system with a ZF linear receiver, it is shown in [31] that the post-processing SNR associated with the mth substream is given by

$$\gamma_m = \frac{E_s}{M_t N_0 [\mathbf{H}_i^H \mathbf{H}_i]^{-1}_{m,m}}, \tag{12.43}$$

where $\mathbf{H}_i = \mathbf{F}_i \mathbf{H}$ is the $M_r \times M_s$ equivalent channel matrix that is simply formed from a subset of M_s columns of $\mathbf{H}$.

The so-called per-stream-received SNR can be approximated as in [17]:

$$\min_{1 \leq m \leq M_s} \gamma_m = \gamma_{min} \geq \lambda^2_{min}(\mathbf{H}_i) \frac{E_s}{N_0 M_s}, \tag{12.44}$$

which is a simple function of the channel. Equation (12.44) shows that the performance of linear receivers with antenna subset selection should improve as the smallest singular value of the equivalent channel increases. In terms of symbol error probability the performance can be calculated from the vector symbol error rate (VSER), which is the probability that at least one substream symbol is in error. Then, the probability of successfully decoded symbol vectors is given by [17]

$$\begin{aligned} P^c_{out} &= 1 - \prod_{m=1}^{M_s} (1 - P_m) \leq 1 - (1 - P_{min})^{M_s} \\ &\approx M_s\, P_{min} \leq M_s\, N_e\, Q\left(\sqrt{\gamma_{min} \frac{d^2_{min}}{2}}\right), \end{aligned} \tag{12.45}$$

where N_e is the average number of nearest neighbors per symbol of the symbol constellation, P_m is the error probability for the mth substream, and d_{min} is the minimum distance of the per-antenna constellation. The bound in (12.45) shows that the receiver performance is simply a function of the minimum SNR of all substreams.

From these results, a selection function can be derived to determine the precoder $\mathbf{F}_i$ from a finite codebook. This selection function is based on the maximization of the minimum singular value $\lambda_{min}(\mathbf{H}_i)$, which from (12.45) represents a suboptimal criteria to minimize the VSER.

It consists in computing $\lambda_{min}(\mathbf{H}_i)$ for each subset of M_s transmit antennas and choosing the subset that optimizes the following criterion:

$$i_{opt} = \arg \max_{1 \leq i \leq \binom{M_t}{}} \lambda^2_{min}(\mathbf{H}_i), \tag{12.46}$$

where $\binom{M_t}{M_s}$ denotes the number of combinations of M_s among M_t antennas.

Concerning the codebook design, let $\mathscr{F}_{M_s}^{(M_t)}$ be the set of $\binom{M_t}{M_s}$ precoders representing the possible transmit antenna subsets. Each precoder is then formed by choosing M_s columns from $\mathbf{I}_{M_t}$. The optimum codebook index i_{opt} is then conveyed back to the transmitter via a feedback channel. Since we have $\binom{M_t}{M_s}$ possible precoders, we need $B = \left\lceil log_2 \binom{M_t}{M_s} \right\rceil$ feedback information bits to perform transmit antenna selection for a fixed M_s and M_t.

Example 12.4 (Codebook Design for Transmit Antenna Selection). For a MIMO system with $M_t = 3$ transmit antennas, we have

$$\mathbf{I}_3 = \left\{ \mathbf{e}_1^{(3)}, \mathbf{e}_2^{(3)}, \mathbf{e}_3^{(3)} \right\} = \begin{bmatrix} 1\,0\,0 \\ 0\,1\,0 \\ 0\,0\,1 \end{bmatrix}, \tag{12.47}$$

where $\mathbf{e}_m^{(3)}$ represents the selection of the mth transmit antenna. It is possible to consider M_s = 1, 2, or 3 and, for each value, we have the possible codebooks:

$$\mathscr{F}_1^{(3)} = \left\{ \begin{bmatrix} 1 \\ 0 \\ 0 \end{bmatrix}, \begin{bmatrix} 0 \\ 1 \\ 0 \end{bmatrix}, \begin{bmatrix} 0 \\ 0 \\ 1 \end{bmatrix} \right\}, \mathscr{F}_2^{(3)} = \left\{ \begin{bmatrix} 1\,0 \\ 0\,1 \\ 0\,0 \end{bmatrix}, \begin{bmatrix} 1\,0 \\ 0\,0 \\ 0\,1 \end{bmatrix}, \begin{bmatrix} 0\,0 \\ 1\,0 \\ 0\,1 \end{bmatrix} \right\}, \text{ and } \mathscr{F}_3^{(3)} = \left\{ \begin{bmatrix} 1\,0\,0 \\ 0\,1\,0 \\ 0\,0\,1 \end{bmatrix} \right\}.$$

Assume that, for each M_s, $\mathscr{F}_{M_s}^{(M_t)}$ has been ordered so that it can be written as

$$\mathscr{F}_{M_s}^{(M_t)} = \left\{ \mathbf{F}_{M_s,1}, \mathbf{F}_{M_s,2}, \ldots, \mathbf{F}_{M_s,\binom{M_t}{M_s}} \right\} \tag{12.48}$$

and each precoder $\mathbf{F}_{M_s,i}$ represents a substream-to-antenna mapping. The columns of each precoder is a *selection diversity vector* as in (12.24).

12.4.2 Multimode Antenna Selection for SM Systems

The *multimode* antenna selection represents an improvement over the transmit antenna selection method described previously. This method, in addition to selecting the transmit antennas, also adapts the number of multiplexed substreams M_s to the current channel, for a fixed total data rate. Since the data rate should be kept constant regardless of M_s, *the number of selected substreams determines the constellation size M to be used for a given channel realization.* This means that the multimode antenna selection is directly associated with adaptive modulation.

Let us rewrite the received signal in (12.42) in the following form:

$$\mathbf{y}[k] = \sqrt{E_s}\,\mathbf{H}\,\mathbf{F}_{M_s,i}\,\mathbf{s}[k] + \mathbf{n}[k]. \tag{12.49}$$

Note that the precoder $\mathbf{F}_{M_s,i}$ is now indexed by M_s and i, the first index determines the number M_s of multiplexed substreams while i associates these substreams with the M_t transmit antennas. Therefore, we are interested in choosing M_s and i such that the VSER is minimized. From (12.45), the VSER can be rewritten as a function of the data rate r_c in bits per Hertz, the number of selected antennas M_s and i leading to the derivation of a new selection function to find the pair $(M_{s\,opt}, i_{opt})$ as [15]

$$\arg \min_{1\le M_s\le M_t,\ 1\le i\le \binom{M_t}{M_s}} N_e(M,r_c)\, Q\left(\sqrt{\min_{1\le m\le M_s} \gamma_m(M_s,i)\frac{d^2_{min}(M_s,r_c)}{2}}\right), \tag{12.50}$$

where $M_s = 1,2,\ldots,M_t$, and $i = 1,2,\ldots,\binom{M_t}{M_s}$. For implementation, $N_e(M_s,r_c)$ and $d^2_{min}(M_s,r_c)$ can be precalculated for all values of M_s.

Example 12.5 (Setting the Parameters of Multimode Antenna Selection). Supposing $M_t = 4$ and $r_c = 8$ bps/Hz, the following transmission modes can be defined:

1. $M_s = 4$ with 4-QAM
2. $M_s = 2$ with 16-QAM
3. $M_s = 1$ with 256-QAM

Then, for the first transmission mode, $N_e(4,8) = 2$ for a 4-QAM constellation, while for the second transmission mode $N_e(2,8) = 3$. This is calculated considering that symbols $0,1,2$, and 3 have four nearest neighbors, symbols $15,6,9$, and 12 have two nearest neighbors, and symbols $4,5,7,8,10,11,13$, and 14 have three nearest neighbors. Averaging over the number of symbols results in

$$N_e(4,8) = \frac{4\times 4 + 2\times 4 + 3\times 8}{16} = 3. \tag{12.51}$$

For the first mode, the minimum squared constellation distance is therefore given by $d^2_{min}(4,8) = 2\,E_s$. This is depicted in Fig. 12.7.

In order to limit the complexity of the multimode selection method, we can rely only on the post-processing SNR. Alternatively, we can additionally consider the minimum distance of the constellation by neglecting the effects of the nearest neighbors on the constellation. In this case, a suboptimal selection function is defined as follows [15]:

$$(M_{s\,opt}, i_{opt}) = \arg \max_{1\le M_s\le M_t,\ 1\le i\le \binom{M_t}{M_s}} \left[\min_{1\le m\le M_s} \gamma_m(M_s,i)\, d^2_{min}(M_s,r_c)\right]. \tag{12.52}$$

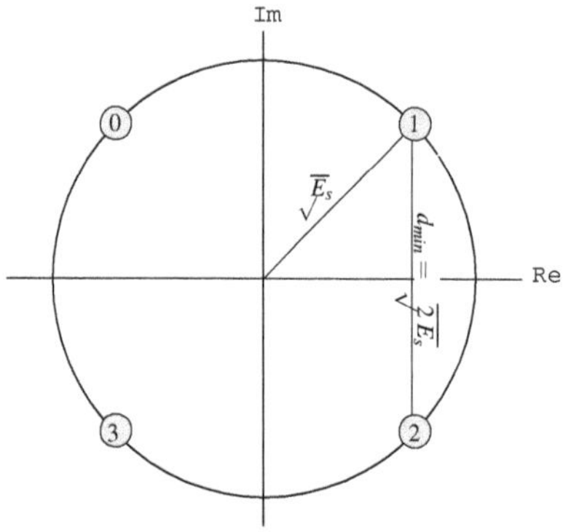

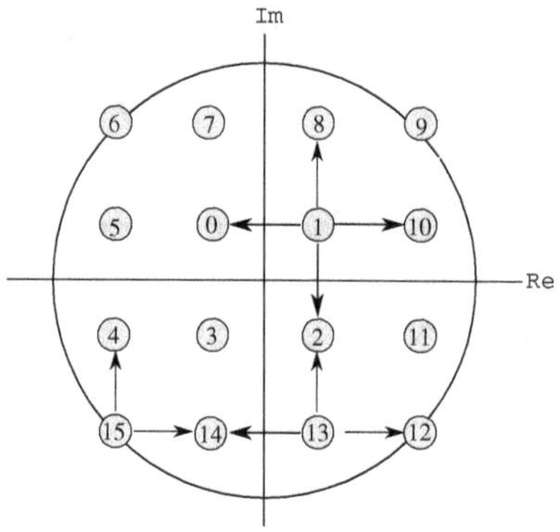

Fig. 12.7 Minimum distance of a 4-QAM constellation and average number of nearest neighbors of a 16-QAM constellation.

Note that a total of $2^{M_s} - 2$ comparisons need to be made to determine the pair $(M_{s\,opt}, i_{opt})$. It is worth mentioning that the effectiveness of this selection criterion is closely related to the accuracy of the bound given in (12.45) for the VSER. In order to further reduce the complexity of the this selection criterion, let us consider the following theorem:

Theorem 1: Let $\lambda_1(\mathbf{H}), \lambda_2(\mathbf{H}), \ldots, \lambda_{M_s}(\mathbf{H})$ be the singular values of $\mathbf{H}$ arranged in decreasing order. Given M_s, the following is true:

$$\lambda_{M_s}^2(\mathbf{H}) \geq \lambda_{M_s}^2(\mathbf{HF}_{M_s,i}) \geq \lambda_{M_t}^2(\mathbf{H}) \tag{12.53}$$

provided that $\mathbf{F}_{M_s,i}$ is full column rank, which is always the case when considering precoding matrices composed of canonical vectors as in Example 12.4. Using the upper bound given by (12.53), a suboptimum selection criterion based on the minimum singular of the equivalent channel can be derived.

This selection criterion consists in solving first for $M_{s\,opt}$ using Theorem 1 and then for i_{opt} that most closely achieves the optimum. This can be resumed in the following selection functions as

$$M_{s\,opt} = \arg \max_{1 \leq M_s \leq M_t} \lambda_{M_s}^2(\mathbf{H}) d_{min}^2(M_s, r_c) \tag{12.54}$$

and then, find the i_{opt} that solves

$$i_{opt} = \arg \max_{1 \leq i \leq \binom{M_t}{M_{s\,opt}}} \lambda_{M_s}^2(\mathbf{HF}_{M_{s\,opt},i}) \tag{12.55}$$

The main advantage of this suboptimum selection criterion is that the singular value decomposition (SVD) of $\mathbf{H}$ needs to be calculated and, at most, $\binom{M_t}{\lfloor M_t/2 \rfloor}$ minimum singular values are needed instead of $2^{M_t} - 1$.

12.4.3 Limited Feedback Unitary Precoding for SM Systems

Despite the great improvements obtained with the multimode antenna selection method, the performance is still highly limited since (i) the columns of $\mathbf{F}_i$ are restricted to the M_s columns of $\mathbf{I}_{M_t}$ and (ii) the size of the codebooks are limited by M_s and M_t. Removing these restrictions about the nature of the precoding matrices as well as the codebook size is a good way to approximate an optimal precoding performance. A better assumption is to consider that each precoder codebook is composed of *unitary matrices* with orthonormal column vectors. This orthonormal structure for the precoder is not restrictive, once the optimum precoder, when full CSI is available, also has the same structure [34].

Working with these more general precoder structures mathematically means that $\mathbf{F}_i \subset \mathscr{U}(M_t, M_s)$, where $\mathscr{U}(M_t, M_s)$ is the set of unitary matrices of dimensions $M_t \times M_s$. Note that, in the case of transmit antenna selection, the codebook contains $\binom{M_t}{M_s}$ matrices, each one being composed of M_s columns of $\mathbf{I}_{M_t}$. Since the columns of the identity matrix belong to the orthonormal set $\mathscr{U}(M_t, M_s)$, transmit antenna selection is a special case (indeed the simplest form) of unitary precoding. The use of unitary precoding along with the problem of quantizing the set of precoders in $\mathscr{U}(M_t, M_s)$ leads to the MIMO-LF unitary precoding technique. In the following is described an application of the QSA approach proposed in Section 12.2 mostly concerning the codebook design criterion.

Still considering a SM system employing a ZF receiver, the SNR associated with the m-th substream follows (12.43) and is given by

$$\gamma_m = \frac{E_s}{M_t N_0 [\mathbf{F}_i^H \mathbf{H}^H \mathbf{H} \mathbf{F}_i]^{-1}_{m,m}}. \tag{12.56}$$

Considering the SNR maximization of the worst-case substream as the optimization criterion, we can use the same bound as in (12.44) for this metric function, given by

$$\gamma_{min} \geq \lambda^2_{min}(\mathbf{H}\mathbf{F}_i) \frac{E_s}{N_0 M_s}. \tag{12.57}$$

Therefore, the selection function which maximizes the minimum SNR is given by

$$i_{opt} = \arg \max_{1 \leq i \leq N} \lambda_{min}(\mathbf{H}\mathbf{F}_i). \tag{12.58}$$

Once the problem of finding a selection function to maximize the minimum SNR is solved, a codebook design criterion has to be found based on the optimum precoder structure belonging to the set $\mathscr{U}(M_t, M_s)$. From [34], this optimum structure is given by

$$\mathbf{F}_{opt} = \mathbf{V}_R^{(M_s)}, \tag{12.59}$$

where $\mathbf{V}_R^{(M_s)}$ is composed of the M_s columns of $\mathbf{V}_R$ corresponding to the M_s greatest eigenvalues, where $\mathbf{V}_R$ is the right singular matrix obtained from the SVD decomposition of the channel $\mathbf{H} = \mathbf{V}_L \Sigma \mathbf{V}_R^H$. The same result is valid when a MMSE filter is used at the receiver and even when the optimization criterion is that of capacity maximization [34].

It is worth noting that most performance metrics only depend on the product $\mathbf{F}^H\mathbf{F}$. Using the fact that right multiplication of $\mathbf{F}$ by an unitary matrix does not change the performance metric, the latter is dependent only on the subspace spanned by the columns of $\mathbf{F}$, and not on its exact formulation. Unitary precoding for SM systems has been addressed in [23] and the main concepts concerning the codebook design can be viewed as extensions of the *Grassmannian line packing* problem for beamforming codebooks [26], which has been discussed in Section 12.3 but considering in the present case an M_s-dimensional subspace packing [24]. Our codebook is assumed to be the set of N matrices from $\mathscr{U}(M_t, M_s)$. The set of all subspaces spanned by matrices in $\mathscr{U}(M_t, M_s)$ is the complex Grassmann manifold, denoted by $G(M_t, M_s)$. The subspace packings in $G(M_t, M_s)$ consist of designing sets of N matrices that maximize the minimum subspace distance. The concept of distance between two subspaces is more complex when considering multidimensional subspaces and some metric distances can be defined depending on the optimization criterion.

For example, suppose that $\mathscr{P}_{F_i}$ is the subspace generated by the columns of $\mathbf{F}_i$, then, the *chordal distance* between $\mathscr{P}_{\mathbf{F}_1}$ and $\mathscr{P}_{\mathbf{F}_2}$ is defined as

$$d_c(\mathbf{F}_1, \mathbf{F}_2) = 1/\sqrt{2}\|\mathbf{F}_1\mathbf{F}_1^H - \mathbf{F}_2\mathbf{F}_2^H\| = \sqrt{M - \sum_{i=1}^{M} \lambda_i^2\{\mathbf{F}_1^H\mathbf{F}_2\}}, \tag{12.60}$$

and the *projection two-norm distance* between the same subspaces is written as

$$d_{proj}(\mathbf{F}_1, \mathbf{F}_2) = \|\mathbf{F}_1\mathbf{F}_1^H - \mathbf{F}_2\mathbf{F}_2^H\|_2 = \sqrt{1 - \lambda_{min}^2\{\mathbf{F}_1^H\mathbf{F}_2\}}. \tag{12.61}$$

As described in Section 12.3.1.3, a packing can be characterized by its minimum distance. In the case of SM systems, the minimum projection two-norm distance is considered, which is given by

$$\delta_{proj}(\mathscr{F}) = \min_{1 \le i \le j \le N} d_{proj}(\mathbf{F}_i, \mathbf{F}_j). \tag{12.62}$$

The goal is to define a codebook design criterion that considers the maximization of the bound for the worst-case SNR given in (12.57). This criterion is defined as follows [23]:

Codebook design criterion for unitary precoders for SM systems: Design the set of precoders $\{\mathbf{F}_i\}_{i=1}^N$ such that the corresponding codebook $\mathscr{F}_N$ maximizes $\delta_{proj}(\mathscr{F}_N)$ between any pair of codeword matrices.

Finding good packings in the Grassmann manifold for arbitrary M_t, M_s, and N, and thus, finding good codebooks is difficult. Fourier approaches exploiting the relationship between non-coherent code design in [18] and Grassmannian subspace packing were used in [24]. We refer the interested reader to this work for further details on the codebook design issue for this specific context.

12.4.4 Performance Results for MIMO-LF Precoding for SM Systems

In this section the presented LF-MIMO methods for SM systems are compared with different techniques for the same scenario assumptions. The comparisons are made in terms of the average VSER, the metric present in almost all of the performance bounds derived in this section. Each plotted VSER curve is an average over 10^6 simulations for each SNR value. The channel is assumed quasi-static, as in Section 12.3.1. The symbols are generated from a M-QAM or M-PSK constellations when the modulation order M is the function of the fixed data rate in bits per second per Hertz (bps/Hz).

Experiment 1 – Transmit antenna selection: In Fig. 12.8 is illustrated a VSER comparison of SM schemes with antenna selection. The systems use 4-QAM constellation to transmit two substreams per channel realization giving a rate of $r_c =$ 4 bps/Hz. The performance gains of 4×2 and 3×2 SM systems employing antenna selection (respectively 8 and 2 dB for average VSER of 10^{-1}) over the 2×2 open-loop V-BLAST scheme are associated with the diversity gains provided by the selection procedure.

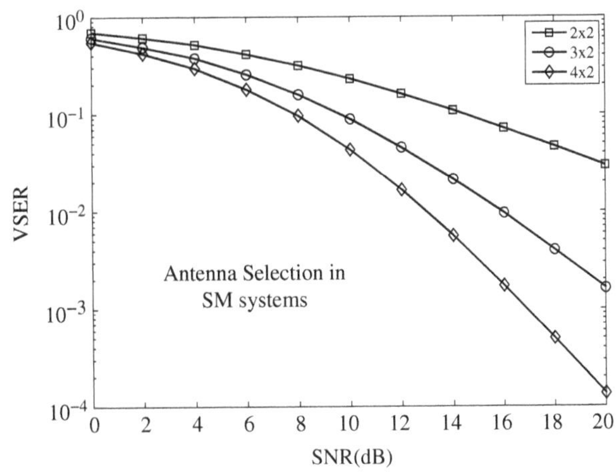

Fig. 12.8 Performance comparison of a 2×2 V-BLAST system with antenna selection.

Experiment 2 – Multimode antenna selection: This technique is an extension of that examined in Experiment 1, the difference being on the additional degree of freedom due to the adaptation of the modulation order M for a fixed data rate r_c. In this experiment, it was considered $M_t = 4$ transmit antennas and $M_r = 4$. For a fixed $r_c = 8$ bps/Hz, the following transmission schemes are used as references for comparisons: (i) $M_s = 4$ with 4-QAM (open-loop), (ii) $M_s = 2$ with 16-QAM (open-loop), and (iii) $M_s = 1$ with 256-QAM and antenna selection. Finally, for the multimode scheme $M_s = 1$, 2, or 4 and modulation orders $M = 4$, 16, or 256 with antenna selection are considered. Figure 12.9 shows that multimode presents the best VSER performance for higher SNRs.

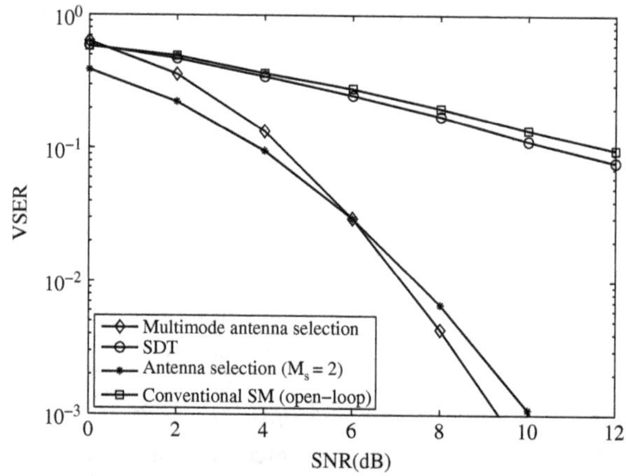

Fig. 12.9 Performance comparison of different 4×4 MIMO schemes with antenna selection and multimode transmission.

Experiment 3 – LF unitary precoding: In this experiment, the LF unitary precoding based on Grassmannian manifold with channel-adaptive techniques such as transmit antenna selection and the optimum MMSE linear precoding proposed in [34] are compared. According to Fig. 12.10, 6-bit unitary precoding presents the best performance over all LF methods considering the same antenna configuration. Note that the difference between optimum MMSE precoding and unitary precoding is only 1 dB for an average VSER of 10^{-3}.

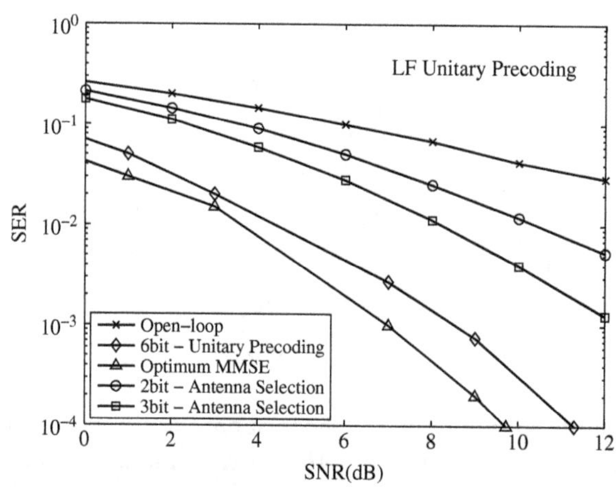

Fig. 12.10 Performance comparison of LF unitary precoding, antenna selection, and optimum MMSE precoding for a 4×4 MIMO system.

12.5 Linear Precoding for Space–Time-Coded Systems

Space–time coding has been considered as an effective MIMO signaling technique to extract transmit diversity gains without any CSI at the transmitter. Despite their great potential, the performance of orthogonal STBCs can be enhanced by adapting the transmission to the channel conditions. In this section, the application of LF transmit antenna selection and unitary precoding methods in orthogonal STBC systems is considered. The received signal in an STBC system can be written as follows:

$$\mathbf{y}[k] = \sqrt{E_s}\,\mathbf{HFS}[k] + \mathbf{n}[k], \tag{12.63}$$

where $\mathbf{S}[k]$ is a space–time codeword represented by a $M_s \times P$ matrix where P is the codeword length. The matrix $\mathbf{S}[k]$ is constructed from uncoded symbols $s_1, \dots, s_R$.

Example 12.6 (3-Antenna OSTBC). The space–time codeword $\mathbf{S}[k]$ for a STBC system employing three transmit antennas is shown below [39]:

$$\mathbf{S}[k] = \begin{bmatrix} s_1 & -s_2 & -s_3 & -s_4 & s_1^* & -s_2^* & -s_3^* & -s_4^* \\ s_2 & s_1 & s_4 & -s_3 & s_2^* & s_1^* & s_4^* & -s_3^* \\ s_3 & -s_4 & s_1 & s_2 & s_3^* & -s_4^* & s_1^* & s_2^* \end{bmatrix}. \tag{12.64}$$

In this example, it is considered $M_s = 3$, $P = 8$, and $R = 4$.

The following section presents some techniques covered by this model when applied to STBCs.

12.5.1 Transmit Antenna Selection for STBC Systems

The transmit antenna selection principle used here is the same as the one presented in Section 12.4.1 for SM systems. We are interested in choosing a subset of M_s among M_t transmit antennas that optimizes some performance metric. The difference is that the performance metric considered when STBC is employed is not the same as those considered in SM systems.

The received SNR in an orthogonal STBC system given by [1], [39]:

$$\gamma = \frac{E_s}{N_0}\frac{\|\mathbf{H}_i\|^2}{M_s}, \tag{12.65}$$

and the error probability is bounded using the Chernoff bound as follows [31]:

$$Prob\{s_r \neq \hat{s}_r|\mathbf{H}\} \leq exp\{-d_{min}^2\|\mathbf{H}_i\|_F^2\}. \tag{12.66}$$

Each s_r is an uncoded symbol which comprises the space–time codeword $\mathbf{S}[k]$ and $\mathbf{H}_i = \mathbf{F}_i\mathbf{H}$ is the equivalent channel after the antenna subset selection. From (12.65) and (12.66) it can be concluded that maximizing the Frobenius norm of the equivalent channel is equivalent to maximizing the effective SNR, i.e., to minimizing the instantaneous probability of error [13].

Considering the same codebook structure as in Section 12.4.1, a selection function can be derived which minimizes the bound on the error probability in (12.66) and maximizes the received SNR in (12.65). The selection function concerning the SNR maximization is then given by

$$i_{opt} = \arg \max_{1 \leq i \leq \binom{M_t}{M}} \|\mathbf{H}_i\|_F^2 = \arg \max_{1 \leq i \leq \binom{M_t}{M}} \|\mathbf{H}\mathbf{F}_i\|_F^2. \tag{12.67}$$

12.5.2 Unitary Linear Precoding for STBC Systems

Similarly to the SM system studied in Section 12.4.3, in this section the design of the precoder codebook for STBC systems with unitary precoding is formalized. As it has been shown in Section 12.5.1, the received SNR (12.65) and the bound on the error probability (12.66) provide us with the same selection function (12.67) for a given codebook. Therefore, the difference between the techniques will be on the precoder structure, which is now given by $\mathbf{F}_{opt} = \mathbf{V}_R^{(M_s)}$, i.e., the same as the one used in unitary precoded SM systems. Consequently, a different codebook design criterion has to be proposed. Again, the codebook design criterion is based on the maximization of the minimum distance between the subspaces generated by the precoding matrices. However, as we have a different selection function, the distance metric used for subspace packing is not the same. In the case of unitary precoding for STBC the chordal distance must be considered, which is given by (12.60) instead of the projection two-norm. The subspace packing can be characterized by its minimum chordal distance, given by

$$\delta_c(\mathscr{F}_N) = \min_{1 \leq i \leq j \leq N} d_c(\mathbf{F}_i, \mathbf{F}_j), \tag{12.68}$$

leading to a new codebook design criterion as stated below:

Codebook design criterion for unitary precoders for STBCs: Design the set of codebook precoders $\{\mathbf{F}_i\}_{i=1}^N$ such that the corresponding codebook $\mathscr{F}_N$ maximizes $\delta_c(\mathscr{F}_N)$ between any pair of codeword matrices.

12.5.3 Performance Results for MIMO-LF OSTBC Systems

Experiment 1 – OSTBC with antenna selection: In this first experiment, an 8-PSK modulation with 3-antenna OSTBC is considered which gives us a data rate of r_c =1.5 bps/Hz. In Fig. 12.11 is illustrated a diversity advantage when one more transmit antenna is available.

Experiment 2 – SBTC G2 with unitary precoding: Now, STBC is considered in a 4×2 MIMO system using 4-QAM constellation,which gives r_c =2 bps/Hz. From

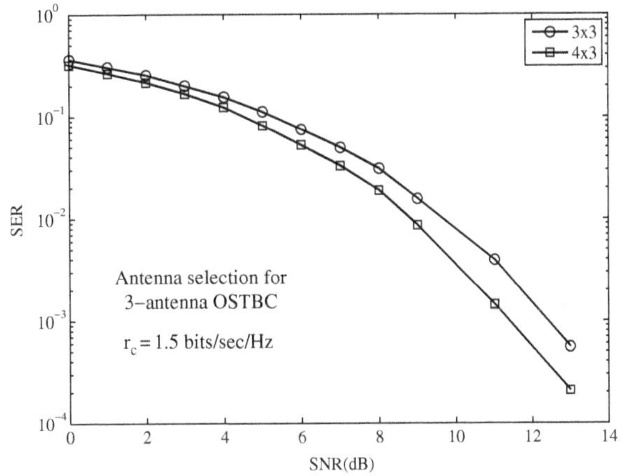

Fig. 12.11 Performance comparison of 3-antenna OSBTC with antenna selection.

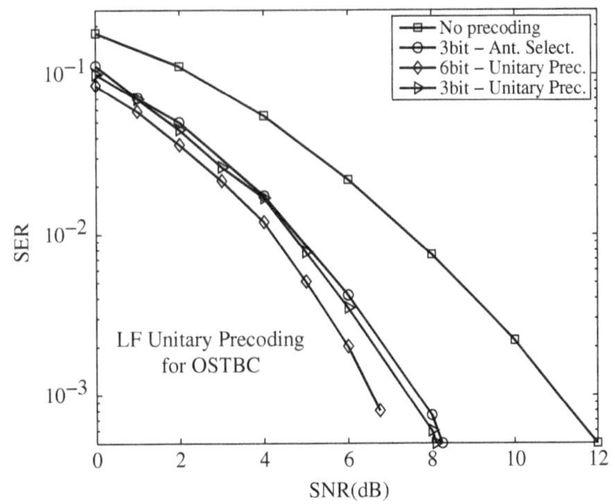

Fig. 12.12 Performance comparison of OSBTC for a 4 × 2 MIMO system with unitary precoding and antenna selection, reproduced with permission from Love, D.J., Heath, R.W. Jr., Limited feedback unitary precoding for orthogonal space–time block codes. IEEE Trans. Signal Process. **53**(1), 64–73 (2005) © 2005 IEEE.

Fig. 12.12, a gain of 3 dB for an average SER of 10^{-3} can be observed. For this system, antenna selection requires $B = \lceil \log_2(6) \rceil = 3$ bits. Note that, by relaxing the constraints concerning the precoding structure and constructing codebooks based on Grassmann manifold (LF unitary precoding), a coding gain of 0.3 dB is obtained over antenna selection with the same 3 bits. Finally, a 6-bit-based codebook provides an array gain of 0.7 dB over antenna selection.

12.6 Tensor-Based Space–Time Precoding (TSTP)

The previous section described the linear precoding methods for STBCs and SM systems relying on the use of limited feedback. Recall that the common feature of these techniques is that they rely on the same linear precoding model given by (12.42), the difference being on the structure assumed by the precoder $\mathbf{F}$, which can be either a transmit antenna selection matrix or a unitary precoding matrix. In this section is proposed a new space–time precoding model that consists of introducing an additional coding matrix **W** in the linear precoding model (12.42). The columns of this matrix are orthogonal codes that can be associated either with the different transmit antennas or with the number of substreams, as will be detailed later. The matrix **W** is called the *space–time precoder*.

The introduction of the orthogonal space–time precoder **W** leads to a tensorial model for the received signal, which can be formulated as a *constrained* PARAFAC model. The conventional PARAFAC tensor model was considered in Chapter 11 for modeling a space–time–frequency multiple-access MIMO system. Here, a modified version of this tensor model is exploited for designing an LF-based MIMO transceiver. For a background on PARAFAC modeling, we refer the interested reader to Chapter 11.

It is worth mentioning that the distinguishing features of the tensor-based space–time precoding (TSTP) approach are

- The TSTP approach covers hybrid schemes mixing spatial multiplexing and transmit diversity, i.e., it allows a selection of multiple transmit antennas for the same substream for obtaining transmit diversity gains.
- Array gains can be achieved by selecting an appropriate structure for the space–time precoding matrix **W**.
- Blind detection is possible from the exploitation of the tensorial structure of the received signal. This avoids the use of bandwidth-consuming training signals for channel estimation.

Note that, thanks to the blind detection capability of the TSTP approach, we overcome the common assumption made in most research studies that perfect channel knowledge is available at the receiver, and the performance of the proposed MIMO-LF techniques can be assessed using an actual blind channel estimate.

Two TSTP systems are now presented. The first one is a *space–time multiplexing (STM)* system consisting of a combination of spatial multiplexing with temporal dispersion coding. The second system covers *space–time spreading (STS)*, where each substream can be transmitted by more than one transmit antenna for achieving transmit spatial diversity. Both TSTP systems will be formulated using the tensor formalism, and a space–time precoder based on transmit antenna selection is presented for each system.

12.6.1 Space–Time Multiplexing (STM) System

In the STM system, a set of M_s among M_t transmit antennas are selected to transmit the M_s data substreams. A different temporal dispersion code with length P symbols is associated with each transmit antenna so that the M_s-transmitted substreams are orthogonal in the space domain. This transmission model describes therefore an orthogonal space-division multiplexing approach, with the added feature of temporal coding and transmit antenna selection.

The received signal model of the STM system can now be stated according to

$$\mathbf{y}[k,p] = \sqrt{E_s}\,\mathbf{HF}D_p(\mathbf{W})\mathbf{s}[k] + \mathbf{v}[k,p], \tag{12.69}$$

where $D_p(\mathbf{W})$ is a diagonal matrix constructed from the pth row of $\mathbf{W}$, of dimension $P \times M_s$, and $\mathbf{F}$ is a transmit antenna selection matrix formed by choosing M_s columns of the identity matrix $\mathbf{I}_{M_t}$ (as in conventional transmit antenna selection for SM systems). Note that the received signal and noise vectors now depend on two indexes, the symbol period index p and the data block index k. Furthermore, one of the differences between the classical space–time block code paradigm and our tensor-based one is in the detection procedure. While in STBC the detection is made for each stream during the P periods, in TSTP it is made in blocks, storing the symbol vectors and their redundancy copies during KP periods.

It is assumed that each data block contains P symbol periods. Note that, for $P = 1$ and $\mathbf{W} = [1, 1, \ldots, 1]$, model (12.69) reduces to the classical linear precoding model for SM systems defined in (12.42).

Regarding the space–time precoder structure, the following two requirements are introduced:

- $\mathbf{W}$ is full rank.
- $\mathbf{W}\mathbf{W}^H = \mathbf{I}_P$.

Both requirements are fulfilled by the following Vandermonde matrix:

$$[\mathbf{W}]_{p,m} \doteq exp[j2\pi(p-1)(m-1)/M_s]. \tag{12.70}$$

With this choice, $\mathbf{W}$ is a full-rank semi-unitary matrix, as required for maximum diversity gain. Moreover, the temporal coding gain P by simple truncation can be controlled using the Vandermonde structure.

12.6.1.1 Tensor Modeling

In order to rewrite (12.69) in the tensorial form, let us define

$$\mathbf{Y}_{\cdot\cdot p} = \left[\mathbf{y}[1,p], \ldots, \mathbf{y}[K,p]\right] \tag{12.71}$$

as an $M_r \times K$ matrix collecting the received signal over K data blocks for a fixed symbol period p of each data block. Define also

$$\mathbf{S} = \left[\mathbf{s}[1], \ldots, \mathbf{s}[K]\right] \tag{12.72}$$

as an $M_s \times K$ matrix collecting the M_s substreams. Note that each data block transmits a different substream. From these definitions, we have

$$\mathbf{Y}_{\cdot\cdot p} = \sqrt{E_s}\,\mathbf{HF}D_p(\mathbf{W})\mathbf{S} + \mathbf{V}_{\cdot\cdot p}, \tag{12.73}$$

where $\mathbf{V}_{..p}$ is the $M_r \times K$ noise signal matrix associated with the pth symbol period. This received signal model can be interpreted as the pth matrix slice of a trilinear PARAFAC decomposition. By comparing (12.73) with (11.4), from which can be concluded that $\mathbf{A}^{(1)}, \mathbf{A}^{(2)}$, and $\mathbf{A}^{(3)}$ correspond to $\mathbf{HF}$, $\mathbf{S}^T$, and $\mathbf{W}$, respectively. Now, stacking row-wise the matrix slices $\mathbf{Y}_{..1}, \ldots, \mathbf{Y}_{..P}$:

$$\mathbf{Y}_2 = \begin{bmatrix} \mathbf{Y}_{..1} \\ \vdots \\ \mathbf{Y}_{..P} \end{bmatrix} = \begin{bmatrix} \mathbf{HF}D_1(\mathbf{W}) \\ \vdots \\ \mathbf{HF}D_P(\mathbf{W}) \end{bmatrix} \mathbf{S} = (\mathbf{W} \diamond (\mathbf{HF}))\mathbf{S} + \mathbf{V}_2 \in \mathbb{C}^{PM_r \times K}, \tag{12.74}$$

where $\diamond$ is the Khatri–Rao (column-wise Kronecker) product and $\mathbf{V}_2$ is a noise matrix formed by concatenating $\mathbf{V}_{..1}, \ldots, \mathbf{V}_{..P}$ in a similar manner as for $\mathbf{Y}_2$. Using (11.5), two other Khatri–Rao factorizations representing the received signal can be obtained by

$$\mathbf{Y}_1 = (\mathbf{S}^T \diamond \mathbf{W})\mathbf{F}^T\mathbf{H}^T + \mathbf{V}_1 \in \mathbb{C}^{KP \times M_r}, \tag{12.75}$$

$$\mathbf{Y}_3 = ((\mathbf{HF}) \diamond \mathbf{S}^T)\mathbf{W}^T + \mathbf{V}_3 \in \mathbb{C}^{M_r K \times P}, \tag{12.76}$$

where the noise matrices $\mathbf{V}_1$ and $\mathbf{V}_3$ have the same dimensions as $\mathbf{Y}_1$ and $\mathbf{Y}_3$, respectively. As described in Chapter 11, $\mathbf{Y}_1$, $\mathbf{Y}_2$, and $\mathbf{Y}_3$ are different matrix "unfoldings" of the received signal tensor. They contain the same information organized in different manners. The same is valid for $\mathbf{V}_1$, $\mathbf{V}_2$, and $\mathbf{V}_3$.

12.6.1.2 Transmit Antenna Selection for the STM System

This section presents a suboptimum per-stream antenna selection method for optimizing the structure of $\mathbf{F}$ in the STM system. The methods consists in allocating the best transmit antenna among M_t for the first substream, then allocating the second best among $M_t - 1$ antennas for the second substream, and so on. Let us suppose that the rth substream is to be transmitted by the ith transmit antenna. For the first substream ($r = 1$), the minimum singular value of the effective channel $\mathbf{h}_i = \mathbf{Hf}_i$, $i = 1, \ldots, M_t$, where $\mathbf{f}_i = \mathbf{e}_i^{M_t}$ is the selection diversity vector from Section 12.3.1.1, is computed. The index corresponds here to the maximum singular value being chosen, following the same idea as in [17]. Denoting $i_{opt}^{(1)}$ the optimum antenna index for the first substream, the selection criterion can be stated as follows:

$$i_{opt}^{(1)} = \arg \max_{1 \le i \le M_t} \lambda_{\min}^2(\mathbf{h}_i). \tag{12.77}$$

After the determination of $i_{opt}^{(1)}$, the same selection procedure is repeated for the second substream but now excluding the index $i_{opt}^{(1)}$ that corresponds to the first selected transmit antenna. In the general case, the selection criterion for the rth substream solves the following problem:

$$i_{opt}^{(r)} = \arg \max_{1 \leq i \leq M_t, \quad i^{(r)} \neq i_{opt}^{(r-1)}} \lambda_{\min}^2(\mathbf{h}_i), \tag{12.78}$$

where $r = 2, \ldots, M_s$. Although this method performs a per-stream transmit antenna selection, the number of feedback bits is the same as for the classical transmit antenna selection method described in Section 12.4.1, and is given by

$$B_{\text{STM}} = \left\lceil log_2 \binom{M_t}{M_s} \right\rceil. \tag{12.79}$$

12.6.2 Space–Time Spreading (STS) System

The STM system presented in the previous section does not cover transmit spatial diversity since each data substream is allocated to only one transmit antenna. In the following, the STS system is presented, in which a given data substream can be allocated to multiple transmit antennas in order to achieve transmit spatial diversity gains. Moreover, different substreams can be allocated to a different number of transmit antennas, which allows a sort of spatial diversity control for each substream. In order to ensure orthogonality in the space–time domain (as in a classical STBC system), each substream is associated with a different orthogonal code, where the orthogonal codes are columns of the space–time precoding matrix **W**.

The received signal model of the STS system is quite similar to that of the STM system, and is given by

$$\mathbf{y}[k,p] = \sqrt{E_s}\,\mathbf{H}D_p(\mathbf{W})\mathbf{F}\mathbf{s}[k] + \mathbf{v}[k,p], \tag{12.80}$$

where **W** is of dimension $P \times M_t$ while **F** is of dimension $M_t \times M_s$. The structure of **W** is the same as that defined in (12.70) for the STM system, except that **W** is now composed of M_s (instead of M_t) orthogonal columns.

On the other hand, the structure of **F** differs from the one assumed in the previous section for the STM system. In the STS system, **F** satisfies the following assumptions:

- The rows of **F** are unit vectors belonging to the canonical set $\{\mathbf{e}_1^{(M_s)}, \ldots, \mathbf{e}_{M_s}^{(M_s)}\}$.
- **F** can be divided into M_s orthogonal submatrices, each one containing α_r equal rows, so that

$$\mathbf{F}^T\mathbf{F} = \text{diag}(\alpha_1, \ldots, \alpha_{M_s}), \quad \text{with} \quad \sum_{r=1}^{M_s} \alpha_r = M_t, \tag{12.81}$$

where α_r is the spatial spreading factor of the rth data substream and defines the number of transmit antennas used to simultaneously transmit the rth substream. As an example, for $M_t = 3$ and $M_s = 2$, a possible configuration for **F** is

$$\mathbf{F} = \begin{bmatrix} 1 & 0 \\ 1 & 0 \\ 0 & 1 \end{bmatrix}. \tag{12.82}$$

In this case, $\alpha_1 = 2$ and $\alpha_2 = 1$. The first substream is transmitted by antennas 1 and 2, while the second substream is transmitted by antenna 3. As it will be discussed later, the goal is to optimize $\mathbf{F}$ for a fixed M_t, M_s, and $\alpha_1, \ldots, \alpha_{M_s}$.

To summarize, it is worth noting that the received signal models (12.69) and (12.80) are quite similar, the main differences being on (i) the position of $\mathbf{F}$ with respect to $\mathbf{W}$ and (ii) the structure of $\mathbf{F}$ which now possibly contain repeated rows.

12.6.2.1 Tensor Modeling

Similarly to the STM system, a matrix-slice representation for the received signal can be obtained as

$$\mathbf{Y}_{\cdot\cdot p} = \sqrt{E_s}\,\mathbf{H}D_p(\mathbf{W})\mathbf{FS} + \mathbf{V}_{\cdot\cdot p}. \tag{12.83}$$

Note the symmetry between models (12.73) and (12.83) with respect to $\mathbf{W}$ and $\mathbf{F}$. This received signal model is also a trilinear PARAFAC decomposition with $\mathbf{A}^{(1)} \to \mathbf{H}$, $\mathbf{A}^{(2)} \to (\mathbf{FS})^T$, and $\mathbf{A}^{(3)} \to \mathbf{W}$. Analogous to (12.75) and (12.76), the three following Khatri–Rao factorizations are obtained to express the received signal tensor:

$$\mathbf{Y}_1 = \big((\mathbf{FS})^T \diamond \mathbf{W}\big)\mathbf{H}^T + \mathbf{V}_1, \tag{12.84}$$

$$\mathbf{Y}_2 = \big(\mathbf{W} \diamond \mathbf{H}\big)\mathbf{FS} + \mathbf{V}_2, \tag{12.85}$$

$$\mathbf{Y}_3 = \big(\mathbf{H} \diamond (\mathbf{FS})^T\big)\mathbf{W}^T + \mathbf{V}_3. \tag{12.86}$$

Again, by comparing (12.75) and (12.76) with (12.84)–(12.86), the basic difference between STM and STS models is in the position of the transmit antenna selection matrix $\mathbf{F}$, which is coupled to the channel matrix $\mathbf{H}$ in the first model and to symbol matrix $\mathbf{S}$ in the second one.

12.6.2.2 Transmit Antenna Selection for the STS System

This section performs a transmit antenna selection on a per-stream basis, but now selecting, for the rth substream, the best *subset* of α_r transmit antennas, $r = 1, \ldots, M_s$. Recall that $\mathbf{F}$ is associated with a unique set $\{\alpha_1, \ldots, \alpha_{M_s}\}$ (up to column permutation), where α_r defines the spatial spreading factor of the rth substream. The method consists of allocating α_1 antennas among M_t to the first substream, then allocating α_2 antennas among $M_t - \alpha_1$ to the second substream, then allocating α_3 antennas among $M_t - \alpha_1 - \alpha_2$ to the third substream, and so on.

Let us suppose that the rth substream is to be transmitted by the ith antenna subset composed of α_r transmit antennas. Let us define

$$N_r = \binom{M_t - \sum_{k=1}^{r-1} \alpha_k}{\alpha_r} \tag{12.87}$$

as the number of possible different antenna subsets for the rth substream. The selection criterion is quite similar to that of the STM system. For the first substream ($r = 1$), the minimum singular value of the effective channel is evaluated by $\bar{\mathbf{H}}_{i_r} = \mathbf{H}\mathbf{F}_{i_r}$, where $i_r = 1, \ldots, N_r$ and $\mathbf{F}_{i_r}$ is formed from a given row of the identity matrix $\mathbf{I}_{M_t}$ repeated α_r times. Let $i_{opt}^{(1)}$ be an index denoting the optimum transmit antenna subset selected to transmit the first substream. For instance, if it is assumed that $M_t = 3$, $M_s = 2$, $\alpha_1 = 2$, and $\alpha_2 = 1$, $\mathbf{F}$ can be chosen from the following codebook:

$$\mathscr{F} = \left\{ \begin{bmatrix} 1 & 0 \\ 1 & 0 \\ 0 & 1 \end{bmatrix}, \begin{bmatrix} 1 & 0 \\ 0 & 1 \\ 1 & 0 \end{bmatrix}, \begin{bmatrix} 0 & 1 \\ 1 & 0 \\ 1 & 0 \end{bmatrix} \right\}. \tag{12.88}$$

A convention can be defined such that $i_{opt}^{(1)}$ specifies an ordered subset of transmit antennas. In this example, the different subsets can be ordered as $\{(1,2),(1,3),(2,3)\}$, to denote the selection of a given pair of transmit antennas. For instance, if $i_{opt}^{(1)} = 1$, the subset $(1,2)$ is selected.

The following selection criterion is used:

$$i_{opt}^{(1)} = \arg \max_{1 \le i \le N_r} \lambda_{\min}^2(\hat{\mathbf{H}}_{i_1}). \tag{12.89}$$

The same selection procedure is repeated for the second substream but now excluding the subset of transmit antennas specified by $i_{opt}^{(1)}$, already selected for the first substream. In the general case, the selection criterion for the rth substream solves the following problem:

$$i_{opt}^{(r)} = \arg \max_{1 \le i \le N_r, \quad i^{(r)} \ne i_{opt}^{(r-1)}} \lambda_{\min}^2(\hat{\mathbf{H}}_{i_r}), \tag{12.90}$$

where $r = 2, \ldots, M_s$.

In the STS system, the number of feedback bits required to perform antenna selection by choosing the mappings $i_{opt}^{(1)}, \ldots, i_{opt}^{(M_s)}$ is given by

$$\mathrm{B_{STS}}(\alpha_1, \ldots, \alpha_{M_s}) = \sum_{r=1}^{M_s} \left\lceil log_2 \binom{M_t - \sum_{k=1}^{r-1} \alpha_k}{\alpha_r} \right\rceil. \tag{12.91}$$

Note that this transmit antenna selection algorithm works sequentially on a per-substream basis. This means that the substream that is first allocated can benefit from all the spatial degrees of freedom and generally presents the best link performance, since the best transmit antenna subset will always be selected for it. The subsequent substreams will be sequentially allocated to the remaining transmit antennas. Since the detection performance is always limited, or upper-bounded, by the worst substream (i.e., the one with the minimum received SNR), it is worth noting that it is important to "schedule" the substreams-to-antenna allocations according to their increasing spatial spreading factor, i.e., $\alpha_1 \leq \cdots \leq \alpha_{M_s}$. In other words, substreams accessing fewer transmit antennas should be allocated first, since they will potentially have a lower SNR at the receiver. This simple ordering policy is adopted here.

12.6.3 Blind Receiver Processing

As already mentioned in previous section, the tensor modeling approach, adopted to model the two TSTP systems in the previous sections, has the attractive feature of allowing a joint blind channel estimation and symbol detection. Consequently, a full exploitation of the TSTP approach consists in performing LF transmit antenna selection using the actual blind channel estimate. That is, the modeling framework used for the STM and STS systems targets the complete transceiver design. On one hand, it models transmitter processing by means of precoding. On the other hand, it leads to a blind receiver processing with channel estimation capabilities.

In the following, a simple blind receiver algorithm that exploits the tensor modeling of the STM system is presented. It is worth to highlight that the receiver for the STS system follows the same algorithm and can be derived in a similar way. The receiver is based on the alternating least-squares (ALS) algorithm [4]. This algorithm has also been used in Chapter 11 for blind detection in the context of a different transceiver model. In our case, this algorithm exploits the two Khatri–Rao factorizations (12.74) and (12.75) for an alternate estimation of the channel and symbol matrices, respectively, through optimization of the following least-squares criteria:

$$\widehat{\mathbf{S}} = \underset{\mathbf{S}}{\operatorname{argmin}} \left\| \mathbf{Y}_2 - \left(\mathbf{W} \diamond (\mathbf{HF}) \right) \mathbf{S} \right\|^2, \tag{12.92}$$

$$\widehat{\mathbf{H}} = \underset{\mathbf{H}}{\operatorname{argmin}} \left\| \mathbf{Y}_1 - \left(\mathbf{S}^T \diamond \mathbf{W} \right) \mathbf{F}^T \mathbf{H}^T \right\|^2. \tag{12.93}$$

Note that $\mathbf{F}$ and $\mathbf{W}$ are known at both transmitter and receiver and the unknown matrices being only $\mathbf{S}$ and $\mathbf{H}$.

It is worth mentioning that the idea of alternating between the estimation of symbol and channel matrices is similar to that proposed in [38]. In our case, uniqueness of the estimated parameters arises due to the exploitation of the tensor structure of the received signal. On the other hand, [38] ensures uniqueness by exploiting the finite-alphabet property of the digitally modulated symbols.

The ALS receiver algorithm is summarized in Algorithm 12.1. The receiver algorithm is composed of two stages. The first operates in an open-loop mode without the presence of the transmit antenna selection matrix, by assuming $M_s = M_t$ and $\mathbf{F} = \mathbf{I}_{M_t}$. The main purpose of this stage is to allow the blind estimation of the full MIMO channel. Based on this channel estimate, the receiver determines the optimum $\mathbf{F}$ and

Algorithm 12.1 Updating ALS Algorithm.

Initialization: $t = 0$; Randomly initialize $\widehat{\mathbf{H}}_{(0)}$;
for all t **do**
repeat
$\widehat{\mathbf{S}}^T_{(t)} \leftarrow \left[\mathbf{W} \diamond (\mathbf{H}_{(t-1)}\mathbf{F})\right]^{\dagger}\mathbf{Y}_2$
$\widehat{\mathbf{H}}^T_{(t)} \leftarrow \left[(\mathbf{S}^T_{(t)} \diamond \mathbf{W})\mathbf{F}^T\right]^{\dagger}\mathbf{Y}_1$
until convergence.
end for

feeds this choice back to the transmitter. In the second stage, the system operates in a closed-loop mode using transmit antenna selection. At the receiver, symbol detection is based on the blind channel estimate performed in the first stage. The same process is then started over for the transmission of subsequent blocks of data.

It is worth mentioning that the closed-loop operation of this receiver assumes that the channel matrix be stationary during the two processing stages. In high-mobility scenarios where the channels are subject to a time-variation, the data block size should be reduced in order to avoid performance degradation due to a possibly outdated channel information at the transmitter.

12.6.4 Simulation Results

In the following, the BER performance of the two TSTP systems are evaluated by means of computer simulations. The objective of the following results is twofold. First, the impact of closed-loop transmission is evaluated, by means of LF transmit antenna selection, on the performance of both STM and STS systems. Second, the "channel-blind" performance of these systems is considered, where the tensor modeling based ALS receiver is used to provide the blind channel estimate to be used in the transmit antenna selection algorithm.

Each plotted BER curve is an average over 5×10^4 Monte Carlo runs. At each run, the transmitted symbols are drawn from a pseudo-random quaternary phase shift keying (QPSK) sequence. The flat-fading channel is assumed to be constant during the transmission of a data block of $N = 50$ transmitted symbols. The BER curves represent the performance averaged over the M_s transmitted substreams and are plotted as a function of an overall SNR measure. For the STM system, this SNR measure is given by (c.f. (12.75))

$$\mathrm{SNR_{dB}} = 10log_{10}\left(\frac{\|(\mathbf{S}^T \diamond \mathbf{W})\mathbf{F}^T\mathbf{H}^T\|^2}{\|\mathbf{V}_1\|^2}\right).$$

For the STS system, the SNR measure obeys a similar expression. At each run, the elements of the noise matrix $\mathbf{V}_1$ are circularly symmetric complex Gaussian random variables with variance $\sigma_v = 10^{(-\mathrm{SNR_{dB}}/10)}$.

In a first experiment, we are interested in evaluating the BER performance of STM and STS systems with channel-blind detection. In order to study the performance gains of an LF-based transceiver design, both open-loop and closed-loop modes are considered. In the open-loop mode, the substreams are randomly assigned to the transmit antennas at each run, while in the closed-loop mode, the LF transmit antenna selection algorithms described in the previous section are applied based on the blindly estimated channel (c.f. Section 12.6.3). Assume that $M_t = 3$, $M_r = 2$, $M_s = 2$, and $P = 3$. For the STS system, $\alpha_1 = 1$ and $\alpha_2 = 2$ are considered, meaning that the first substream is transmitted by a single transmit antenna while the second substream is spread across two transmit antennas to achieve transmit spatial diversity.

Figure 12.13(a) depicts the results for the STM system while Fig. 12.13(b) is valid for the STS system. First of all, in both figures it can be observed that the use of closed-loop transmission by means of LF transmit antenna selection leads to substantial performance gains over open-loop transmission. For a target BER of 10^{-3}, the closed-loop STM system has an SNR gain of more than 4 dB. For the same target BER, the closed-loop STS system has an SNR gain of nearly 6 dB over its open-loop counterpart. From Fig. 12.13(a) and (b), the closed-loop performance of STM and STS systems can be compared. According to these figures, the STM system achieves a BER of 10^{-3} with an SNR of 11 dB, approximately, while the STS system achieves the same BER with less than 9 dB of SNR. Such an SNR gain provided by the STS system with respect to the STM one corroborates the spatial diversity gains achieved with STS system due to spatial spreading of the second substream which is not present in the STM system.

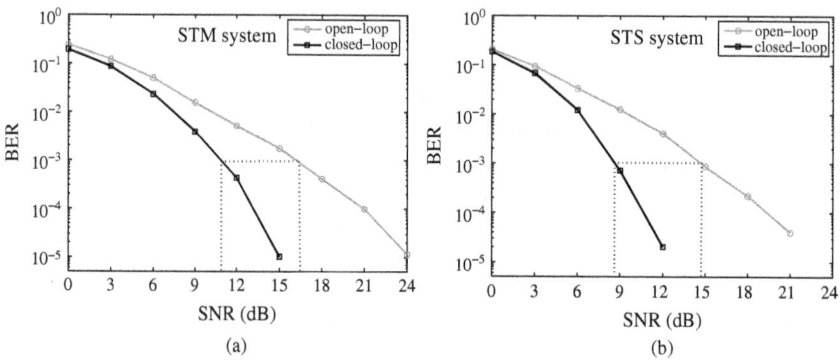

Fig. 12.13 BER performance of the (a) STM system with $M_t = 3$ and $M_r = 2$ and (b) STS system with $M_t = 3$ and $M_r = 2$ using open-loop and closed-loop transmission.

In the next experiment, the closed-loop performance of the channel-blind STS system is evaluated under three different transmit schemes. Assume that $M_t = 4$, $M_r = 3$, and $P = 3$. In the first scheme, $M_s = 2$ and $(\alpha_1, \alpha_2) = (1, 3)$. In the second scheme, $M_s = 2$ and $(\alpha_1, \alpha_2) = (2, 2)$. The third scheme considers $M_s = 3$ and $(\alpha_1, \alpha_2, \alpha_3) = (1, 1, 2)$. Note that both schemes 1 and 2 transmit two substreams

Fig. 12.14 Performance of three different STS systems with $M_t = 4$ and $M_r = 3$.

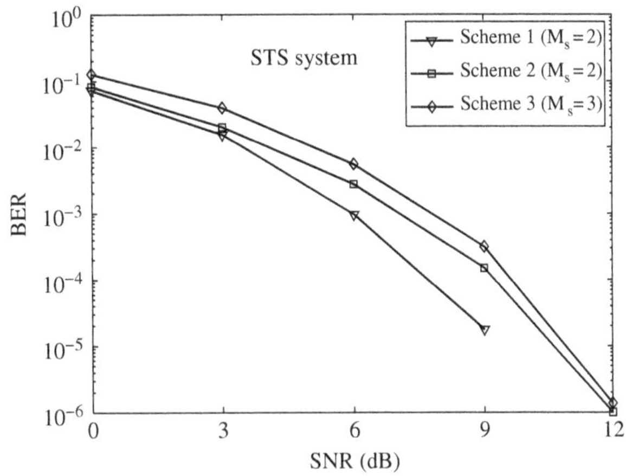

but they differ in partitioning of the transmit antennas. While scheme 1 allocates one antenna to the first substream and three antennas to the second substream, the scheme 2 allocates two antennas to each substream. In scheme 3, an additional substream has to be transmitted, thus leading to a higher spectral efficiency. According to Fig. 12.14, the performance degrades when going from scheme 1 to scheme 3, which is in accordance with our expectations. Note that the transmit spatial diversity gain of the second substream is higher in scheme 1 compared to scheme 2, since the second substream is spread across three transmit antennas in scheme 1 while only two transmit antennas are used in scheme 2 for the second substream. This explains the performance gain of scheme 1 over scheme 2. On the other hand, scheme 3 has the worst performance since an additional substream is present and less spatial degrees of freedom are available for its transmission in this case.

As previously emphasized, a distinguishing feature of the TSTP approach to transceiver design based on tensor modeling is that a blind estimate of the MIMO

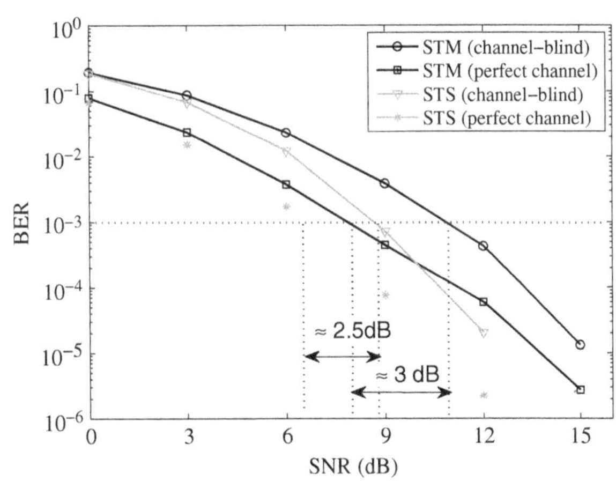

Fig. 12.15 Blind channel estimate versus perfect channel knowledge: Impact on the performance of STM and STS systems when operating in closed loop.

channel estimation can be performed at the receiver, which is subsequently used for LF transmit antenna selection. Figure 12.15 compares the closed-loop performance of STM and STS systems using either blind channel estimate or perfect channel knowledge. For both systems, it is assumed that $M_t = 3$, $M_r = 2$, $M_s = 2$, and $P = 3$. It is worth noting that, for a target BER of 10^{-3}, the channel-blind STM system presents an SNR loss of nearly 3 dB compared to the perfect channel case. Regarding the STS system, and for the same target BER, the gap between channel-blind and perfect channel-based transceivers is approximately 2.5 dB. Such a loss of performance is expected and is naturally due to the channel estimation errors of the blind receiver algorithm.

12.7 Conclusions and Research Directions

Channel-adaptive MIMO techniques exploiting LF channel information have attracted significant interest over the past few years, due to their great potential in enhancing link reliability while coping with practical requirements of upcoming fourth generation (4G) wireless communication systems. This chapter has provided an overview of the state-of-the-art on the main LF methods used in wireless transceiver design. A roadmap to the MIMO-LF research has been presented by showing how the most popular MIMO techniques can benefit from the CSI at the transmitter. Some quantization issues that are important to LF-based transceiver designs have been briefly presented. Some codebook design criteria for beamforming, SM systems, and STBC strategies has also been discussed. The performance of competing MIMO-LF techniques have been evaluated by means of simulation experiments. In this chapter, a new tensor modeling approach was proposed to MIMO-LF transceiver design. Based on the TSTP approach, two MIMO systems were presented namely STM and STS systems and an LF-based transmit antenna selection algorithm has been proposed for each system. Due to the blind channel estimation capability of the TSTP approach, a channel-blind-based LF transmit adaptation can be performed. This allowed us to access the performance of STM and STS systems using actual blind channel estimate.

In this chapter, we have restricted ourselves to a time-invariant channel for simplicity reasons, since our main objective is to understand the basic concepts associated with this research area. It is worth mentioning, however, that in more practical time-varying channel conditions, LF methods can present a significant performance degradation. Another practical issue affecting the performance of LF communications is the presence of errors and delays in the feedback control channel. These issues deserve further investigation and should be addressed in future works. In this sense, the adoption of an adequate time-varying channel model is of key importance. Regarding the proposed TSTP modeling approach, we should mention that, although we have evaluated the impact of actual blind channel estimation on the closed-loop performance of the STM and STS systems, a more complete and realistic performance evaluation should take into account the impact of errors and delays on the feedback channel. The TSTP modeling approach can also be generalized in order to cover more MIMO transmit schemes. For instance, both STM and STS systems do not allow different substreams to share the same transmit antenna for transmission. In this context, it would be interesting to incorporate this feature into these systems since more degrees of freedom would be available to design optimized closed-loop MIMO transceivers.

For future work, it remains to evaluate the link performance of MIMO-LF transceivers (including the TSTP one) when retransmission schemes and outer-coding/interleaving are taken into account. Finally, we highlight that the study of all these LF methods in a multiuser scenario is of fundamental importance. In this context, the investigation of new *multiuser feedback* methods that take into account users' quality-of-service requirements in the feedback design is a growing research topic. Further research also includes the integration of LF methods into a cross-layer design framework as well as the adaptation of LF transceiver design to OFDM systems, where transceiver complexity and feedback overhead issues are crucial.

References

1. Alamouti, S.: A simple transmit diversity technique for wireless communications. IEEE J. Sel. Areas Commun. **16**(8), 1451–1458 (1998)
2. de Almeida, A.L.F.: Tensor modeling and signal processing for wireless communication systems. Ph.D. thesis, University of Nice Sophia-Antipolis (2008)
3. Barg, A., Nogin, Y.D.: Bounds on packings of spheres in the grassmann manifold. IEEE Trans. Inf. Theory **48**(9), 2450–2454 (2002)
4. Bro, R.: Multi-way analysis in the food industry: Models, algorithms and applications. Ph.D. thesis, University of Amsterdam, Amsterdam (1998)
5. Conway, J.H., Hardin, R.H., Sloane, N.J.A.: Packing lines, planes, etc., packing in grassmannian spaces. Exp. Math. **5**, 139–159 (1996)
6. El Gamal, H., Damen, M.O.: Universal space-time coding. IEEE Trans. Inf. Theory **49**(5), 1097–1119 (2003)
7. Foschini, G.J.: Layered space-time architecture for wireless communications in a fading environment when using multiple antennas. Bell Labs Tech. J. **1**(2), 41–59 (1996)
8. Foschini, G.J., Gans, M.J.: On limits of wireless communications when using multiple antennas. Wireless Pers. Commun. **6**(3), 311–335 (1998)
9. Gershman, A.B., Sidiropoulos, N.D.: Space-Time Processing for MIMO Communications. The Johns Hopkins University Press, London (1996)
10. Gersho, A., Gray, R.M.: Vector Quantization and Signal Compression. Kluwer, Boston, MA (1992)
11. Golden, G., Foschini, G., Valenzuela, R., Wolniansky, P.: Detection algorithm and initial laboratory results using the V-BLAST space-time communications architecture. IEE Electron. Lett. **35**(7), 14–15 (1999)
12. Goldsmith, A., Jafar, S.A., Jindal, N., Vishwanath, S.: Capacity limits of MIMO channels. IEEE J. Sel. Areas Commun. **21**(5), 684–702 (2003)
13. Gore, D.A., Heath Jr., R.W., Paulraj, A.J.: Transmit selection in spatial multiplexing systems. IEEE Commun. Lett. **6**(11), 491–493 (2002)
14. Hassibi, B., Hochwald, B.M.: High-rate codes that are linear in space and time. IEEE Trans. Inf. Theory **48**(7), 1804–1824 (2002)
15. Heath, R.W., Love, D.J.: Multimode antenna selection for spatial multiplexing systems with linear receivers. IEEE Trans. Signal Process. **53**, 3042–3056 (2005)

16. Heath, R., Paulraj, A.: A simple scheme for transmit diversity using partial channel feedback. In: Proc. IEEE Assilomar Conf. on Signals and Systems, vol. 2 (1998)
17. Heath, R.W., Paulraj, A.: Antenna selection for spatial multiplexing systems based on minimum error rate. In: Proc. IEEE Int. Conf. Commun., vol. 2, pp. 570–574 (2001)
18. Hochwald, M., Marzetta, T., Richardson, T., Sweldens, W., Urbanke, R.: Systematic design of unitary space-time constellations. IEEE Trans. Inf. Theory **46**, 1962–1973 (2000)
19. Holma, H., Toskala, A.: WCDMA for UMTS: Radio Access for Third Generation Mobile Communications. Wiley, West Sussex (2002)
20. Larsson, E.G., Stoica, P.: Space-Time Block Coding for Wireless Communications. Cambridge University Press, New York (2003)
21. Love, D.J., Heath, R.W.: Equal gain transmission in multiple-input multiple-output wireless systems. IEEE Trans. Commun. **51**(7), 1102–1110 (2003)
22. Love, D.J., Heath, R.W.: Limited feedback unitary precoding for orthogonal space-time block-codes. IEEE Trans. Signal Process. **53**(1), 64–73 (2005)
23. Love, D.J., Heath, R.W.: Limited feedback unitary precoding for spatial multiplexing systems. IEEE Trans. Inf. Theory **51**(8), 2967–2976 (2005)
24. Love, D.J., Heath, R.W.: Multimode precoding for MIMO wireless systems. IEEE Trans. Signal Process. **53**(10), 3674–3687 (2005)
25. Love, D.J., Heath, R.W., Santipach, W., Honig, M.L.: What is the value of limited feedback for MIMO channels? IEEE Commun. Mag. **42**(10), 54–59 (2004)
26. Love, D.J., Heath, R.W., Strohmer, T.: Grassmannian beamforming for multiple-input multiple-output wireless systems. IEEE Trans. Inf. Theory **49**(10), 2735–2747 (2003)
27. Mukkavilli, K.K., Sabharwal, A., Erkip, E., Aazhang, B.: On beamforming with finite rate feedback in multiple antenna systems. IEEE Trans. Inf. Theory **49**(10), 2562–2579 (2003)
28. Murthy, C., Rao, B.: Quantization methods for equal gain transmission with finite rate feedback. IEEE Trans. Signal Process. **55**(1), 233–245 (2007)
29. Murthy, C., Zheng, J., Rao, B.: Performance of quantized equal gain transmission with noisy feedback channels. IEEE Trans. Signal Process. **56**(6), 2451–2460 (2008)
30. Narula, A., Lopez, M., Trott, M., Wornell, G.: Efficient use of side information in multiple-antenna data transmission over fading channels. IEEE J. Sel. Areas Commun. **16**(27), 1423–1436 (1998)
31. Paulraj, A., Nabar, R., Gore, D.: Introduction to Space-Time Wireless Communications. Cambridge University Press, Cambridge (2003)
32. Sampath, H., Paulraj, A.: Linear precoding for space-time coded systems with known fading correlations. IEEE Commun. Lett. **6**, 239–241 (2002)
33. Sampath, H., Stoica, P., Paulraj, A.: Generalized linear precoder and decoder design for MIMO channels using the weighted MMSE criterion. IEEE Trans. Commun. **49**(12), 2198–2206 (2001)
34. Scaglione, A., Stoica, P., Barbarossa, S., Giannakis, G.B., Sampath, H.: Optimal designs for space-time linear precoders and decoders. IEEE Trans. Signal Process. **50**(5), 1051–1064 (2002)
35. Simon, M.K., Alouini, M.: Digital Communications over Fading Channels. Wiley, New York, NY (2004)
36. Skoglund, M., Jongren, G.: On the capacity of a multiple-antenna communication link with channel side information. IEEE J. Sel. Areas Commun. **21**, 395–405 (2003)
37. Sollenberger, N.R.: Diversity and automatic link transfer for a TDMA wireless access link. In: Proc. IEEE GLOBECOM, vol. 1, pp. 532–536 (1993)
38. Talwar, S., Viberg, M., Paulraj, A.J.: Blind separation of synchronous co-channel digital signals using an antenna array Part I: Algorithms. IEEE Trans. Signal Process. **44**(5), 1184–1197 (1996)
39. Tarokh, V., Seshadri, N., Calderbank, A.R.: Space-time codes for high data rate wireless communications: Performance criterion and code construction. IEEE Trans. Inf. Theory **44**(2), 744–765 (1998)

40. Telatar, I.E.: Capacity of multi-antenna gaussian channels. Eur. Trans. Telecomm. **10**(6), 585–595 (1999)
41. Titus, K.Y.L.: Maximum ratio transmission. IEEE Trans. Commun. **47**(10), 1458–1461 (1999)
42. Tse, C.H., Yip, K.W., Ko, T.S.: Performance tradeoffs maximum ratio transmission and switched-transmit diversity. In: Proc. IEEE 11th Symp. Pers. Ind. Mob. Radio Commun. (PIMRC), vol. 2, pp. 1485–1489 (2000)
43. Xia, P., Zhou, S., Giannakis, G.B.: Achieving the welch bound with difference sets. IEEE Trans. Inf. Theory **51**, 1990–1907 (2005)
44. Zheng, L., Tse, D.N.C.: Diversity and multiplexing: A fundamental tradeoff in multiple-antenna channels. IEEE Trans. Inf. Theory **49**(5), 1073–1096 (2003)
45. Zhou, S., Giannakis, G.B.: Optimal transmitter eigen-beamforming and space-time block coding based on channel mean feedback. IEEE Trans. Signal Process. **50**(10), 2599–2613 (2002)
46. Zhou, S., Giannakis, G.B.: Optimal transmitter eigen-beamforming and space-time block coding based on channel correlations. IEEE Trans. Inf. Theory **49**, 1673–1690 (2003)

Index